Embryology and Birth Defects

Embryology and Birth Defects

Edited by Jimmy Taylor

hayle
medical

New York

Hayle Medical,
750 Third Avenue, 9th Floor,
New York, NY 10017, USA

Visit us on the World Wide Web at:
www.haylemedical.com

ISBN: 978-1-63241-494-6

Trademark Notice: Registered trademark of products or corporate names are used only for explanation and identification without intent to infringe.

Cataloging-in-Publication Data

Embryology and birth defects / edited by Jimmy Taylor.
    p. cm.
Includes bibliographical references and index.
ISBN 978-1-63241-494-6
1. Embryology, Human. 2. Abnormalities, Human. I. Taylor, Jimmy.
QM611 .E43 2018
612.62--dc23

Table of Contents

Preface

The branch of biology which studies the development of fetus is known as embryology. It also studies the birth defects and disorders found before birth. Embryology is becoming an important area for research as it focuses on certain diseases and mutations that can be detected and treated during pregnancy. Different approaches, evaluations, methodologies and advanced studies on embryology and birth defects have been included in this book. For someone with an interest and eye for detail, this book covers the most significant topics in this field.

This book is a comprehensive compilation of works of different researchers from varied parts of the world. It includes valuable experiences of the researchers with the sole objective of providing the readers (learners) with a proper knowledge of the concerned field. This book will be beneficial in evoking inspiration and enhancing the knowledge of the interested readers.

In the end, I would like to extend my heartiest thanks to the authors who worked with great determination on their chapters. I also appreciate the publisher's support in the course of the book. I would also like to deeply acknowledge my family who stood by me as a source of inspiration during the project.

Editor

1

Sperm-Associated Antigen 6 (SPAG6) Deficiency and Defects in Ciliogenesis and Cilia Function: Polarity, Density, and Beat

Maria E. Teves[1], Patrick R. Sears[3], Wei Li[1], Zhengang Zhang[1,4], Waixing Tang[5], Lauren van Reesema[1], Richard M. Costanzo[6], C. William Davis[7], Michael R. Knowles[7], Jerome F. Strauss III[1,2], Zhibing Zhang[1,2]*

1 Department of Obstetrics and Gynecology, Virginia Commonwealth University, Richmond, Virginia, United States of America, 2 Department of Biochemistry and Molecular Biology, Virginia Commonwealth University, Richmond, Virginia, United States of America, 3 Cystic Fibrosis Center, University of North Carolina, Chapel Hill, North Carolina, United States of America, 4 Department of Infectious Diseases, Tongji Medical College, Huazhong University of Science and Technology, Wuhan, Hubei, China, 5 Department of Otorhinolaryngology, University of Pennsylvania, Philadelphia, Pennsylvania, United States of America, 6 Department of Physiology and Biophysics, Virginia Commonwealth University, Richmond, Virginia, United States of America, 7 Department of Cell & Molecular Physiology of Medicine, University of North Carolina, Chapel Hill, North Carolina, United States of America

Abstract

SPAG6, an axoneme central apparatus protein, is essential for function of ependymal cell cilia and sperm flagella. A significant number of Spag6-deficient mice die with hydrocephalus, and surviving males are sterile because of sperm motility defects. In further exploring the ciliary dysfunction in Spag6-null mice, we discovered that cilia beat frequency was significantly reduced in tracheal epithelial cells, and that the beat was not synchronized. There was also a significant reduction in cilia density in both brain ependymal and trachea epithelial cells, and cilia arrays were disorganized. The orientation of basal feet, which determines the direction of axoneme orientation, was apparently random in Spag6-deficient mice, and there were reduced numbers of basal feet, consistent with reduced cilia density. The polarized epithelial cell morphology and distribution of intracellular mucin, α-tubulin, and the planar cell polarity protein, Vangl2, were lost in Spag6-deficient tracheal epithelial cells. Polarized epithelial cell morphology and polarized distribution of α-tubulin in tracheal epithelial cells was observed in one-week old wild-type mice, but not in the Spag6-deficient mice of the same age. Thus, the cilia and polarity defects appear prior to 7 days post-partum. These findings suggest that SPAG6 not only regulates cilia/flagellar motility, but that in its absence, ciliogenesis, axoneme orientation, and tracheal epithelial cell polarity are altered.

Editor: Yulia Komarova, University of Illinois at Chicago, United States of America

Funding: This research was supported by NIH grant HD076257, Virginia Commonwealth University Presidential Research Incentive Program (PRIP) and Massey Cancer Award (to ZZ), NIH grants HD37416 (JFS), and HL071798 (MRK). The authors also thank Pamela J. Gigliotti for technical support with the tissue processing at the VCU Biological Macromolecule Core Facility, supported, in part, with the funding from NIH-NCI Cancer Center Core Grant 5P30CA016059. The authors declare no competing financial interests. The funders had no role in study design, data collection and analysis, decision to publish, or preparation of the manuscript.

Competing Interests: The authors have declared that no competing interests exist.

* Email: zzhang4@vcu.edu

Introduction

Mammalian SPAG6 is the orthologue of PF16, a component of the central apparatus of the "9+2" axoneme of the green algae model organism, Chlamydomonas reinhardtii [1]. In Chlamydomonas, PF16 protein is present along the length of the flagella, and immunogold labeling localizes the PF16 protein to a single microtubule of the central pair. Mutations in the Chlamydomonas PF16 gene cause flagellar paralysis, and PF16 is believed to be involved in C1 central microtubule stability and flagellar motility [2]. In addition to Chlamydomonas reinhardtii, SPAG6/PF16 has been shown to regulate flagellar motility in other models, including trypanosomes, Plasmodium, and Giardia [3], [4,5].

Gene targeting has been used to create mice lacking SPAG6 [6]. Approximately 50% of Spag6-deficient animals died from hydrocephalus before adulthood, and males surviving to maturity were infertile. Even though an abnormal axoneme ultrastructure

was discovered in the Spag6-deficient sperm [6], cilia of brain ependymal cells and trachea epithelial cells from the mutant mice contained "9+2" axonemes that appeared to be grossly intact [7]. However, brain ependymal cells of Spag6-deficient mice are functionally defective since hydrocephalus develops.

In further characterizing the cilia abnormalities of Spag6-deficient mice, we discovered that ciliary beat frequency was significantly reduced. The mutant mice also had fewer trachea and ependymal cilia, and these cilia were arrayed in a random fashion on the cell surface. The central pair orientation differed significantly between cilia of the Spag6-deficient mice, reflecting the random orientation of basal feet. The polarized epithelial cell morphology and distribution of α-tubulin and planar cell polarity protein, Vangl2, were lost in Spag6-deficient tracheal epithelial cells. These findings suggest that mouse SPAG6 has multiple functions: it regulates cilia/flagellar motility through the central pair apparatus; but also plays a role in ciliogenesis, axoneme orientation, and cell polarity.

Materials and Methods

Spag6 and Spag16L mutant mice

Spag6 and Spag16L mutant mice were generated previously in our laboratory [6,8]. All animal work was approved by Virginia Commonwealth University's Institutional Animal Care & Use Committee (protocol #AM10297 and AD10000167) in accordance with Federal and local regulations regarding the use of non-primate vertebrates in scientific research.

High-speed video analysis of ciliary beat frequency

Ciliary beat frequency was assessed with the Sisson–Ammons video analysis (SAVA) system (Ammons Engineering, Mt. Morris, MI) [9]. Tracheas from wild type and Spag6-deficient mice (3 weeks old) were removed and video movies were taken within five minutes with an Nikon Eclipse TE-2000 inverted microscope (×40 phase-contrast objective) equipped with an ES-310 Turbo monochrome high-speed video camera (Redlake, San Diego, CA) set at 125 frames per second. The ciliary beat pattern was evaluated on slow-motion playbacks.

Transmission electron microscopy

For transmission electron microscopy (TEM), the samples (from 3 week old mice) were cut into small sections (2×2 mm) and fixed in 2.5%glutaraldehyde (PH = 7. 4) for 6–8 hours at 4°C. They were washed and post fixed in 2% OsO4 for 1 hour, at 4°C. The tissue was dehydrated through ascending series of ethanol concentrations and embedded in araldite CY212. Semi thin sections (1 µm) were cut and stained with toluidine blue. Ultra-thin sections (60–70 nm) were cut and stained with uranyl acetate and alkaline lead citrate.

Scanning electron microscopy

Specimens (from 3 week old mice) were fixed with 1.5% glutaraldehyde and 1.5% paraformaldehyde in 0.1 M sodium phosphate buffer, pH 7.3 for 3 hours at room temperature and postfixed for two hours in 2% osmium tetroxide in 0.1 sodium phosphate buffer. After dehydration in graded ethanol, samples for scanning electron microscopy (SEM) were dried in a critical-point dryer (Polaron, Watford, UK), mounted on stubs, and coated with gold-palladium in a cool sputter coater (Fisons Instruments Uckfield, UK). The specimens were examined using a scanning electron microscope DSM 960 (Zeiss Oberkochen, Germany).

Histology

H&E and Periodic acid-Schiff (PAS) staining on mouse trachea (from 1 and 3 week old mice) were carried out using standard procedures. 5 µm sections were cut for experiments.

Immunofluorescence staining of brain and trachea

Brain and trachea from wild-type and Spag6-mutant mice (3 week old) were fixed with 4% paraformaldehyde in 0.1 M PBS (pH 7.4), and 5 µm paraffin sections were made. For the immunofluorescence, the method described by Tsuneoka was used [10]. The sections were incubated with an anti-Vangl2 or anti-acetylated tubulin primary antibody at 4°C for overnight. Slides were washed with PBS and incubated for 1 hour at room temperature with Alexa 488-conjugated anti-mouse IgG secondary antibody (1:500; Jackson ImmunoResearch Laboratories) or Cy3-conjugated anti-rabbit IgG secondary antibody (1:5000; Jackson ImmunoResearch Laboratories). Following secondary antibody incubation, the slides were washed again three times in PBS, mounted using VectaMount with 4′, 6-diamidino-2-pheny-lindole (DAPI) (Vector Laboratories, Burlingame, CA), and sealed with a cover slip. Images were captured by confocal laser-scanning microscopy (Leica TCS-SP2 AOBS).

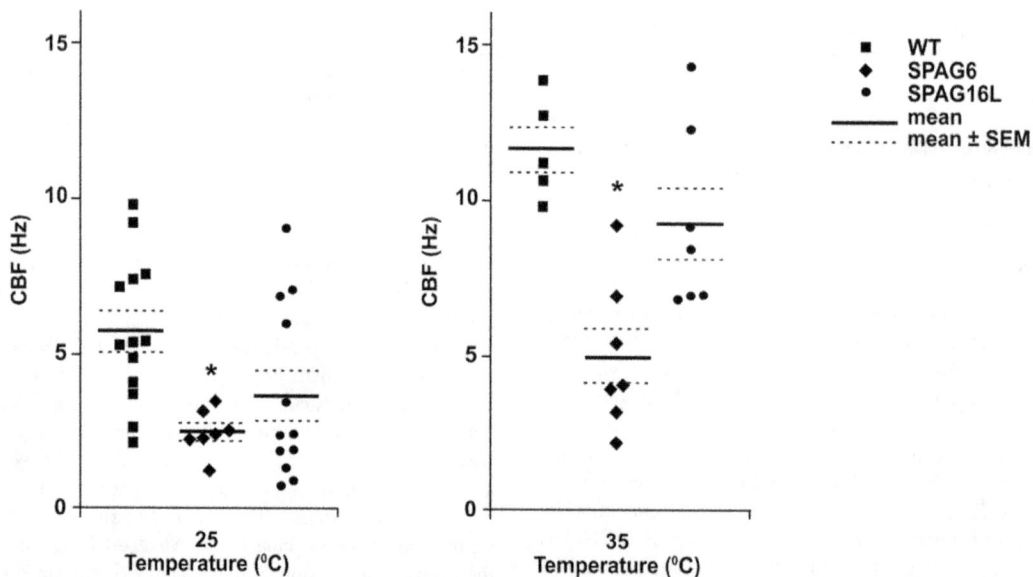

Figure 1. Trachea ciliary beat frequency (CBF) is dramatically decreased in Spag6-mutant mice. Graph showing ciliary beat frequency for wild-type, Spag6, and Spag16L mutant mice. The mean CBF in Spag6-deficient mice was significantly lower than that in the wild type and Spag16L mice at both 25°C (n = 13, 7 and 12 for wild-type, Spag6 mutant, and Spag16L mutant mice respectively) and 35°C (n = 5, 7 and 7 for wild-type, Spag6 mutant, and Spag16L mutant mice respectively). *p<0.05. ANOVA was conducted to determine significant difference.

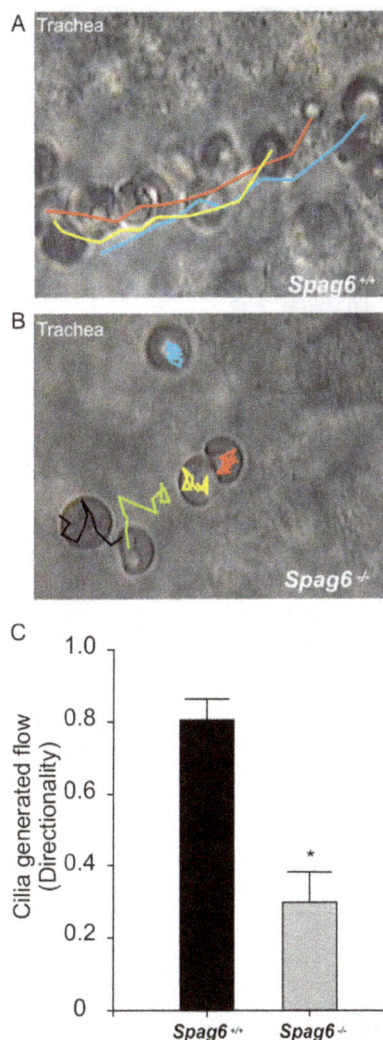

Figure 2. Cilia-generated flow is significantly reduced in *Spag6*-deficient tracheal epithelium. A) Longitudinal view of tracheal epithelia from wild-type mouse showing the tracking of movement of blood cells. B) Longitudinal view of tracheal epithelia from *Spag6* knockout mouse showing the tracking of movement of blood cells. C) Cilia generated flow was quantified by analyzing the directionality of movement of blood cells. * Significant differences (p<0.05) vs. wild-type. Data are presented as mean ± SEM. The colors indicates movement track of individual blood cells.

Video-microscopy

Tracheas were collected from three week old wild-type and *Spag6* knockout mice. Trachea sections were placed luminal side down on a coverslip containing some blood drops in 37°C PBS. Cilia movement and blood cells flows were observed with differential interference contrast microscopy using an inverted microscope (Nikon) equipped with a 100 X oil immersion objective. Images were recorded at 30 frames per second with SANYO color CCD, Hi-resolution camera (VCC-3972, Sanyo Electric Co, Japan) and Pinnacle Studio HD (Ver. 14.0, Pinnacle Systems, Inc., Mountain View, CA, USA) software. Several randomly selected areas were imaged for each sample. Quantification of blood cells directionality was performed with ImageJ software and plugin MTrackJ (NIH). 200 blood cells were tracked for each sample. Directionality was defined as the net displace-

ment achieved divided by the total distance traveled. A directionality of 1 indicated the blood cell moved in a straight line, while a directionality of 0 represents a random movement approach. Data represent mean ± SEM of three mice for each genotype.

Western blot analysis

Equal amounts of protein (50 μg/lane) were heated to 95°C for 10 minutes in sample buffer, loaded onto 10% sodium dodecyl sulfate-polyacrylamide gels, electrophoretically separated, and transferred to polyvinylidene difluoride membranes (Millipore, Billerica, MA). Membranes were blocked (Tris-buffered saline solution containing 5% nonfat dry milk and 0.05% Tween 20 (TBST)) and then incubated overnight with indicated antibodies at 4°C. After washing in TBST, the blots were incubated with second antibodies for 1 hour at room temperature. After washing, the proteins were detected with Super Signal chemiluminescent substrate (Pierce, Rockford, IL).

Quantitative analysis of basal foot orientation

A reference line was drawn for each image. For each basal foot, a vector connecting the center of the basal body and the protrusion of the basal foot was drawn. The angle between this vector and the reference line was measured manually using ImageJ software (NIH). Five images were analyzed from each mouse, and three wild-type and three mutant mice were used for the analysis. The mean angle was calculated for each cell using Oriana 4.0 software (Kovach Computing Services). The mean angle was defined as mean ciliary direction (shown as 0° in each circular plot graph). Deviation from the mean angle was calculated for all of the basal feet analyzed. Deviation angles of the basal feet were pooled and plotted on a circular graph using Oriana 4.0 software.

Statistical methods

Significant difference of axoneme number and basal feet number between wild-type and *Spag6*-deficient mice was calculated using student t test. Significant difference of CBF among wild-type, *Spag6*, and *Spag16L*-deficient mice was calculated using ANOVA. * = significant at 0.05. Statistical analysis of tracheal epithelial cell basal body rootlet orientation was carried out with Oriana 4.0 (Kovach Computing Services) circular statistics software.

Results and Discussion

A recent study reported that the mouse has two copies of the *Spag6* gene, the one previously studied on chromosome 16, which is proposed to have evolved from the parental isoform, *Spag6-BC061194*, which is located on chromosome 2 [11]. Even though the amino acid sequences of the two SPAG6 proteins are 97% identical, the nucleotide sequences of the two *Spag6* genes are significantly different. We confirmed that the *Spag6*-deficient mouse we created and studied retains the *Spag6*-BC061194 gene, so that the phenotypes we have described represent the solely the impact of the loss of the "evolved" *Spag6* gene (data not shown).

Ciliary beat frequency (CBF) of tracheal cilia was measured in *Spag6*-deficient and littermate wild-type mice. Compared to the wild-type mice, baseline CBF was significantly reduced in the *Spag6*-deficient mice at both room temperature (25°C) and 35°C (Fig. 1). Of note, mutation of the *Spag16L* gene, which encodes a central apparatus protein, SPAG16L, that interacts with SPAG6, does not cause CBF abnormalities or uncoordinated cilia beat [12], despite the fact that *Spag16L* mutant mice are infertile due

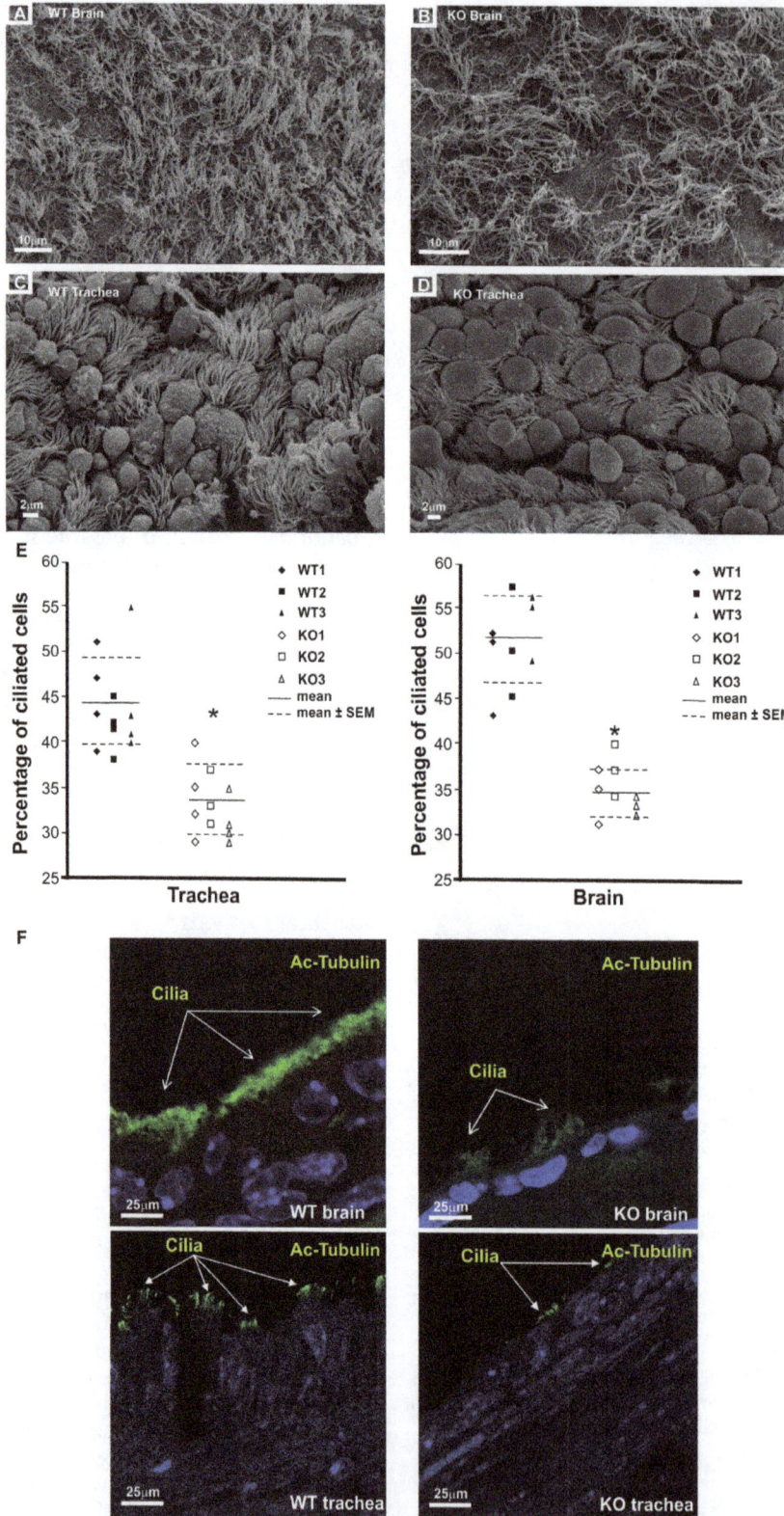

Figure 3. Analyses of cilia in the trachea epithelial cells and brain ependymal cells by scanning electronic microscopy and immunofluorescence staining. Representative images from SEM analyses from wild-type and *Spag6*-deficient mice. Cilia of the wild type mice were well-organized in both brain ependymal cells (A) and trachea epithelial cells (C). In contrast, there was a dramatic reduction in cilia density in both brain ependymal cells (B) and trachea epithelial cells (D) of the *Spag6*-deficient mice, and the cilia were disorganized. To calculate the percentage of ciliated cells, cells with cilia and total cells were counted from three (brain) or four (trachea) SEM images from each mouse, and ratio was calculated (E). Three wild-type and three *Spag6*-deficient mice were analyzed. * Significant differences vs. wild-type (p<0.05). Brain and trachea

sections from wild type and *Spag6*-deficient mice were examined by immunofluorescence staining using an antibody targeting acetylated tubulin. In the wild type mice, the cilia-containing signal was continuously observed along the surface of the epithelial cells. However, in the *Spag6*-deficient mice, the signal was discontinuous (F).

to a severe sperm motility defect [8]. Thus, SPAG6 has functional roles in tracheal cilia that are distinct from those of SPAG16L.

Tracheal ciliary beating was also observed by video microscopy. Consistent with the CBF results, cilia from wild-type mice beat at a faster rate, and the beat was coordinated, with all the cilia beating in the same direction at a specific time point (Video S1). The metachronal beating resulted in a directional flow as shown by the movement of particles/blood cells (Video S2 and Fig. 2A). However, cilia from *Spag6*-deficient mice beat at a much slower rate, and the beating was largely uncoordinated. At specific time points, some cilia beat in one direction, but others in an opposite direction (Video S3). Significantly reduced directed flow was observed as the blood cells collected at the beginning of the tracheal tubes (Video S4, Fig. 2B and Fig. 2C).

Scanning electron microscopy (SEM) was carried out on three (3 week old) wild-type and three *Spag6*-deficient mice of the same age to examine cilia orientation. Tracheal and ependymal cilia in the wild-type animals are anchored to the cell surface in organized arrays (Fig. 3A, 3C). In contrast, cilia arrays of *Spag6*-deficient mice were disorganized (Fig. 3B, 3D, Figure S1C and Figure S1D). In addition, there was a dramatic reduction in cilia density in both brains and tracheas of the *Spag6*-deficient mice, and the percentage of ciliated cells was significantly lower in the mutant mice (Fig. 3E). Cilia in these tissues were further examined by immunofluorescence staining using an antibody to acetylated tubulin. In wild-type mice, the cilia signal was continuous along the surface of the epithelial cells, and extended away from the cell surface into the lumen (Fig. 3F, left panel). However, in the *Spag6*-deficient mice, the signal was discontinuous, and the signal extended to a lesser extent from cell surface than in wild-type tissues (Fig. 3F, right panel).

Transmission electron microscopy (TEM) was conducted to examine the orientation of the central pair microtubules in four wild-type and four *Spag6*-deficient mice. In wild-type animals, orientation of the two central microtubules of all the cilia in the ependymal cells (Fig. 4A) and tracheal epithelial cells (Fig. 4C) was consistent, as shown by the similar orientation of lines connecting the two microtubules in all the axonemes. In the *Spag6*-deficient mice, the axoneme structure appeared normal, but the orientation of the two central microtubules was random; lines connecting central microtubules pointed to one direction in some axonemes, while the lines pointed to a different direction in others (Fig. 4B, and Fig. 4D). To compare the cilia number in the brain ependymal and trachea epithelial cells of wild-type and *Spag6*-deficient mice, the axoneme number was counted from ten TEM images randomly selected from each group. The *Spag6*-deficient mice had significantly lower axoneme numbers than that in the wild-type mice (p<0.05, Fig. 4E).

The ciliary beat orientation is determined by the orientation of the basal feet [13]. The basal feet were examined in the brain ependymal and trachea epithelial cells of wild-type and the *Spag6*-deficient mice by TEM. In the wild-type mice, the basal feet were present in both brain ependymal cells (Fig. 5A) and trachea epithelial cells (Fig. 5C), and they were organized in a similar orientation. However, in the *Spag6*-deficient mice, even though basal feet were morphologically intact, the orientation was random. Like the orientation of the central microtubules in the *Spag6*-deficient mice, some basal feet pointed to one direction, others pointed in different direction (Fig. 5B, and Fig. 5D). The

basal feet number was also counted from the TEM images, and the number was significantly reduced in both brain ependymal (p<0.05) and trachea epithelial cells (p<0.05) of the *Spag6*-deficient mice (Fig. 5E). To determine if the organization of basal feet in *Spag6*-mutant mice is significantly different compared to wild-type animals, basal foot orientation of tracheal epithelial cells was analyzed in three *Spag6* mutant mice (Figure S2A) and three wild-type mice (Figure S2B). Statistical analysis demonstrated that there was significant difference in the r_{cell} metric between the mutant and wild-type mice (Fig. 5F), suggesting that intracellular planar polarity was lost in the mutant mice.

To investigate tissue-level cell polarity, histological sections of tracheas from three (3 week old) wild type and three *Spag6*-deficient mice were examined by light microscopy. H&E staining revealed that in the wild-type mice, two or three rows of nuclei were present in the pseudostratified columnar epithelium lining of the trachea, and the nuclei were oval in shape and oriented in a basal/apical distribution (Fig. 6A). However, the *Spag6*-deficient epithelial cells did not form the pseudostratified columnar morphology. Only one row of nuclei was present, and the cells lay relatively flat along the basement membrane, with most cells having round nuclei (Fig. 6B). The *Spag6*-deficient tracheal epithelial cells not only lose the polarized morphology, the polarized distribution of mucin was also absent. In the wild-type mice, PAS staining demonstrated that mucin was localized in apical region of epithelial cells (Fig. 6C). However, this pattern was never seen in the epithelial cells of *Spag6*-deficient mice. In contrast, mucin was present throughout the cytoplasm (Fig. 6D). The polarized morphology of wild-type tracheal epithelial cells was also observed in low magnification TEM images, and mucin was observed on the trachea surface (Fig. 6E). However, the polarized pattern was lost in the mutant mice, and mucin was not detected on the surface of the tracheal epithelial cells (Fig. 6F).

The localization of the planar cell polarity protein, Vangl2, was examined in the tracheas of three wild-type and three *Spag6*-deficient mice by immunofluorescence staining. Even though there was no difference in total expression level of the protein in trachea/lung between wild-type and the *Spag6*-deficient mice by Western blot analysis (Figure S3), it appears that in wild-type trachea epithelial cells from three-week old mice, Vangl2 signal was more intense in the apical region (Figure S4A). This polarized distribution was not evident in the *Spag6*-deficient epithelial cells (Figure S4B).

Immunofluorescence staining was also conducted on tracheas from three week old wild-type and age-matched *Spag6*-deficient mice using an anti-α-tubulin antibody. In the trachea of wild-type mice, cilia were intensely stained. Inside the epithelial cells, a strong signal was visualized in the apical regions (Figure S4C). However, in the *Spag6*-deficient mice, the signal was evenly distributed throughout the whole cell body (Figure S4D). The microtubule distribution pattern is consistent with that of the planar cell polarity (PCP) proteins, suggesting that polarized microtubules might contribute to the localization of PCP proteins.

One-week mice were analyzed, when the mutant mice did not show obvious abnormalities related to reduced ciliary motility, such as hydrocephalus. As in the three-week old wild-type mice, epithelial cells in the one-week old wild-type mice are polarized as shown by H&E staining (Fig. 7A), and immunofluorescence staining using an anti-acetylated tubulin antibody revealed a

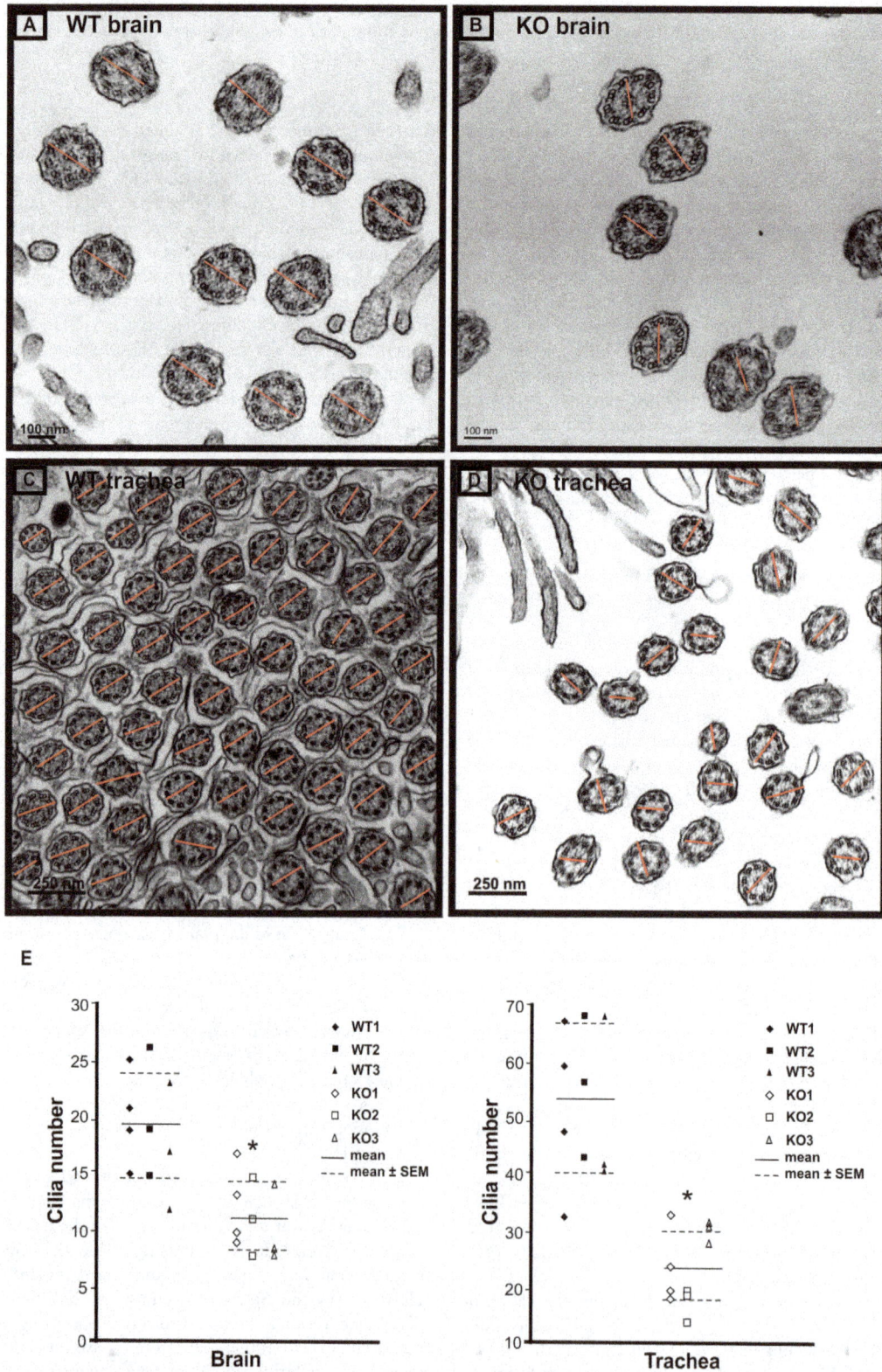

Figure 4. Examination of rotational polarity of ciliary axoneme of brain ependymal cells and trachea epithelial cells by transmission electronic miscroscopy. Axoneme cross-sectional images were taken with a transmission electron microscope. The rotational polarity of each axoneme was evaluated by the angle of the line connecting the central pair. Notice that the orientation of the lines in wild type mice is similar (A: brain; C: trachea). while the orientation of the lines in the *Spag6*-deficient mice varies among axonemes (B: brain; D: trachea). E. Average axoneme number counted from ten images randomly selected from each group. Three or four images were counted from each mouse, and three wild-type and three mutant mice were examined. Horizontal lines represent the means and SEMs. * p<0.05.

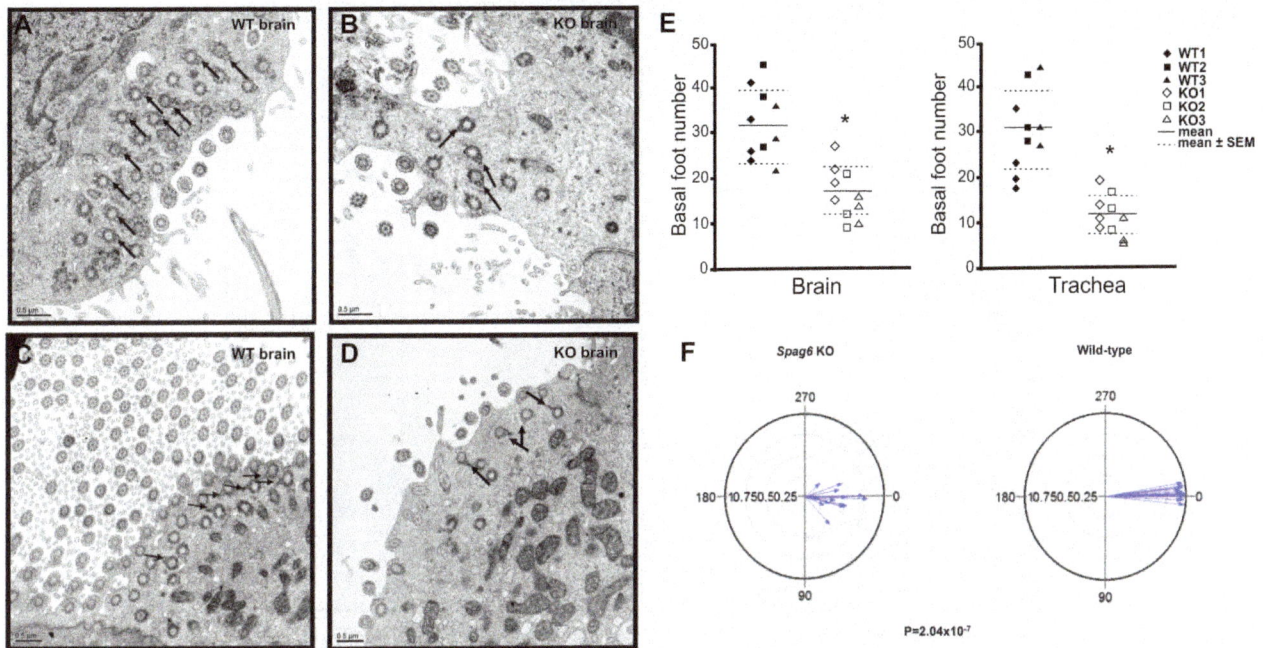

Figure 5. Basal feet polarity of brain ependymal cells and trachea epithelial cells was lost in the *Spag6*-deficient mice. Basal body images were taken with a transmission electron microscope. Notice that the basal feet point to the same orientation in the wild type animals (A: brain; C: trachea). However, they point to different orientation in the *Spag6*-deficient mice (B: brain; D: trachea). The number of basal body in the *Spag6*-deficient mice was significantly reduced in both brain and trachea. The arrows point to the basal feet. E. Average basal feet number counted from ten images randomly selected from each group. Horizontal lines represent means and SEMs. Three or four images were counted from each mouse, and three wild-type and three mutant mice were examined. * p<0.05. F. Circular plots of tracheal epithelial cell basal feet orientation in *Spag6*-deficient (left) and wild-type mice (right). For each mouse, basal foot orientation from five images was analyzed. For each image, the angel for one basal foot orientation was set as 0° (or 360°), angels of the rest basal feet were measured. Each plot represents the combined data from three mice as shown in Figure S2 (* p<0.001 between the two groups).

strong cilia signal (Fig. 7C). However, the epithelial cells of the *Spag6*-deficient mice did not show the polarized morphology. Compared to the wild-type mice, cilia number is dramatically reduced (Fig. 7B, 7D). The polarized distribution of α-tubulin is obvious in the wild-type mice (Figure S5A), but not in the *Spag6*-deficient mice (Figure S5B). These findings indicate that the cilia number and orientation defects are present earlier than 7 days post-partum.

It has been reported that a feedback loop generated by fluid flow contributes to cilia polarization [14,15]. Indeed, our findings of disorganization and reduced number of cilia in ependymal and tracheal epithelial cells in *Spag6*-mutants, and alterations in basal body alignment are similar to those previously described in other mutant mice as a consequence of reduced ciliary motility. Studies involving Jhy$^{lacZ/lacZ}$ mice showed disorganization and altered axonemal structure of the ependymal cilia. However, the hydrocephalus appeared to be unrelated to abnormal brain development or patterning [16]. Observations in *ktu*-mutant mice revealed that the PCP protein, Vangl1, localized asymmetrically in ependymal and tracheal epithelial cells, while the alignment of basal bodies only differed from wild-type mice in brain ependymal cells, suggesting that ciliary motility was required in the alignment of brain ependymal cells, but not for airway cilia [17].

There are other possible mechanisms, in addition to ciliary motility defects, as causal factors of the phenotypes in SPAG6-mutant mice. In multi-ciliated cells, basal bodies are replicated deep within the cytoplasm, and their apical movement and docking are thought to involve regulated actin assembly [18,19] and vesicle trafficking [20]. Indeed, actin is enriched at the apical

surface of ciliated epithelial cells [21,22]. Disruption of the actin cytoskeleton blocks basal body migration and ciliogenesis [19]. In this case, the ciliogenesis defect is associated with the failure of basal body docking at the apical plasma membrane. *Spag6*-deficient cells may have a disrupted actin cytoskeleton affecting basal body docking, which traps some basal bodies inside the cytoplasm where they are degraded, with the result that fewer cilia develop in the brain ependymal and tracheal epithelial cells.

A recent study demonstrated that silencing of the *Spag6* gene in Xenopus larvae gives rise to disruption of orientation of basal bodies, suggesting that the planar cell polarity mechanism might be involved [14]. PCP refers to the polarization of a field of cells within the plane of a cell sheet [23]. It is a downstream branch of Wnt signaling [24]. This form of polarization is required for diverse cellular processes in vertebrates, including convergent extension (CE) [25] and the establishment of PCP in epithelial tissues and ciliogenesis [26,27]. In multi-ciliated cells, planar polarity is present in two distinct models, termed rotational polarity and tissue-level polarity [28,29]. The former refers to the alignment of the basal bodies within each multi-ciliated cell, and the latter to the coordination of many multi-ciliated cells across the tissue. SEM and high magnification TEM studies clearly demonstrated that *Spag6*-deficient epithelial cells in the brain and trachea lost rotational polarity. Alternatively, SPAG6 may cause ciliogenesis defects through a role in basal bodies. Pearson et al. reported that SPAG6 was present in newly assembled basal bodies [30], and SPAG6 localizes to the center of the transition zone at the site of central pair assembly [31].

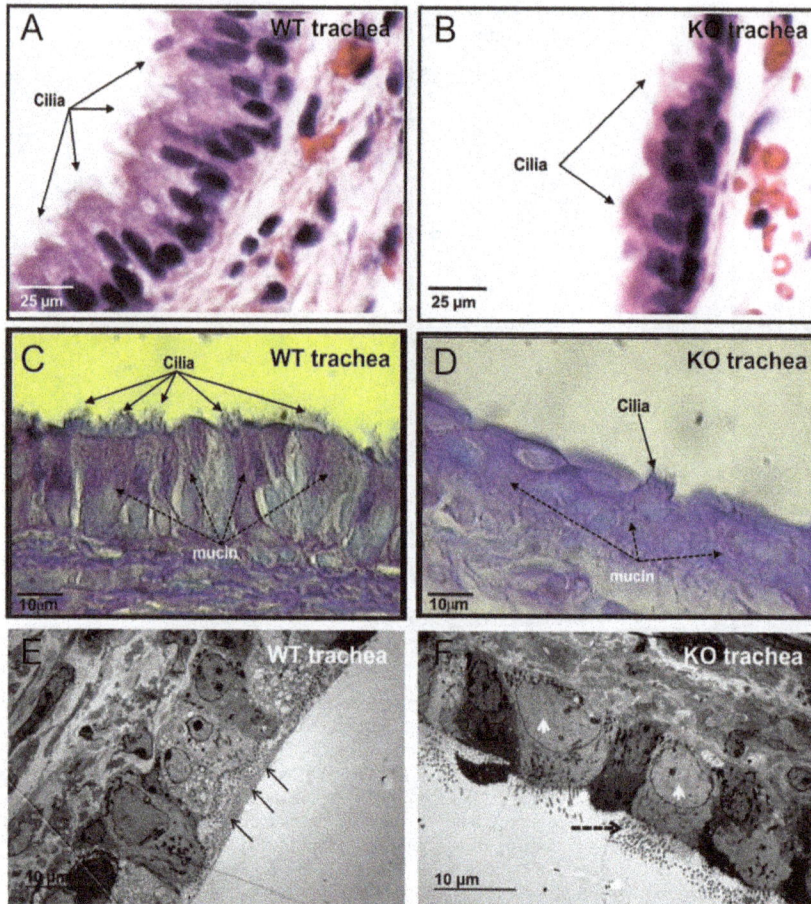

Figure 6. Examination of trachea epithelial cell polarity in three-week old mice. H&E stained tissue demonstrating that two or three rows of nuclei were seen in the pseudostratified columnar epithelium lining the trachea of the wild-type mice, and the nuclei were oval shape and the two poles were at basal/apical distribution (Fig. 6A). In contrast, in the *Spag6*-deficient mice, the cells lie flat along the basement membrane, most cells had round nuclei (Fig. 6B). PAS staining demonstrated that mucin was localized in apical region of epithelial cells (Dashed arrows in Fig. 6C). However, this pattern was never seen in the epithelial cells of *Spag6*-deficient mice, mucin was present through the whole cytoplasm (Fig. 6D). The above-mentioned differences between wild-type and *Spag6*-deficient mice were confirmed by TEM with low magnification. The wild-type epithelial cells show polarized pattern, and mucin was found along the surface in all the three mice analyzed (arrows in Fig. 6E), where the cilia axonemes were located. However, in the mutant mice, the epithelial cells lost this pattern, and the cells look larger than those in the wild-type mice (arrow heads), and no mucin was found in any of the three mice analyzed (Fig. 6F). Three wild-type and three mutant mice were analyzed and the results were similar.

Several recent studies revealed that the PCP signaling cascade is a central regulator of the orientation of cilia-mediated fluid flow. Disruption of core PCP genes, including the Dishevelled (Dvl1), Celsr2 and Celsr3 resulted in a randomization of rotational polarity [32,33,15]. PCP signaling also controls the tissue-level polarity of multi-ciliated cells, PCP proteins, van Gogh-like 2 (Vangl2) and Frizzled are in this case [29]. A more recent study indicated that PCP proteins, including Vangl1, Vangl2, Prickle2 (PK2), Dishevelled1 (Dvl1) and Dvl2 localize asymmetrically to the tracheal epithelial cell cortex [34].

Little is known about the regularity of orientation of basal bodies in multi-ciliated cells. Previous studies of multi-ciliated cells suggested that microtubules attached to the basal feet link basal bodies to one another, and also to the apical junctions [35]. The classic planar polarity in the Drosophila wing epithelial cells is also associated with sub-apical microtubules [36,37]. These microtubules are planar polarized, with their plus ends enriched at the distal face of cells, where Dvl1 and Frizzled localize. It appears that the apical microtubule network is an upstream regulator of

PCP signaling [38,39]. It is suggested that a similar planar polarized web of microtubules may also influence planar polarity of basal bodies. In addition, basal bodies in multi-ciliated cells make complex connections to both actin and cytokeratin networks, and these may be involved in polarization [40]. We have previously shown that SPAG6 is a microtubule binding protein [1]. It may play a role in stabilizing the microtubule system. In the absence of SPAG6, sub-apical microtubule stability might be affected, which could result in disruption of polarized PCP distribution, causing the basal bodies to lose their polarized localization.

Defects in mammalian cilia or flagella motility/function caused by mutations in central apparatus genes appear to depend upon the genetic background and cellular context. For instance, mutation of the *Pcdp1* gene results in several phenotypes commonly associated with primary ciliary dyskinesia. Homozygous mutants on a C57BL/6J background develop severe hydrocephalus and mainly die within the first week of life. However, on other genetic backgrounds (129S6/SvEvTac), mice

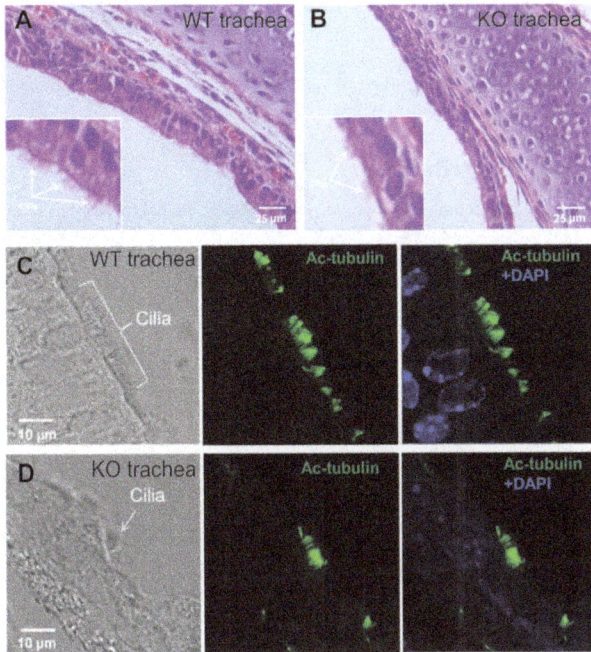

Figure 7. Examination of trachea epithelial cell polarity in one-week old mice. H&E stained tissue demonstrating the polarized pattern of epithelial cells in the trachea of wild type mice (Fig. 7A), but not in the *Spag6*-deficient mice (Fig. 7B). Acetylated tubulin signal is abundant along the tracheal epithelial cells in the wild-type mice (Fig. 7C), the signal is dramatically reduced in the *Spag6*-deficient mice (Fig. 7D). Three wild-type and three mutant mice were analyzed and similar results were observed.

develop either mild or no hydrocephalus with survival to adulthood. The respiratory epithelial cilia have a normal ultrastructure, but beat with reduced frequency. Interestingly, the male mice are infertile, producing sperm with no visible flagella, suggesting that the mechanisms regulating the biogenesis of cilia and flagella are likely to be different [41]. Tracheal epithelial cilia from *Spef2*-deficient mice beat at lower frequency and have a normal 9+2 axonemal structure without apparent defects in the dynein arms, but epididymal sperm lack recognizable axonemal structures [42]. SPAG16L-null mice show no evidence of cilia dysfunction, such as hydrocephalus, sinusitis, and bronchial infection [8], and have tracheal epithelial cells with motile cilia [12]. However, males are infertile due to severe sperm motility defects, even though the sperm have a normal axoneme ultrastructure [8]. In the *Spag17*-mutant mouse, the rapid neonatal demise is associated with a profound respiratory phenotype characterized by immotile cilia and defects in the 9+2 axonemal structure [12]. This is not observed in cilia from knockouts of the *Spag6* and *Spag16* genes. Although the genetic background of mutant mice may significantly influence the phenotypes, the observations summarized above suggest that specific central pair genes may have unique roles in different cell types.

In conclusion, our studies demonstrate that SPAG6 deficiency causes multiple abnormalities in the function of cilia and ciliated cells, including defects in sperm flagellar motility; ciliogenesis; ciliary beat; axoneme orientation; cell morphology and polarity.

Supporting Information

Figure S1 Analysis of cilia in the trachea epithelial cells and brain ependymal cells by scanning electronic microscopy. Tracheas and brains from wild type and Spag6-deficient mice were processed for SEM. Notice that cilia in the brains (A) and trachea (C) of the wild-type animals sit on the cell surface in a highly ordered state. However, cilia in the ependymal cells (B) and trachea (D) of Spag6-deificent mice appeared to be disordered on the cell surface. Fig. S1 shows cilia in the trachea epithelial cells and brain ependymal cells by scanning electronic microscopy with high magnification.

Figure S2 Circular plots of tracheal epithelial cell basal foot orientation in three individual *Spag6*-deficient (upper) and three individual wild-type mice (lower). Five TEM images were randomly selected from each mouse and basal foot orientations were measured. Arrow direction represents the mean vector of cilium orientation per cell; arrow length is the length of the mean vector, with longer arrows indicating stronger coordination of orientation. r_{cell} is the length of mean vector and describes rotational orientation.

Figure S3 Analysis of Vangl2 protein expression level in the lung/trachea by Western blotting. Lungs/tracheas from three wild-type and three *Spag6*-deficient mice were homogenized and Western blotting was performed with anti-Vangl2 antibody, the membrane was striped and re-probed with an anti-actin antibody as a loading control. There was no difference in Vangl2 protein expression level between the wild-type and *Spag6*-deficient mice.

Figure S4 Examination of Vangl2 and α–tubulin localization in trachea epithelial cells of three-week old mice. The distribution of the PCP protein, Vangl2, and, α-tubulin was examined by immunofluorescence staining. More intense signal was detected in the apical regions in wild-type trachea epithelial cells (arrows in A for Vangl2 and arrowheads in C for α-tubulin). These proteins appeared to be distributed evenly throughout the cytoplasm in cells from *Spag6* mutant mice (dashed arrows in B). In the trachea of wild-type mice, cilia were also intensively stained by an anti-α-tubulin antibody (arrows in C and D).

Figure S5 Examination of α-tubulin localization in trachea epithelial cells in one-week old mice. Distribution of α-tubulin is polarized in the wild-type mice (arrowheads in upper panel). However, the polarized pattern is not seen in the *Spag6*-deficient mice (lower panel), where α-tubulin is evenly distributed throughout the cytoplasm.

Video S1 Trachea ciliary beat observed by video microscopy. Cilia from wild-type mice (*Spag6* knockout littermate) beat at a fast rate, and the beat was coordinated, with all the cilia beating in the same direction at a specific time point.

Video S2 Airway clearance in wild-type mice. Real time video showing the efficiency of ciliated epithelium in moving particles (blood cells) in the trachea. Arrows indicate the direction flow.

Video S3 Cilia from *Spag6*-deficient mice beat at much slower rate, and the beating is largely uncoordinated. At specific time points, some cilia beat in one direction, while others beat in the opposite direction.

Video S4 *Spag6*-deficient mice fail to clear particles from the airway. Uncoordinated cilia from tracheal epithelium failed to generate blood cell flow. Arrows indicate the presence of blood cells stacked at the beginning of tracheal tube. Video is shown in real time.

Acknowledgments

We thank Dr. Eszter K. Vladar in Dr. Jeffrey Axelrod's laboratory at Department of Pathology, Stanford University for assistance with circular statistics. We thank Dr. Bruce. K. Rubin, Professor and Chair of Department of Pediatric, Virginia Commonwealth University for his comments and edits. We thank Ling Zhang, associate professor of Wuhan University of Science and Technology for assistance with statistics on cilia and basal foot number.

Author Contributions

Conceived and designed the experiments: ZZ. Performed the experiments: MET WL PRS ZGZ WT LVR ZZ. Analyzed the data: CWD MRK JFS ZZ. Contributed reagents/materials/analysis tools: RMC MRK JFS ZZ. Wrote the paper: ZZ JFS MET.

References

1. Sapiro R, Tarantino LM, Velazquez F, Kiriakidou M, Hecht NB, et al. (2000) Sperm antigen 6 is the murine homologue of the Chlamydomonas reinhardtii central apparatus protein encoded by the PF16 locus. Biol Reprod 62: 511–518.
2. Smith EF, Lefebvre PA (1996) PF16 encodes a protein with armadillo repeats and localizes to a single microtubule of the central apparatus in Chlamydomonas flagella. J Cell Biol 132: 359–370.
3. Branche C, Kohl L, Toutirais G, Buisson J, Cosson J, et al. (2006) Conserved and specific functions of axoneme components in trypanosome motility. J Cell Sci 119(Pt 16): 3443–3455.
4. Straschil U, Talman AM, Ferguson DJ, Bunting KA, Xu Z, et al. (2010) The Armadillo repeat protein PF16 is essential for flagellar structure and function in Plasmodium male gametes. PLoS One 5: e12901.
5. House SA, Richter DJ, Pham JK, Dawson SC (2011) Giardia flagellar motility is not directly required to maintain attachment to surfaces. PLoS Pathog 7: e1002167.
6. Sapiro R, Kostetskii I, Olds-Clarke P, Gerton GL, Radice GL, et al. (2002) Male infertility, impaired sperm motility, and hydrocephalus in mice deficient in sperm-associated antigen 6. Mol Cell Biol 22: 6298–6305.
7. Zhang Z, Tang W, Zhou R, Shen X, Wei Z, et al. (2007) Accelerated mortality from hydrocephalus and pneumonia in mice with a combined deficiency of SPAG6 and SPAG16L reveals a functional interrelationship between the two central apparatus proteins. Cell Motil Cytoskeleton 64(5): 360–376.
8. Zhang Z, Kostetskii I, Tang W, Haig-Ladewig L, Sapiro R, et al. (2006) Deficiency of SPAG16L causes male infertility associated with impaired sperm motility. Biol Reprod 74(4): 751–759.
9. Sisson JH, Stoner JA, Ammons BA, Wyatt TA (2003) All-digital image capture and whole field analysis of ciliary beat frequency. J Microsc 211: 103–111.
10. Tsuneoka M, Nishimune Y, Ohta K, Teye K, Tanaka H, et al. (2006) Expression of Mina53, a product of a Myc target gene in mouse testis. Int J Androl 29: 323–330.
11. Qiu H, Gołas A, Grzmil P, Wojnowski L (2013) Lineage-specific duplications of Muroidea Faim and Spag6 genes and atypical accelerated evolution of the parental Spag6 gene. J Mol Evol 77(3): 119–29.
12. Teves ME, Zhang Z, Costanzo RM, Henderson SC, Corwin FD, et al. (2013) Spag17 is Essential for Motile Cilia Function and Neonatal Survival. Am J Respir Cell Mol Biol 48(6): 765–772.
13. Kunimoto K, Yamazaki Y, Nishida T, Shinohara K, Ishikawa H, et al. (2012) Coordinated ciliary beating requires Odf2-mediated polarization of basal bodies via basal feet. Cell 148: 189–200.
14. Mitchell B, Jacobs R, Li J, Chien S, Kintner C (2007) A positive feedback mechanism governs the polarity and motion of motile cilia. Nature 447: 97–101.
15. Guirao B, Meunier A, Mortaud S, Aguilar A, Corsi JM, et al. (2010) Coupling between hydrodynamic forces and planar cell polarity orients mammalian motile cilia. Nat Cell Biol 12: 341–350.
16. Appelbe OK, Bollman B, Attarwala A, Triebes LA, Muniz-Talavera H, et al. (2013) Disruption of the mouse Jhy gene causes abnormal ciliary microtubule patterning and juvenile hydrocephalus. Dev Biol 382(1): 172–85.
17. Matsuo M, Shimada A, Koshida S, Saga Y, Takeda H (2013) The establishment of rotational polarity in the airway and ependymal cilia: analysis with a novel cilium motility mutant mouse. Am J Physiol Lung Cell Mol Physiol 304(11): L736–45.
18. Dawe HR, Farr H, Gull K (2007) Centriole/basal body morphogenesis and migration during ciliogenesis in animal cells. J Cell Sci 120: 7–15.
19. Boisvieux-Ulrich E, Lainé MC, Sandoz D (1990) Cytochalasin D inhibits basal body migration and ciliary elongation in quail oviduct epithelium. Cell Tissue Res 259: 443–454.
20. Sorokin SP (1968) Reconstructions of centriole formation and ciliogenesis in mammalian lungs. J Cell Sci 3: 207–230.
21. Park TJ, Haigo SL, Wallingford JB (2006) Ciliogenesis defects in embryos lacking inturned or fuzzy function are associated with failure of planar cell polarity and Hedgehog signaling. Nat Genet 38: 303–311.
22. Pan J, You Y, Huang T, Brody SL (2007) RhoA-mediated apical actin enrichment is required for ciliogenesis and promoted by Foxj1. J Cell Sci 120: 1868–1876.
23. Goodrich LV, Strutt D (2011) Principles of planar polarity in animal development. Development 138: 1877–1892.
24. Wallingford JB, Mitchell B (2011). Strange as it may seem: the many links between Wnt signaling, planar cell polarity, and cilia. Gene Dev 25: 201–213.
25. Ybot-Gonzalez P, Savery D, Gerrelli D, Signore M, Mitchell CE, et al. (2007) Convergent extension, planar-cell-polarity signalling and initiation of mouse neural tube closure. Development 134: 789–799.
26. Dworkin S, Jane SM, Darido C (2011) The planar cell polarity pathway in vertebrate epidermal development, homeostasis and repair. Organogenesis 7: 202–208.
27. Wallingford JB (2010) Planar cell polarity signaling, cilia and polarized ciliary beating. Curr Opin Cell Biol 22: 597–604.
28. Mirzadeh Z, Han YG, Soriano-Navarro M, García-Verdugo JM, Alvarez-Buylla A (2010) Cilia organize ependymal planar polarity. J Neurosci 30: 2600–2610.
29. Mitchell B, Stubbs JL, Huisman F, Taborek P, Yu C, et al. (2009) The PCP pathway instructs the planar orientation of ciliated cells in the Xenopus larval skin. Curr Biol 19: 924–929.
30. Pearson CG, Giddings TH Jr, Winey M (2009) Basal body components exhibit differential protein dynamics during nascent basal body assembly. Mol Biol Cell 20: 904–914.
31. Kilburn CL, Pearson CG, Romijn EP, Janet B, Meehl JB, et al. (2007) New Tetrahymena basal body protein components identify basal body domain structure. J Cell Biol 178: 905–912.
32. Park TJ, Mitchell BJ, Abitua PB, Kintner C, Wallingford JB (2008) Dishevelled controls apical docking and planar polarization of basal bodies in ciliated epithelial cells. Nat Genet 40: 871–879.
33. Tissir F, Qu Y, Montcouquiol M, Zhou L, Komatsu K, et al. (2010) Lack of cadherins Celsr2 and Celsr3 impairs ependymal ciliogenesis, leading to fatal hydrocephalus. Nat Neurosci 13(6):700–707.
34. Vladar EK, Bayly RD, Sangoram AM, Scott MP, Axelrod JD (2012) Microtubules enable the planar cell polarity of airway cilia. Curr Biol 22(23): 2203–2212.
35. Sandoz D, Chailley B, Boisvieux-Ulrich E, Lemullois M, Laine MC, et al. (1988) Organization and functions of cytoskeleton in metazoan ciliated cells. Biol Cell 63: 183–193.
36. Eaton S, Wepf R, Simons K (1996) Roles for Rac1 and Cdc42 in planar polarization and hair outgrowth in the wing of Drosophila. J Cell Biol 135:1277–1289.
37. Turner CM, Adler PN (1998) Distinct roles for the actin and microtubule cytoskeletons in the morphogenesis of epidermal hairs during wing development in Drosophila. Mech Dev 70: 181–192.
38. Shimada Y, Yonemura S, Ohkura H, Strutt D, Uemura T (2006) Polarizedtransport of Frizzled along the planar microtubule arrays in Drosophila wing epithelium. Dev Cell 10: 209–222.
39. Hannus M, Feiguin F, Heisenberg CP, Eaton S (2002) Planar cell polarization requires Widerborst, a B′ regulatory subunit of protein phosphatase 2A. Development 129: 3493–3503.
40. Chailley B, Nicolas G, Lainé MC (1989) Organization of actin microfilaments in the apical border of oviduct ciliated cells. Biol Cell 67: 81–90.
41. Lee L, Campagna DR, Pinkus JL, Mulhern H, Wyatt TA, et al. (2008) Primary ciliary dyskinesia in mice lacking the novel ciliary protein Pcdp1. Mol Cell Biol 28(3): 949–957.
42. Sironen A, Kotaja N, Mulhern H, Wyatt TA, Sisson JH, et al. (2011) Loss of SPEF2 function in mice results in spermatogenesis defects and primary ciliary dyskinesia. Biol Reprod 85(4): 690–701.

2

Maternal Condition but Not Corticosterone Is Linked to Offspring Sex Ratio in a Passerine Bird

Lindsay J. Henderson[1,2]*, Neil P. Evans[2], Britt J. Heidinger[2,3], Aileen Adams[2], Kathryn E. Arnold[2,4]

1 Department of Neurobiology, Physiology and Behavior, University of California Davis, Davis, California, United States of America, 2 College of Medical, Veterinary & Life Sciences, The University of Glasgow, Glasgow, United Kingdom, 3 Department of Biological Sciences, North Dakota State University, Fargo, North Dakota, United States of America, 4 Environment Department, The University of York, York, United Kingdom

Abstract

There is evidence of offspring sex ratio adjustment in a range of species, but the potential mechanisms remain largely unknown. Elevated maternal corticosterone (CORT) is associated with factors that can favour brood sex ratio adjustment, such as reduced maternal condition, food availability and partner attractiveness. Therefore, the steroid hormone has been suggested to play a key role in sex ratio manipulation. However, despite correlative and causal evidence CORT is linked to sex ratio manipulation in some avian species, the timing of adjustment varies between studies. Consequently, whether CORT is consistently involved in sex-ratio adjustment, and how the hormone acts as a mechanism for this adjustment remains unclear. Here we measured maternal baseline CORT and body condition in free-living blue tits (*Cyanistes caeruleus*) over three years and related these factors to brood sex ratio and nestling quality. In addition, a non-invasive technique was employed to experimentally elevate maternal CORT during egg laying, and its effects upon sex ratio and nestling quality were measured. We found that maternal CORT was not correlated with brood sex ratio, but mothers with elevated CORT fledged lighter offspring. Also, experimental elevation of maternal CORT did not influence brood sex ratio or nestling quality. In one year, mothers in superior body condition produced male biased broods, and maternal condition was positively correlated with both nestling mass and growth rate in all years. Unlike previous studies maternal condition was not correlated with maternal CORT. This study provides evidence that maternal condition is linked to brood sex ratio manipulation in blue tits. However, maternal baseline CORT may not be the mechanistic link between the maternal condition and sex ratio adjustment. Overall, this study serves to highlight the complexity of sex ratio adjustment in birds and the difficulties associated with identifying sex biasing mechanisms.

Editor: Alexandre Roulin, University of Lausanne, Switzerland

Funding: LJH, KEA, and NPE received PhD funding from Natural Environment Research Council (award number NE/H526886/1, http://www.nerc.com/). KEA was also supported by a Royal Society University Research Fellowship (https://royalsociety.org/). The funders had no role in study design, data collection and analysis, decision to publish, or preparation of the manuscript.

Competing Interests: The authors have declared that no competing interests exist.

* Email: lindsayhenderson@hotmail.com

Introduction

Maternal quality and natal conditions can affect the survival and reproductive potential of offspring in a sex-specific manner [1,2]. Therefore mothers breeding in favourable conditions are predicted to gain fitness benefits from investing in the sex that will benefit most from those conditions, and vice versa [2,3]. In agreement with sex allocation theory there are both experimental and correlative studies that demonstrate that brood sex ratio adjustment is associated maternal condition and natal conditions in avian species [4,5,6,7,8]. However, the replication of results has proved difficult, with outcomes differing between years and studies [7,9]. Moreover, predicting the direction of a sex ratio bias has been problematic, with evidence of no bias or the opposing bias from that expected in empirical studies [10,11,12]. Overall, the variety of avian life histories, extended parental care and the array of factors that could influence the benefits of sex ratio manipulation cause the prediction of sex ratio adjustment in birds to be complex [13]. In this case, identifying a mechanism of sex ratio adjustment would offer insight into the potential costs of manipulation and may improve predictions of when sex ratio adjustment is expected to occur [14].

In birds females are the heterogametic sex (producing Z- and W-bearing ova), therefore it has been suggested that primary sex ratio adjustment (occurring prior to laying) could be under maternal control [15]. The steroid hormone corticosterone (CORT) has been proposed to play a role in brood sex ratio adjustment, as it is modulated in response to factors implicated in sex ratio adjustment, including maternal body condition, environmental conditions and partner quality [5,8,16,17,18]. Baseline CORT concentrations fluctuate within individuals in response to the prevalent conditions, to maintain homeostasis and energy-balance through their effects on behaviour and physiological processes [19]. Due to this, elevated baseline CORT has been associated with poor body condition [5,17,20,21], inclement environmental conditions and incompatible breeding partners in birds [22,23,24,25]. Therefore it would be expected to be associated with investment in the sex whose survival and reproductive success is least affected by poor developmental conditions [14,17]. In agreement with this hypothesis correlative and experimental studies have found a link between elevated CORT and female biased brood sex ratios, in species where males are the larger sex and therefore may be more sensitive to poor natal conditions than females [5,16,17]. In addition, in the Gouldian finch (*Erythrura gouldiae*) genetically incompatible breeding pairs have an 84% greater mortality of daughters compared to compatible pairs. Elevated CORT in females constrained to mate with genetically incompatible partners, has been correlatively and causally linked to the adaptive overproduction of male offspring [18,26].

The mechanism by which CORT could influence the sex of offspring is currently unknown, but could potentially act at the pre- or post-laying stage. A number of hypothesis have been posited including, differential segregation of the sex chromosomes during meiosis [27] or selective reabsorption of ova dependent upon sex, which could result in laying gaps or infertile eggs due to selective fertilization of ova based on sex [14]. Alternatively, yolk CORT concentrations can influence hatching success [28], nestling growth [29] and nestling survival [17,30], thus could affect brood sex ratio through early embryo death or sex-specific nestling mortality. Finally, mothers with elevated CORT may provide inferior parental care, thus negatively impacting offspring quality and survival [17]. The methods that have been used to experimentally elevate CORT and investigate the effects upon sex ratio do not provide unequivocal evidence to establish the mechanism through which CORT acts. For example, CORT has been experimentally elevated within baseline and above baseline levels for prolonged periods (days) during egg laying [16,17,26]. This method could have knock-on physiological effects, such as changes in glucose availability, which has also been implicated in brood sex ratio adjustment [31]. This methodology could also influence maternal care and yolk concentrations [17], which could result in sex ratio adjustment after laying. Interestingly, studies provide evidence for both of these hypotheses as there is evidence of causal relationship between maternal baseline CORT and brood sex ratio both at laying [5,11,16] and at fledging [17]. Alternatively, to address whether CORT directly affects offspring sex, other studies have elevated CORT transiently during sex determining meiosis [11,12]. However, these studies have used pharmacological CORT concentrations and have resulted in a offspring sex ratio bias opposite to that predicted from sex allocation theory [11,12]. Ultimately, there remains a lack of consensus as to how maternal CORT may cause brood sex ratio adjustment.

There have been a number of studies that present evidence of brood sex ratio adjustment in the blue tit (*Cyanistes caeruleus*) [32,33,34,35,36]. Specifically, studies have examined whether females paired with attractive males, can increase their fitness by investing in sons rather than daughters, as sons may inherit their fathers attractiveness [34]. Although there is some correlative and causal evidence of a link between paternal attractiveness and male-biased brood sex ratio in blue tits, the results have proved difficult to replicate and have varied between years [33,36]. As maternal baseline CORT has been shown to be higher in females with unattractive mates [37,38], CORT could be the mechanism through which females could adjust brood sex ratio in response to male attractiveness [18]. There is also evidence for assortative mating in blue tits [39], thus females paired with high quality attractive partners may also be of superior quality. However, to date the relationship between maternal condition, maternal baseline CORT and brood sex ratio has not been investigated in blue tits.

In this study, a free-living population of blue tits were monitored for three years to assess the relationship between maternal baseline CORT, maternal condition and brood sex ratio. To identify the potential mechanisms of sex ratio adjustment, both primary and secondary sex ratio were established. In addition, laying gaps were recorded to identify evidence of potential reabsorption of ova, and un-hatched eggs were analysed to test for sex-biased fertilization or embryo death. Nestling mass and growth were also measured to investigate whether maternal baseline CORT and/or maternal condition had sex-specific effects on nestlings, therefore indicating the potential benefits of brood sex ratio adjustment. In the final year of the study a field-based experiment was conducted to investigate whether transient elevation of maternal CORT at sex determining meiosis during egg laying influenced brood sex ratio. In order to ensure CORT concentrations were elevated during sex determination we acutely elevated CORT above baseline levels but within the physiological range for this species. We predicted that reduced maternal condition and elevated maternal CORT would be associated with reduced nestling quality and a female biased sex ratio, as males are the larger sex and may be more sensitive to poor conditions.

The main aims of this study were to investigate whether; 1) endogenous maternal baseline CORT and maternal body condition were correlated with primary or secondary brood sex ratio, 2) maternal condition and/or baseline CORT concentrations were related to nestling mass and growth rate, 3) maternal baseline CORT and body condition were correlated and 4) experimental elevation of maternal CORT during egg laying influenced brood sex ratio and offspring quality.

Methods

Empirical study

Blue tits breeding in nest boxes in oak-dominated woodland around Loch Lomond, Scotland (56° 13′ N, 4° 13′ W) were studied for three years from April to June 2008–2010. Nest boxes were monitored regularly from the onset of nest building to establish the date the first egg was laid (lay date). Nests were then checked every second day and eggs were counted to establish clutch size. In addition, as blue tits lay one egg per day [40], this allowed laying gaps to be identified. When eggs were found to be warm and no new eggs had been laid on 2 consecutive visits, incubation was deemed to have started and mothers were left undisturbed for 10 days. Nests were then visited every day to establish hatching date, when >50% of eggs had hatched this was considered day 1. All un-hatched eggs and dead nestlings were collected for molecular sexing (see Molecular sex identification).

To measure maternal baseline CORT, birds were caught during provisioning, on day 5–7 after hatching. Mothers were captured on the nest by blocking the entrance hole, and a small blood sample was obtained (about 80–100 µl) after puncture of the brachial vein. All samples were collected within 3 minutes of initial blockage of the nest box entrance. Blood samples were immediately stored on ice and separated through centrifugation within 2 h of collection. Occasionally researchers had to wait for a short period near the nest for birds to enter the nest box, when this occurred the duration of the time at the nest was noted. CORT samples were considered to be baseline because time spent at the nest before capture, time between sampling and blockage of the nest box entrance and time of day were not related to maternal CORT (GLM: $n = 79$, time at nest; $t = 1.26$, $P = 0.21$, sampling time; $t = -0.01$, $P = 0.99$ and time of day; $t = -1.25$, $P = 0.22$). The plasma portion of the sample was removed and stored at $-20°C$ until assay. Circulating CORT concentrations were measured using a double antibody radioimmunoassay (for full details see [41]). CORT was measured in three assays for which the detection limit was 0.03 ng/ml and the intra-and inter-assay variation was 9±2% and 10±5% respectively.

Baseline CORT was measured during provisioning rather than egg-laying for two reasons, i) previous studies that have found a link between maternal baseline CORT and brood sex ratio, have blood sampled mothers post egg-laying [5,16], and ii) baseline CORT concentrations did not differ significantly between females measured during laying or provisioning (figure 1, females matched based on lay date, n = laying: 9, provisioning: 10, t-Test: $t = -1.08$, $P = 0.30$).

To establish body condition, mothers were weighed to the nearest 0.05 g with a Pesola spring balance and wing length was measured. Maternal condition was established by taking the residuals from a linear regression of mass (g) and wing length (mm). Parental birds were sexed based on presence/absence of a brood patch and aged based on plumage characteristics [42] and all birds captured were fitted with a uniquely numbered aluminium ring (British Trust for Ornithology). As some mothers were measured for body condition but not baseline CORT sample sizes differ between analyses.

Experimental manipulation of maternal CORT

To deliver CORT non-invasively to breeding blue tits, CORT solution was injected into mealworms that were placed in nest boxes during egg-laying. Sex determining meiotic division occurs 2–4 hrs before ovulation [43], with oviposition occurring approximately 24 hrs later [44,45]. As blue tits lay in the early hours of the morning [46], sex determination is expected to occur during the night, approximately 28 hrs before oviposition. Therefore, mothers were fed mealworms injected with exogenous CORT or without (Control) in the evening to coincide with sex determination to establish whether maternal CORT influenced the sex ratio at laying. We used a transient method of CORT elevation to control for knock-on physiological effects that could result from chronic elevation of CORT.

Crystalline CORT (Sigma) was dissolved directly into peanut oil (Sigma) through sonification. To ensure the CORT was dissolved evenly in the peanut oil, the solution was sonicated before each

use. Mealworms were injected with 20μl of peanut oil containing one of the following concentrations of CORT: (1) Control, no CORT or (2) CORT, 0.3 mg/ml. Hence, mothers received 0 or 6μg CORT per mealworm. We injected solution into mealworms with a 25-μl Hamilton syringe using a 26-gauge ½-inch needle. Prior to injection, mealworms were kept at −20°C to reduce movement during injection. The needle was inserted ventrally, into the anterior abdomen, between two segments. If fluid leaked from the mealworm after injection, it was not used. We used mealworms of approximately 20 mm and 0.1 g in size. This methodology was validated under lab conditions where blue tits fed 6μg CORT injected mealworms had significantly higher CORT 10 minutes after ingestion than blue tits fed Control mealworms (CORT: $n = 4$, mean±SE, 39.9±4.0 ng/ml, Control: $n = 4$, 3.1±0.95 ng/ml, t-Test: $t = 5.54$, $P = 0.001$). CORT concentrations did not differ significantly between groups 30 mins after ingestion (t-Test: $t = 0.96$, $P = 0.41$). The circulating CORT concentrations achieved after spiked mealworm consumption were within 1 SE of the mean concentrations ($n = 5$, mean±SE, 28.2±11.2 ng/ml) found in blue tits after subjection to a standardized capture and restraint procedure [47]. Therefore, CORT concentrations were elevated within the natural range for this species. For full validation methods please see Materials S1.

In 2010 prior to the onset of nest building, small plastic cups (40×20 mm) were secured on the inside of all nest boxes to later provide a tray for mealworms. Nest boxes were monitored weekly to establish the onset of nest building. When nests were found to be half to fully constructed they were checked daily for eggs. When the first egg was laid, the nest was randomly assigned by the toss of a coin to the CORT or Control group. Beginning that day, a CORT spiked (6 μg) or Control mealworm was placed into the plastic tray every evening between 17:30 and 19:30, until no more eggs had been laid on 2 consecutive visits. This time was chosen because female blue tits have been found to roost as early as 19:00 during egg laying in Scotland [48]. Mothers received their first mealworm to coincide with the sex determination of their third egg. The mean clutch size was 10.9±1.5, consequently on average>80% of eggs laid were manipulated. If we assume a 0.5 sex ratio for the two un-manipulated eggs, with a clutch size of 10, a consistent female sex bias caused by our experiment of>0.1 in the manipulated eggs, would cause a brood sex ratio of <0.42.

To check that the treatment targeted female rather than male blue tits, a hide was erected close to a sub-sample of nest boxes during egg laying (n = CORT: 4, Control: 3). The nests were then monitored after the mealworm was placed in the nest until sunset and then checked the following morning before 06:00. This was to record activity at the nest during this time and to establish if the mealworm had been consumed during the night. The mealworm was consumed by 06:00 for each nest box checked. On only one occasion a bird was recorded to enter the nest on more than one occasion before roosting. For all other nests only one bird was recorded to enter the nest and not leave. As female rather than male breeding blue tits roost in the nest box during laying [48], when only one bird was seen entering and not leaving the nest before sunset, it was assumed to be the female. Therefore our observations suggest that only mothers consumed the mealworms between 19:00–06:00. The progress of all manipulated nests were followed as previously described, and nestlings weighed and sexed as below.

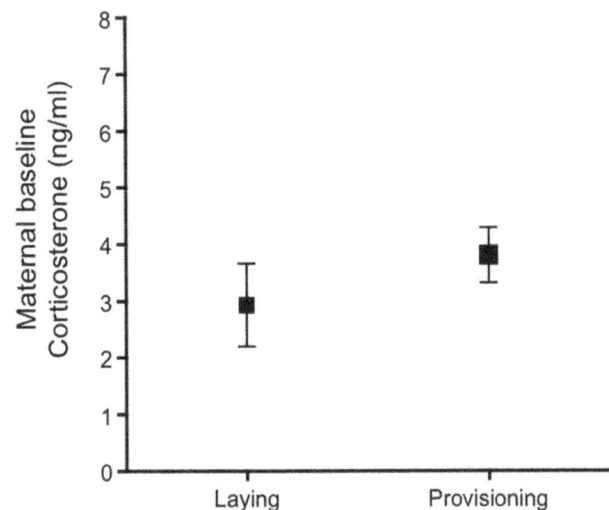

Figure 1. Maternal baseline CORT did not significantly differ when measured during egg laying versus nestling provisioning (n=laying: 9, provisioning: 10). Graph shows mean±SE.

Nestling condition

All individually marked nestlings were weighed to the nearest 0.01 g with a digital balance on day 4, 6, 8, 10 and 14 after hatching. Nestling growth rate was calculated individually as mass gain day^{-1} from day 4–14 after hatching. On day 14 chicks were fitted with a uniquely numbered aluminium ring (British Trust for Ornithology).

Molecular sex identification

At the age of 14 days, nestlings were blood sampled to provide DNA for molecular sex identification. All nestlings that were collected dead before this time and un-hatched eggs where development was evident were also sexed. A salt extraction [49] or Qiagen DNeasy kits were employed for DNA extraction. Primers were P2/P8 [50]. PCR amplification was carried out in a total volume of 10μl. The final reaction conditions were as follows: 0.8μM of each primer, 200μM of each dNTP, target DNA, 0.35 units GoTaq polymerase (Promega), 2μM (5×) GoTaq Flexi Buffer (Promega) and 2μM of 25 mM $MgCl_2$. Thermal cycling was carried out in a Biometra UnoII: 94°C/2 min, 30 cycles of (49°C/40 s, 72°C/40 s and 94°C/30 s), 49°C/1 min, 72°C/ 5 min. PCR products were separated by electrophoresis on a 2% agarose gel stained with ethidium bromide.

Where possible both the primary sex ratio (Pri), i.e. the sex ratio of all eggs laid, and the secondary sex ratio (Sec), i.e. the sex ratio of all nestlings that fledged was calculated for each nest. However, some of the un-hatched eggs collected showed no evidence of development, therefore they could not be sexed and are hereon referred to as 'unviable' eggs. In this case, the nests were included in primary sex ratio analysis if all the remaining offspring were sexed. For some nests it was not possible to calculate primary sex ratio as dead nestlings or eggs were lost prior to sexing, usually because the mother removed them from the nest. In this case only secondary sex ratio was calculated. Due to this samples sizes of primary and secondary sex ratio differ. Overall 91.6% of eggs laid were sexed ($n = 1511$) from 150 un-manipulated broods. Nests were not included in the analysis if any eggs were accidentally broken or molecular sexing was not successful for an individual egg or chick.

Ethical note

In order to minimise disruption to parents and nestlings a number of precautions were followed. All adult birds were captured and blood sampled within 15 minutes of initial disturbance at the nest, on the majority of occasions birds were disturbed for less than 10 minutes. Furthermore, nestlings were held on heat pads to prevent them from getting chilled and never disturbed for longer than 30 minutes. During the experimental study, nest visits were carried out as quickly as possible to minimise disturbance, and because maternal CORT was elevated within the natural range recorded for this species, the physiological stress mothers were subjected to was comparable to concentrations commonly experienced by the birds. All work was conducted under licence from the UK Home Office and was subject to review by the University of Glasgow ethics committee. Fieldwork was conducted on land leased by The University of Glasgow. This study did not involve endangered or protected species.

Statistical analysis

Generalized Linear Models (GLMs) with binomial errors and no explanatory terms were used to test if brood sex ratios differed from a binomial distribution [36,51]. To investigate whether year, maternal body condition, maternal baseline CORT, lay date, and maternal age explained variation in brood sex ratio, GLMs with a binomial error structure and a logit link function, weighted by brood size were used [51]. Overdispersion was not deemed a problem as the residual mean deviance (residual deviance/residual d.f.) was always less than 1.5 [51]. All CORT data were square

root transformed because of non-normality. GLMs were used to establish whether maternal condition, lay date and age explained variance in maternal baseline CORT. GLMs with a binomial error structure were employed to establish whether the occurrence of laying gaps and unviable eggs (present:1, absent:0) were related to year, experimental treatment, maternal baseline CORT and maternal body condition. General Linear Mixed Models (GLMMs), with brood ID as the random factor, were employed to assess whether nestling mass on day 14 after hatching and nestling growth rate were explained by year, sex, lay date, brood size, maternal baseline CORT and maternal condition. All models were run twice, once with maternal CORT and once with maternal condition. In addition, all two-way interactions with year or sex, and the interaction between maternal body condition and baseline CORT were included in full models.

For the experimental study in 2010, GLMs were used to establish whether maternal condition, maternal CORT and lay date differed between control, CORT and un-manipulated broods. GLMs with a Poisson error structure were used to compare clutch size and number fledged between treatment groups. Similar to the empirical study, Generalized Linear Models with a binomial error structure and a logit link function, weighted by brood size were used to investigate whether the treatment group affected brood sex ratio [51]. GLMMs were used to establish whether nestling mass on day 14 after hatching and nestling growth rate were affected by the treatment.

All models were optimised using backward elimination of non-significant terms. Model validations were applied where appropriate and the underlying statistical assumptions of normality and homogeneity of variance were verified. All statistical analyses were conducted using R version 2.8.0, and the nlme library was used for GLMMs. For raw data please see Data S1.

Results

Population brood sex ratio

Population primary and secondary brood sex ratio did not depart from a binomial distribution (Pri: $z = 1.37$, $P = 0.17$, Sec: $z = 0.68$, $P = 0.49$). Furthermore, population brood sex ratios did not differ between years (Pri: $z = -0.60$, $P = 0.55$, Sec: $z = -1.40$, $P = 0.16$).

When the sex of both un-hatched eggs and nestlings that died before fledging was analysed, there was no indication of sex-biased offspring mortality (Yates' corrected: $n =$ M: 37, F: 30, $X^2 = 0.73$, $P = 0.39$). The number of mothers that exhibited laying gaps did not differ between years ($z = -0.55$, $P = 0.58$). However, the number of mothers that laid unviable eggs (where development was not detected) was significantly lower in 2010 compared with the other two years, with 55%, 38% and 18% of broods containing unviable eggs in 2008–2010 respectively ($z = -3.21$, $P = 0.001$).

Maternal baseline corticosterone, body condition and brood sex ratio

Maternal body condition was significantly correlated with both primary and secondary brood sex ratio in a year-specific manner (table 1). In 2010 only, mothers in superior condition had more male biased broods (figure 2a, Pri: $z = 3.12$, $P = 0.002$, Sec: $z = 3.19$, $P = 0.001$). Maternal CORT (figure 2b, Pri: $z = 0.30$, $P = 0.77$, Sec: $z = 0.51$, $P = 0.61$), maternal age and lay date were not related to brood sex ratio.

Maternal body condition and CORT did not predict whether mothers exhibited laying gaps (CORT: $z = -0.99$, $P = 0.32$, Condition: $z = 0.38$, $P = 0.70$). Also, whether mothers laid unviable eggs was not associated with maternal body condition

Table 1. The results of GLMs investigating whether maternal body condition influenced a) primary ($n = 89$) and b) secondary ($n = 88$) brood sex ratio.

Factor	d.f. Effect	SE	z	P
a) Primary				
Year	2	0.24	0.21	0.84
Maternal condition	1	0.19	−0.17	0.86
Age	1	0.17	−0.04	0.97
Lay date	1	0.02	0.64	0.52
Maternal condition *Year	2	0.38	2.75	**0.006**
b) Secondary				
Year	2	0.25	−0.40	0.69
Maternal condition	1	0.20	−0.16	0.87
Age	1	0.16	−0.05	0.96
Lay date	1	0.03	0.55	0.58
Maternal condition *Year	2	0.37	2.81	**0.005**

or CORT (CORT: $z = -0.09$, $P = 0.93$, Condition: $z = 0.17$, $P = 0.86$).

Maternal baseline corticosterone, body condition and nestling condition

In all years, mothers in superior body condition had heavier and faster growing offspring than mothers in poor condition (table 2, figure 3ab). Maternal body condition did not influence nestling mass or growth in a sex-specific manner (Mass: Body Condition × Sex; $t = -1.00$, $P = 0.32$ and Growth: Body Condition × Sex: $t = -0.65$, $P = 0.52$). Male nestlings were consistently heavier and grew at a faster rate than their female siblings in all years (table 2, figure 3ab). Nestling mass and growth rate did not differ between the years (table 2).

Maternal baseline CORT was negatively correlated with nestling mass, but not nestling growth (figure 3c, Mass; $t = -2.18$, $P = 0.03$ and Growth; $t = -1.60$, $P = 0.12$). Maternal baseline CORT did not explain variation in nestling mass or growth in a sex-specific manner (Mass: Sex × CORT; $t = -0.90$, $P = 0.37$, Growth: Sex × CORT; $t = -1.59$, $P = 0.11$).

Maternal baseline corticosterone and body condition

In 2008 and 2009, maternal CORT was significantly higher than 2010 (figure 4, $t = -4.39$, $P < 0.001$). However, maternal body condition was significantly lower in 2009 compared with the other two years (figure 4, $t = -2.00$, $P = 0.048$). Maternal body condition did not explain variation in maternal baseline CORT ($t = -1.02$, $P = 0.31$).

Experimental study

There were 48 nests manipulated and monitored until fledging in 2010 ($n = $ Control 24, CORT 24). In addition, we matched all experimental nests by date with 24 un-manipulated control nests. Lay date ($t = 0.33$, $P = 0.74$), clutch size ($z = 0.22$, $P = 0.83$), number fledged ($z = -0.32$, $P = 0.74$), maternal body condition ($t = 0.56$, $P = 0.58$) and maternal CORT (figure 5, $t = -1.76$, $P = 0.09$) did not differ between treatments. In addition, the number of mothers that exhibited laying gaps ($z = -0.56$, $P = 0.58$) and laid unviable eggs ($z = -1.49$, $P = 0.14$) did not differ between the treatments.

Primary and secondary brood sex ratios did not differ between treatments (figure 6, Pri: $z = -0.47$, $P = 0.64$, Sec: $z = -0.05$, $P = 0.96$). There was a marginally non-significant trend that the interaction term Treatment x Maternal body condition predicted offspring sex ratio (Pri: $z = 1.86$, $P = 0.06$, Sec: $z = 1.80$, $P = 0.07$). When the treatments were analysed individually, maternal body condition was positively correlated with male biased broods in the un-manipulated group only (Pri: Un-manipulated, $z = 3.02$, $P = 0.002$, Control, $z = 0.40$, $P = 0.69$, CORT, $z = -0.07$, $P = 0.95$; Sec: Un-manipulated, $z = 3.08$, $P = 0.002$, Control, $z = 0.56$, $P = 0.56$, CORT, $z = -0.24$, $P = 0.81$).

Nestling mass and growth rate did not differ between treatments (Mass: $t = -0.89$, $P = 0.38$ and Growth: $t = -0.49$, $P = 0.63$), and were not affected by treatment in a sex-specific manner (Mass: Sex × Treatment; $t = 0.002$, $P = 0.99$ and Growth: Sex × Treatment; $t = 0.68$, $P = 0.50$). Similar to un-manipulated nests, male nestlings from manipulated broods grew at a significantly faster rate than their female siblings ($t = 5.12$, $P < 0.001$) and were heavier ($t = 8.47$, $P < 0.001$).

Discussion

Our results suggest that maternal CORT is not related to brood sex ratio adjustment in the blue tit. Furthermore, unlike previous studies maternal baseline CORT was not correlated with maternal condition [5,17]. In agreement with this, exogenous elevation of maternal CORT during egg laying did not influence offspring sex ratio. However, mothers in superior condition produced male biased broods in one year of the study, and both maternal body condition and baseline CORT were associated with indices of nestling quality. Specifically, maternal CORT was negatively correlated with nestling mass, but not growth rate, whereas, maternal condition was positively correlated with both nestling mass and growth rate. This result suggests that maternal condition is more closely related to offspring quality than baseline CORT. Although male nestlings were heavier and grew at a faster rate than female nestlings, maternal condition and CORT concentrations did not influence nestling condition in a sex-specific manner.

Studies that have found a link between maternal CORT and brood sex ratio have also found maternal CORT to be associated with factors that may influence the adaptive significance of sex ratio adjustment, i.e. maternal condition [5,17], environmental

A

B

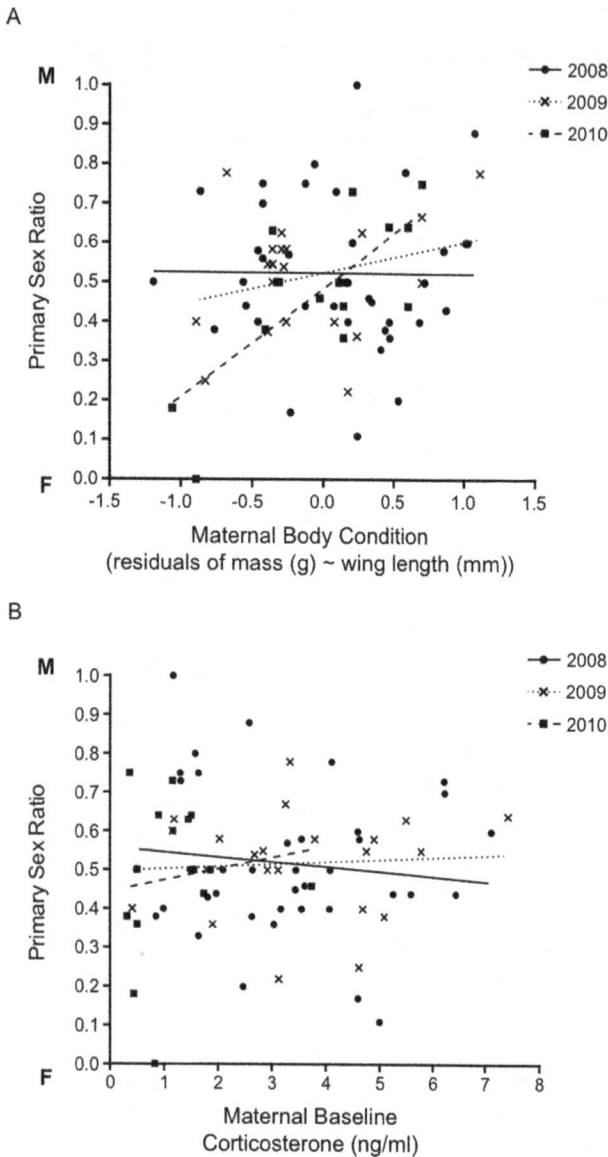

Figure 2. The relationship between primary brood sex ratio and a) maternal body condition (*n*=89) and b) maternal baseline CORT (*n*=76) from 2008–2010. M indicates a male and F indicates a female biased brood sex ratio.

conditions [8,52] and mate attractiveness [18,38]. In the present study however, maternal condition was linked to brood sex ratio in one year, but was not correlated with maternal CORT in any year. Furthermore, a previous study that found a correlation between maternal CORT and sex ratio manipulation, also found sex-specific effects of elevated maternal CORT upon nestling mass [17]. However, our results do not support sex-specific effects of maternal CORT upon nestling condition. Therefore, in blue tits maternal CORT may not be indicative of circumstances that might favour sex ratio adjustment, and thus may not be expected to influence brood sex ratio.

Our results provide evidence that CORT may not be directly linked to sex ratio adjustment consistently across bird species. Previous studies that present evidence of a link between maternal CORT and offspring sex ratio differ in the timing of sex ratio adjustment, and therefore the potential mechanisms employed. For example, there is evidence of a pre-laying mechanism in peafowl (*Pavo cristatus*), Japanese quail (*Coturnix coturnix japonica*), Gouldian finch and white-crowned sparrows (*Zonotrichia leucophrys*) [5,16,18,26,38], as maternal baseline CORT was found to be correlated and causally linked to the primary sex ratio in these species. However, in the European starling (*Sturnus vulgaris*), exogenous elevation of maternal CORT during egg laying was associated with secondary brood sex ratio adjustment through male nestling mortality [17]. The lack of a relationship between maternal CORT, brood sex ratio and nestling mortality in this study and the contrasting findings of previous studies highlight the need for additional research to establish the generality of hormonal mechanisms in sex ratio manipulation and the timing of these adjustments.

Importantly, there were limitations of the methods employed in this study. Although previous studies have found a significant relationship between maternal baseline CORT measured post-laying and brood sex ratio [5,16], it would have been preferable to measure CORT during egg laying to coincide with the sex determination of chicks. Furthermore, although CORT concentrations did not differ between breeding stages in our population, and there is evidence that CORT concentrations are consistent between breeding stages within individuals [53,54], this could not be ascertained in this study. Therefore, the CORT concentrations measured during brood rearing may not have reflected those experienced by mothers during egg laying. However, it is important to note that the probability of nest desertion is high in wild birds when they are blood sampled during early breeding stages [55]. This is an important limitation for studies of wild birds,

Table 2. The results of GLMs investigating whether maternal body condition influenced a) nestling mass and b) nestling growth (broods = 59).

Factor	d.f. Effect	SE	*t*	*P*
a) Nestling mass				
Year	2	0.23	1.11	0.27
Sex	1	0.05	10.97	**<0.001**
Maternal condition	1	0.17	3.09	**0.003**
b) Nestling growth				
Year	2	0.10	−0.51	0.61
Sex	1	0.02	5.06	**<0.001**
Brood size	1	0.01	−2.06	**0.04**
Maternal condition	1	0.07	2.37	**0.02**

A

B

C

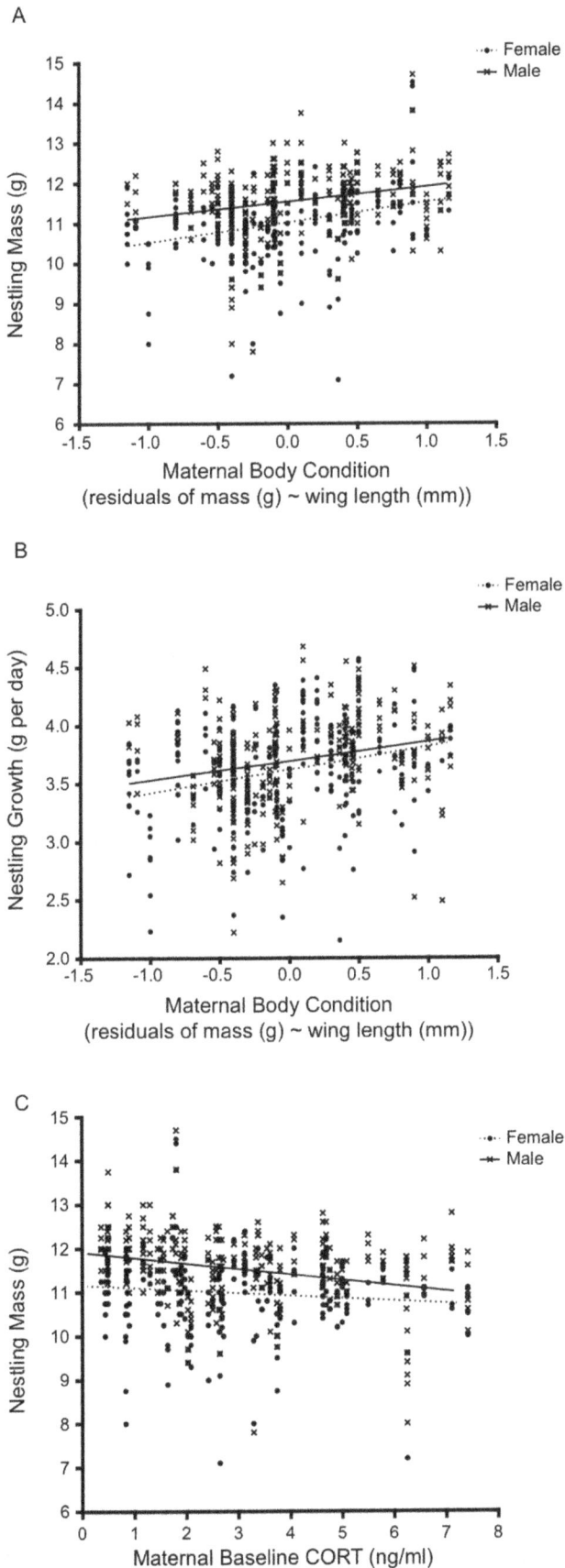

Figure 3. The relationship between a) maternal body condition and nestling mass on day 14 after hatching, b) maternal body condition and nestling growth rate, and c) maternal baseline CORT and nestling mass on day 14 after hatching from 2008–2010 (*n*=chicks: 520, broods: 57).

especially in species such as the blue tit, where re-nesting is not possible due to the short duration of abundant food that is essential for breeding [40].

The method used in this study to elevate maternal CORT was non-invasive, and elevated CORT concentrations above baseline to stress-induced levels for a transitory period during sex determining meiosis. As mothers began the treatment once they had laid their first egg, only ~80% of each clutch was manipulated (out of mean brood size of 10). Our results showed that the primary and secondary brood sex ratios produced after the CORT treatment were 0.47 and 0.45 respectively. Therefore the mean brood sex ratio did not deviate>0.1 from parity, when taking into account that the first two eggs were not manipulated. Due to the limitations of our manipulation we would not have been able to identify a change of <0.1 in brood sex ratio. However, an adjustment of <0.1 would be an extremely small change in brood sex ratio that would be unlikely to affect maternal fitness. Also, as the study was field based it was not possible to monitor hatching, therefore we could not establish the laying order of chicks. Previous studies provide evidence that birds can manipulate the sex of offspring adaptively dependent upon the order that they are laid [56]. Thus future studies that take laying order in account would be insightful.

There is evidence of both a correlative and causal link between maternal condition and brood sex ratio from a wide range of avian species [4,5,57,58]. However, there are also studies that have found no such relationship [9,59,60,61]. In addition, studies that have measured maternal condition and offspring sex ratio over multiple years in birds are rare, and where contrasting patterns between years have been found convincing biological explanations are lacking [9,62]. There is evidence to suggest that the link between maternal condition and offspring sex ratio is influenced by the prevalent conditions. In red deer (*Cervus elaphus*) the tendency of dominant females to produce more male offspring disappeared as population density increased [52], which has been suggested to have been caused by increased mortality of male foetuses as conditions became less favourable [63]. In our study, maternal condition was related to offspring sex ratio in one year of our study (2010). In 2010 there were significantly fewer mothers that laid unviable eggs compared with the two other years. Thus if the unviable eggs in the previous two years were male this may have obscured the effect of maternal condition upon sex ratio. However, the incidence of unviable eggs was not linked to maternal condition, and when the unviable eggs were considered male and the data re-analysed, maternal condition remained non-significantly correlated with brood sex ratio in two out of the three years (maternal condition × year: $z = 2.72$, $P = 0.01$). Therefore it is unlikely that sex-biased early embryo death or fertilization of ova obscured sex ratio adjustment in these years.

Interestingly, the results of our experiment showed that our manipulation negated the correlation between maternal body condition and brood sex ratio in 2010. In both the CORT and control groups there was no correlation between maternal body condition and brood sex ratio, but there was a positive correlation between body condition and brood sex ratio in un-manipulated broods. A recent study has shown that CORT manipulation during laying can influence the relationship between maternal condition and brood sex ratio. For example, exogenous CORT

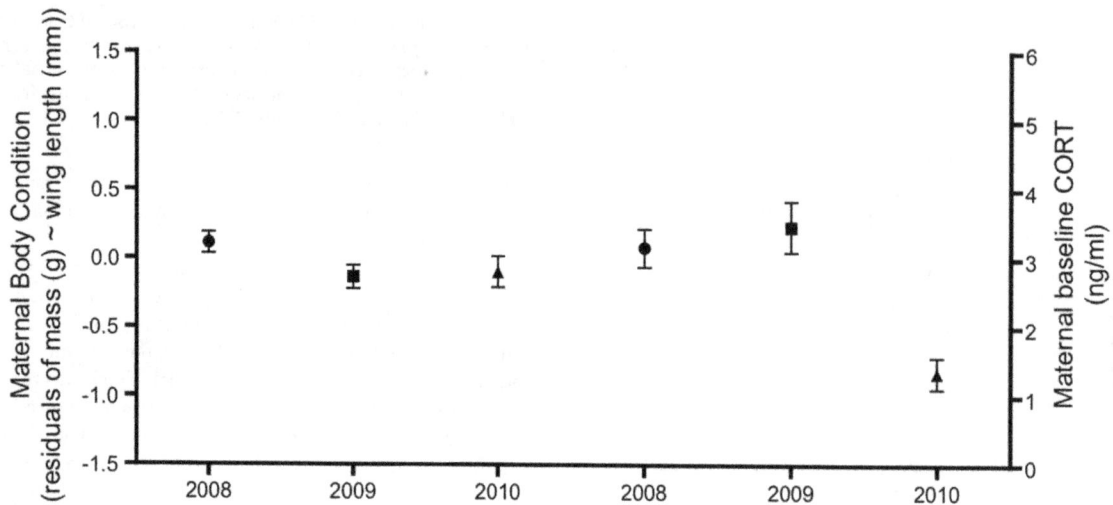

Figure 4. Inter-annual variation in maternal baseline CORT (*n*=76) and maternal body condition (*n*=89) from 2008–2010.

treatment caused a negative relationship between maternal body mass and brood sex ratio in chickens, whereas in the control group there was a positive correlation between body mass and sex ratio [64]. Overall, these results provide evidence of the context-dependence of the relationship between maternal condition and brood sex ratio.

Variation in the breeding conditions between years could influence the fitness benefits of sex ratio adjustment in relation to maternal condition. In the great tit (*Parus major*) a closely related species, natal conditions influence lifetime reproductive success more strongly in male compared with female birds [65]. Therefore, mothers in superior condition may derive fitness benefits from investing in sons only when breeding conditions are good. In our study, the year that brood sex ratio was linked to maternal condition (2010) was characterized by lower maternal CORT and relatively good maternal condition, most likely due to high food availability and favourable weather conditions in this year compared with the other years [66]. However, manipulative studies are required to provide convincing evidence of a link between maternal condition and brood sex ratio adjustment in this species.

Mothers in superior condition had heavier, faster growing offspring, but these effects were not sex specific. Therefore, although male nestlings grew at a faster rate and were heavier than females, maternal condition did not influence growth or mass more strongly in sons compared with daughters. In spite of this, mothers may have improved their fitness by investing in sons when they were in superior condition, as improved nestling mass and growth during the nestling phase can have beneficial long-term effects for male but not female birds. For example in the great tit, improved nestling mass close to fledging was linked to greater reproductive success in sons but not daughters [67]. Unfortunately, it was not possible in our study to investigate the effects of maternal condition upon the future reproductive success of offspring, as few nestlings were re-captured in subsequent years.

Our finding that the relationship between maternal condition and brood sex ratio adjustment differed between years is consistent with previous studies in the blue tit. Evidence in support of a relationship between paternal attractiveness and offspring sex ratio varies both between years and populations [32,33,34,35,36]. This may suggest that brood sex ratio manipulation is non-adaptive or constrained by chromosomal sex determination, or that the

optimal sex ratio varies across years dependent upon the prevalent conditions [36]. However, as previous studies have shown that temporal change in maternal body condition during breeding predicts offspring sex ratio rather than absolute values [68,69], the lack of relationship in two years of our study may have been because we employed a single measure of maternal body condition.

Ultimately, their large brood size, limited size dimorphism (~5%) and variation in extra-pair paternity and thus male reproductive success [70,71], may suggest that there is weak selection on sex ratio adjustment in the blue tit compared with other species. However, to establish whether the contrasting relationships between parental quality and brood sex ratio across years are significant, further knowledge concerning the fitness benefits of sex ratio adjustment in relation to parental quality across multiple years and environmental conditions are required.

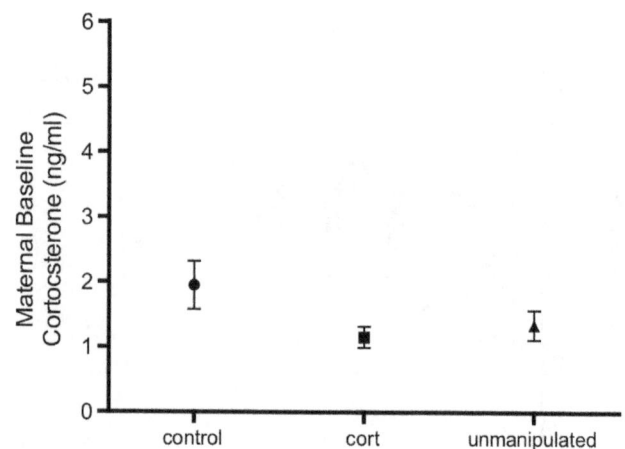

Figure 5. Maternal baseline CORT between experimental treatments. CORT: mothers fed CORT spiked mealworms during egg laying (*n*=15), Control: mothers fed mealworms (*n*=13) and un-manipulated mothers (*n*=17).

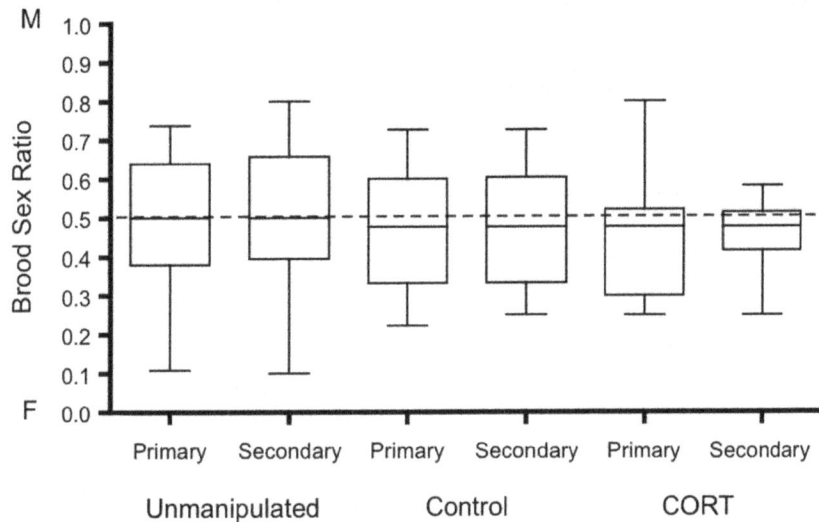

Figure 6. Primary and secondary brood sex ratios between experimental treatments. CORT: mothers fed CORT spiked mealworms during egg laying (*n* = 24), Control: mothers fed mealworms (*n* = 24) or un-manipulated mothers (*n* = 24). Graph shows median and interquartile range. M indicates a male and F indicates a female biased brood sex ratio. Dashed line denotes a 50:50 brood sex ratio.

Conclusions

This study does not provide evidence that maternal baseline CORT plays a role in brood sex ratio adjustment in the blue tit. Our results do provide some evidence that maternal condition was linked to offspring sex ratio adjustment, and that this relationship was context-dependent. Further studies that manipulate maternal condition and/or natal conditions and investigate the effects upon the lifetime reproductive success of offspring would be valuable. Overall, this study serves to highlight the complexity of sex ratio adjustment in birds and the difficulties associated with identifying sex biasing mechanisms.

Supporting Information

Data S1 Combined raw data files.

References

1. Jones KS, Nakagawa S, Sheldon BC (2009) Environmental sensitivity in relation to size and sex in birds: meta-regression analysis. American Naturalist 174: 122–133.
2. Trivers RL, Willard DE (1973) Natural Selection of Parental Ability to Vary the Sex Ratio of Offspring. Science 179 90–92.
3. Charnov EL (1982) The theory of sex allocation. Princeton: Princeton University Press.
4. Nager RG, Monaghan P, Griffiths R, Houston DC, Dawson R (1999) Experimental demonstration that offspring sex ratio varies with maternal condition. Proceedings of the National Academy of Sciences 96: 570–573.
5. Pike T, Petrie M (2005) Maternal body condition and plasma hormones affect offspring sex ratio in peafowl. Animal Behaviour 70: 745–751.
6. Stauss M, Segelbacher G, Tomiuk J, Bachmann L (2005) Sex ratio of *Parus major* and *P. caeruleus* broods depends on parental condition and habitat quality. Oikos 109: 367–373.
7. Ewen JG, Cassey P, Moller AP (2004) Facultative primary sex ratio variation: a lack of evidence in birds? Proceedings of the Royal Society B: Biological Sciences 271: 1277–1282.
8. Komdeur J, Magrath MJ, Krackow S (2002) Pre-ovulation control of hatchling sex ratio in the Seychelles warbler. Proceedings of the Royal Society B-Biological Sciences 269: 1067–1072.
9. Radford AN, Blakey JK (2000) Is variation in brood sex ratios adaptive in the great tit (Parus major)? Behavioral Ecology 11: 294–298.
10. West SA (2009) Conditional Sex Allocation I: Basic Scenarios. Sex Allocation: Princeton University Press.

Materials S1 Supporting information regarding the validation of the non-invasive technique to manipulate maternal CORT concentrations.

Acknowledgments

We thank S. Bairner, G. Casasole, J. Desjardins, C. Fischbacher, C. Foote, E. Forbes, K. Mikolajczak, L. Mills, G. Ortolani, S. Parsche, E. Pooley, Y. Roggia, J. Sciberras, V. Smith, F. Svendsen, R. Vetter and A. Whyte for their help with fieldwork. We also thank L. Fleming for her assistance with genetic sex determination.

Author Contributions

Conceived and designed the experiments: LJH NPE KEA. Performed the experiments: LJH KEA BJH. Analyzed the data: LJH. Contributed reagents/materials/analysis tools: NPE KEA AA. Wrote the paper: LJH.

11. Gam AE, Mendonça MT, Navara KJ (2011) Acute corticosterone treatment prior to ovulation biases offspring sex ratios towards males in zebra finches Taeniopygia guttata. Journal of Avian Biology 42: 253–258.
12. Pinson SE, Parr CM, Wilson JL, Navara KJ (2011) Acute corticosterone administration during meiotic segregation stimulates females to produce more male offspring. Physiological and Biochemical Zoology 84: 292–298.
13. Komdeur J, Pen I (2002) Adaptive sex allocation in birds: the complexities of linking theory and practice. Philosophical Transactions of the Royal Society B: Biological Sciences 357: 373–380.
14. Pike TW, Petrie M (2003) Potential mechanisms of avian sex manipulation. Biological Reviews 78: 553–574.
15. Oddie K (1998) Sex discrimination before birth. Trends in ecology & evolution 13: 130–131.
16. Bonier F, Martin PR, Wingfield JC (2007) Maternal corticosteroids influence primary offspring sex ratio in a free-ranging passerine bird. Behavioral Ecology 18: 1045–1050.
17. Love OP, Chin EH, Wynne-Edwards KE, Williams TD (2005) Stress hormones: A link between maternal condition and sex-biased reproductive investment American Naturalist 166: 751–766.
18. Pryke SR, Rollins LA, Buttemer WA, Griffith SC (2011) Maternal stress to partner quality is linked to adaptive offspring sex ratio adjustment. Behavioral Ecology 22: 717–722.
19. Wingfield JC (2005) The concept of allostasis: Coping with a capricious environment. Journal of Mammalogy 86: 248–254.
20. Kitaysky AS, Wingfield JC, Piatt JF (1999) Dynamics of food availability, body condition and physiological response in breeding black-legged kittiwakes. Functional Ecology 13: 577–585.

21. Schoech SJ, Mumme RL, Wingfield JC (1997) Corticosterone, reproductive status, and body mass in a cooperative breeder, the Florida scrub-jay (Aphelocoma coerulescens). Physiological Zoology 70: 68–73.

22. Buck CL, O'Reilly KM, Kildaw SD (2007) Interannual variability of Black-legged Kittiwake productivity is reflected in baseline plasma corticosterone. General and Comparative Endocrinology 150: 430–436.

23. Kitaysky AS, Kitaiskaia EV, Wingfield JC, Piatt JF (2001) Dietary restriction causes chronic elevation of corticosterone and enhances stress response in red-legged kittiwake chicks. Journal of Comparative Physiology B 171: 701–709.

24. Kitaysky AS, Piatt JF, Wingfield JC (2007) Stress hormones link food availability and population processes in seabirds. Marine Ecology Progress Series 352: 245–258.

25. Marra PP, Holberton RL (1998) Corticosterone levels as indicators of habitat quality: effects of habitat segregation in a migratory bird during the non-breeding season. Oecologia 116: 284–292.

26. Pryke SR, Rollins LA, Griffith SC, Buttemer WA, Sockman K (2014) Experimental evidence that maternal corticosterone controls adaptive offspring sex ratios. Functional Ecology.

27. Rutkowska J, Badyaev AV (2008) Meiotic drive and sex determination: molecular and cytological mechanisms of sex ratio adjustment in birds. Philosophical Transactions of the Royal Society B: Biological Sciences 363: 1675–1686.

28. Saino N, Romano M, Ferrari RP, Martinelli R, Møller AP (2005) Stressed mothers lay eggs with high corticosterone levels which produce low-quality offspring. Journal of Experimental Zoology Part A: Comparative Experimental Biology 303A: 998–1006.

29. Hayward LS, Wingfield JC (2004) Maternal corticosterone is transferred to avian yolk and may alter offspring growth and adult phenotype. General and Comparative Endocrinology 135: 365–371.

30. Cyr N, Romero LM (2007) Chronic stress in free-living European starlings reduces corticosterone concentrations and reproductive success. General and Comparative Endocrinology 151: 82–89.

31. Bennett NC, Bateman PW, Lemons PR, Cameron EZ (2008) Experimental alteration of litter sex ratios in a mammal. Proceedings of the Royal Society B: Biological Sciences 275: 323–327.

32. Delhey K, Peters A, Johnsen A, Kempenaers B (2007) Brood sex ratio and male UV ornamentation in blue tits (Cyanistes caeruleus): correlational evidence and an experimental test. Behavioral Ecology and Sociobiology 61: 853–862.

33. Griffith SC, Örnborg J, Russell AF, Andersson S, Sheldon BC (2003) Correlations between ultraviolet coloration, overwinter survival and offspring sex ratio in the blue tit. Journal of Evolutionary Biology 16: 1045–1054.

34. Sheldon BC, Andersson S, Griffith SC, Örnborg J, Sendecka J (1999) Ultraviolet colour variation influences blue tit sex ratios. Nature 402: 874–877.

35. Dreiss A, Richard M, Moyen F, White J, Møller AP, et al. (2006) Sex ratio and male sexual characters in a population of blue tits, Parus caeruleus. Behavioral Ecology 17: 13–19.

36. Korsten PC, Lessells KM, Mateman AC, Van der Velde M, Komdeur J (2006) Primary sex ratio adjustment to experimentally reduced male UV attractiveness in blue tits. Behavioral Ecology 17: 539–546.

37. Griffith SC, Pryke SR, Buttemer WA (2011) Constrained mate choice in social monogamy and the stress of having an unattractive partner. Proceedings of the Royal Society B: Biological Sciences.

38. Pike TW, Petrie M (2005) Offspring sex ratio is related to paternal train elaboration and yolk corticosterone in peafowl Biology letters 1: 204–207.

39. Hunt S, Cuthill IC, Bennett AT, Griffiths R (1999) Preferences for ultraviolet partners in the blue tit. Animal Behaviour 58: 809–815.

40. Perrins CM (1979) British Tits. London: Collins.

41. Henderson LJ, Heidinger BJ, Evans NP, Arnold KE (2013) Ultraviolet crown coloration in female blue tits predicts reproductive success and baseline corticosterone. Behavioral Ecology 24: 1299–1305.

42. Svensson L (1992) Identification guide to european passerines. Stockholm: BTO.

43. Olsen MW, Fraps RM (1950) Maturation changes in hen ovum. Journal of Experimental Zoology 114: 475–487.

44. Romanoff AJ (1960) The Avian Embryo. New York: Macmillian.

45. Sturkie PD (2000) Sturkie's Avian Physiology. San Diego: Academic Press.

46. Perrins CM (1996) Eggs, egg formation and the timing of breeding. Ibis 138: 2–15.

47. Wingfield JC, Smith JP, Farner DS (1982) Endocrine Responses of White-Crowned Sparrows to Environmental Stress. The Condor 84: 399–409

48. Pendlebury CJ, Bryant DM (2005) Night-time behaviour of egg-laying tits. Ibis 147: 342–345.

49. Nicholls JA, Double MC, Rowell DM, Magrath RD (2000) The evolution of cooperative and pair breeding in thornbills Acanthiza (Pardalotidae). Journal of Avian Biology 31: 165–176.

50. Griffiths R, Double MC, Orr K, Dawson RJ (1998) A DNA test to sex most birds. Molecular Ecology 7: 1071–1075.

51. Wilson K, Hardy ICW (2002) Statistical analysis of sex ratios: an introduction. In: Hardy ICW, editor. Sex ratios: concepts and research methods. Cambridge: Cambridge University Press.

52. Kruuk LEB, Clutton-Brock T, Rose KE, Guinness FE (1999) Population density effects sex ratio vraiation in red deer. Nature 399: 459–461.

53. Wingfield J, Farner DS (1976) Avian Endocrinology: Field Investigations and Methods. The Condor 78: 570–573.

54. Wingfield JC, Farner DS (1978) The annual cycle of plasma irLH and steroid hormones in feral populations of the white-crowned sparrow, Zonotrichia leucophrys gambelii Biology of Reproduction 19: 1046–1056.

55. Criscuolo F (2001) Does blood sampling during incubation induce nest desertion in the female common eider Somateria mollissima. Marine Ornithology 29: 47–50.

56. Badyaev AV (2002) Sex-biased hatching order and adaptive population divergence in a passerine bird. Science 295: 316–318.

57. Clout MN, Elliott GP, Robertson BC (2002) Effects of supplementary feeding on the offspring sex ratio of kakapo: a dilemma for the conservation of a polygynous parrot. Biological Conservation 107: 13–18.

58. Kilner R (1998) Primary and secondary sex ratio manipulation by zebra finches. Animal Behaviour 56: 155–164.

59. Koenig WD, Dickinson JL (1996) Nestling Sex-Ratio Variation in Western Bluebirds. The Auk 113: 902–910.

60. Leech DI, Hartley IR, Stewart IRK, Griffith SC, Burke T (2001) No effect of parental quality or extrapair paternity on brood sex ratio in the blue tit (Parus caeruleus). Behavioral Ecology 12: 674–680.

61. Whittingham LA, Dunn PO, Nooker JK (2005) Maternal influences on brood sex ratios: an experimental study in tree swallows. Proceedings of the Royal Society B: Biological Sciences 272: 1775–1780.

62. Korpimäki E, May CA, Parkin DT, Wetton JH, Wiehn J (2000) Environmental- and parental condition-related variation in sex ratio of kestrel broods. The Journal of Avian Biology 31: 128–134

63. Hardy ICW (2002) Sex ratios: concepts and research methods. Cambridge: Cambridge University Press.

64. Aslam MA, Groothuis TGG, Smits MA, Woelders H (2014) Effect of corticosterone and hen body mass on primary sex ratio in laying hen (Gallus gallus), using unincubated eggs. Biology of Reproduction.

65. Wilkin TA, Sheldon BC (2009) Sex Differences in the Persistence of Natal Environmental Effects on Life Histories. Current Biology 19: 1998–2002.

66. Henderson LJ, Evans NP, Heidinger BJ, Herborn KA, Arnold KE (In Prep.) Are glucocorticoids associated with fitness? Linking foraging conditions, corticosterone and reproductive success in the blue tit, Cyanistes caeruleus.

67. Tilgar V, Mänd R, Kilgas P, Mägi M (2010) Long-term consequences of early ontogeny in free-living Great Tits, Parus major. Journal of Ornithology 151: 61–68.

68. Cameron EZ, Linklater WL (2007) Extreme sex ratio variation in relation to change in condition around conception. Biology Letters 3: 395–397.

69. Goerlich VC, Dijkstra C, Boonekamp JJ, Groothuis TGG (2010) Change in body mass can overrule the effects of maternal testosterone on primary offspring sex ratio of first eggs in homing pigeons. Physiological and Biochemical Zoology 83: 490–500.

70. Charmantier A, Blondel J, Perret P, Lambrechts MM (2004) Do extra-pair paternities provide genetic benefits for female blue tits Parus caeruleus? Journal of Avian Biology 35: 524–532.

71. Gullberg A, Tegelström H, Gelter HP (1992) DNA fingerprinting reveals multiple paternity in families of great and blue tits (Parus major and P. caeruleus). Hereditas 117: 103–108.

3

Ras GTPase Activating Protein Colra1 Is Involved in Infection-Related Morphogenesis by Regulating cAMP and MAPK Signaling Pathways through CoRas2 in *Colletotrichum orbiculare*

Ken Harata, Yasuyuki Kubo*

Laboratory of Plant Pathology, Graduate School of Life and Environmental Sciences, Kyoto Prefectural University, Kyoto, Japan

Abstract

Colletotrichum orbiculare is the causative agent of anthracnose disease on cucurbitaceous plants. Several signaling pathways, including cAMP–PKA and mitogen-activating protein kinase (MAPK) pathways are involved in the infection-related morphogenesis and pathogenicity of *C. orbiculare*. However, upstream regulators of these pathways for this species remain unidentified. In this study, *ColRA1*, encoding RAS GTPase activating protein, was identified by screening the *Agrobacterium tumefaciens*-mediated transformation (AtMT) mutant, which was defective in the pathogenesis of *C. orbiculare*. The *coira1* disrupted mutant showed an abnormal infection-related morphogenesis and attenuated pathogenesis. In *Saccharomyces cerevisiae*, Ira1/2 inactivates Ras1/2, which activates adenylate cyclase, leading to the synthesis of cAMP. Increase in the intracellular cAMP levels in *coira1* mutants and dominant active forms of *CoRAS2* introduced transformants indicated that Colra1 regulates intracellular cAMP levels through CoRas2. Moreover, the phenotypic analysis of transformants that express dominant active form *CoRAS2* in the *comekk1* mutant or a dominant active form *CoMEKK1* in the *coras2* mutant indicated that CoRas2 regulates the MAPK CoMekk1–Cmk1 signaling pathway. The CoRas2 localization pattern in vegetative hyphae of the *coira1* mutant was similar to that of the wild-type, expressing a dominant active form of *RFP–CoRAS2*. Moreover, we demonstrated that bimolecular fluorescence complementation (BiFC) signals between Colra1 and CoRas2 were detected in the plasma membrane of vegetative hyphae. Therefore, it is likely that Colra1 negatively regulates CoRas2 in vegetative hyphae. Furthermore, cytological analysis of the localization of Colra1 and CoRas2 revealed the dynamic cellular localization of the proteins that leads to proper assembly of F-actin at appressorial pore required for successful penetration peg formation through the pore. Thus, our results indicated that Colra1 is involved in infection-related morphogenesis and pathogenicity by proper regulation of cAMP and MAPK signaling pathways through CoRas2.

Editor: Zhengyi Wang, Zhejiang University, China

Funding: Grants-in-Aid for scientific Research from the Ministry of Education, Culture, Sports, Science and Technology (Grant number 24248009). The funders had no role in study design, data collection and analysis, decision to publish, or preparation of the manuscript.

Competing Interests: The authors have declared that no competing interests exist.

* Email: y_kubo@kpu.ac.jp

Introduction

Colletotrichum orbiculare is the causative agent of cucumber anthracnose disease. The infection process is initiated by the recognition of an appropriate surface. A series of changes in the morphology upon recognizing the appropriate signals, including the formation of a specialized infection structure called appressorium, is important for the successful infection of the host plants.

Several signal-transduction related genes associated with these morphological changes have been characterized in *C. orbiculare* [1]. It also has been shown that the MAPK and cyclic AMP (cAMP) signaling pathways are linked to infection-related morphological changes in this fungus. A yeast MAPK kinase (MAPKK) kinase *STE11* homolog, *CoMEKK1*, is involved in the appressorium development in *C. orbiculare* [2]. *C. orbiculare* mutant with the adenylate cyclase encoding gene *cac1* shows

defectiveness in conidial germination [3], however, no upstream regulators of cAMP and MAPK signaling pathways have been identified in *C. orbiculare*.

Ras is the prototypical member of the small GTP binding protein superfamily that plays a pivotal role in proliferation and differentiation in eukaryotic cells. In *Magnaporthe oryzae*, MoRas1 and MoRas2 interact with Mst50, which interacts directly with the mitogen-activated proteins Mst11 and Mst7 [4]. In *Fusarium graminearum*, FgRas2 is involved in hyphal growth and pathogenicity [5]. In *C. trifolii*, CtRas is involved in conidial germination and hyphal growth [6]. Ras acts as a molecular switch that exists both in the active (GTP-bound) and the inactive (GTPase-bound) states. Cycling between these two states is aided by the interaction of a GTPase activating protein and guanine-nucleotide exchange factors. In *Saccharomyces cerevisiae*, the Ras activity is controlled positively by the guanine nucleotide exchange factor (GEF) Cdc25 and negatively by the GTPase activating proteins Ira1 and Ira2 (GAPs) [7]. In *S. cerevisiae*, the RAS GTPase activating proteins Ira1 and Ira2 inactivate Ras1 and Ras2, which in turn activate adenylate cyclase, promoting the synthesis of cAMP from ATP. Ira1 and Ira2 are important factors for controlling cAMP production. The cAMP signaling pathway plays a pivotal role in transducing environmental cues during cell development in *Magnaporthe oryzae* [8]. Moreover, the cAMP signaling pathway of *C. orbiculare* plays a critical role in regulating conidial germination and pathogenicity [9]. Adenylate cyclase gene *MAC1* encodes a Ras-association domain [10], [11]. In terms of protein structure, the Ras protein putatively interacts with Mac1 in *M. oryzae*. It was shown that cyclase-associated protein Cap1 of *M. oryzae* directly interacts with Mac1 and plays a role in the activation of Mac1, which may function downstream of Ras [12]. In *Ustilago maydis*, *RAS2* promotes filamentous growth through a MAP kinase cascade, however, the function of *RAS2* in the cAMP signal transduction remains unknown [13]. Therefore, direct evidences for the involvement of the Ras protein in the cAMP signal transduction are not sufficient in phytopathogenic fungi.

Because Ras proteins transduce signals from the cell surface to various intracellular effectors, they are located and function only at the plasma membrane [14], [15]. Conversely, data provided by the yeast GFP fusion localization database [16] and other published data [17], [18] indicate that Ras2, Cyr1, Cdc25, and Ira proteins mainly localize at the internal membranes of the endoplasmic reticulum and mitochondrial membranes, but only marginally at the plasma membrane. Recently, it was reported that active Ras accumulates mainly in the plasma membrane and nucleus when the cells are grown on medium containing glucose, whereas it accumulates mainly in the mitochondria in glucose-starved cells [19]. Moreover, PKA activity causes Ira2 to move away from the mitochondria [19]. Therefore, it is considered that the behavior of Ras proteins and Ras GTPase activating proteins and their localization depends on various growth conditions. In plant pathogenic fungi, subcellular dynamics of Ras proteins and Ras GTPase activating proteins during infection and the location of this interaction remain unclear.

In this study, we identified a novel *S. cerevisiae* homolog gene *CoIRA1*, encoding the RAS GTPase-activating protein (RAS-GAP) of *C. orbiculare* and characterized the *CoIRA1* function in relation to the activation of cAMP-PKA and MAPK signaling pathways. By phenotypic analysis of the *coira1* mutant, cytological analysis of CoRas2 localization and BiFC assay, we showed that CoIra1 regulates these signaling pathways through CoRas2 as a negative regulator. Conclusively, we presented CoIra1 is involved in infection-related morphogenesis by regulating cAMP and MAPK signaling pathways through CoRas2 in *C. orbiculare*.

Materials and Methods

Fungal and bacterial strains

The 104-T (MAFF240422) *C. orbiculare* (Berk. & Mont.) Arx [syn. *C. lagenarium* (pass.); Ellis & Halst.] strain was used as the wild-type. All the *C. orbiculare* strains used in this study are listed in Table 1 and were cultured on PDA media (3.9% [w/v] PDA; Difco laboratories, Detroit) at 24°C. One shop TOP10 chemically competent *E. coli* cultured in Luria-Bartani media [20] at 37°C was used as a host for gene manipulation. When required, the supplement kanamycin was added to the medium at 50 μg/ml. *A. tumefaciens* C58C1 cultured in Luria-Bartani media at 28°C was used to transform *C. orbiculare* by AtMT.

Fungal transformation

For the fungal transformation, we used an AtMT protocol that was previously described [21] with slight modifications. The hygromycin-resistant transformants were selected on the PDA medium containing 100 μg/ml of hygromycin B (Wako Chemicals, Osaka, Japan), 50 μg/ml of cefotaxim (Wako Chemicals, Osaka, Japan), and 50 μg/ml of spectinomycin (Wako Chemicals, Osaka, Japan). The bialaphos-resistant transformants were selected on an SD medium containing 10 μg/mL of bialaphos (Meiji Seika Kaisha, Ltd., Tokyo, Japan), 100 μg/ml of cefotaxim, and 100 μg/ml of spectinomycin. The sulfonylurea-resistant transformants were selected on an SD medium containing 4 μg/ml of chlorimuron-ethyl (Chem Service West Chester, PA, USA.), 100 μg/ml of cefotaxim, and 100 μg/ml of spectinomycin.

Genomic DNA blot analysis

The total DNA from the mycelia of *C. orbiculare* was isolated, and a DNA blot analysis was performed using a previously described method [22]. DNA digestion, gel electrophoresis, labeling of probes, and hybridization were performed according to the manufacturer's instructions following standard methods [20]. DNA probes were labeled with DIG-dUTP using a BcaBESTTM DIG labeling kit (Takara Bio, Ohtsu, Japan). Hybridized DNA was detected with anti-Digoxygenin antibody Fab fragments conjugated to alkaline phosphatase (Roche Diagnostics, Tokyo, Japan). Light emission from the enzymatic dephosphorylation of the CDP-Star Detection Reagent (GE Healthcare, Tokyo, Japan) was detected using the Fujifilm LAS-1000 Plus Gel Documentation System (Fujifilm, Tokyo).

Construction of the *CoIRA1* gene replacement vector and *CoIRA1* complementation vector

To replace the *CoIRA1* gene with the hygromycin-resistance gene, we constructed a *CoIRA1* gene-replacement vector pBIG4MRBrev-coira1. We first amplified the upstream region of the *CoIRA1* gene, the hygromycin-resistance gene, and the downstream region of the *CoIRA1* gene by PCR using the primer pairs CoIRA1F1A–CoIRA1R2D, CoIRA1hphF1C–CoIRA1hphR1D, and CoIRA1F2C–CoIRA1R1B, respectively. The primers used are listed in Table S1. Next, the pBIG4MRBrev-coira1 vector was constructed using the GeneART seamless cloning and assembly kit (Life Technologies, Carlsbad, California USA) with the amplified product and the *A. tumefaciens* binary vector pBIG4MRBrev. This vector, which contains a bialaphos resistance gene, was used as the gene replacement plasmid.

To perform a complementation assay of the *coira1* mutant, we constructed the CoIRA1 complementation vector pBIG4MRBrev–CoIRA1. We first amplified the upstream region of the *CoIRA1* gene, the middle region of the *CoIRA1* gene, and the downstream region of the *CoIRA1* gene by PCR using the primer pairs CoIRA1F3A–CoIRA1R4D, CoIRA1F5D–CoIRA1R5C, and CoIRA1F4C–CoIRA1R4D, respectively. Next, the pBIG4MRBrev–CoIRA1 vector was constructed with the amplified product, and the *A. tumefaciens* binary vector pBIG4MRBrev was constructed using the GeneART seamless cloning and assembly kit. This vector, which contains a bialaphos resistance gene, was used as the gene replacement plasmid.

Construction of dominant active and negative *CoRAS1* vectors

To construct a dominant active form of the *CoRAS1* vector, we amplified a 1.4-kb fragment containing the upstream region of the *CoRAS1* gene, a 2.0-kb fragment containing the downstream

Table 1. Fungal strains used in this study.

Strain	Description	References
WT	Wild-type strain of *Colletorichum orbiculare*	This study
Dl1-1	*coira1* disruptant of WT	This study
Dl1-2	*coira1* disruptant of WT	This study
El1-1	*ColRA1* ectopic transformant WT	This study
Cl1-1	Dl1-1 complemented with *ColRA1*	This study
Cl1-2	Dl1-2 complemented with *ColRA1*	This study
DC1	*cac1* disruptant of WT	Yamauchi et al. 2004
DARS1	WT transformed with a dominant active form *CoRAS1*	This study
DARS2	WT transformed with a dominant active form *CoRAS2*	This study
iDNRA1	Dl1-1 transformed with a dominant negative form *CoRAS1*	This study
iDNRA2	Dl1-1 transformed with a dominant negative form *CoRAS2*	This study
DMK1	*comekk1* disruptant of WT	Sakaguchi et al. 2010
DRS2-1	*coras2* disruptant of WT	This study
DRS2-2	*coras2* disruptant of WT	This study
CRS2-1	DRS2-1 complemented with *CoRAS2*	This study
CRS2-2	DRS2-2 complemented with *CoRAS2*	This study
DRS2/DAMK1	DRS2 transformed with a dominant active form *CoMEKK1*	This study
DMK/DARS2	DMK1 transformed with a dominant negative form *CoRAS2*	This study
DRS2/RFP-RS2	DRS2 transformed with *RFP-CoRAS2*	This study
WT/RFP-RS2	WT transformed with *RFP-CoRAS2*	This study
WT/RFP-DARS2	WT transformed with RFP-*DACoRAS2*	This study
iRFP-RS2	Dl1-1 transformed with *RFP-CoRAS2*	This study
Vc-RS2	WT transformed with *VENUS*(1–158aa)*-CoRAS2*	This study
IRA1-Vn	WT transformed with *ColRA1-VENUS*(159aa–238aa)	This study
Vc-RS2/IRA1-Vn	IRA-Vn transformed with *VENUS*(1–158aa)*-CoRAS2*	This study
IRA1-VENUS	WT transformed with *ColRA1-VENUS*	This study
RFP-RS2/IRA1-VENUS	WT/RFP-RS2 transformed with *ColRA1-VENUS*	This study
RFP-DARS2/IRA1-VENUS	WT/RFP-DARS2 transformed with *ColRA1-VENUS*	This study
LA/IRA1-VENUS	IRA1-VENUS transformed with LifeAct-*RFP*	This study
WT/LA	WT with LifeAct-*RFP*	This study
iLA	Dl1-1 transformed with LifeAct-*RFP*	This study
DPS1	*pks1* disruptant of WT	Takano et al. 1995
DSD1	*ssd1* disruptant of WT	Tanaka et al. 2007

region of the *CoRAS1* gene, and the linearized pBIG4MRBrev vector by PCR using the primer pairs CoRAS1F1A–CoRAS1^{G17V}R1B, CoRAS1^{G17V}F1A–CoRAS1R1B, and CoRAS1pBIF1A–R1B. Next, the pBIG4MRBrev–CoRAS1^{G17V} vector was constructed with the amplified product using the GeneART seamless cloning and assembly kit. To construct a dominant negative form of the *CoRAS1* vector, we amplified a 1.4-kb fragment containing the upstream region of the *CoRAS1* gene, a 2.0-kb fragment containing the downstream region of the *CoRAS1* gene, and the linearized pBIG4MRBrev vector by PCR using the primer pairs RAS1^{S22N}F1A–RAS1R1B, RAS1F1A–RAS1^{S22N}R1B, and CoRAS1pBIF1A–R1B, respectively. Next, the pBIG4MRBrev–CoRAS1^{S22N} vector was constructed with the

amplified product using the GeneART seamless cloning and assembly kit.

Construction of the *CoRAS2* gene replacement, complementation, dominant active or negative vectors

To replace the *CoRAS2* gene for a hygromycin-resistance gene, we constructed the *CoRAS2* gene-replacement vector pBIG4MRBrev-coras2. We first amplified the upstream region of the *CoRAS2* gene, the hygromycin-resistance gene, the downstream region of the *CoRAS2* gene, and the linearized pBIG4MRBrev vector by PCR using the primer pairs CoRAS2F1A–CoRAS2R2D, CoRAS2hphF1C–CoRAS2hphR1D, CoRAS2F2C–CoRAS2R1B, and CoRAS2pBIF1–R1, respectively. Next, the pBIG4MRBrev–coras2 vector was constructed with the amplified product using the GeneART seamless cloning and assembly kit. To perform a complementation assay for the *coras2* mutant, we constructed the *CoRAS2* complementation vector pBIG4MR-Brev–CoRAS2. We amplified a 3.3-kb fragment containing the *CoRAS2* gene and the linearized pBIG4MRBrev vector by PCR using the primer pairs CoRAS2F1A–R1B and CoRAS2pBIF1A–R1B, respectively. Next, the pBIG4MRBrev–CoIRA1 vector was constructed with the amplified product using the GeneART seamless cloning and assembly kit.

To construct a dominant active form of the *CoRAS2* vector, we amplified a 1.4-kb fragment containing the upstream region of the *CoRAS2* gene, a 1.9-kb fragment containing the downstream region of the *CoRAS2* gene, and the linearized pBIG4MRBrev vector by PCR using the primer pairs CoRAS2F1A–CoR-AS2^{Q65L}R1B, CoRAS2^{Q65L}F1A–CoRAS2R1B, and CoRAS2p-BIF1A–R1B, respectively. Next, the pBIG4MRBrev–CoR-AS2^{Q65L} vector was constructed with the amplified product using the GeneART seamless cloning and assembly kit. To construct a dominant negative form of the CoRas2 vector, we amplified a 1.2-kb fragment containing the upstream region of the *CoRAS2* gene, a 2.2-kb fragment containing the downstream region of the *CoRAS2* gene, and the linearized pBIG4MRBrev vector by PCR using the primer pairs CoRAS2F1A–CoRAS2-^{G19A}R1B, CoRAS2^{G19A}F1A–CoRAS2R1B, and CoRAS2p-BIF1A–R1B, respectively. Next, the pBIG4MRBrev–CoR-AS2^{G19A} vector was constructed with the amplified product using the GeneART seamless cloning and assembly kit.

Construction of the *CoIRA1*–VENUS vector

To construct the pBITEF–VENUS vector, we amplified a 0.9-kb fragment containing the TEF promoter and the upstream region of the *GFP* gene, a 0.6-kb fragment containing the downstream region of the *GFP* gene, and the linearized pBIG4MRBrev vector by PCR using the primer pairs TEFF1–VENUSR1, VENUSF1–glyGFPR1, and pBIG4VENUSF1–R1, respectively. Next, the pBIG4MRBrev–TEF–VENUS vector was constructed with the amplified product using the GeneART seamless cloning and assembly kit. To construct the CoIRA1–VENUS vector, we amplified the *VENUS* gene, the hygromycin-resistance gene, and the linearized pBIG4MRBrev–CoIRA1 vector by PCR using the primer pairs glyGFPF1–GFPR1, VENUShphF1–VENUSR1, and pBICoIRA1–VENUSF1–CoIR-ApBIcomR1–GFP, respectively. Next, the pBIG4MRBrev–CoIRA1–VENUS–hph vector was constructed with the amplified product using the GeneART seamless cloning and assembly kit.

Construction of the *RFP* fused *CoRAS2* and the *RFP* fused *CoRAS2*^{Q65L} vectors

To construct the *RFP* fused *CoRAS2* vector, we amplified the *RFP* gene and linearized pBIG4MRBrev–CoRAS2 vector by PCR using the primer pairs RFPF1–glyRFPR1 and CoRAS2p-BIF1–R1, respectively. Next, the pBIG4MRBrev–RFP–CoRAS2 vector was constructed with the amplified product using the GeneART seamless cloning and assembly kit. To construct the *RFP*–fused *CoRAS2*^{Q65L} vector, we amplified the *RFP* gene and linearized pBIG4MRBrev–CoRAS2^{Q65L} vector using the primer pairs RFPF1–glyRFPR1 and CoRAS2pBIF1–R1, respectively. Next, the pBIG4MRBrev–RFP–CoRAS2^{Q65L} vector was constructed with the amplified product using the GeneART seamless cloning and assembly kit.

Construction of the VENUS–N (1–158 aa) fused CoIRA1 and VENUS–C (159–238 aa) fused CoRAS2 vectors for BiFC assays

To construct the *VENUS*-N fused *CoIRA1* vector, we amplified the *VENUS*–N fragment, the hygromycin-resistance gene, and the linearized pBIG4MRBrev–CoIRA1 vector by PCR using the primer pairs glyGFPF1–αGFPR1, VENUShphF1–VENUSR1, and pBICoIRA1–VENUSF1–CoIRApBIcomR1–GFP respectively. Next, the pBIG4MRBrev–CoIRA1–nVENUS–hph vector was constructed with the amplified product using the GeneART seamless cloning and assembly kit. To construct the *VENUS*–C fused *CoRAS2* vector, we amplified the *VENUS*–C fragment and linearized pBIG4MRBrev–CoRAS2 vector by PCR using the primer pair BGFPF2–glyGFPR1 and CoRAS2PBIcompF3–c-CoRAS2PBIcompR3, respectively. Next, the pBIG4MRBrev–cVENUS–CoRAS2 vector was constructed with the amplified product using the GeneART seamless cloning and assembly kit.

Intracellular cAMP measurements

Mycelia were collected from three-day old PSY liquid cultures and frozen in liquid nitrogen. All samples were lyophilized for 24 h and weighed. For every 10 mg of mycelia, 200 µl of ice-cold 6% trichloroacetic acid was added. The precipitate was removed by centrifugation at 2000×g for 15 min at 4°C, the supernatant was transferred to a new tube, and the TCA was extracted four times with five volumes of water-saturated ether. The concentration of cAMP was determined using the cAMP Biotrak Enzyme immunoassay system (GE Health Life Science, UK) according to the manufacturer's instructions.

Western blot

The total protein was isolated from vegetative hyphae using a previously described method [23]. The protein separated on SDS–PAGE gels was transferred onto a PVDF membrane using an Xcell SureLock Mini-Cell. The phosphorylation activation of Maf1 and Cmk1 MAPK kinase was detected using a PhosphoPlus p44/42 MAP kinase antibody kit (Cell Signaling Technology). Alkaline Phosphatase-conjugated secondary antibody and light emission from the enzymatic dephosphorylation of the CDP-Star Detection Reagent (GE Healthcare, Tokyo, Japan) was detected using the Fujifilm LAS-1000 Plus Gel Documentation System (Fujifilm, Tokyo). Anti-actin antibodies (Wako, Japan) were used at a 1:1000 dilution for Western blot analysis.

Pathogenicity tests

An inoculation assay on cucumber cotyledons (*Cucumis sativus* L. "Suyo") was performed as described by Tsuji et al. (1997) [24]. The conidia of *C. orbiculare* were obtained from seven-day old cultures, and drops of 10-µl conidial suspension (1×10^5 conidia per ml) were added on the surface of cucumber cotyledons at different locations. To assess invasive growth ability, 10-µl drops of spore suspension were added on wounded sites that were created by scratching the leaf surface with a sterile toothpick. After inoculation, the cotyledons were incubated at 24°C for seven days.

Light microscopy

For appressorium formation and penetration assays *in vitro*, conidia were harvested from seven-day old PDA cultures and suspended in distilled water. The conidial suspension, adjusted to 1×10^5 conidia per ml, was placed on a multiwell glass slide (eight-well multi-test slide; ICN Biomedicals, Aurora, OH, U.S.A.) and incubated in humid boxes at 24°C. Germlings were observed by a Nikon ECLIPSE E600 microscope with differential interference contrast optics (Nikon, Tokyo, Japan). To observe the formation of infectious hyphae in cucumber leaves, the conidial suspension was inoculated on the abaxial surface of cucumber cotyledons and incubated at 24°C for three days. Then, the inoculation site was cut off and stained with 0.1% lactophenol-aniline blue. VENUS and RFP fluorescence was observed by a Carl ZEISS Axio Imager M2 microscope (Zeiss, Gottingen, Germany) with 470 and 595 nm of excitation wavelength, respectively.

Results

Identification of an *IRA1/2* homolog, *CoIRA1*, in *C. orbiculare*

CoIRA1 was first identified as a mutant gene in the *Agrobacterium tumefaciens*-mediated transformation (AtMT) T-DNA mutant AA4510 of *C. orbiculare*, which shows an attenuated pathogenicity on cucumber leaves. DNA flanking regions, adjacent to the inserted plasmid, were isolated from the mutant AA4510 by thermal asymmetrical interlaced-polymerase chain reaction (TAIL-PCR) and the amplified products were subsequently sequenced. The TAIL-PCR result showed that the T-DNA fragment was inserted probable 10-bp downstream from the translational origin of the gene ENH81573, based on *C. orbiculare* genomic information [25]. This gene putatively encodes a 2255-amino acid protein with a predicted RAS GTPase-activating protein (RASGAP) domain (Figure S1A). We named this gene *CoIRA1*.

A blast search of the *CoIRA1* fungal genome homologs in non-redundant protein database of the National Center for Biotechnology Information indicated that the derived amino acid sequence from this gene is significantly similar to that of the *IRA1/2* homolog in *S. cerevisiae* and filamentous fungi, including *C. graminicola*, *Neurospora crassa*, and *M. oryzae* (Figure S1B).

CoIRA1 is required for infection-related morphogenesis

To analyze the function of *CoIRA1*, the disruption vector pBIG4MRBrevcoira1 was designed to replace the wild-type *CoIRA1* gene with the hygromycin phosphotransferase (*hph*) gene, by double crossover homologous recombination. Successful replacement of the targeted gene in the transformants was confirmed by genomic DNA blot analysis (Figure S2). A single 3.4-kb band was detected in the wild type, whereas a single 5.6-kb band was detected in *coira1* mutants, as expected from such a targeted gene replacement event (Figure S2). In ectopic transformants, the 3.4-kb band and several additional bands were detected, indicating ectopic insertion events.

The hyphal growth of the *coira1* mutant was similar to that of the wild-type, but the *coira1* mutant formed a rather dark colony compared with the wild-type (Figure S3). The *coira1* mutant also showed reduced conidiation compared with the wild-type. To investigate infection-related morphogenesis in the *coira1* mutant, we observed conidial germination and appressorium development on a glass slide, and infection hyphae development on cellulose membranes. In the wild-type, *coira1* mutant, ectopic transformants, and *CoIRA1* reintroduced transformants, approximately 80% of the conidia germinated and formed darkly melanized appressoria after 24 h, but the frequency of abnormal appressorium formation in the *coira1* mutant was higher compared with the wild-type, as observed on glass slides (Figures 1A and B). The appressorium turgor pressure is required to penetrate the plant surface mechanically during infection [26], therefore, we evaluated the appressorium turgor in the *coira1* mutant using a cytorrhysis assay [27]. The proportion of the collapsed appressoria at each glycerol concentration in the *coira1* mutant was similar to that of

the wild-type (Figure S4), indicating that Colra1 is not involved in the generation of the appressorium turgor pressure. The wild-type, ectopic transformants, and *CoIRA1* reintroduced transformants formed normal infection-hyphae that penetrated the cellulose membranes with high frequency. The *coira1* mutant effectively formed infection hyphae as did the wild-type. Interestingly, the *coira1* mutant hyphae had a bulbous shape, which was quite different from those of the wild-type (Figures 2A and B). These results indicated that *CoIRA1* is involved in the progression of normal morphogenesis during infection.

CoIRA1 is involved in the pathogenicity of the host plant

To investigate whether the pathogenicity of *coira1* mutants is attenuated, conidial suspensions were inoculated onto cucumber cotyledons. The pathogenicity of the *coira1* mutant was found to be attenuated during the infection of the cucumber plants compared with that of the wild-type (Figure 3A). Moreover, when conidial suspensions were applied directly on wounded sites, the *coira1* mutant caused the formation of smaller lesions compared with the wild-type (Figure S5). Microscopic observations of the infection process of the *coira1* mutant showed that it formed normal darkly melanized appressoria and infection hyphae, however, its frequency of infection hyphae formation was lower than that of the wild-type (Figures 3B and C). These data indicated that during the infection of the cucumber, the *coira1* mutants had a defective infection-related morphogenesis. In *C. orbiculare*, we showed that the *ssd1* mutant that are defective in proper fungal cell walls constitution failed to penetrate into host epidermal cells due to the increased defence reaction of the host plant with rapidly induced callose deposition at at the attempted penetration site from appressoria [28]. Therefore, to investigate whether the observed low frequency of the infection hyphae formation in the *coira1* mutant was induced by a host defense mechanism, we monitored the callose deposition under the appressorium of the *coira1* mutant on the cucumber cotyledons and found that the extent and frequency of the deposition was similar to that of the wild-type (Figure S6A). Moreover, the *coira1* mutant caused the formation of smaller lesions than the wild-type on heat-shocked cucumber cotyledons (Figure S6B), which impaired the host defense responses, indicating that the induction of plant defense responses is not involved in the attenuated pathogenicity observed for *coira1* mutants.

CoIRA1 is involved in cAMP signaling

In *S. cerevisiae*, Ira1/2 inactivates Ras1/2, which in turn activates adenylate cyclase, leading to the synthesis of cAMP from ATP [29]. We therefore examined whether the appressorium morphogenesis of the *coira1* mutants is affected by exogenous cAMP signals. In the presence of 10 mM cAMP, the frequency of abnormal appressorium formation observed in the *coira1* mutant was higher compared with that of the wild-type (Figures 4A and B). This data suggested that Colra1 is involved in cAMP signaling pathway during the process of the appressorium formation. We identified two Ras genes and named *CoRAS1* (ENH84705) and *CoRAS2* (ENH80898). The comparative amino acid sequence analysis of Ras proteins showed that *CoRAS1* was 90% identical to *M. oryzae* MoRAS1 and *CoRAS2* was 80% identical to *M. oryzae* MoRAS2. To better understand relationship between Colra1 and Ras proteins, we set up a hypothesis that the wild-type expressing a dominant active form *CoRAS1* and *CoRAS2* increases abnormal appressorium formation compared with the wild-type in the presence of excess cAMP, while the *coira1* mutant expressing a dominant active form *CoRAS1* and *CoRAS2* decrease abnormal appressorium compared with the *coira1* mutant in the presence of

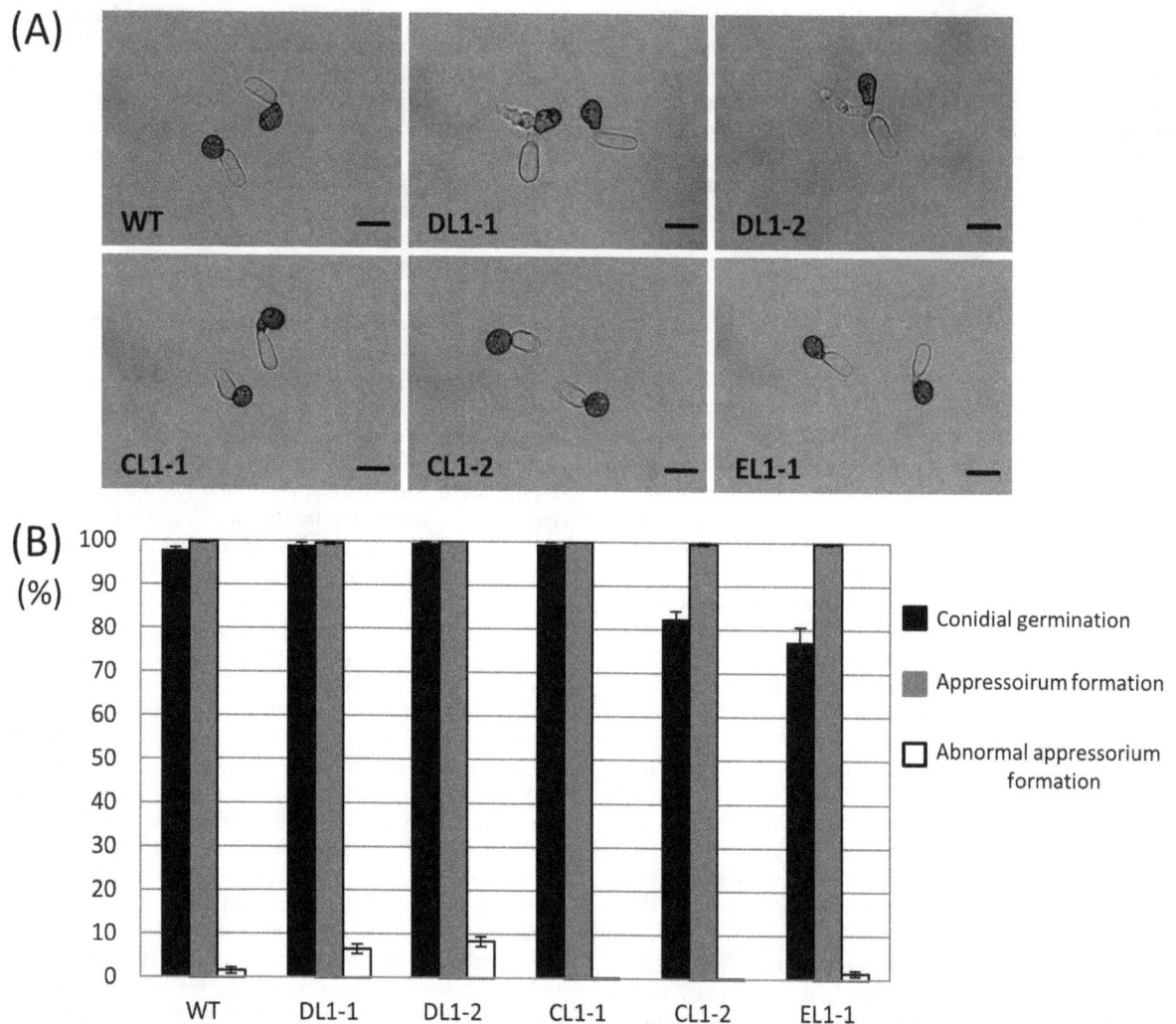

Figure 1. Appressorium formation in *coira1* mutants of *C. orbiculare* on glass slides. (A) Conidial suspensions of each strain in distilled water were incubated on multiwell glass slides at 24°C for 24 h. WT, the wild-type 104-T; DL1-1 and DL1-2, the *coira1* mutants; CL1-1, the Co*IRA1*-complemented transformant of DL1-1; CL1-2, the Co*IRA1*-complemented transformant of DL1-2; EL1-1, ectopic strain. Scale bar, 10 μm. (B) Percentages of conidial germination, appressorium formation, and abnormal appressorium formation in the *C. orbiculare* WT and *coira1* mutants on multiwell glass slides. Approximately 100 conidia of each strain were observed per well on multiwell slide glass. Three replicates were examined. Three independent experiments were conducted, and standard errors are shown. Black bar, conidial germination; gray bar, appressorium formation; white bar, abnormal appressorium formation.

excess cAMP. Therefore, we generated these transformants. The Ras protein shows a high degree of amino acid conservation. On the basis of the human Ras gene mutation information, we constructed the dominant active forms CoRas1 (*RAS1*^G17V^) and CoRas2 (*RAS2*^Q65L^) alleles by replacing Gln-65 with Leu and Gly-17 with Val, respectively. Moreover, based on *C. trifolii* and *Candida albicans* Ras gene mutation information, we constructed the dominant negative forms CoRas1 (*RAS1*^S22N^) and CoRas2 (*RAS2*^G19A^) alleles by replacing Ser-22 with Asn and Gly-19 with Ala, respectively [6], [30]–[32]. Next, we generated the dominant active and negative forms of Co*RAS1* and Co*RAS2* of *C. orbiculare*. DARS1 (Dominant Active form Co*RAS1*^G17V^) is the wild-type transformant, expressing the dominant active form Co*RAS1*^G17V^ and DARS2 (Dominant Active form Co*RAS2*^Q65L^) is the wild-type transformant, expressing the dominant active form Co*RAS2*^Q65L^. On the other hand, iDNRS1 (the *coira1*/Dominant Negative form Co*RAS1*^S22N^) is the *coira1* transformant, expressing the dominant negative form Co*RAS1*^S22N^ and iDNRS2 (the

coira1/Dominant Negative form Co*RAS2*^G19A^) is the *coira1* transformant, expressing the dominant negative form Co*RAS2*^G19A^.

DARS1 and DARS2 showed normal appressorium formation on the glass slides in distilled water (Figure 4) and normal penetration hyphae similar to the wild-type on the cellulose membranes (Figure S7), however, DARS2 caused the formation of smaller lesions compared with the wild-type and DARS1 on the cucumber cotyledons (Figure S8). In the presence of 10 mM cAMP, DARS2 formed abnormal appressoria, similar to those observed for the *coira1* mutants. In contrast, iDNRS2 suppressed the abnormal appressorium formation compared with that of the *coira1* mutants (Figure 4). These data indicated that CoIra1 could control the cAMP signaling pathway through CoRas2 during the process of appressorium formation. We also checked the intracellular cAMP accumulation in the *coira1* mutant and the dominant active and negative Co*RAS1* and Co*RAS2* introduced transformants in vegetative hyphae whether the cAMP signaling

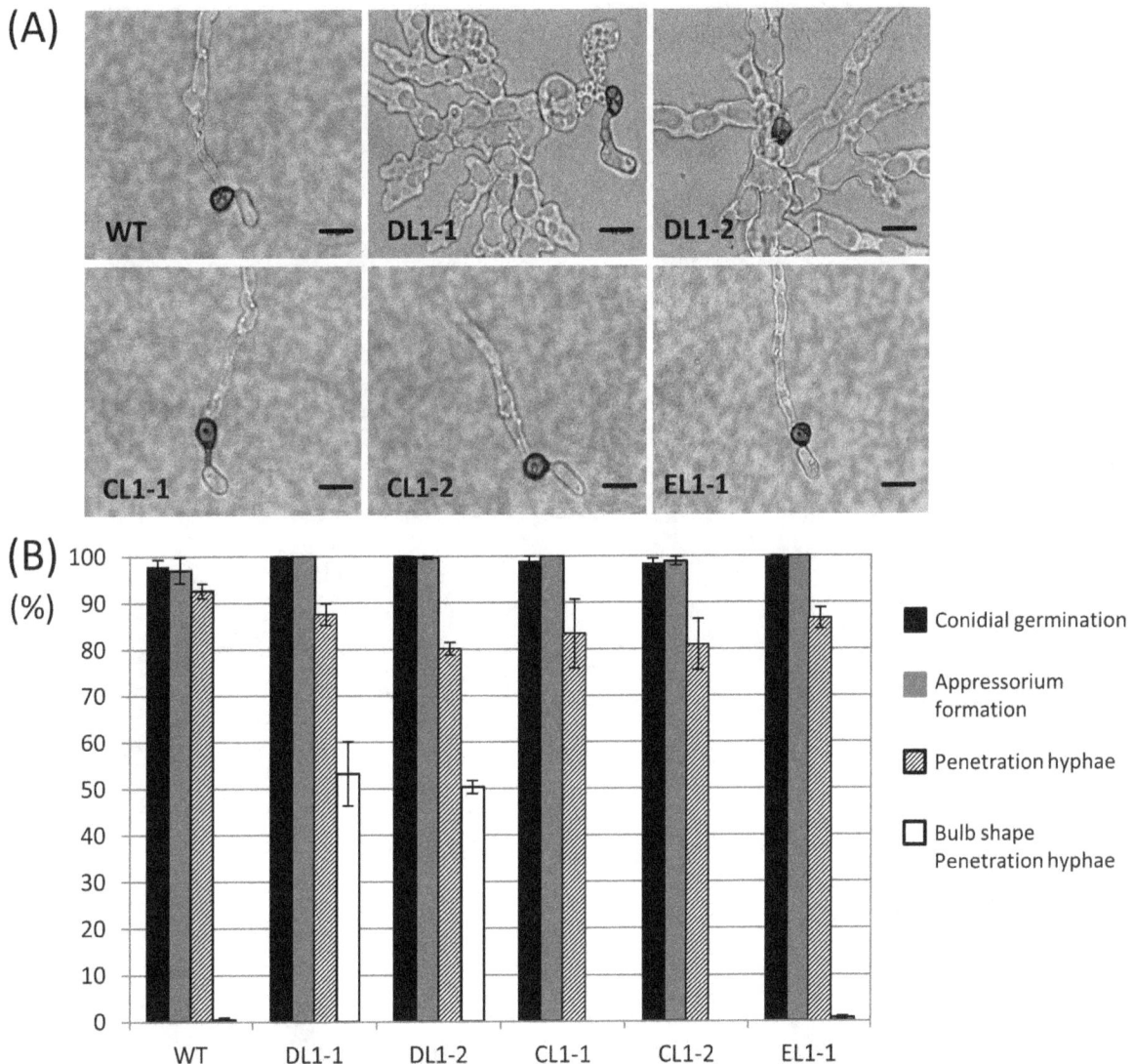

Figure 2. Penetration hyphae formation of the *coira1* mutants of *C. orbiculare* on cellulose membranes. Conidial suspensions of each strain in distilled water were incubated on cellulose membranes at 24°C for 48 h. WT, the wild-type 104-T; DL1-1 and DL1-2, the *coira1* mutant; CL1-1, the *CoIRA1*-complemented transformant of DL1-1; CL1-2, the *CoIRA1*-complemented transformant of DL1-2; EL1-1, the ectopic strain. Scale bar, 10 μm. (B) Percentages of conidial germination, appressorium formation, penetration hyphae formation, and bulb-shaped penetration-hyphae formation of *C. orbiculare* WT and *coira1* mutants on cellulose membranes. Approximately 200 conidia of each strain were observed on cellulose membranes. Three replicates were examined. Three independent experiments were conducted, and standard errors are shown. Black bar, conidial germination; gray bar, appressorium formation; slash bar, penetration hyphae; white bar, bulb-shape penetration formation.

pathway was regulated by those genes. The *coira1* mutants accumulated higher levels of cAMP compared with the wild-type (Figure 5). Moreover, intracellular cAMP levels in the DARS1 and DARS2 mutants were similar to those of the *coira1* mutant. These data indicated that intracellular cAMP levels in the vegetative hyphae were controlled by CoIra1, CoRas1, and CoRas2.

CoRAS2 is involved in conidial germination and pathogenicity

CoIra1 regulates intracellular cAMP levels through Ras proteins. Therefore, to analyze the functional roles of *CoRAS1* and *CoRAS2*, we aimed to generate *coras1* and *coras2* mutants by AtMT. Whereas we successfully developed *coras2* mutants, the generation of *coras1* mutants was challenging, indicating that *CoRAS1* could be an essential gene in *C. orbiculare*. To investigate

whether *CoRAS2* is involved in infection-related morphogenesis, we observed conidial germination and appressorium formation on glass slides. Strikingly, conidial germination was not observed on glass slides for *coras2* mutants, indicating that *CoRAS2* is involved in conidial germination (Figure S9). To investigate whether the pathogenicity of the *coras2* mutant was attenuated during the infection of the cucumber cotyledons, conidial suspensions were inoculated onto the cucumber cotyledons. The *coras2* mutant was defective in pathogenesis to the cucumber cotyledons, forming small speck lesion (Figure S10). Microscopic observation revealed that most conidia in the *coras2* mutant were defective in conidial germination as *in vitro* condition, indicating that the attenuated pathogenicity of this mutant was due to a defect in conidial germination and that small speck lesion was apparently caused by small numbers of appressorium forming conidia.

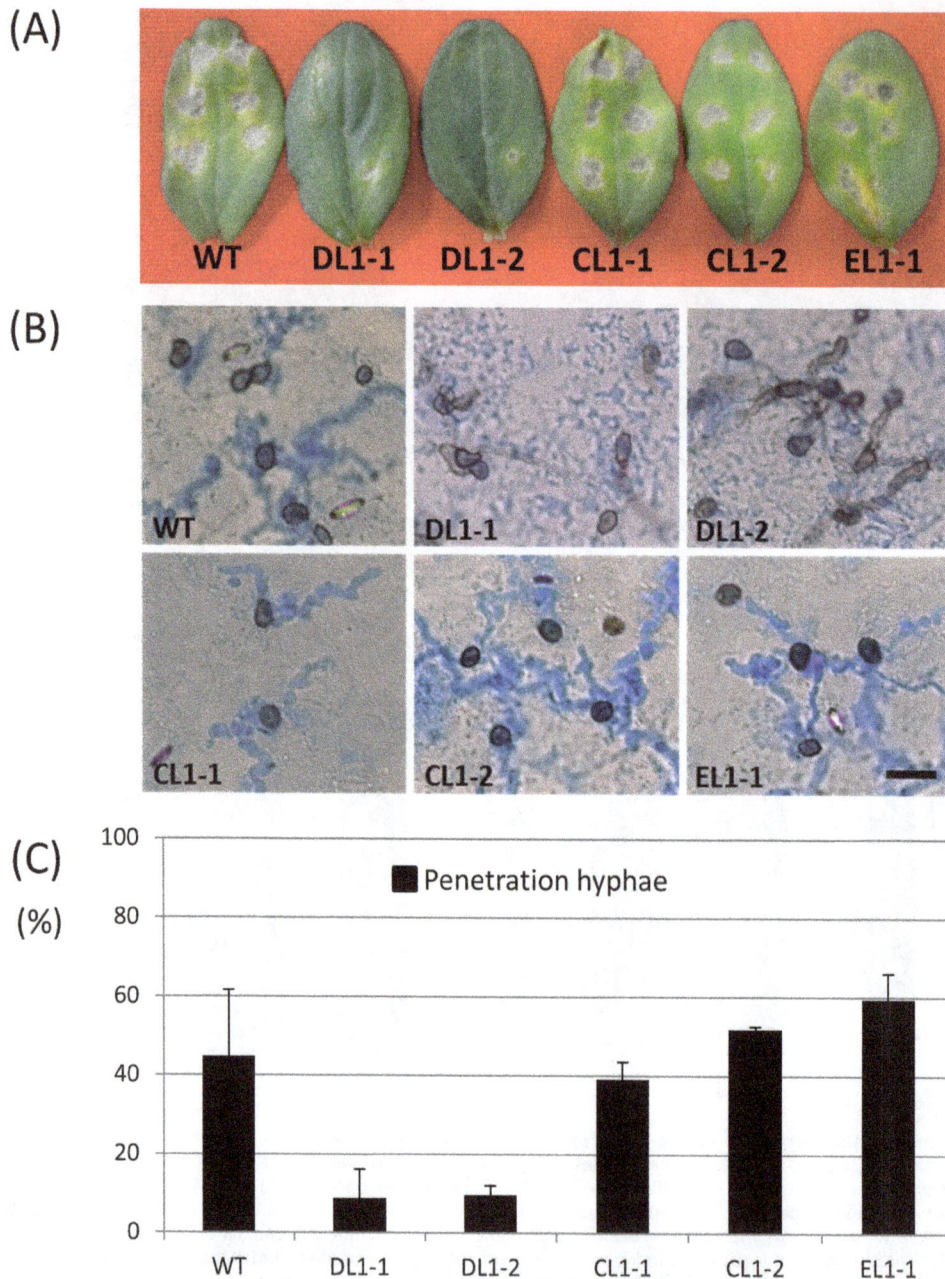

Figure 3. Pathogenicity assay and penetration ability of *coira1* mutants of *C. orbiculare* on the cucumber cotyledons. (A) Conidial suspensions of each strain (10 μl) placed on detached cucumber cotyledons and leaves incubated at 24°C for three days. The figure shows the leaves after incubation with the following strains: WT, the wild-type 104-T; DL1-1 and DL1-2, the *coira1* mutant; CL1-1, the *ColRA1*-complemented transformant of DL1-1; CL1-2, the *ColRA1*-complemented transformant of DL1-2; EL1-1, the ectopic strain. (B) Penetration hyphae development of each strain on the cucumber cotyledons. Conidial suspensions (10 μl) were applied to the abaxial surface of the cucumber cotyledons and incubated at 24°C for 72 h. Scale bar, 20 μm. (C) Percentage of penetration hyphae of the *coira1* mutants on the abaxial surface of cucumber cotyledons. Approximately 100 appressorium were observed per incubated site. Three replicates were examined. Three independent experiments were conducted, and standard errors are shown. Black bar, penetration hyphae. Scale bar, 20 μm.

CoRas2 is an upstream regulator of the MAPK signaling pathway

From previous reports, the MAPK CoMekk1–Cmk1 signaling pathway has been shown to play pivotal roles in conidial germination and appressorium formation in several *Colletotrichum* speciese [2], [33]. In *C. orbiculare*, the Ste11 homolog *CoMEKK1* encodes a Ras-association domain. Thus, to analyze whether CoRas2 is an upstream regulator of CoMekk1, we generated a

comekk1 transformant DMK1/DARS2 expressing a dominant active form of the *CoRAS2* allele. In addition, we generated the *coras2* mutant DRS2/DAMK1 expressing a dominant active form of the *CoMEKK1* allele. A normal conidial germination and appressorium formation was observed on the glass slide for the DRS2/DAMK1 mutant (Figure 6). On the other hand, the defective conidial germination of the *coras2* mutant remained in the DMK1/DARS2 mutant, showing a similar frequency of

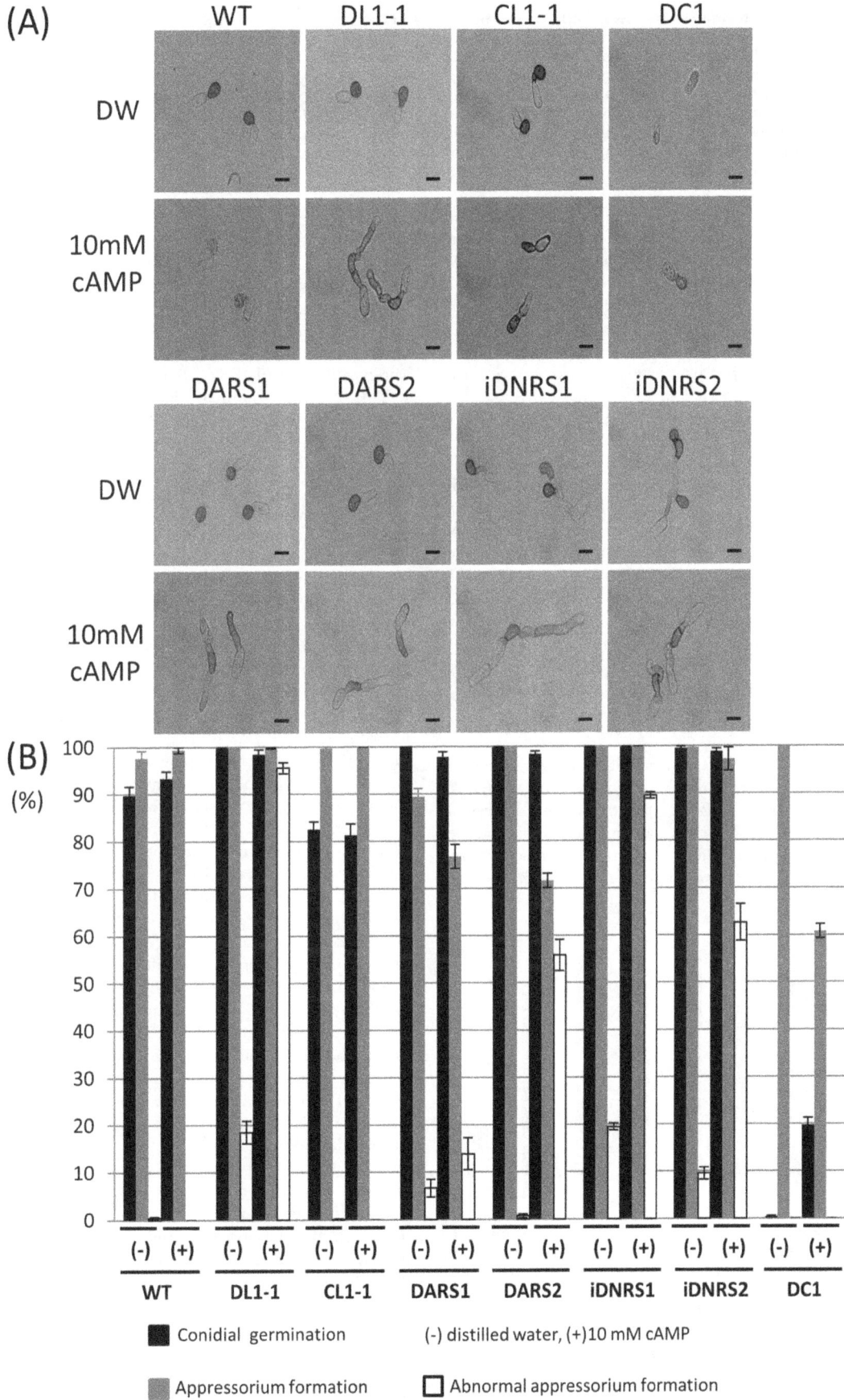

(A)

(B)

Conidial germination   (-) distilled water, (+)10 mM cAMP

Appressorium formation   ☐ Abnormal appressorium formation

Figure 4. Appressorium formation by *coira1* mutants on glass slide in the presence of 10-mM cAMP. (A) Conidial suspensions of each strain in distilled water or 10-mM cAMP were incubated on the multiwell glass slide at 24°C for 24 h. WT, the wild-type 104-T; DL1-1, the *coira1* mutant; CL1-1, the *CoIRA1*-complemented transformant of DL1-1; DC1, the *cac1* mutant; DARS1, WT transformed with a dominant active form *CoRAS1*; DARS2, WT transformed with a dominant active form *CoRAS2*; iDNRS1, DL1-1 transformed with a dominant negative form *CoRAS1*; iDNRS2, DL1-1 transformed with a dominant negative form *CoRAS2*. Scale bar, 10 μm. (B) Percentages of conidial germination, appressorium formation, and abnormal appressorium formation of *C. orbiculare* on multiwell glass slides in the presence of 10-mM cAMP. Approximately 100 conidia of each strain were observed on multiwell glass slides. Three replicate experiments were examined. Three independent experiments were conducted, and standard errors are shown. Black bar, conidial germination; gray bar, appressorium formation that includes normal appressorium and abnormal appressorium; white bar, abnormal appressorium formation. (–) distilled water, (+) 10-mM cAMP.

conidial germination as that in the *coras2* mutant. These data suggested that *CoRAS2* is an upstream regulator of the MAPK CoMekk1–Cmk1 signaling pathway.

The MAP kinase Cmk1 is activated through threonine/tyrosine phosphorylation catalyzed by MAPKK, which at the same time is activated through serine phosphorylation catalyzed by CoMekk1. To determine whether the phosphorylation of Cmk1 in the *coira1* and *coras2* mutants was affected in the vegetative hyphae, we investigated the Thr–Gln–Tyr residue phosphorylation of Cmk1 using an anti-TpEY specific antibody. The phosphorylation levels of Cmk1 in the *coira1* mutant and DARS2 were higher compared with those in the wild-type, whereas phosphorylation levels of Cmk1 in iDNRS2 were lower compared with those in the *coira1* mutant, which was similar to those of the wild-type (Figure 7). These data indicated that CoIra1 negatively regulates the phosphorylation of Cmk1 through CoRas2. Interestingly, the phosphorylation level of Cmk1 in DARS1 was higher compared with that in the wild-type, indicating that CoRas1 positively regulates the phosphorylation of MAPK Cmk1.

CoRas2 localization in the *coira1* mutant was similar to the active CoRas2 localization pattern in vegetative hyphae

To analyze the cellular CoRas2 localization, a red fluorescent protein (RFP) gene was fused to the C-terminus of the *CoRAS2* gene. In *S. cerevisiae* and *C. albicans*, the Ras protein conserves a CAAX motif (C, cysteine; A, aliphatic amino-acids; X; methionine or serine), which is important for post-translational modification, including farnesylation and palmitoylation, to ensure specific

membrane localization and function [29], [34]. We assumed that the *RFP gene* fused C-terminus of *CoRAS2* gene may block the localization and function of CoRas2. Thus, the *RFP* gene fused to the N-terminus of *CoRAS2* gene was constructed, and the *RFP–CoRAS2* gene controlled under the *CoRAS2* native promoter was transformed into the *coras2* mutant (DRS2/RFP–RS2) and the wild-type (WT/RFP–RS2). Both DRS2/RFP–RS2 and WT/RFP–RS2 retained normal infection-related morphogenesis and pathogenicity. The signals of RFP–CoRas2 were detected within vesicle-like structures in conidia (Figure 8). During the process of premature appressoria formation, signals of the RFP–CoRas2 were localized mainly in subcellular compartments resembling vacuole-like structures in conidia, but these signals were not detected in matured appressoria. In vegetative hyphae, RFP–CoRas2 proteins showed no specific localization and their signals were uniformly distributed throughout the entire cell (Figure 8). To analyze the cellular localization of the active form of CoRas2, the *RFP* gene was fused to the N-terminus of the dominant active form of the *CoRAS2* gene. The *RFP–DACoRAS2* gene controlled under the native *CoRAS2* promoter was transformed into the wild-type (WT/RFP–DARS2). During the process of appressoria formation, the signal pattern of the RFP–DACoRas2 was similar to that of the RFP–CoRas2 (Figure S11), however, the signals of the RFP–DACoRas2 in vegetative hyphae were detected predominantly in the plasma membrane, unlike those of native CoRas2. To analyze whether CoIra1 negatively regulates CoRas2, we generated an iRFP–RS2 transformant expressing *RFP–CoRAS2* in the *coira1* mutant and observed the cellular localization of CoRas2 in this mutant. During appressoria formation, the signals of RFP-CoRas2 in the *coira1* mutant were detected as vesicle like structures as well as RFP-CoRas2 and RFP-DACoRas2. In vegetative hyphae, the signals of RFP-CoRas2 in the *coira1* mutant were detected at the plasma membrane as well as RFP-DACoRas2 localization (Figure 9). Engagement of CoIra1 in the regulation of CoRas2 during appressorium formation was not directly suggested by these data from the viewpoint of the cellular localization of Ras protein, whereas it was suggested that CoIra1 functions as a negative regulator for CoRas2 in vegetative hyphae. Furthermore, to analyze whether CoIra1 regulates the CoRas2 in the plasma membrane in vegetative hyphae, we generated the Vc–RS2/IRA–Vn transformant, expressing the C-terminal domain (159–238) *VENUS* that was fused with *CoRAS2* and the N-terminal domain (1–158) of *VENUS* fused with *CoIRA1* in the wild-type for bimolecular fluorescence complementation (BiFC) assays [35]. BiFC fluorescence was detected at the plasma membrane in vegetative hyphae. This result indicated that CoIra1 regulates CoRas2 at the plasma membrane in vegetative hyphae (Figure 10).

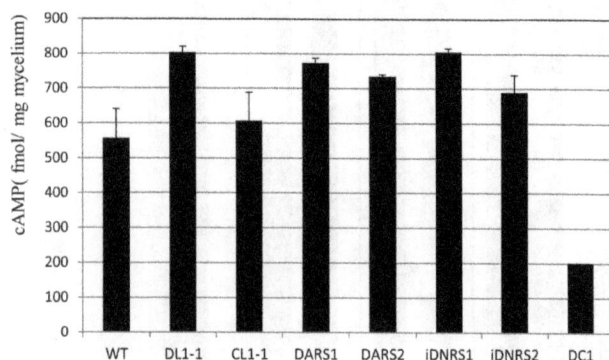

Figure 5. Intracellular cAMP levels in the *coira1* mutant. Intracellular cAMP levels were measured in tissue collection from three-day old liquid cultures of each strain. WT, the wild-type 104-T; DL1-1, the *coira1* mutant; CL1-1, the *CoIRA1*-complemented transformant of DL1-1; DARS1, WT transformed with a dominant active form *CoRAS1*; DARS2, WT transformed with a dominant active form *CoRAS2*; iDNRS1, DI1-1 transformed with a dominant negative form *CoRAS1*; iDNRS2, DI1-1 transformed with a dominant negative form *CoRAS2*; DC1, the *cac1* mutant. Three independent experiments were conducted, and standard errors are shown.

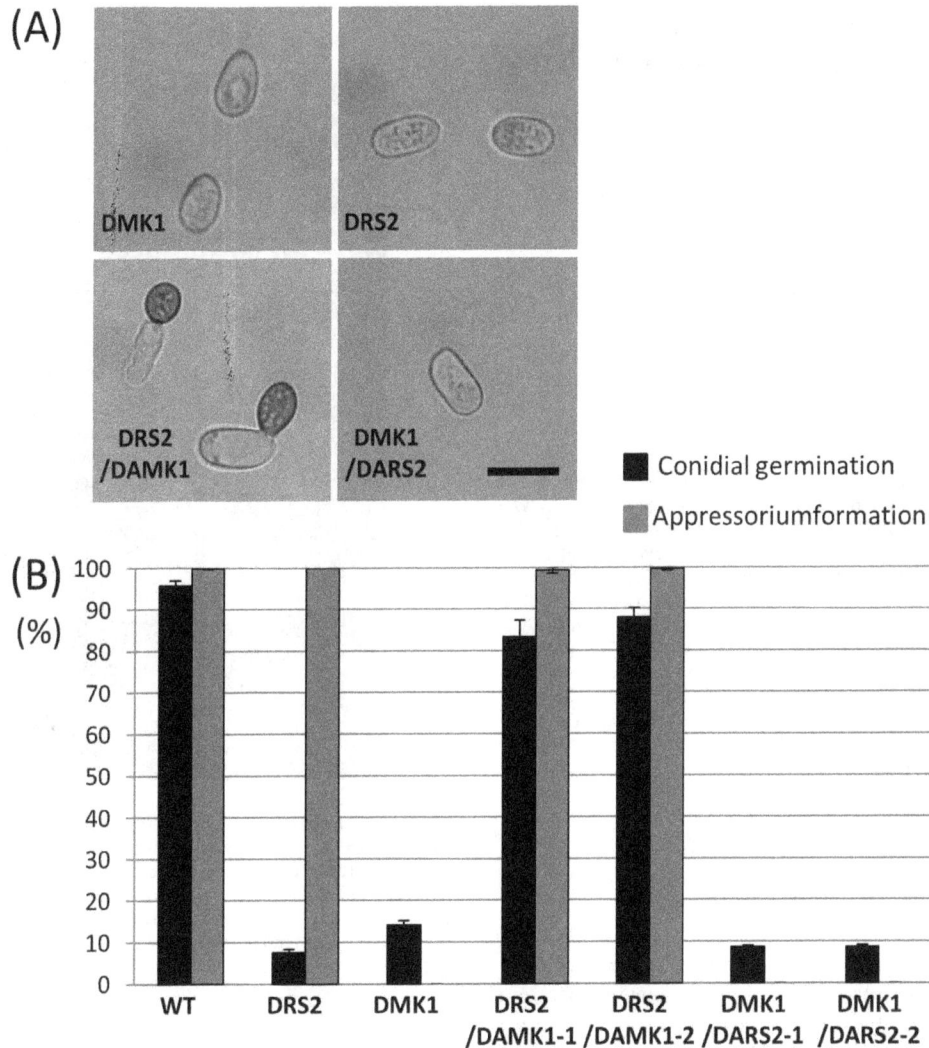

Figure 6. Appressorium formation assay of DRS2/DAMK1 and DMK1/DARS2 on the glass slides. (A) Conidial suspensions of each strain in distilled water incubated on multiwell glass slides at 24°C for 24 h. DRS2, the *coras2* mutant; DMK1, the *comekk1* mutant; DRS2/DAMK1, DRS2 transformed with a dominant active form *CoMEKK1*; DMK1/DARS2, DMK1 transformed with a dominant active form *CoRAS2*. Scale bar, 10 μm. (B) Percentages of conidial germination and appressorium formation in *C. orbiculare* on multiwell glass slides. Approximately 100 conidia of each strain were observed per well on the multiwell slide glass. Three replicates were examined. Three independent experiments were conducted, and standard errors are shown. Black bar, conidial germination; gray bar, appressorium formation.

CoIra1 colocalizes with CoRas2 in pregerminated conidia and appressoria that initiate the differentiation of infection hyphae

To elucidate the cellular localization of CoIra1, the *VENUS* fluorescence gene was fused to the C-terminal of the *CoIRA1* gene [36]. *VENUS* fused with *CoIRA1*, expressing under a native *CoIRA1* promoter was transformed into the wild-type (IRA1–VENUS). CoIra1–VENUS was detected in a vesicle-like structure of conidia (Figure 11). In the germinated conidia, CoIra1–VENUS was detected at the tip of the germ tubes and distributed uniformly in the conidial cells. In developing appressoria, CoIra1–VENUS did not show any specific localization, and uniform signals were distributed throughout the appressoria except the presumable lipid bodies, similar to those on cucumber leaves. In developing infection hyphae, CoIra1–VENUS showed a uniform distribution either in the initial or late infection hyphae in cucumber cotyledons (Figure S12). Then, to elucidate the

intracellular colocalization of CoIra1 and CoRas2, we generated the RFP–RS2/IRA1–VENUS and RFP–DARS2/IRA1–VENUS transformants by introducing the *RFP* fused with *CoRAS2* and *VENUS* fused with *CoIRA1* genes into the wild-type, and the *RFP* fused with a dominant active form *CoRAS2* and *VENUS* fused with *CoIRA1* genes into the wild-type, respectively. RFP–CoRas2 and CoIra1–VENUS colocalized at vesicle-like structures in conidia and germinated conidia (Figure 11), however, the RFP–CoRas2 and CoIra1–VENUS did not show specific colocalization in developing appressoria, similar to RFP–DARS2/IRA1–VENUS (Figure S11). Surprisingly, after 48 h of conidial incubation on glass slides, RFP–CoRas2 and CoIra1–VENUS were colocalized in a vesicle-like structures in appressoria (Figure 12). Conclusively, CoIra1 colocalizes with CoRas2 in pregerminated conidia and in appressoria that initiate differentiation of infection hyphae.

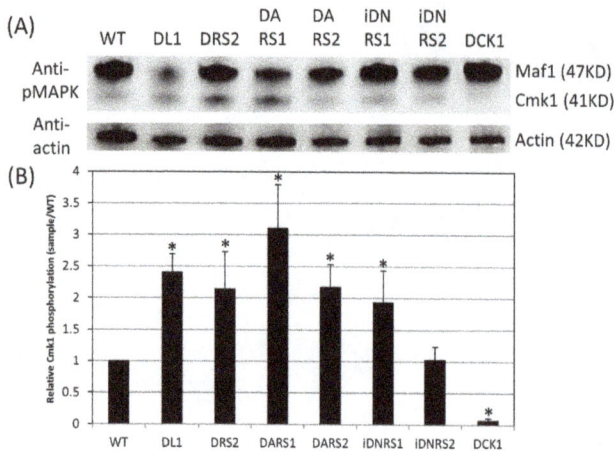

(A)

(B)

Figure 7. The phosphorylation of MAPK Cmk1 in the *coira1* mutant. (A) The total protein isolated from mycelia of each strain. WT; the wild-type, DL1; the *coira1* mutant, DRS2; the *coras2* mutant, DARS1, WT transformed with a dominant active form *CoRAS1*, DARS2; WT transformed with a dominant active form *CoRAS2*, iDNRS1; the *coira1* mutant transformed with a dominant negative form *CoRAS1*, iDNRS2; the *coira1* mutant transformed with a dominant negative form *CoRAS2*, DCK1; the *cmk1* mutant The anti-phospho p44/42 MAPK antibody detected a 41-KD Cmk1 and 47-KD Maf1. The anti-actin antibody detected a 42-KD actin. (B) Relative activity of MAPK Cmk1 phosphorylation of each mutant was calculated by comparison of signal intensity with that of the wild-type, normalized by actin signal. The quantitative analysis of phosphorylated Cmk1 was performed by four replicated experiments. Asterisk represents significant differences between the wild type and each mutant. (Student's t test: *indicate $P<0.05$).

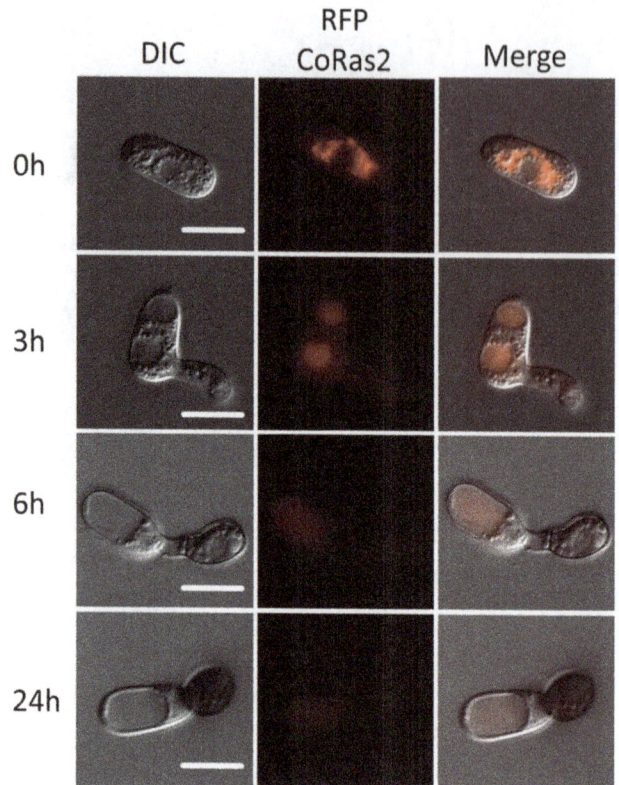

Figure 8. Localization of a functional RFP–CoRas2 fusion protein in *C. orbiculare* during conidial germination and appressorium formation. Conidial suspensions of the DRS2/RFP–RS2strain in distilled water were incubated in glass slides at 24°C for 0 h, 3 h, 6 h, and 24 h. After incubation, RFP fluorescence was observed by fluorescence microscopy. DRS2/RFP–RS2, the *coras2* mutant expressing *RFP–CoRAS2*. Scale bar, 10 μm.

Colra1 regulates the assembly of actin at the appressorium pore

In *M. oryzae*, it has been reported that proper assembly of the F-actin network in the appressorium pore is required for host infection and MAPK *Mst12*, the *C. orbiculare* Cst1 homolog, is known to be involved in the proper assembly of the F-actin network in the appressorium pore [37], [38]. Moreover, it has been reported that cAMP signaling is involved in the remodeling of the actin structure in *S. cerevisiae* [39]. Our data showed that CoIra1 regulated the cAMP and MAPK signaling pathways (Figure 5 and 7). To elucidate whether CoIra1 colocalize with F-actin, we generated the LA/IRA1–VENUS transformant by introducing Lifeact–*RFP* into it. In *C. orbiculare*, Lifeact, which binds to F-actin, forms vesicle-like structures in pregerminated conidia and then Lifeact–*RFP* mainly localizes in the appressorium pore during appressoria development. We examined the localization of CoIra1–VENUS and Lifeact–*RFP* in pregerminated conidia and appressorium pores. CoIra1–VENUS was colocalized with Lifeact–*RFP* in pregerminated conidia, however, CoIra1–VENUS was not clearly colocalized with Lifeact–RFP in the appressorium pore (Figure 13). To elucidate whether CoIra1 is involved in the assembly of the F-actin in the appressorium pore, we generated the iLA transformant by introducing Lifeact–*RFP* into the *coira1* mutant. In *C. orbiculare*, the wild-type showed a specific assembly of F-actin in the appressorium pore on the cucumber cotyledons, however, the frequency of this assembly in the *coira1* mutant was lower than that in the wild-type (Figure 14). These data indicated that CoIra1 is involved in the assembly of the F-actin in the appressorium pore.

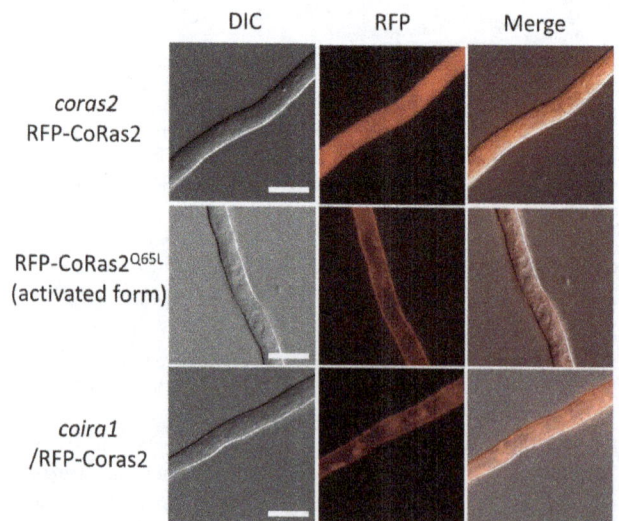

Figure 9. CoRas2 localization was regulated by Colra1 in vegetative hyphae. Conidia harvested from each strain were observed on glass slides by fluorescent microscopy. coras2/RFP–CoRAS2, the *coras2* mutant expressing the *RFP–CoRAS2*; RFP–CoRas2^Q65L, the wild-type strain expressing *RFP–CoRAS2*^Q65L; and coira1/RFP–CoRas2, the *coira1* mutant expressing *RFP–CoRAS2*. Scale bar, 10 μm.

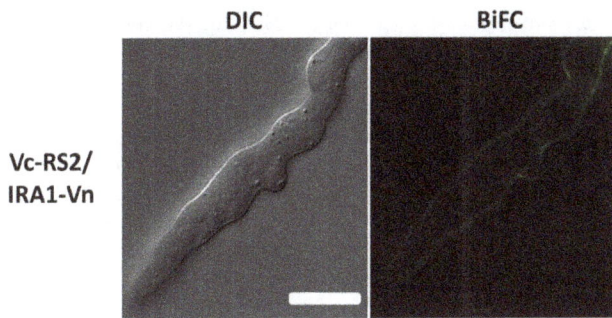

Figure 10. BiFC assays for Colra1 and CoRas2 interactions in vegetative hyphae. Conidial suspensions of Vc-CoRas2/Colra1-n transformant in liquid PSY medium were incubated at 28°C for 24 h and BiFC fluorescence was observed by fluorescent microscopy in vegetative hyphae. Vc–CoRas2/Colra1–Vn; the wild-type strain expressing Vc–CoRAS2 and Vn–CoIRA1. Scale bar, 10 μm.

Discussion

Colra1 is involved in the crosstalk between cAMP and MAPK signaling pathways through CoRas2 in C. orbiculare

In *S. cerevisiae*, the activation of the cAMP-dependent pathway causes cells to undergo unipolar growth, which is coupled with an elongated growth that is controlled by the filamentous MAPK pathway [40]. In *C. orbiculare*, cAMP–PKA signal transduction is involved in conidial germination, and the MAPK cascade is involved in appressorium development [1]. In *S. cerevisiae*, an increase in the intracellular levels of RAS–GTP against RAS–GDP is observed in *ira1* and *ira2* mutants, activating various target effectors [41]. Our data indicated that intracellular cAMP levels and the phosphorylation of MAPK Cmk1 in the *coira1* mutant, DARS1, and DARS2 was higher than that in the wild-type. Interestingly, the phosphorylation of Cmk1 in the *coras2* mutant was higher compared with the wild-type, although CoRas2 is an upstream regulator of CoMekk1–Cmk1. In *M. oryzae*, the intracellular cAMP levels of the dominant active MEK $Mst7^{S212D}_{T216E}$ strain are lower than those of the wild-type [42]. Recently, the functional analysis of the adenylate cyclase-associated protein encoding *CAP1*, the ortholog of yeast *Srv2*, Cap1 may play a role in the feedback inhibition of MoRas2 signaling when Pmk1 MAP kinase is activated [12]. In yeast and phytopathogenic fungi, cAMP signaling is intimately associated with a MAPK homologous with PMK1 for regulating various developmental and plant-infection processes [43]–[45]. We assumed that the constitutively active forms CoRas1 and CoRas2 induce excessive activation of the cAMP–PKA and MAPK CoMekk1–Cmk1 signaling pathways, resulting in the interruption of the cAMP–PKA and MAPK signaling pathways (Figure 15). Interestingly, the level of cAMP and the phosphorylation of Cmk1 in DARS1 was higher than that in the DARS2; however, the pathogenicity of DARS1 was not attenuated during the infection of cucumber leaves (Figure S11). Therefore, CoRas1 may be involved in the cAMP–PKA and MAPK CoMekk1–Cmk1 signaling pathways in vegetative hyphae. To understand Ras-mediated signaling pathways clearly, further functional analysis of the relationship between CoRas1 and CoRas2 during infection is required.

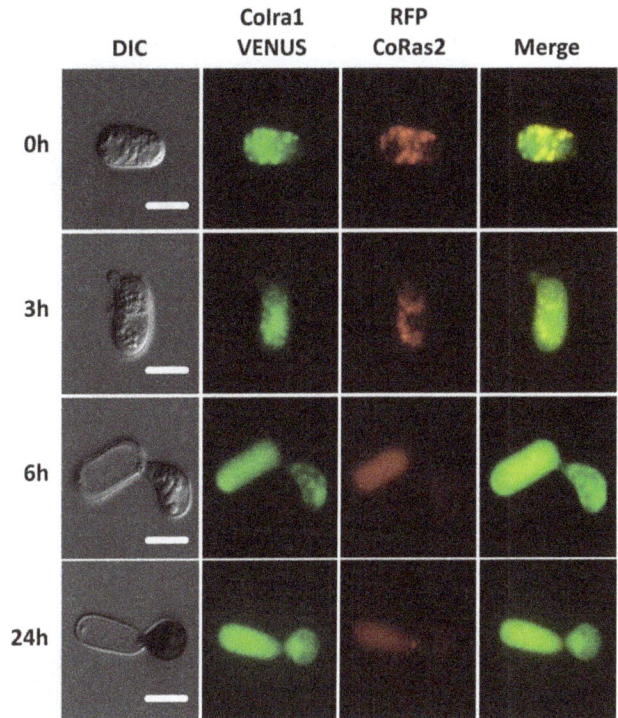

Figure 11. Assay for colocalization of Colra1 and CoRas2. Conidial suspensions of the RFP–RS2/IRA1–VENUS strain were incubated on glass slides at 24°C for 0 h, 3 h, 6 h, and 24 h and observed by fluorescent microscopy. RFP–RS2/IRA1–VENUS, the wild-type strain expressing CoIRA1–VENUS and RFP–CoRAS2. Scale bar, 10 μm.

Colra1 interacts with CoRas2 and negatively regulates CoRas2

Ras serves as a molecular switch, coupling activated membrane receptors to the downstream signaling molecule, by alternating between the GTP-bound (active) and the GDP bound (inactive) conformations. In *S. cerevisiae*, Ira1/2 negatively regulates Ras1/2; therefore, the ira1/2 mutants accumulate high amounts of Ras–GTP compared with the wild-type [7]. In the present study, the localization pattern of CoRas2 in the *coira1* mutant was similar to that in the active form CoRas2 in vegetative hyphae. Moreover, BiFC assays supported that CoIra1 regulates CoRas2 at the plasma membrane of vegetative hyphae. It is likely that CoIra1 negatively regulates CoRas2 in vegetative hyphae. In *Aspergillus fumigatus*, RasA localizes in plasma membranes in hyphae where it associates with and stimulates targeted effectors, resulting in the

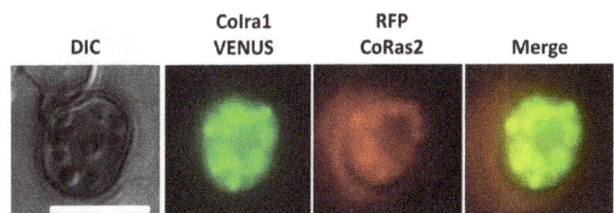

Figure 12. Assay for the colocalization of Colra1 and CoRas2 in appressoria at 48 h after inoculation on cucumber leaves. Conidial suspensions of RFP–RS2/IRA1-strain were incubated in cucumber leaves at 24°C for 48 h and observed under fluorescent microscopy. RFP–RS2/IRA1–VENUS, the wild-type strain expressing CoIRA1–VENUS and RFP–CoRAS2. Scale bar, 10 μm.

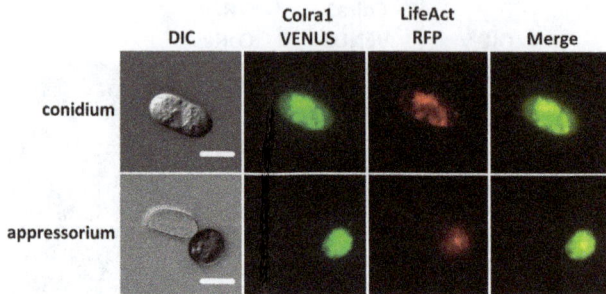

Figure 13. Assay for colocalization of Colra1 and F-actin. Conidial suspensions of the LA/IRA1–VENUS strain were incubated on glass slides at 24°C for 0 h and 24 h and observed under fluorescent microscopy. LA/IRA1–VENUS, the wild-type strain expressing *ColRA1–VENUS* and Lifeact–*RFP*. Scale bar, 10 μm.

regulation of polarized morphogenesis [46]. We considered that CoIra1 negatively regulates CoRas2, which is required for the proper morphogenesis of vegetative hyphae. However, this morphogenesis in the *coira1* mutant and the dominant active form CoRas2 expressed strain was similar to that in the wild-type. In *C. orbiculare*, the conidia in the MAPK *cmk1* disruption mutant fails to germinate, however, by adding 0.1-% yeast extract, conidial germination efficiency can be restored [33]. Thus, there is a strong possibility that a nutrient-specific signaling pathway regulates conidial germination in *C. orbiculare*. Therefore, we speculated the presence of a bypass pathway that regulates vegetative hyphae morphogenesis independent of the CoIra1–CoRas2 signaling pathway in the presence of nutrients.

ColRA1 regulates proper infection-related morphogenesis

Our data indicated that the pathogenicity of the *coira1* mutant was attenuated during the infection of the cucumber cotyledons.

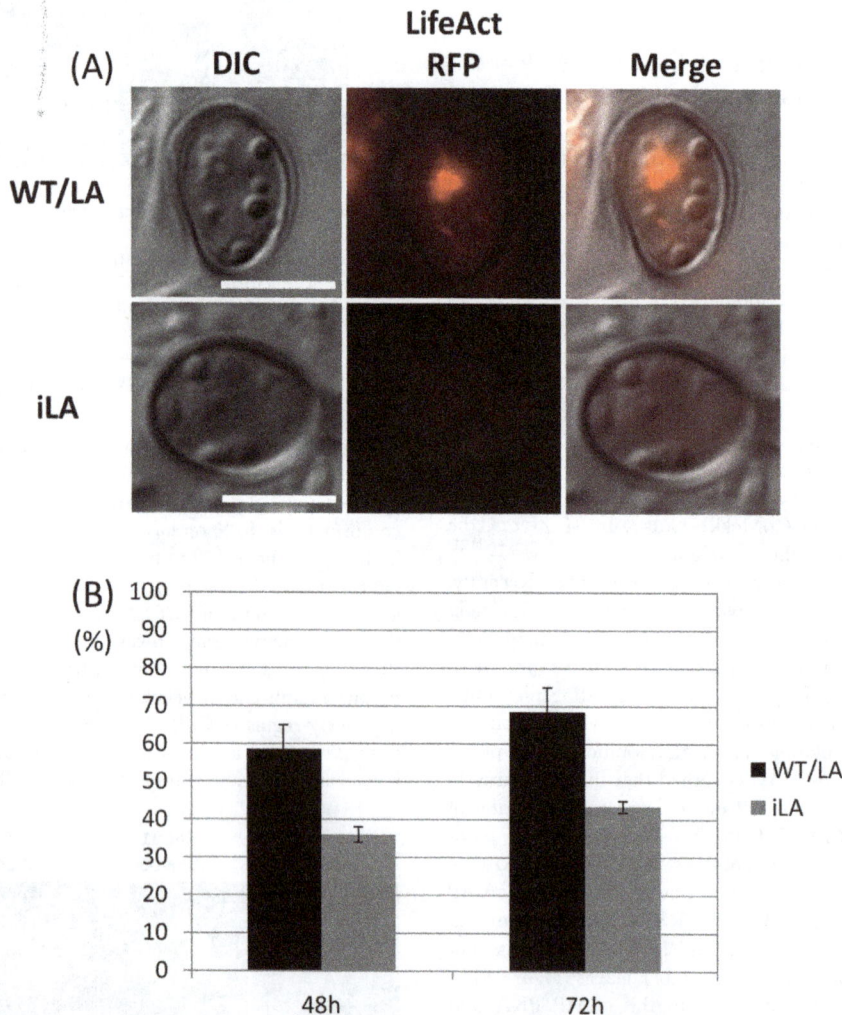

Figure 14. Assembly of F-actin in the appressorium pores of the *coira1* mutant. (A) Micrograph of F-actin organization in the appressorium pore visualized by the expression of Lifeact–RFP in the wild-type and in the *coira1* mutant. Conidial suspensions (10 μl) of each strain were applied to the abaxial surface of the cucumber cotyledons and incubated at 24°C for 48 h. WT/RA, the wild type expressing Lifeact–RFP; iRA, the *coira1* mutant expressing Lifeact–RFP (B) Percentage of the assembly of F-actin in the appressorium pore of the *coira1* mutants on the abaxial surface of cucumber cotyledons. Approximately 100 appressoria of each strain were observed per incubated site for 48 h, 72 h post-inoculation. Two replicates were examined. Three independent experiments were conducted, and standard errors are shown. Black bar, WT/RA; gray bar, Ira; WT/RA, the wild type expressing LifeAct-RFP; iRA, the *coira1* mutant expressing LifeAct-RFP.

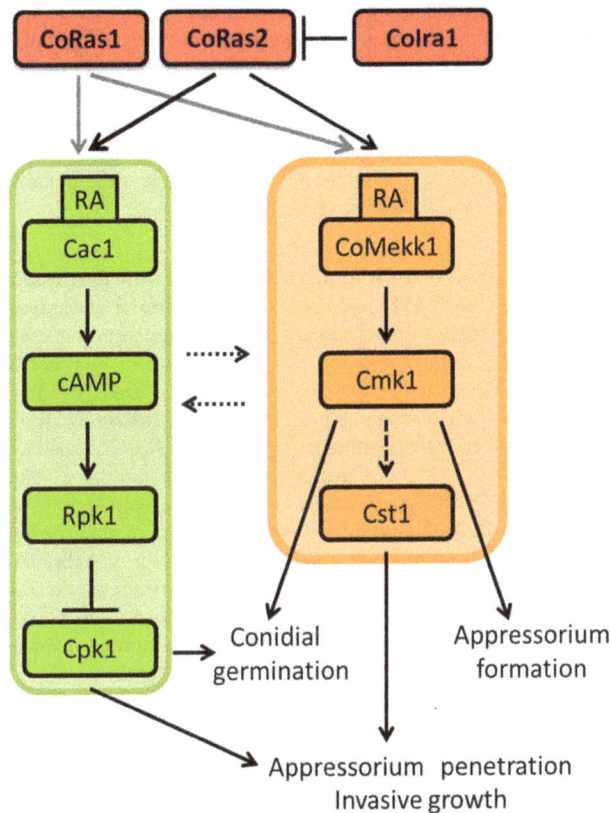

Figure 15. The hypothetical model for CoIra1 and CoRas1/2 mediated signaling transduction pathway in *C. orbiculare*. The cAMP-PKA signaling pathway is involved in conidial germination, appressorium penetration and invasive hyphae formation. The MAPK CoMekk1–Cmk1 signaling pathway is involved in conidial germination and appressorium formation. The CoRas2 localization pattern in the *coira1* mutant was similar to that in DARS2. Moreover, BiFC assays supported that CoIra1 interacted with CoRas2 in the plasma membrane. Therefore, CoIra1 negatively regulates CoRas2. The *coira1* mutant and DARS2 significantly induced abnormal appressorium formation and the frequency of abnormal appressorium formation in iDNRS2 was lower compared with that in the *coira1* mutant. Moreover, intracellular cAMP levels in the *coira1* mutant and DARS2 was high compared with those in the wild type. Therefore, CoIra1 regulates cAMP-PKA signaling pathway through CoRas2. The conidia of the *coras2* mutant failed to germinate, whereas DRS2/DAMK1 restored the phenotype of the *coras2* mutant. Therefore, CoRas2 is an upstream regulator of the MAPK CoMekk1–Cmk1 signaling pathway. Interestingly, the phosphorylation of MAPK Cmk1 in the *coras2* mutant was higher compared with that in the wild-type, although CoRas2 is an upstream regulator of CoMekk1–Cmk1. Therefore, that CoIra1 may be a key factor for regulating the crosstalk between the cAMP–PKA and CoMekk1–Cmk1 MAPK signaling pathway through CoRas2. Intracellular cAMP levels in DARS1 were higher compared with those in the wild-type in vegetative hyphae. However, DARS1 showed lower sensitivity to exogenous cAMP in appressorium development compared with DARS2. Moreover, the intensity of pathogenesis in DARS1 was similar to that in the wild-type. Therefore, CoRas1 could be involved in the cAMP–PKA signaling pathway in vegetative hyphae but not during infection related morphogenesis.

Plants have evolved with a variety of defense mechanisms against attacking phytopathogenic fungi. These include the deposition of cell wall reinforcement components (papillae), hypersensitive cell death, and the synthesis of an antimicrobial secondary metabolite. However, callose deposition was not observed following attempted penetration from the appressoria of the *coira1* mutant. Further-

more, the pathogenesis of the *coira1* mutant was not restored during the infection of heat-shocked cucumber cotyledons. Therefore, there is a small possibility that the observed reduction of pathogenicity in the *coira1* mutant was caused by the plant defense response. In *C. orbiculare*, the generation of the appressorium turgor pressure is required for mechanical penetration of the plant surface [47], however, appressorial turgor of the *coira1* mutant was not affected, as shown by the cytorrhysis assay [27]. The pathogenicity of the *coira1* mutant was attenuated compared with that of the wild-type during infection of wounded leaves, thereby resulting in the reduction in the development of hyphae causing infection in the host plants. In *C. orbiculare*, the catalytic subunit of PKA Cpk1 and adenylate cyclase Cac1 is involved in infectious growth in plants [3]. We postulate that excessive cAMP levels in the *coira1* mutant would affect the process of the infection hyphae development in the host plant. The vegetative hyphae of the *coira1* mutant and the wild-type showed normal morphology; however, the *coira1* mutant formed bulb-shaped hyphae on the cellulose membranes. In response to nitrogen starvation, diploid *S. cerevisiae* cells undergo pseudohyphal growth, which is enhanced by the expression of the dominant active allele *RAS2* [48]. Further investigation has revealed that the pseudohyphal growth is controlled by both Kss1 MAPK and cAMP–PKA signaling [49]. Similarly, *C. albicans* strains carrying the activating $RAS1^{V13}$ allele formed more abundant hyphae in a shorter time than that in the wild-type strain. Thus, activated Ras is a key factor for regulating hyphal morphogenesis in fungi. In this study, the molecular mechanism of bulb-shape penetration hyphae caused by the *coira1* mutation remains unclear. Therefore, further functional analysis of CoIra1 will be helpful for understanding the regulation of penetration hyphae morphogenesis for the requirement of proper infection of the host plant.

CoIra1 and CoRas2 is involved in the appressoria-mediated differentiation of infection hyphae development

CoRas2 and CoIra1 did not show specific colocalization in developing appressoria. Consistently, CoIra1 and the active form CoRas2 did not colocalize in developing appressoria. Therefore, during appressorium formation, CoRas2 may be regulated by other factors different from CoIra1. In *C. orbiculare*, 48 to 72 h post-inoculation is a crucial phase for appressorium-mediated penetration. Our data indicated that CoIra1 and CoRas2 colocalized in a vesicle-like structure in the appressoria 48 h post-inoculation. Moreover, in *M. oryzae*, the correct organization of F-actin in the appressorium pore is important for penetration peg development [37]. Our data also indicated that CoIra1 is involved in the assembly of the F-actin in the appressorium pore. Therefore, we assume that CoIra1 could be involved in the F-actin organization in appressoria, which is required for penetration peg emergence. However, it is likely that CoIra1 does not directly regulate F-actin, because it does not colocalize with the F-actin in the appressoria pore. In *M. oryzae*, NADPH oxidases regulate the septin-mediated assembly of F-actin in the appressorium pore [50]. Moreover, in *C. neoformans*, the GTP-bound form of Ras1 interacts with Rho–GEF Cdc24 that mediates the activation of Cdc42 and Rac proteins, and Cdc42 is involved in cytokinesis and bud morphogenesis through the organization of septin proteins [51]. In *Epichloë festucae*, NoxA is activated by a small GTPase RacA [52]. Therefore, in *C. orbiculare*, Ras–Cdc42–septin proteins and the Ras–Rac–Nox proteins signaling pathway may regulate the assembly of F-actin in the appressorium pore. Our future work will identify interaction factors between CoIra1 and

CoRas2, elucidating those that are involved in the assembly of F-actin in the appressorial pore.

Supporting Information

Figure S1 Organization of the *CoIRA1* gene in *C. orbiculare*. (A) Schematic representation of *CoIRA1*. Exons are indicated by gray boxes. The predicted RASGAP domains are indicated by slashed boxes. Eleven exons of *CoIRA1* are indicated by a gray square. Ten introns of *CoIRA1* are indicated by a black bar among 11 exons. (B) RASGAP domain in CoIra1. Amino acid sequence alignment of the predicted *CoIRA1* gene product with homologs from *Saccharomyces cerevisiae IRA1*, *IRA2*, and *Magnaporthe oryzae* (M.o.). Identical amino acids are indicated by a black background, similar residues are indicated by a gray background, and gaps introduced for alignments are indicated by a hyphen. The predicted RASGAP domain is indicated by a black line.

Figure S2 Gene disruption of the *CoIRA1* of *C. orbiculare*. (A) *CoIRA1* gene disruption by homologous recombination with the *CoIRA1* disruption vector in which a *hph* fragment was inserted into the *CoIRA1* gene. Through double crossover, an *Eco*RV fragment of approximately 3.4 kb containing wild-type *CoIRA1* was predicted to be replaced by a fragment of approximately 5.6 kb containing the *hph* fragment. (B) *CoIRA1* gene disruption was confirmed by Southern blot analysis. Genomic DNAs from the wild-type 104-T and transformants were digested with *Eco*RV and probe with an upstream 1.0-kb fragment of the *CoIRA1* gene. WT, wild-type; dis1-5, *coira1* mutants; ec1-2, ectopic strains.

Figure S3 Hyphal growth and conidia number of the *coira1* mutant on PDA. (A) Each strain was grown on the PDA medium at 24°C for five days and the number of conidia harvested from a 9-cm PDA plate at 5 days after incubation at 24°C.

Figure S4 Appressorium cytorrhysis assay for the *coira1* mutant. Appressoria formed on the multiwell slides were exposed to glycerol solutions ranging from 0 M to 4 M and the percentage of collapsed spherical appressorium was counted. Approximately 200 conidia were observed per well. Three replicates were examined. Three independent experiments were conducted, and standard errors are shown. Solid line, the wild-type; dotted line, the *coira1* mutant.

Figure S5 Pathogenicity assay of the *coira1* mutant in wounded leaves. Conidial suspensions of each strain were inoculated on wounded sites on the cotyledon of the cucumber prepared by scratching the leaves with a sterile toothpick. The leaves were incubated at 24°C for seven days. Strains: WT, wild-type 104-T; DL1, the *coira1* mutant; CL1, the *CoIRA1*-complemented transformant of DL1; DPS1, the *pks1* mutant.

Figure S6 Host defense response was not induced by the penetration of the *coira1* mutant. (A) Quantification of papilla formation at sites of attempted penetration by appressorium in *C. orbiculare*. At three days, leaf epidermal strips inoculated with each strain was stained with Aniline blue to reveal the papilla and observed with epi-fluorescence microscopy. Strains: WT, wild-type; DL1, the *coira1* mutant; CL1, the *CoIRA1*-complemented transformant of DL1; DSD1, the *ssd1*

mutant. At least 200 appressoria were counted for each strain and standard deviations were calculated from three replicated experiments. (B) Pathogenicity assay of the *coira1* mutant on heat-shock cotyledons after the heat treatment at 50°C for 30 s, cucumber cotyledons were inoculated with conidial suspensions. Strains: WT, wild-type; DL1, *coira1* mutant; CL1, the *CoIRA1*-complemented transformant of DL1; DSD1, the *ssd1* mutant. Controls were not exposed to heat shock.

Figure S7 Penetration hyphae formation of a dominant active form *CoRAS1* and *CoRAS2* introduced transformants on cellulose membranes. Conidial suspensions of each strain in distilled water were incubated on cellulose membranes at 24°C for 48 h. WT, the wild-type 104-T; DARS1, WT transformed with a dominant active form *CoRAS1*; DARS2, WT transformed with a dominant active form *CoRAS2*. Scale bar, 10 μm. (B) Percentages of penetration hyphae formation, and bulb-shaped penetration-hyphae formation of *C. orbiculare* WT, DARS1 and DARS2 on cellulose membranes. Approximately 200 conidia of each strain were observed on cellulose membranes. Three replicates were examined. Three independent experiments were conducted, and standard errors are shown. black bar, penetration hyphae; gray bar, bulb-shape penetration hyphae formation.

Figure S8 Pathogenicity assay of a dominant active form *CoRAS1* and *CoRAS2* introduced transformants on cucumber cotyledons. Conidial suspensions of each strain were inoculated on the detached cucumber cotyledons, and the leaves were incubated at 24°C for seven days. WT, the wild-type; DARS1, WT transformed with a dominant active form *CoRAS1*; DARS2, WT transformed with a dominant active form *CoRAS2*.

Figure S9 Appressorium formation in *coira1* mutants of *C. orbiculare* on glass slides. (A) Conidial suspensions of each strain in distilled water were incubated on multiwell glass slides at 24°C for 24 h. WT, wild-type; DRS2-1 and DRS2-2, the *coras2* mutant; CRS2-1, the *CoRAS2*-complemented transformant of DRS2-1; CR2-2, the *CoRAS2*-complemented transformant of DRS2-2. Scale bar, 10 μm. (B) Percentages of conidial germination, appressorium formation in *C. orbiculare* WT and *coras2* mutants on multiwell glass slides. Approximately 100 conidia of each strain were observed per well on multiwell glass slides. Three replicates were examined. Three independent experiments were conducted, and standard errors are shown. Black bar, conidial germination; gray bar, appressorium formation.

Figure S10 Pathogenicity assay of *coras2* mutants of *C. orbiculare* on the cucumber cotyledons. Conidial suspensions of each strain were placed on detached cotyledons of the cucumber, and the leaves were incubated at 24°C for seven days. Shown are the leaves after incubation with the following strains: WT, wild-type 104-T; DRS2-1 and DSR2-2, the *coras2* mutant; CRS2-1, the *CoRAS2*-complemented transformant of DRS2-1; the CRS2-2, *CoRAS2*-complemented transformant of DRS2-2.

Figure S11 Assay for colocalizations of CoIra1 and active form CoRas2. Conidial suspensions of RFP–DARS2/IRA1–VENUS strain were incubated on glass slides at 24°C for 0 h, 3 h, 6 h and 24 h and observed by fluorescent microscopy. RFP–DARS2/IRA1–VENUS, the wild-type strain expressing

RFP fused with a dominant active form *CoRAS2* and *CoIRA1*–VENUS. Scale bar, 10 μm.

Figure S12 Localization of a functional CoIra1–VENUS fusion protein in *C. orbiculare* in initial and late infection hyphae in the cucumber leaves. The wild-type strain expressing *CoIRA1*–VENUS was inoculated on cucumber leaves and incubated at 48 h, 72 h and CoIra1–VENUS was observed using fluorescent microscopy.

Table S1 PCR primers used in this study.

Acknowledgments

We are grateful to Dr. Y. Takano (Kyoto University) for providing the *cac1* and *maf1* mutants of *C. orbiculare*. We thank Dr. G. Tsuji (Kyoto Prefectural University) for providing the Lifeact–RFP vector.

Author Contributions

Conceived and designed the experiments: KH YK. Performed the experiments: KH. Analyzed the data: KH YK. Contributed reagents/materials/analysis tools: KH YK. Contributed to the writing of the manuscript: KH YK.

References

1. Kubo Y, Takano Y (2013) Dynamics of infection-related morphogenesis and pathogenesis in *Colletotrichum orbiculare*. J Gen Plant Pathol 79: 233–242. doi:10.1007/s10327-013-0451-9.
2. Sakaguchi A, Tsuji G, Kubo Y (2010) A yeast *STE11* homologue *CoMEKK1* is essential for pathogenesis-related morphogenesis in *Colletotrichum orbiculare*. Mol Plant Microbe Interact 23: 1563–1572. doi:1 0.1094/MPMI-03-10-0051.
3. Yamauchi J, Takayanagi N, Komeda K, Takano Y, Okuno T (2004) cAMP-PKA signaling regulates multiple steps of fungal infection cooperatively with Cmk1 MAP kinase in *Colletotrichum lagenarium*. Mol Plant Microbe Interact 17: 1355–1365. doi:10.1094/MPMI.2004.17.12.1355.
4. Park G, Xue C, Zhao S, Kim Y, Orbach M, et al. (2006) Multiple upstream signals converge on the adaptor protein Mst50 in *Magnaporthe grisea*. Plant Cell 18: 2822–2835. doi:10.1105/tpc.105.038422.
5. Bluhm BH, Zhao X, Flaherty JE, Xu JR, Dunkle LD (2007) *RAS2* regulates growth and pathogenesis in *Fusarium graminnearum*. Mol Plant Microbe Interact 20: 627–636. doi:10.1094/MPMI-20-6-0627.
6. Ha YS, Memmott SD, Dickman MB (2003) Function analysis of Ras in *Colletotrichum trifolii*. FEMS Microbiol Lett 26: 315–321.
7. Harashima T, Anderson S, Yates JR 3rd, Heitman J (2006) The kelch proteins Gbp1 and Gpb2 inhibit Ras activity via association with the yeast RasGAP neurofibromin homologs Ira1 and Ira2. Mol Cell 22: 819–830. doi:10.1016/j.molcel.2006.05.011.
8. Lee YH, Dean RA (1993) cAMP regulates infection structure formation in the plant pathogenic fungus *Magnaporthe grisea*. Plant Cell 5: 693–700. doi:10.1105/tpc.5.6.693.
9. Takano Y, Komeda K, Kojima K, Okuno T (2001) Proper regulation of cyclic AMP-dependent protein kinase is required for growth, conidiation, and appressorium formation in the anthracnose fungus *Colletotrichum lagenarium*. Mol Plant Microbe Interact 14: 1149–1157. doi:10.1094/MPMI.2001.14.10.1149.
10. Adachi K, Hamer JE (1998) Divergent cAMP signaling pathways regulate growth and pathogenesis in the rice blast fungus *Magnaporthe grisea*. Plant Cell 10: 1361–1374. doi:10.1105/tpc.10.8.1361.
11. Choi W, Dean RA (1997) The adenylate cyclase gene *MAC1* of *Magnaporthe grisea* controls appressorium formation and other aspects of growth and development. Plant Cell 9: 1973–1983. doi:10.1105/tpc.9.11.1973.
12. Zhou X, Zhang H, Li G, Shaw B, Xu JR (2012) The Cyclase-associated protein Cap1 is important for proper regulation of infection-related morphogenesis in *Magnaporthe oryzae*. PLOS Pathogen 8: e1002911. doi:10.1371/journal.-ppat.1002911.
13. Lee N, Kronstad JW (2002) ras2 controls morphogenesis, pheromone response, and pathogenicity in the fungal pathogen *Ustilago maydis*. Eukaryot Cell 1: 954–966. doi:10.1128/EC.1.6.954-966.2002.
14. Gibbs JB (1991) Ras C-terminal processing enzymes-new drug targets? Cell 65: 1–4.
15. Magee T, Marshall C (1999) New insights into the interaction of Ras with the plasma membrane. Cell 98: 9–12. doi:10.1016/S0092-8674(00)80601-7.
16. Ghaemmaghami S, Huh WK, Bower K, Howson RW, Belle A, et al. (2003) Global analysis of protein expression in yeast. Nature 16: 737–741. doi:10.1038/nature02046.
17. Belotti F, Tisi R, Paiardi C, Groppi S, Martegani E (2011) PKA-dependent regulation of Cdc25 RasGEF localization in budding yeast. FEBS Lett 15: 3914–3920. doi:10.1016/j.febslet.2011.10.032.
18. Belotti F, Tisi R, Paiardi C, Rigamonti M, Groppi S, et al. (2012) Localization of Ras signaling complex in budding yeast. Biochim Biophys Acta 1823: 1208–1216. doi:10.1016/j.bbamcr.2012.04.016.
19. Broggi S, Martegani E, Colombo S (2013) Live-cell imaging of endogenous Ras-GTP shows predominant Ras activation at the plasma membrane and in the nucleus in the *Saccharomyces cerevisiae*. Int J Biochem Cell Biol 45: 384–394. doi:10.1016/j.biocel.2012.10.013.
20. Sambrook J, Fritsch EF, Maniatis T (1989) Molecular Cloning: A laboratory Manual. Cold Spring Harbor Laboratory Press, Cold Spring Harbor, NY.
21. Tsuji G, Fujii S, Fujihara N, Hirose C, Tsuge S, et al. (2003) Agrobacterium tumefaciens-mediated transformation for random insertional mutagenesis in

Colletotrichum lagenarium. J Gen Plant Pathol 69: 230–239. doi:10.1007/s10327-003-0040-4.
22. Takano Y, Kubo Y, Kawamura C, Tsuge T, Furusawa I (1997) The *Alternaria alternata* melanin biosynthesis gene restores appressorial melanization and penetration of cellulose membranes in the melanin-deficient albino mutant of *Colletotrichum lagenarium*. Fungal Genet Biol 21: 131–140. doi:10.1006/fgbi.1997.0963.
23. Bruno KS, Tenjo F, Li L, Hamer JE, Xu JR (2004) Cellular localization and role of kinase activity of PMK1 in *Magnaporthe grisea*. Eukaryot Cell 3: 1525–1532. doi:10.1128/EC.3.6.1525-1532.2004.
24. Tsuji G, Takeda T, Furusawa I, Horino O, Kubo Y (1997) Carpropamid, an anti-rice blastfungicide, inhibits scytalone dehydratase activity and appressorial penetration in *Colletotrichum lagenarium*. Pestic Biochem Physiol 57: 211–219.
25. Gan P, Ikeda K, Irieda H, Narusaka M, O'Connel RJ, et al. (2013) Comparative genomic and transcriptomic analyses reveal the hemibiotrophic stage shift of *Colletotrichum* fungi. New Phytol 197: 1236–1249. doi:10.1111/nph.12085.
26. Bechinger C, Giebel KF, Schnell M, Leiderer P, Deising HB, et al. (1999) Optical measurements of invasive forces exerted by appressoria of a plant pathogenic fungus. Science 17: 1896–1899. doi:10.1126/science.285.5435.1896.
27. Howard RJ, Ferrari MA, Roach DH, Money NP (1991) Penetration of hard substrates by a fungus employing enormous turgor pressures. Proc Natl Acad Sci U S A 15: 11281–11284.
28. Tanaka S, Ishihana N, Yoshioka H, Huser A, O'Connell R, et al. (2009) The *Colletotrichum orbiculare SSD1* mutant enhances *Nicotiana benthamiana* basal resistance by activating a mitogen-activated protein kinase pathway. Plant Cell 8: 2517–2526 doi:10.1105/tpc.109.068023.
29. Tamanoi F (2011) Ras signaling in yeast. Genes Cancer 2: 210–215. doi:10.1177/1947601911407322.
30. Seeburg PH, Colby WW, Capon DJ, Goeddel DV, Levinson AD (1984) Biological properties of human c-Ha-ras1 genes mutated at codon 12. Nature 312: 71–75. doi:10.1038/312071a0.
31. Der CJ, Finkel T, Cooper GM (1986) Biological and biochemical properties of human rasH genes mutated at codon 61. Cell 17: 167–176.
32. Feng Q, Summers E, Guo B, Fink G (1999) Ras signaling is required for serum-induced hyphal differentiation in *Candida albicans*. J Bacteriol 181: 6339–6346.
33. Takano Y, Kikuchi T, Kubo Y, Hamer JE, Mise K, et al. (2000) The *Colletotrichum lagenarium* MAP kinase gene CMK1 regulates diverse aspects of fungal pathogenesis. Mol Plant Microbe Interact 13: 374–383. doi:10.1094/MPMI.2000.13.4.374.
34. Piispanen AE, Bonnefoi O, Carden S, Deveau A, Bassilana M Hogan DA (2011) Roles of Ras1 membrane localization during *Candida albicans* hyphal growth and farnesol response. Eukaryot Cell 10: 1473–1484. doi:10.1128/EC.05153-11.
35. Herrera F, Tenreiro S, Miller-Fleming L, Outeiro TF (2011) Visualization of cell-to-cell transmission of mutant huntingtin oligomers. PLoS Curr 3: RRN1210. doi:10.1371/currents.RRN1210.
36. Nagai T, Ibata K, Park ES, Kubota M, Mikoshiba K, et al. (2002) A variant of yellow fluorescent protein with fast and efficient maturation for cell-biological applications. Nat Biotechnol 20: 87–90. doi:10.1038/nbt0102-87.
37. Dagdas YF, Yoshino K, Dagdas G, Ryder LS, Bielska E, et al (2012) Septin-mediated plant cell invasion by the rice blast fungus, *Magnaporthe oryzae*. Science 336: 1590–1595. doi:10.1126/science.1222934.
38. Tsuji G, Fujii S, Tsuge S, Shiraishi T, Kubo Y (2003) The *Colletotrichum lagenarium* Ste12-like gene *CST1* is essential for appressorium penetration. Mol Plant Microbe Interact 16: 315–325. doi.org/10.1094/MPMI.2003.16.4.315.
39. Gourlay CW, Ayscough KR (2006) Actin-induced hyperactivation of the Ras signaling pathway leads to apoptosis in *Saccharomyces cerevisiae*. Mol Cell Biol 17: 6487–6501. doi:10.1128/MCB.00117-06.
40. Vinod PK, Sengupta N, Bhat PJ, Venkatesh KV (2008) Integration of global signaling pathways, cAMP-PKA, MAPK and TOR in the regulation of FLO11. PLoS One 3: e1663. doi:10.1371/journal.pone.0001663.
41. Phan VT, Ding VW, Li F, Chalkley RJ, Burlingame A, et al. (2010) The RasGAP proteins Ira2 and neurofibromin are negatively regulated by Gpb1 in

yeast and ETEA in humans. Mol Cell Biol 30: 2264–2279. doi:10.1128/MCB.01450-08.

42. Zhao X, Kim Y, Park G, Xu JR (2005) A mitogen-activated protein kinase cascade regulating infection-related morphogenesis in *Magnaporthe grisea*. Plant Cell 17: 1317–1329. doi:doi.org/10.1105/tpc.104.029116.

43. Cherkasova VA, McCully R, Wang Y, Hinnebusch A, Elion EA (2003) A novel functional link between MAP kinase cascades and the Ras/cAMP pathway that regulates survival. Curr Biol 13: 1220–1226. doi:10.1016/S0960-9822(03)00490-1.

44. Kaffarnik F, Müller P, Leibundgut M, Kahmann R, Feldbrügge M (2003) PKA and MAPK phosphorylation of Prf1 allows promoter discrimination in *Ustilago maydis*. EMBO J 22: 5817–5826. doi:10.1093/emboj/cdg554.

45. Lee N, D'Souza CA, Kronstad JW (2003) Of smuts, blasts, mildews, and blights cAMP signaling in phytopathogenic fungi. Ann Rev Phytopathol 41: 399–427. doi:10.1146/annurev.phyto.41.052002.095728.

46. Fortwendel JR, Juvvadi PR, Rogg LE, Asfaw YG, Burns KA, et al. (2012) Plasma membrane localization is required for RasA-mediated polarized morphogenesis and virulence of *Aspergillus fumigatus*. Eukaryot Cell 11: 966–977. doi:10.1128/EC.00091-12.

47. Fujihara N, Sakaguchi A, Tanaka S, Fujii S, Tsuji G, et al. (2010) Peroxisome biogenesis factor PEX13 is required for appressorium-mediated plant infection by the anthracnose fungus *Colletotrichum orbiculare*. Mol Plant Microbe Interact 23: 436–445. doi:10.1094/MPMI-23-4-0436.

48. Gimeno CJ, Ljungdahl PO, Styles CA, Fink GR (1992) Unipolar cell divisions in the yeast *S. cerevisiae* lead to filamentous growth: regulation by starvation and *RAS*. Cell 68: 1077–1090.

49. Pan X and Heitman J (1999) Cyclic AMP-dependent protein kinase regulates pseudohyphal differentiation in *Saccharomyces cerevisiae*. Mol Cell Biol 19: 4847–4887.

50. Ryder LS, Dagdas YF, Mentlak TA, Kershaw MJ, Thornton CR, et al. (2013) NADPH oxidases regulate septin-mediated cytoskeletal remodeling during plant infection by the rice blast fungus. Proc Natl Acad Sci USA 110: 3179–3184. doi:10.1073/pnas.1217470110.

51. Ballou ER, Selvig K, Narloch JL, Nichols CB, Alspaugh JA (2013) Two Rac paralogs regulate polarized growth in the human fungal pathogen *Cryptococcus neoformans*. Fungal Genet Biol 47: 58–75. doi:10.1016/j.fgb.2013.05.006.

52. Tanaka A, Takemoto D, Hyon GS, Park P, Scott B (2008) NoxA activation by the small GTPase RacA is required to maintain a mutualistic symbiotic association between *Epichloë festucae* and perennial ryegrass. Mol Microbiol 68: 1165–1178. doi:10.1111/j.1365-2958.2008.06217.

4

Association of Aminoacyl-tRNA Synthetases Gene Polymorphisms with the Risk of Congenital Heart Disease in the Chinese Han Population

Min Da[1][9], Yu Feng[1][9], Jing Xu[2][9], Yuanli Hu[1], Yuan Lin[3], Bixian Ni[3], Bo Qian[1], Zhibin Hu[3], Xuming Mo[1]*

1 Department of Cardiothoracic Surgery, The Affiliated Nanjing Children's Hospital of Nanjing Medical University, Nanjing, Jiangsu, P.R. China, **2** Department of Thoracic and Cardiovascular Surgery, The First Affiliated Hospital of Nanjing Medical University, Nanjing, Jiangsu, P.R. China, **3** Department of Epidemiology and Biostatistics, State Key Laboratory of Reproductive Medicine, Nanjing Medical University, Nanjing, Jiangsu, P.R. China

Abstract

Aminoacyl-tRNA synthetases (ARSs) are in charge of cellular protein synthesis and have additional domains that function in a versatile manner beyond translation. Eight core ARSs (*EPRS, MRS, QRS, RRS, IRS, LRS, KRS, DRS*) combined with three nonenzymatic components form a complex known as multisynthetase complex (MSC).We hypothesize that the single-nucleotide polymorphisms (SNPs) of the eight core ARS coding genes might influence the susceptibility of sporadic congenital heart disease (CHD). Thus, we conducted a case-control study of 984 CHD cases and 2953 non-CHD controls in the Chinese Han population to evaluate the associations of 16 potentially functional SNPs within the eight ARS coding genes with the risk of CHD. We observed significant associations with the risk of CHD for rs1061248 [G/A; odds ratio (OR) = 0.90, 95% confidence interval (CI) = 0.81–0.99; $P = 3.81 \times 10^{-2}$], rs2230301 [A/C; OR = 0.73, 95%CI = 0.60–0.90, $P = 3.81 \times 10^{-2}$], rs1061160 [G/A; OR = 1.18, 95%CI = 1.06–1.31; $P = 3.53 \times 10^{-3}$] and rs5030754 [G/A; OR = 1.39, 95%CI = 1.11–1.75; $P = 4.47 \times 10^{-3}$] of *EPRS* gene. After multiple comparisons, rs1061248 conferred no predisposition to CHD. Additionally, a combined analysis showed a significant dosage-response effect of CHD risk among individuals carrying the different number of risk alleles ($P_{trend} = 5.00 \times 10^{-4}$). Compared with individuals with "0–2" risk allele, those carrying "3", "4" or "5 or more" risk alleles had a 0.97-, 1.25- or 1.38-fold increased risk of CHD, respectively. These findings indicate that genetic variants of the *EPRS* gene may influence the individual susceptibility to CHD in the Chinese Han population.

Editor: Yong-Gang Yao, Kunming Institute of Zoology, Chinese Academy of Sciences, China

Funding: This work was supported in part by National Natural Science Foundation of China (grant numbers 81370277 and 81300128); Jiangsu Provincial Special Program of Medical Science (grant number BL2013013); Ph.D. Programs Foundation of Ministry of Education of China (grant number 20123234120015); and Jiangsu Natural Science Foundation (grant number BK20131025). The funders had no role in study design, data collection and analysis, decision to publish, or preparation of the manuscript.

Competing Interests: The authors have declared that no competing interests exist.

* Email: mohsuming15@sina.com

[9] These authors contributed equally to this work.

Introduction

Congenital heart disease(CHD) is the most common human birth defect and the leading cause of perinatal mortality, with an incidence of approximately 6–8 per 1000 live births or even higher [1,2,3]. With the advances in surgical techniques, the prognosis of children with complicated and uncomplicated CHDs continues to improve, but the reported incidence remains unchanged [4]. The etiology of CHD is complex and possibly includes the interaction of inherited factors and environmental exposures [5,6,7]. A multitude of research studies have identified both chromosomal abnormality and gene mutations as causation for the syndromic heart malfunction [8]. However, the origin of non-syndromic CHD, which accounts for most of all congenital cardiac abnormalities, is waiting to be uncovered further.

Over the past decades, plenty of genes have been identified as candidates to be responsible for CHD [9,10,11]. However, aminoacyl-tRNA synthetases (ARSs) that seemed to be in charge

of only cellular protein synthesis were overlooked. ARSs catalyze the attachment of amino acids to their cognate tRNAs with high fidelity [12,13]. Recent research has shown that eukaryote ARSs, distinguished from their prokaryotic counterparts, have additional domains and motifs such as glutathione S-transferase (GST), WHEP domains, leucine zipper domains, and α-helicalappendices that function beyond translation [14] and may link with a variety of human diseases, such as cancer, neuronal pathologies, autoimmune disorders, and disrupted metabolic conditions [13,15]. Recently, the nontranslational functions of vertebrate ARSs have been associated with cytoplasmic forms and nuclear and secreted extracellular forms that impact cardiovascular development pathways [16].

Eight core aminoacyl-tRNA synthetases (ARSs), bifunctional glutamyl-prolyl-tRNA synthetase (*EPRS*), isoleucyl-tRNA synthetase (*IRS*), leucyl-tRNA synthetase (*LRS*), methionyl-tRNA synthetase (*MRS*), glutaminyl-tRNA synthetase (*QRS*), lysyl-tRNA synthetase (*KRS*), aspartyl-tRNA synthetase (*DRS*), and arginyl-tRNA synthetase (*RRS*), form a macromolecular protein complex with three auxiliary factors, designated ARS-interacting multi-functional protein 1 (AIMP1), AIMP2 and AIMP3. This complex is known as the multisynthetase complex (MSC). The MSC may act as a depot for ARSs, which could be subsequently released from the macromolecular complexes to participate in auxiliary tasks beyond translation [17], generate a channel for the delivery of tRNAs [18,19] and help the proofreading of newly synthesized nuclear tRNAs in the nucleus [20].

According to the expressed sequence tags (EST) profile in the public database UniGene (http://www.ncbi.nlm.nih.gov/UniGene), all eight of the core ARS coding genes were expressed in human heart tissues, with transcripts ranging from 44 to 502 per million (**Figure S1**). Thus, it is plausible that changes in the core ARSs may affect heart development and are related to the occurrence of CHD. However, to date, no research has reported a relation between the genetic variants of the core ARS genes and CHD susceptibility.

To determine the effect of genetic variants in the core ARS genes on CHD development, we conducted a case-control study by investigating the genotype frequency distribution of the 16 potential functional polymorphisms in the eight members of the MSC.

Materials and Methods

Ethics Statement

This study was approved by the institutional review board of Nanjing Medical University and adhered to the tenets of the Declaration of Helsinki. The design and performance of the current study involving human subjects were clearly described in a research protocol. All participants and/or their parents were voluntary and completed the informed consent in writing before taking part in this research.

Study populations

The case-control analysis included 984 affected children with sporadic CHD and 2953 unrelated non-CHD controls. All subjects were genetically unrelated ethnic Han Chinese. Subjects for the study were consecutively recruited from the Affiliated Nanjing Children's Hospital of Nanjing Medical University and the First Affiliated Hospital of Nanjing Medical University, Nanjing, China, from March 2009 to December 2011. All CHD patients were diagnosed based on echocardiography, with some diagnoses further confirmed by cardiac catheterization and/or surgery. Potential study subjects were initially surveyed with a brief questionnaire at clinics to determine whether they were willing to participate in a research study; we then conducted a face-to-face interview to obtain demographic information. Cases that had clinical features of developmental syndromes, multiple major developmental anomalies or known chromosomal abnormalities were excluded. The exclusion criteria also included a positive family history of CHD in a first-degree relative (parents, siblings and children), maternal diabetes mellitus, phenylketonuria, maternal teratogen exposure (e.g., pesticides and organic solvents), and maternal therapeutic drug exposure during the intrauterine period. Controls were non-CHD outpatients from the same geographic areas. They were recruited from the hospitals listed above during the same time period. Controls with congenital anomalies or cardiac disease were excluded. For each participant, approximately 2 ml of whole blood was obtained to extract genomic DNA for genotyping analysis.

SNP selection and genotyping

Eight ARSs (*EPRS, MRS, QRS, RRS, IRS, LRS, KRS, DRS*) that formed MSC were selected. For each ARS-coding gene, we first used the public HapMap single nucleotide polymorphism (SNP) database (phase II+ III Feb 09, on NCBI B36 assembly and dbSNP b126) to search for SNPs that localized within gene regions, with MAF≥0.05, in the Chinese Han population. Then, a web-based analysis tool was used to predict the function of these SNPs (http://snpinfo.niehs.nih.gov/snpinfo/snpfunc.htm). Finally, a total of 27 potentially functional SNPs were selected in 8 ARS-coding genes. We next conducted linkage disequilibrium (LD) analysis by the Haploview 4.2 software, and only one SNP was selected in the case of multiple SNPs in the same haplotype block ($r^2 > 0.8$). Eighteen (rs1061160, rs1061248, rs2230301 and rs5030754 in *EPRS*; rs508904 in *MRS*; rs193466, rs2305737 and rs244903 in *RRS*; rs1058751, rs10820966 and rs556155 in *IRS*; rs10988 in *LRS*; rs2233805 and rs3784929 in *KRS*; rs2164331, rs309142, rs309143 and rs6738266 in *DRS*) of 27 SNPs remained. Two SNPs (rs6738266 and rs2164331) were excluded due to primer design failure.

Genomic DNA was isolated from leukocyte pellets of venous blood by proteinase K digestion, followed by phenol-chloroform extraction and ethanol precipitation. Nanodrop and DNA electrophoresis were used to check the quality and quantity of DNA samples before genotyping. The genotyping was performed by Illumina Infinium BeadChip (Illumina, Inc.). All SNPs were successfully genotyped with call rates >95% (**Table 1**).

Statistical analyses

The differences between the CHD patients and control subjects were evaluated in the distributions of demographic characteristics, selected variables, and frequencies of genotypes of the 16 polymorphisms using Student's t-test (for continuous variables) or the χ^2 test (for categorical variables). The χ^2 test determined the Hardy-Weinberg equilibrium of the genotype distribution of polymorphisms in the control group. LD between SNPs was evaluated using Haploview 4.2.Odds ratios (ORs) and 95% confidence intervals (CIs) were estimated by logistic regression analyses in the additive model to estimate the associations between the variants genotypes and risk of CHD. Chi-square-based Q-test was applied to test the heterogeneity of associations between subgroups, and the heterogeneity was considered significant when $P < 0.05$. All statistical analyses were performed using the Statistical Analysis System software (v.9.1.3; SAS Institute, Cary, NC, USA). All tests were two-sided, and $P < 0.05$ was considered significant.

Results

An overview of the study design using a flowchart was performed as shown in **Figure 1**. We systematically investigated the association of potentially functional SNPs with CHD susceptibility in 984 cases and 2953 controls in a Chinese population. There were no statistically significant differences for the distributions of age and gender between cases and controls ($P = 0.261$ and $P = 0.832$, respectively). Among the 984 CHD patients, 312 had atrial septal defect (ASD), 585 were diagnosed with ventricular septal defect (VSD), and 87 were diagnosed with ASD combined with VSD.

The genotype distributions of the 16 SNPs and the associations with CHD risk are summarized in **Table 2**. The observed genotype frequencies of these SNPs were in agreement with Hardy-Weinberg equilibrium in the controls (P value from 0.16 to 1.00) except rs10988 ($P = 0.04$). Among the 16 SNPs, significant

Table 1. Primary information for 16 functional SNPs in ARS-coding genes.

Gene	ARS	Chr. (cytoband)	SNP	Position (bp)[a]	Location	Predicted function[b]	MAF[c]	Allele[c]	HWE[d]	Genotyping call rate (%)
EPRS	Glutamyl-prolyl-tRNA synthetase	1q41	**rs1061248**	219968681	3'UTR	miRNA[e]	0.378	G/A	0.63	99.5
			rs1061160	219981426	exon	Splice sites[f]	0.366	G/A	0.76	99.8
			rs5030754	219983387	exon	Splice sites	0.012	G/A	0.82	99.9
			rs2230301	220024283	exon	Splice sites, nsSNP[g]	0.073	A/C	0.90	99.9
DRS	Aspartyl-tRNA synthetase	2q21.3	rs309143	135956608	intron	TFBS[h]	0.207	A/G	0.40	99.8
			rs309142	1395957754	intron	TFBS	0.488	A/G	0.29	99.8
LRS	Leucyl-tRNA synthetase	5q32	rs10988	146120433	exon	nsSNP	0.280	G/A	0.04	99.9
RRS	Arginyl-tRNA synthetase	5q34	rs244903	168486505	exon	Splice sites, nsSNP	0.146	A/G	0.35	99.7
			rs193466	1684865598	intron	TFBS	0.305	A/G	0.58	99.7
			rs2305737	168519256	3'UTR	Splice sites, miRNA	0.146	C/A	0.71	99.8
IRS	Isoleucyl-tRNA synthetase	9q22.31	rs1058751	92210634	3'UTR	miRNA	0.122	A/T	0.22	99.5
			rs556155	92223355	exon	nsSNP	0.195	A/G	0.25	99.9
			rs10820966	92293147	intron	TFBS	0.078	A/T	0.16	99.9
MRS	Methionyl-tRNA synthetase	12q13.3	rs508904	57488311	intron	TFBS	0.289	A/G	0.70	99.8
KRS	Lysyl-tRNA synthetase	16q23.1	rs3784929	75643129	intron	TFBS	0.159	G/A	0.66	99.8
			rs2233805	75647244	intron	TFBS	0.049	A/G	1.00	99.9

[a]Derived from the UCSC Genome Browser on Human Feb. 2009 (GRCh37/hg19) Assembly (http://genome.ucsc.edu/);
[b]Derived from an online tool-SNPinfo (http://snpinfo.niehs.nih.gov/snpfunc.html);
[c]Major/minor allele;
[d]Hardy-Weinberg equilibrium test among controls;
[e]miRNA: microRNA
[f]Splice sites: Exonic splicing enhancer (ESE) or exonic splicing silencer (ESS) binding sites;
[g]nsSNP: non-synonymous polymorphisms.
[h]TFBS: Transcription factor binding sites;

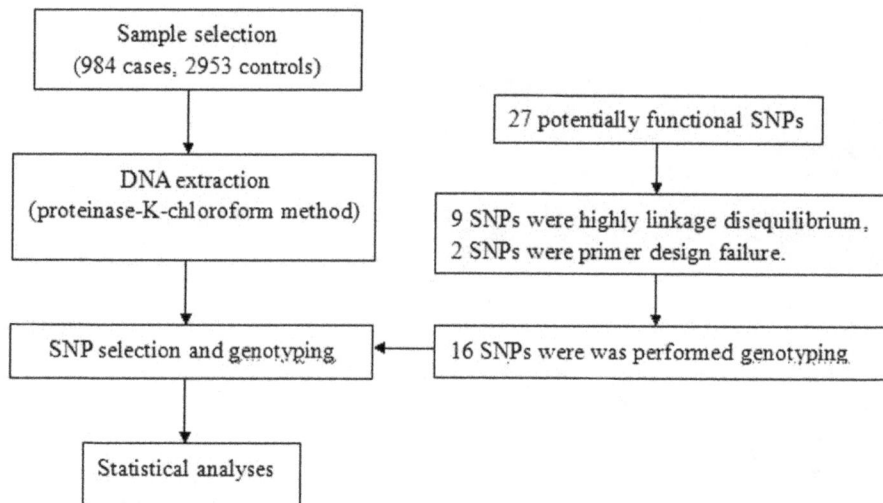

Figure 1. Study design procedures for association of ARSs gene polymorphisms with the risk of congenital heart disease in the Chinese Han population.

associations were observed between 4 SNPs (rs1061248, rs1061160, rs5030754 and rs2230301) and CHD risk following a logistic regression analysis in the additive model. All four SNPs were in the *ERPS* gene. The G allele of rs1061248 and the A allele of rs2230301 were associated with a decreased risk of CHD [additive model: odds ratio (OR) = 0.90, 95% confidence interval (CI) = 0.81–0.99, $P = 3.81 \times 10^{-2}$; and OR = 0.73, 95%CI = 0.60–0.90, $P = 3.53 \times 10^{-3}$, respectively]; however, the G allele of rs1061160 and the G allele of rs5030754 were associated with an increased risk of CHD (OR = 1.18, 95%CI = 1.06–1.31, $P = 1.28 \times 10^{-3}$; and OR = 1.39, 95%CI = 1.11–1.75, $P = 4.47 \times 10^{-3}$, respectively). We further calculated P values for the false discovery rate to perform multiple comparisons. After comparisons, we found that rs2230301, rs5030754 and rs1061160 correlated with CHD risk, whereas rs1061248 lost its significant association with the risk of CHD. In contrast, no obvious evidence of a significant association between the other 12 SNPs and CHD risk was found.

We have listed the results of the genotypic association analysis in **Table 3**. In dominant genetic model, for rs1061160 and rs5030754 polymorphisms, AG+AA and AG+GG genotypes were associated with an increased risk of CHD compared with the GG genotype, respectively(OR = 1.25, 95%CI = 1.07–1.46; OR = 1.44, 95%CI = 1.14–1.82). For rs2230301 polymorphism, AC+CC genotypes were associated with a decreased risk of CHD compared with the AA genotype(OR = 0.73, 95% CI = 0.59–0.91).

Additionally, we performed haplotype analysis (**Table 4**). As shown, the haplotype "GAAA" (combination of risk alleles of the four SNPs) was associated with an increased risk of CHD, whereas the protective allele combination "AGGC" was associated with a decreased risk of CHD. In the stratification analysis, we further evaluated the associations of the four SNPs in *EPRS* with CHD risk in subgroups stratified by gender and specific CHD phenotypes. As shown in **Table 5**, similar effects were observed among the subgroups.

We also conducted a combined analysis of the four promising SNPs to test their joint effects on CHD risk. There was a significant dosage-response effect among individuals carrying the different number of risk alleles and CHD risk ($P_{\text{trend}} = 5.00 \times 10^{-4}$). Compared with individuals with "0–2" risk allele, those carrying "3", "4" or "5 or more" risk alleles had a 0.97- (95% CI = 0.69–1.37), 1.25- (95% CI = 1.05–1.50) or 1.38-fold (95% CI = 1.14–1.68) increased risk of CHD, respectively (**Table 6**).

Discussion

In this study, we systematically investigated the association of potentially functional SNPs in ARS-coding genes of the MSC with CHD susceptibility in 984 cases and 2953 controls in a Chinese population. We observed significant association of four SNPs (rs1061248, rs1061160, rs5030754 and rs2230301) in the *EPRS* gene with the risk of CHD, and the risk remarkably accelerated in the individuals who carried more risk alleles. Although ASD and VSD represent the most common congenital heart malfunctions, the accurate pathogenesis is poorly understood. Based on previous research, the ARS-coding genes of MSC take part in diverse functional activities, and some of them have been proven to be crucial for heart development and proper functioning. Few studies have linked the variants of MSC genes to congenital heart disease. To our knowledge, we provide the first evidence that SNPs in *EPRS*, one of the core coding genes in MSC, may modulate the process of CHD.

Some ARSs in MSC have been demonstrated to have a close correlation with cardiovascular development. Glutamyl-prolyl-tRNA synthetase (*EPRS*) is a bifunctional enzyme that could translationally suppress vascular endothelial growth factor-A (VEGF-A) to regulate angiogenesis [21] and seems to act as a key gatekeeper of inflammatory gene translation [22]. Lysyl-tRNA synthetase (*KRS*) is secreted to trigger pro-inflammatory response [23] and plays a key role via Ap4A as an important signaling molecule in the transcriptional activity of microphthalmia transcription factor(MITF) [24], which has been demonstrated to be necessary in heart growth [25]. Glutaminyl-tRNA synthetase (*QRS*) can bind and inhibit the apoptotic activity of apoptosis signal-regulating kinase 1 (ASK1) [26], which has been demonstrated to be a new intracellular regulator of p38 MAPK activation in cardiac myogenic differentiation [27]. Han and colleagues [28] reported that Leucyl-tRNA synthetase (*LRS*) acts as a vital

Table 2. Summary of associations between 16 SNPs of MSC genes with congenital heart disease.

Chr. (cytoband)	Gene	SNP	Allele[a]	Case[b] (N=984)	Control[b] (N=2953)	MAF[c] Cases	MAF[c] Controls	HWE[d]	Additive model OR(95%CI)	Additive model P	P_{FDR}[e]
1q41	EPRS	rs1061248	G/A	217/486/271	741/1459/744	0.47	0.50	0.63	0.90 (0.81–0.99)	3.81×10^{-2}	0.152
		rs1061160	G/A	184/486/311	465/1402/1083	0.44	0.40	0.76	1.18 (1.06–1.31)	1.82×10^{-3}	0.029
		rs5030754	G/A	1/113/869	6/241/2704	0.06	0.04	0.82	1.39 (1.11–1.75)	4.47×10^{-3}	0.023
		rs2230301	A/C	3/114/865	20/440/2490	0.06	0.08	0.90	0.73 (0.60–0.90)	3.53×10^{-3}	0.028
2q21.3	DRS	rs309143	A/G	38/308/634	100/927/1921	0.20	0.19	0.40	1.03 (0.91–1.17)	6.40×10^{-1}	1.024
		rs309142	A/G	172/499/310	526/1472/949	0.43	0.43	0.29	1.01 (0.91–1.12)	9.11×10^{-1}	0.911
5q32	LRS	rs10988	G/A	47/340/596	120/1041/1789	0.22	0.22	0.04	1.02 (0.90–1.16)	7.32×10^{-1}	0.901
5q34	RRS	rs244903	A/G	15/213/754	50/622/2270	0.12	0.12	0.35	1.01 (0.87–1.18)	9.06×10^{-1}	0.966
		rs193466	A/G	74/399/509	229/1162/1553	0.28	0.28	0.58	1.02 (0.91–1.14)	7.72×10^{-1}	0.505
		rs2305737	C/A	20/270/691	60/744/2145	0.16	0.15	0.71	1.10 (0.95–1.26)	2.12×10^{-1}	0.565
9q22.31	IRS	rs1058751	A/T	0/150/828	11/412/2516	0.08	0.07	0.22	1.04 (0.86–1.27)	6.71×10^{-1}	0.976
		rs556155	A/G	16/206/760	34/627/2291	0.12	0.12	0.25	1.03 (0.88–1.21)	6.79×10^{-1}	0.905
		rs10820966	A/T	10/213/760	40/681/2230	0.12	0.13	0.16	0.91 (0.77–1.06)	2.21×10^{-1}	0.505
12q13.3	MRS	rs508904	A/G	105/442/433	301/1268/1380	0.33	0.32	0.70	1.07 (0.96–1.20)	2.01×10^{-1}	0.643
16q23.1	KRS	rs3784929	G/A	33/302/644	95/849/2005	0.19	0.18	0.66	1.08 (0.95–1.23)	2.40×10^{-1}	0.480
		rs2233805	A/G	0/31/951	1/106/2844	0.02	0.02	1.00	0.86 (0.57–1.29)	4.63×10^{-1}	0.823

[a]Major/minor allele;
[b]Variant homozygote/Heterozygote/Wild type homozygote;
[c]Minor allele frequency among cases/controls;
[d]Hardy-Weinberg equilibrium test among controls.
[e]Multiple comparisons P values for false discovery rate.

Table 3. Summary of genotypic association analysis of four SNPs of the *EPRS gene* with congenital heart disease.

SNP	Genotype	Cases	Controls	OR (95%CI)	P
rs1061248	GG	271	744	1.00	–
	AG	486	1459	0.91 (0.77–1.09)	3.11×10^{-1}
	AA	217	741	**0.80 (0.65–0.99)**	**3.74×10^{-2}**
	AG+AA	703	2200	0.88 (0.75–1.03)	1.15×10^{-1}
rs1061160	GG	311	1083	1.00	–
	AG	486	1402	**1.21 (1.03–1.42)**	**2.35×10^{-2}**
	AA	164	465	**1.38 (1.11–1.70)**	**3.07×10^{-3}**
	AG+AA	650	1867	**1.25 (1.07–1.46)**	**4.54×10^{-3}**
rs5030754	GG	869	2704	1.00	–
	AG	113	241	**1.46 (1.15–1.85)**	**1.72×10^{-3}**
	AA	1	6	0.52 (0.06–4.31)	5.44×10^{-1}
	AG+GG	114	247	**1.44 (1.14–1.82)**	**2.51×10^{-3}**
rs2230301	AA	865	2490	1.00	–
	AC	114	440	**0.75 (0.60–0.93)**	**8.99×10^{-3}**
	CC	3	20	0.43 (0.13–1.46)	1.76×10^{-1}
	AC+CC	117	460	**0.73 (0.59–0.91)**	**4.90×10^{-3}**

mediator for amino acid signaling to mTORC1, and the latter has been found to be related to the normal development of cardiovascular tissue [29].

Human *EPRS*, the largest polypeptide from the complex, is a bifunctional enzyme in which the two domains exhibiting each catalytic activity are linked by three tandem WHEP motifs [30]. *EPRS* contains 29 exons and 28 introns. In response to interferon-γ (IFNγ), *EPRS* is phosphorylated and released from its residence in the MSC. MSC then forms another multi-component complex, known as IFN-γ–activated inhibitor of translation (GAIT), with other regulatory proteins at a 3′UTR region that is involved in the translational silencing of target transcripts, such as VEGF-A [31,32,33]. As documented in many studies, VEGF-A shares a close relationship with CHD, and both the increased and decreased expression of VEGF-A during heart development can result in various CHD [34,35,36]. The SNP rs2230301, a missense SNP located at the 23rd exon of the *EPRS* gene, may act as a part of the exonic splicing enhancer based on the online tool SNPinfo [37]. The missense mutation would change the sequence of *EPRS* and may lead to protein misfolding and malfunction. We used a web-based analysis tool to predict the potential function of the

SNPs, and rs2230301 was predicted to be a missense variant that may result in an amino acid alteration from aspartic acid (Asp) to glutamic acid (Glu) (http://snpinfo.niehs.nih.gov/snpinfo/snpfunc.htm). The NCBI database confirmed the results (http://www.ncbi.nlm.nih.gov/). However, the predicted results differed from the *in-silico* analysis. To further validate the function of this variant, some functional studies should be performed in some follow-up studies. The SNP rs1061248 is located at the 3′ regulatory region of the *EPRS* gene with a predicted function as a MicroRNA-binding site. Considering its potentially functional role, it is likely that this polymorphism might alter miRNA binding, thereby modulating the biological function of *EPRS*. The two synonymous SNPs rs1061160 and rs5030754 were localized on the seventh exon and the eleventh exon, respectively. Recently, a synonymous SNP was reported to alter the function of the protein in certain circumstances [38].

Several limitations of the present study need to be addressed. First, we did not replicate the results in additional individuals; this may contribute to potential false positive errors. The present analysis was restricted to individuals of Chinese Han descent, and therefore, the findings may not hold true for individuals of other

Table 4. The haplotypic association of the four SNPs of the *EPRS* gene with congenital heart disease.

Haplotype[a]	case (%)	control (%)	OR (95%CI)	P
AGGA	905 (45.99)	2835 (48.0)	1.00 (referent)	
GAGA	740 (37.60)	2075 (35.13)	1.12 (0.99–1.25)	5.34×10^{-2}
GGGC	100 (5.08)	370 (6.26)	0.85 (0.67–1.07)	1.62×10^{-1}
GGGA	84 (4.27)	259 (4.39)	1.02 (0.79–1.31)	9.04×10^{-1}
GAAA	113 (5.74)	253 (4.28)	**1.40 (1.11–1.77)**	**4.91×10^{-3}**
AGGC	20 (1.02)	107 (1.81)	**0.59 (0.36–0.95)**	**3.00×10^{-2}**
Others	6 (0.30)	7 (0.12)	2.69 (0.90–8.01)	7.65×10^{-2}

[a]SNP order: rs1061248, rs1061160, rs5030754 and rs2230301.

Table 5. Stratified analysis on the associations between four SNPs in *EPRS* with congenital heart disease.

Characteristics	rs1061248					rs1061160				
	Case [a]	Control [a]	OR (95%CI) [b]	P [b]	P [c]	Case [a]	Control [a]	OR (95%CI) [b]	P [b]	P [c]
Gender										
male	105/241/138	472/856/464	0.87 (0.76–1.00)	4.90×10^{-2}	0.527	91/237/160	286/844/667	1.16 (1.00–1.33)	4.79×10^{-2}	0.751
female	112/245/133	269/603/280	0.93 (0.80–1.09)	3.74×10^{-1}		93/249/151	179/558/416	1.20 (1.03–1.40)	1.83×10^{-2}	
Diagnostic groups										
ASD	71/139/98	741/1459/744	0.84 (0.71–0.99)	4.27×10^{-2}	0.483	68/146/98	465/1402/1083	1.26 (1.07–1.49)	6.44×10^{-3}	0.489
VSD	126/299/154	741/1459/744	0.91 (0.80–1.03)	1.41×10^{-1}		104/294/184	465/1402/1083	1.16 (1.02–1.32)	2.21×10^{-2}	
ASD/VSD	20/48/19	741/1459/744	1.03 (0.76–1.39)	8.71×10^{-1}		12/46/29	465/1402/1083	1.03 (0.76–1.40)	8.52×10^{-1}	

Characteristics	rs5030754					rs2230301				
	Case [a]	Control [a]	OR(95%CI) [b]	P [b]	P [c]	Case [a]	Control [a]	OR(95%CI) [b]	P [b]	P [c]
Gender										
male	1/58/429	4/135/1659	1.58 (1.16–2.15)	3.90×10^{-3}	0.211	3/64/421	13/280/1504	0.83 (0.63–1.09)	1.77×10^{-1}	0.260
female	0/55/440	2/106/1045	1.18 (0.84–1.65)	3.37×10^{-1}		0/50/444	7/160/986	0.65 (0.47–0.90)	9.74×10^{-3}	
Diagnostic groups										
ASD	0/42/270	6/241/2704	1.62 (1.15–2.27)	5.54×10^{-3}	0.119	0/31/280	20/440/2490	0.59 (0.41–0.86)	5.93×10^{-3}	0.281
VSD	1/67/516	6/241/2704	1.40 (1.07–1.85)	1.56×10^{-2}		2/76/506	20/440/2490	0.83 (0.65–1.06)	1.37×10^{-1}	
ASD/VSD	0/4/83	6/241/2704	0.53 (0.19–1.43)	2.08×10^{-1}		1/7/79	20/440/2490	0.62 (0.31–1.21)	1.62×10^{-1}	

[a] Major/minor allele;
[b] Calculated by additive model;
[c] P for heterogeneity.

Table 6. Combined effects of rs1061248, rs1061160, rs1061160, and rs2230301 on CHD.

Number of risk alleles[a]	Case (%)	Control (%)	OR (95% CI)[b]	p^b
0–2	262 (26.95)	939 (31.96)	1.00	
3	50 (5.14)	184 (6.26)	0.97 (0.69–1.37)	8.79×10^{-1}
4	384 (39.51)	1098 (37.37)	1.25 (1.05–1.50)	1.37×10^{-2}
≥5	276 (28.40)	717 (24.40)	1.38 (1.14–1.68)	1.20×10^{-3}
Trend				5.00×10^{-4}

[a]rs1061248 G, rs1061160 A, rs5030754 A, and rs2230301 A were assumed as risk alleles;
[b]Calculated by additive model.

races and ethnicities. Additionally, the limited sample size may contribute to the failed validation in the stratified analysis concerning the association between the SNPs and CHD. We performed the statistic power analysis of the significant SNPs in the studied population. The powers of three SNPs (rs1061248, rs5030754, and rs2230301) are lower than 0.6 because the sample size of our study is relatively small (984 CHD cases and 2953 non-CHD controls) and the effects of our target common SNPs are weak. Further replication of the association signal in an independent cohort for the four SNPs would support the conclusions. Therefore, the results are required to be further replicated by well-designed studies in additional large-scale Chinese Han populations.

In conclusion, we conducted a case-control study to investigate the role of genetic variants in ARS-coding genes of MSC in the development of CHD in a Chinese population. We observed that four SNPs (rs1061248, rs1061160, rs5030754, and rs2230301) in

the *EPRS* gene may confer susceptibility to sporadic CHD and that the risk significantly increased with the number of risk alleles. However, further studies with functional evaluations are warranted to elucidate the potentially biological mechanisms of these polymorphisms in the development of CHD.

Supporting Information

Figure S1 Expressed sequence tags (EST) profile of the 8 core ARSs coding genes.

Author Contributions

Conceived and designed the experiments: XM. Performed the experiments: MD YF JX. Analyzed the data: YH YL BN BQ. Contributed reagents/materials/analysis tools: ZH. Contributed to the writing of the manuscript: MD YF JX XM.

References

1. Hoffman JI, Kaplan S (2002) The incidence of congenital heart disease. J Am Coll Cardiol 39: 1890–1900.
2. Botto LD, Correa A, Erickson JD (2001) Racial and temporal variations in the prevalence of heart defects. Pediatrics 107: E32.
3. Rosamond W, Flegal K, Friday G, Furie K, Go A, et al. (2007) Heart disease and stroke statistics–2007 update: a report from the American Heart Association Statistics Committee and Stroke Statistics Subcommittee. Circulation 115: e69–171.
4. Gatzoulis MA (2004) Adult congenital heart disease: a cardiovascular area of growth in urgent need of additional resource allocation. Int J Cardiol 97 Suppl 1: 1–2.
5. van der Bom T, Zomer AC, Zwinderman AH, Meijboom FJ, Bouma BJ, et al. (2011) The changing epidemiology of congenital heart disease. Nat Rev Cardiol 8: 50–60.
6. Jenkins KJ, Correa A, Feinstein JA, Botto L, Britt AE, et al. (2007) Noninherited risk factors and congenital cardiovascular defects: current knowledge: a scientific statement from the American Heart Association Council on Cardiovascular Disease in the Young: endorsed by the American Academy of Pediatrics. Circulation 115: 2995–3014.
7. Huhta J, Linask KK (2013) Environmental origins of congenital heart disease: the heart-placenta connection. Semin Fetal Neonatal Med 18: 245–250.
8. Pierpont ME, Basson CT, Benson DW Jr, Gelb BD, Giglia TM, et al. (2007) Genetic basis for congenital heart defects: current knowledge: a scientific statement from the American Heart Association Congenital Cardiac Defects Committee, Council on Cardiovascular Disease in the Young: endorsed by the American Academy of Pediatrics. Circulation 115: 3015–3038.
9. Wessels MW, Willems PJ (2010) Genetic factors in non-syndromic congenital heart malformations. Clin Genet 78: 103–123.
10. Wolf M, Basson CT (2010) The molecular genetics of congenital heart disease: a review of recent developments. Curr Opin Cardiol 25: 192–197.
11. Garg V (2006) Insights into the genetic basis of congenital heart disease. Cell Mol Life Sci 63: 1141–1148.
12. Schimmel P (1987) Aminoacyl tRNA synthetases: general scheme of structure-function relationships in the polypeptides and recognition of transfer RNAs. Annu Rev Biochem 56: 125–158.
13. Park SG, Ewalt KL, Kim S (2005) Functional expansion of aminoacyl-tRNA synthetases and their interacting factors: new perspectives on housekeepers. Trends Biochem Sci 30: 569–574.
14. Guo M, Yang XL, Schimmel P (2010) New functions of aminoacyl-tRNA synthetases beyond translation. Nat Rev Mol Cell Biol 11: 668–674.
15. Park SG, Schimmel P, Kim S (2008) Aminoacyl tRNA synthetases and their connections to disease. Proc Natl Acad Sci U S A 105: 11043–11049.
16. Guo M, Schimmel P (2013) Essential nontranslational functions of tRNA synthetases. Nat Chem Biol 9: 145–153.
17. Ray PS, Arif A, Fox PL (2007) Macromolecular complexes as depots for releasable regulatory proteins. Trends Biochem Sci 32: 158–164.
18. Simos G, Grosshans H, Hurt E (2002) Nuclear export of tRNA. Results Probl Cell Differ 35: 115–131.
19. Hopper AK, Phizicky EM (2003) tRNA transfers to the limelight. Genes Dev 17: 162–180.
20. Nathanson L, Deutscher MP (2000) Active aminoacyl-tRNA synthetases are present in nuclei as a high molecular weight multienzyme complex. J Biol Chem 275: 31559–31562.
21. Ray PS, Fox PL (2007) A post-transcriptional pathway represses monocyte VEGF-A expression and angiogenic activity. EMBO J 26: 3360–3372.
22. Mukhopadhyay R, Jia J, Arif A, Ray PS, Fox PL (2009) The GAIT system: a gatekeeper of inflammatory gene expression. Trends Biochem Sci 34: 324–331.
23. Park SG, Kim HJ, Min YH, Choi EC, Shin YK, et al. (2005) Human lysyl-tRNA synthetase is secreted to trigger proinflammatory response. Proc Natl Acad Sci U S A 102: 6356–6361.
24. Lee YN, Nechushtan H, Figov N, Razin E (2004) The function of lysyl-tRNA synthetase and Ap4A as signaling regulators of MITF activity in FcepsilonRI-activated mast cells. Immunity 20: 145–151.
25. Tshori S, Gilon D, Beeri R, Nechushtan H, Kaluzhny D, et al. (2006) Transcription factor MITF regulates cardiac growth and hypertrophy. J Clin Invest 116: 2673–2681.
26. Aizenshtat LI, Rozenblyum YZ (1975) The OPD-1 trial frame for children. Biomed Eng (NY) 8: 177–178.
27. Choi TG, Lee J, Ha J, Kim SS (2011) Apoptosis signal-regulating kinase 1 is an intracellular inducer of p38 MAPK-mediated myogenic signalling in cardiac myoblasts. Biochim Biophys Acta 1813: 1412–1421.
28. Han JM, Jeong SJ, Park MC, Kim G, Kwon NH, et al. (2012) Leucyl-tRNA synthetase is an intracellular leucine sensor for the mTORC1-signaling pathway. Cell 149: 410–424.
29. Malhowski AJ, Hira H, Bashiruddin S, Warburton R, Goto J, et al. (2011) Smooth muscle protein-22-mediated deletion of Tsc1 results in cardiac

hypertrophy that is mTORC1-mediated and reversed by rapamycin. Hum Mol Genet 20: 1290–1305.

30. Rho SB, Lee JS, Jeong EJ, Kim KS, Kim YG, et al. (1998) A multifunctional repeated motif is present in human bifunctional tRNA synthetase. J Biol Chem 273: 11267–11273.

31. Arif A, Jia J, Moodt RA, DiCorleto PE, Fox PL (2011) Phosphorylation of glutamyl-prolyl tRNA synthetase by cyclin-dependent kinase 5 dictates transcript-selective translational control. Proc Natl Acad Sci U S A 108: 1415–1420.

32. Sampath P, Mazumder B, Seshadri V, Gerber CA, Chavatte L, et al. (2004) Noncanonical function of glutamyl-prolyl-tRNA synthetase: gene-specific silencing of translation. Cell 119: 195–208.

33. Ray PS, Jia J, Yao P, Majumder M, Hatzoglou M, et al. (2009) A stress-responsive RNA switch regulates VEGFA expression. Nature 457: 915–919.

34. Zhao W, Wang J, Shen J, Sun K, Zhu J, et al. (2010) Mutations in VEGFA are associated with congenital left ventricular outflow tract obstruction. Biochem Biophys Res Commun 396: 483–488.

35. Ackerman C, Locke AE, Feingold E, Reshey B, Espana K, et al. (2012) An excess of deleterious variants in VEGF-A pathway genes in Down-syndrome-associated atrioventricular septal defects. Am J Hum Genet 91: 646–659.

36. Vannay A, Vasarhelyi B, Kornyei M, Treszl A, Kozma G, et al. (2006) Single-nucleotide polymorphisms of VEGF gene are associated with risk of congenital valvuloseptal heart defects. Am Heart J 151: 878–881.

37. Xu Z, Taylor JA (2009) SNPinfo: integrating GWAS and candidate gene information into functional SNP selection for genetic association studies. Nucleic Acids Res 37: W600–605.

38. Kimchi-Sarfaty C, Oh JM, Kim IW, Sauna ZE, Calcagno AM, et al. (2007) A "silent" polymorphism in the MDR1 gene changes substrate specificity. Science 315: 525–528.

5

Genome-Wide Association Analysis of Tolerance to Methylmercury Toxicity in *Drosophila* Implicates Myogenic and Neuromuscular Developmental Pathways

Sara L. Montgomery[1], Daria Vorojeikina[1], Wen Huang[2], Trudy F. C. Mackay[2], Robert R. H. Anholt[2], Matthew D. Rand[1]*

1 Department of Environmental Medicine, University of Rochester School of Medicine and Dentistry, Rochester, New York, United States of America, 2 Department of Biological Sciences, Genetics Program, and W. M. Keck Center for Behavioral Biology, North Carolina State University, Raleigh, North Carolina, United States of America

Abstract

Methylmercury (MeHg) is a persistent environmental toxin present in seafood that can compromise the developing nervous system in humans. The effects of MeHg toxicity varies among individuals, despite similar levels of exposure, indicating that genetic differences contribute to MeHg susceptibility. To examine how genetic variation impacts MeHg tolerance, we assessed developmental tolerance to MeHg using the sequenced, inbred lines of the *Drosophila melanogaster* Genetic Reference Panel (DGRP). We found significant genetic variation in the effects of MeHg on development, measured by eclosion rate, giving a broad sense heritability of 0.86. To investigate the influence of dietary factors, we measured MeHg toxicity with caffeine supplementation in the DGRP lines. We found that caffeine counteracts the deleterious effects of MeHg in the majority of lines, and there is significant genetic variance in the magnitude of this effect, with a broad sense heritability of 0.80. We performed genome-wide association (GWA) analysis for both traits, and identified candidate genes that fall into several gene ontology categories, with enrichment for genes involved in muscle and neuromuscular development. Overexpression of glutamate-cysteine ligase, a MeHg protective enzyme, in a muscle-specific manner leads to a robust rescue of eclosion of flies reared on MeHg food. Conversely, mutations in *kirre*, a pivotal myogenic gene identified in our GWA analyses, modulate tolerance to MeHg during development in accordance with *kirre* expression levels. Finally, we observe disruptions of indirect flight muscle morphogenesis in MeHg-exposed pupae. Since the pathways for muscle development are evolutionarily conserved, it is likely that the effects of MeHg observed in *Drosophila* can be generalized across phyla, implicating muscle as an additional hitherto unrecognized target for MeHg toxicity. Furthermore, our observations that caffeine can ameliorate the toxic effects of MeHg show that nutritional factors and dietary manipulations may offer protection against the deleterious effects of MeHg exposure.

Editor: Dennis C. Ko, Duke University, United States of America

Funding: Funding by P30 ES001247, T32 ES07026 (http://www.niehs.nih.gov/) National Institute of Environmental Sciences (MDR, SLM), GM045146, R21 ES021719, (http://www.nigms.nih.gov/) National Institute of General Medical Sciences (RRHA, TFCM), and NCSU Initiative for Biological Complexity, North Carolina State University (WH). The funders had no role in study design, data collection and analysis, decision to publish, or preparation of the manuscript.

Competing Interests: The authors have declared that no competing interests exist.

* Email: matthew_rand@urmc.rochester.edu

Introduction

Methylmercury (MeHg) is a potent environmental neurotoxin that presents a risk to human health. MeHg exposures occur predominantly through dietary intake of fish species that harbor elevated levels of the organometal. Historic accidental MeHg poisonings in Minamata, Japan (1950's) and Iraq (1970's) demonstrated that the neurotoxic effects of MeHg result primarily from fetal exposures [1,2]. In congenital Minamata disease, MeHg-exposed pregnant women with little to no neurological signs give birth to children with a range of severe clinical manifestations akin to cerebral palsy, including mental retardation, ataxia and motor deficits, growth retardation, speech and auditory deficits [3]. Clinical cases of Minamata disease, and limited samples of human fetal brain histopathology associated with them, have consolidated the notion that the developing nervous system is a target tissue for MeHg toxicity [4–6].

Large-scale epidemiologic studies that have investigated outcomes of prenatal MeHg exposure through seafood diets have yielded incongruent results with respect to neurological deficits [7,8]. Subsequent studies have explored genetic predisposition and nutritional modifiers as factors that confer tolerance or susceptibility to MeHg among populations and in individuals [9–11]. In addition to neurotoxicity, recent population studies have identified MeHg effects on cardiovascular factors (e.g., heart rate variability and blood pressure) [12,13] and the immune system [14,15]. While less well studied, there is evidence that overall fetal and infant growth rates are inversely related to prenatal MeHg exposure [16,17]. Given the prevailing focus on neural-specific mechanisms, the extent to which other developing organ systems are affected by MeHg during development has not been fully explored.

MeHg distributes rapidly and ubiquitously in living tissues and demonstrates an exceptionally high affinity for biological thiols, including glutathione (GSH) and protein thiols [18,19]. As a result, the potential molecular and cellular pathways perturbed by MeHg during development are numerous. Thus, natural variation in phenotypic outcomes of MeHg exposure (e.g. tolerance or susceptibility) is the consequence of MeHg interaction with multiple gene products. However, few studies investigating MeHg mechanisms have focused on genetic variation as the underlying framework for variation in MeHg susceptibility. Conducting genome-wide association (GWA) analyses for MeHg susceptibility is challenging in human populations as both the extent of early developmental exposure and its postnatal manifestations are difficult to quantify, and environmental exposure in human populations cannot be controlled precisely. Thus, genome-wide studies on the genetic underpinnings of variation in MeHg exposure are best performed in a model system, where exposure conditions, environmental growth conditions and genetic background can be controlled and effects of MeHg exposure precisely quantified. Fundamental insights based on evolutionarily conserved biological processes can then be extrapolated to human populations.

Drosophila melanogaster presents an excellent genetic model system for quantitative genetic analyses of complex traits, and has resulted in the identification of genetic networks that underlie several stress responses, such as starvation resistance, chill coma recovery, startle behavior, oxidative stress sensitivity, and exposure to alcohol [20–23]. The recent establishment of the *Drosophila melanogaster* Genetic Reference Panel (DGRP), consisting of 205 wild-derived inbred fly strains with fully sequenced genomes, enables GWA studies in a population where all genetic variants are known [24,25]. Linkage disequilibrium decays rapidly within *Drosophila* [24] and the limited population structure in the DGRP can be corrected for by taking into account segregating inversions and genomic relatedness [25]. In addition, the Drosophila model allows for rapid assessment of candidate genes through functional analyses of mutants.

Here, we have used the DGRP lines to perform a GWA analysis for variation in tolerance/susceptibility to MeHg toxicity during development. We measured the development of flies exposed to MeHg during larval and pupal stages by scoring eclosion (adult hatching) as a phenotypic endpoint. In addition, we examined development on MeHg food supplemented with caffeine, a previously identified dietary modifier of MeHg toxicity in fly development [26]. We find significant genetic variation in tolerance to MeHg as well as in modulation of MeHg toxicity by caffeine. Gene network and gene ontology analyses reveal, among others, an enrichment of genes related to development of muscle and the neuromuscular junction. We present a characterization of pupal MeHg phenotypes and functional analyses in mutant and transgenic flies that confirm a role for muscle development as a target for MeHg toxicity.

Materials and Methods

Drosophila Stocks

The following lines were obtained from the Bloomington *Drosophila* Stock Center (Indiana University): *Mi{MIC}kirre*[MI07148], (#41549), *Mi{MIC}kirre*[MI00678] (#41463) and *P{EP}-kirre*[G1566] (#32593), *Mef2*-RFP (#26882), $y^1 w^{67c23}$ (#6599), w^{1118}(#5905), *Mef2-Gal4* (#27390), and the entire *Drosophila melanogaster* Genetic Reference Panel (DGRP). The DGRP is a

set of fully sequenced inbred lines generated by 20 generations of full-sib mating of progeny of wild-caught females from the Raleigh, NC population [24]. *UAS-GCLc5* and *UAS-GCLc 6* were kindly provided by Dr. William C. Orr, Southern Methodist University, Dallas TX [60]. All stocks were maintained at 25°C with 60% humidity and reared on cornmeal-molasses-agar culture medium.

MeHg tolerance assays

Eclosion on MeHg food was assayed as previously described [27,61]. Briefly, 30–50 first instar larvae (mixed sexes) were seeded on MeHg-containing media (0–15 μM). Assays for each treatment condition were performed on n = 150–300 larvae in three vials with 50 larvae each. On day 13 after larvae seeding, the number of eclosed adult flies were counted and expressed as percent of larvae applied to the food. To assess overall developmental tolerance on MeHg media, an eclosion index was calculated by normalizing the mean eclosion rate for each MeHg concentration to the eclosion on 0 μM MeHg for each strain. An overall eclosion index was then generated by summing the normalized percent eclosion values obtained on the 5, 10, and 15 μM MeHg treatments for each strain. For example, if a strain exhibits 95% eclosion on 5 μM MeHg, 50% eclosion on 10 μM MeHg and 5% eclosion on 15 μM MeHg, the eclosion index would be 95+50+5 = 150. Some DGRP strains demonstrated less than 60% eclosion on 0 μM MeHg (9 lines total), exhibiting poor baseline eclosion behavior. These lines were omitted from the analyses, for a total of 167 DGRP lines.

Parallel assays assessing the modulating effects of caffeine on eclosion rates were also determined on 0 μM and 10 μM MeHg with and without 2 mM caffeine (LKT Laboratories; St. Paul MN). This concentration of caffeine was previously characterized as a sub-toxic and tolerance-promoting dose in two wild-type *Drosophila* strains [26]. Eclosion values were normalized to the 0 μM MeHg condition for each strain. A caffeine difference index was determined for each line by subtracting the normalized eclosion rate on 10 μM MeHg alone from that on 10 μM MeHg+ 2 mM caffeine. Lines exhibiting 0% eclosion on 10 μM MeHg and 10 μM+2 mM caffeine were omitted from GWA analysis, leaving 139 lines for GWA analysis.

Quantitative genetic analyses

We performed a mixed effects model analysis of variance (ANOVA) using the model $Y = \mu + D + L + D \times L + \varepsilon$ to partition phenotypic variation, where Y is the response variable, μ is the overall mean, D is the fixed effect for dose, and L and $D \times L$ are the random effects for line and dose by line interaction. Significance of the effects was tested using type III F tests implemented in SAS Proc Mixed (SAS Institute). Broad sense heritability was calculated as $H^2 = \dfrac{\sigma_L^2 + \sigma_{DXL}^2}{\sigma_L^2 + \sigma_{DXL}^2 + \sigma_\varepsilon^2}$. For the effect of caffeine, because there was only a single dose of MeHg, a simpler model was fitted as $Y = \mu + L + \varepsilon$, and $H^2 = \dfrac{\sigma_L^2}{\sigma_L^2 + \sigma_\varepsilon^2}$.

Genome-wide association analysis

We performed genome-wide association analysis on line means of the phenotype using the DGRP analysis portal (http://dgrp2. gnets.ncsu.edu; [24,25]). Briefly, the hatching index or caffeine index was used to fit a mixed model for each variant in the form of $Y = \mu + M + g + \varepsilon$, where Y is the line means adjusted for *Wolbachia* infection and five major inversion polymorphisms (*In(2L)t*, *In(2R)NS*, *In(3R)Y*, *In(3R)P*, *In(3R)Mo*) in the DGRP, μ is

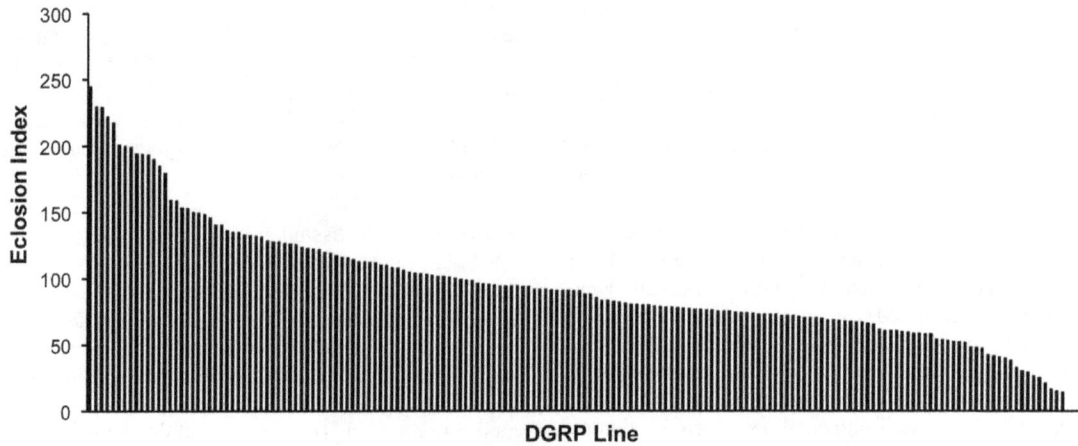

Figure 1. Genetic variation in MeHg tolerance during development. 176 DGRP lines were assayed in triplicate for eclosion on media containing 0, 5, 10 and 15 μM MeHg. A cumulative index (Eclosion Index) was generated by summing the percent eclosion on 5, 10 and 15 μM MeHg food for each strain (see methods). The histogram represents a rank ordering of the eclosion index for each of the DGRP lines.

the overall population mean, M is the effect of DNA variant being tested, and g is a polygenic component with covariance between lines determined by their genomic relationship [25]. For Wolbachia infection adjustment we fit a linear model with the infection status and major inversion genotypes as covariates and the raw phenotypes as the response variable. Residuals from this linear model were used as inputs for GWAS. We performed 2,180,555 tests for association for the MeHg treatment alone, restricting the analyses to variants for which at least 6 lines contain the minor allele (for a minor allele frequency, MAF, >3.6%). We performed 2,357,353 tests for association for the MeHg+caffeine treatment, restricting the analysis to variants for which at least 4 lines contained the minor allele (MAF>2.9%).

GeneMANIA network analysis

Polymorphism-based single marker analysis for MeHg and MeHg+caffeine was used to identify top candidate genes found to be associated with developmental tolerance to MeHg. Candidate genes were uploaded to the GeneMANIA prediction server (www.genemania.org), a web interface to identify networks of gene functions associated with a query list of genes [39]. Functional networks were derived using automatic query-dependent gene weighting and biological function-based gene ontology (GO) weighting to identify interactions based on co-expression, co-localization, genetic, and physical interactions of query and non-query genes related to biological integration networks. Outputs from GeneMANIA were constructed in tabular form and in graphical form using a Cytoscape plugin v3.1.0 (Cytoscape). Functional enrichment is based on the GO categories and is reported as Q-values of a false discovery rate (FDR) using a corrected hypergeometric test for enrichment. Coverage ratios for

the number of annotated genes in the displayed network versus the number of genes with that annotation in the genome are also reported. Q-values are derived using the Benjamini-Hochberg procedure. Categories are displayed up to a Q-value cutoff of 0.1.

Muscle phenotype characterization

Mef2>RFP L1 larvae were seeded onto 0, 10, and 15 μM MeHg media and monitored until pupae formation at 25°C. Stage 6–12 pupae were selected based on established developmental landmarks such as the appearance of green in Malpighian tubules, body color, eye color, bristles development and wing color [62]. Pupae were dissected from their case and positioned on a Superfrost microscope slide (VWR International; Radnor, PA) for fluorescent reporter imaging of the indirect flight muscles (IFMs) within the thorax. Phosphate-buffered saline (PBS) was added drop-wise to each dissected pupae to avoid desiccation. Fluorescent microscopy was performed with a Leitz Orthoplan 2 microscope (Leica Microsystems; Buffalo Grove, IL) equipped with a SPOT Insight QE 4.2 Camera (SPOT Imaging Solutions; Sterling Heights, MI) and imaging software using a 4× objective. Images were assembled in Microsoft PowerPoint.

Functional validation of candidate genes

Functional validation of candidate genes related to muscle development was performed with eclosion assays using *kirre*^MI07148, *kirre*^MI00678 and *kirre*^G1566 mutants (Bloomington *Drosophila* Stock Center, Indiana University). Eclosion on MeHg food (0–15 μM) was assayed for the *kirre* mutants and the corresponding y^1w^{67c23} control strain. Experiments overexpressing *UAS-GCLc5* and *UAS-GCLc6* were conducted by using the

Table 1. ANOVA table for MeHg tolerance.

Source of Variation	Df	MS (Type III)	F	P value	σ² (SE)
Line	172	0.0909	12.42	<0.0001	0.0299 (0.0035)
Error	346	0.0073			0.0073 (0.0006)

Phenotype = hatching rate for each replicate at 5, 10, 15 μM MeHg normalized to the line means of hatching rate at 0 μM MeHg. Residuals are weighted by the square of number of flies assayed.

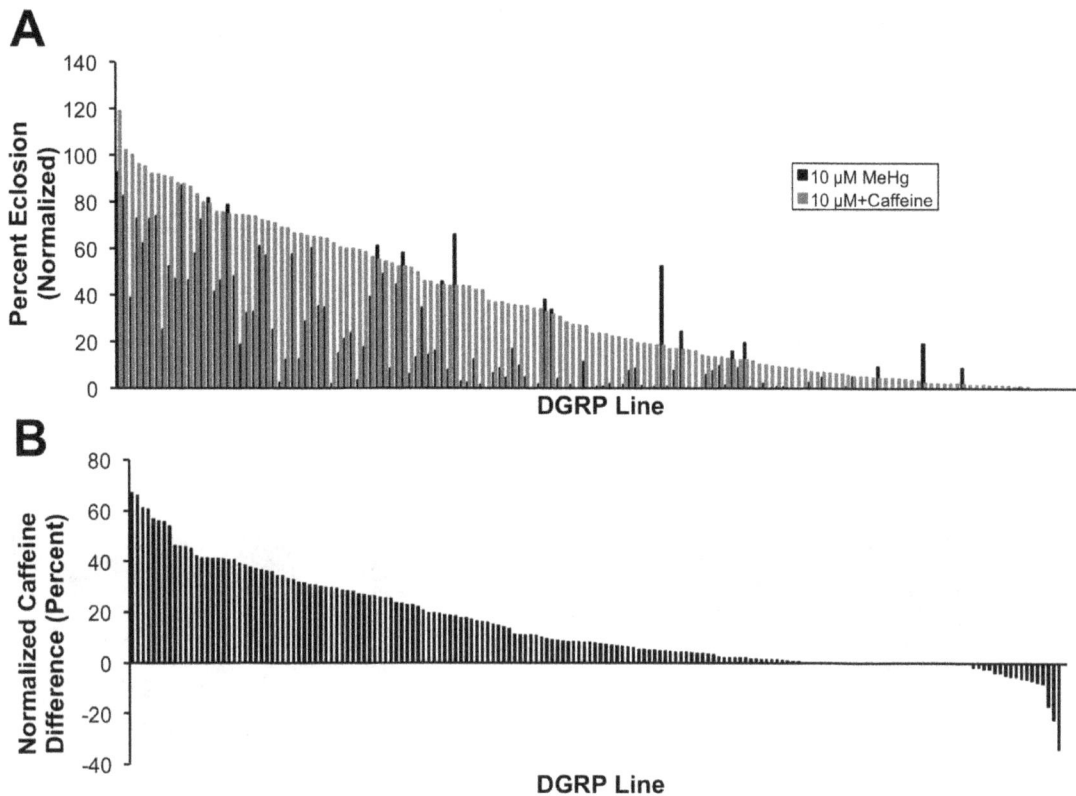

Figure 2. Caffeine effect on MeHg tolerance exhibits genetic variation. (A) Eclosion rates of DGRP lines were determined on 0 μM and 10 μM MeHg food with and without addition of 2 mM caffeine. The histogram is a rank ordered expression of eclosion rates on the MeHg+caffeine condition (gray bars) paired with the respective line on MeHg alone (black bars). (B) A caffeine difference index was determined for each DGRP line by subtracting the normalized eclosion rate on 10 μM MeHg from that on 10 μM MeHg+2 mM caffeine. Positive values indicate a beneficial effect of caffeine and negative values indicate a detrimental effect of caffeine relative to 10 μM MeHg alone. Lines showing 0% eclosion on both MeHg alone and MeHg+caffeine were omitted from the analyses leaving 139 lines for GWA analyses.

muscle-specific *Mef2-Gal4* driver. Female *Mef2-GAL4* flies were bred with male *UAS-GCLc5*, *UAS-GCLc6* or *w^{1118}* control flies, and progeny assayed for eclosion on MeHg media (0–25 μM). Rates of eclosion for indicated strains at each MeHg concentration are expressed in proportions (% eclosion). Assays were performed to achieve n = 150 larvae. Statistical consideration of differences between experimental and control fly strains are therefore comparisons of proportion values and not of continuous values. Since error determinations in proportion values become restricted at the edges (*i.e.* near 100%), an analysis of variance (ANOVA) was not used. Statistical analyses of eclosion assays were therefore done using a pairwise 2-tailed z-test, treating each MeHg concentration categorically. *p*-values of less than 0.01 were considered significant.

Gene expression

GCLc and *kirre* gene expression was measured by quantitative real-time PCR (qRT-PCR) of RNA extracts isolated from first instar larvae or staged pupae. For the indicated genotypes, RNA was extracted by pooling 15–20 larvae or 20–25 pupae. The tissue was homogenized and RNA extracted with Trizol (Invitrogen; Grand Island, NY). qRT-PCR quantification was performed on a Bio-Rad CFZ Connect Real-Time PCR Detection System using CFX Manager software. cDNA synthesis and reverse transcription was performed in a Bio-Rad iScript SYBR Green one-step

Table 2. ANOVA table for caffeine effect.

Source of Variation	Df	MS (Type III)	F	P value	σ^2 (SE)
Dose	2	66.4455	10788	<0.0001	Fixed
Line	172	0.1983	32.21	<0.0001	0.0155 (0.0027)
Dose × Line	344	0.0711	11.55	<0.0001	0.0231 (0.0019)
Error	3	0.0062			0.0062 (0.0003)

Phenotype = (hatching rate for each replicate at 10 μM MeHg + Caffeine normalized to the line means of hatching rate at 0 μM MeHg) – (line means of hatching rate at 10 μM MeHg normalized to the line means of hatching rate at 0 μM MeHg). Residuals are weighted by the square of number of flies assayed.

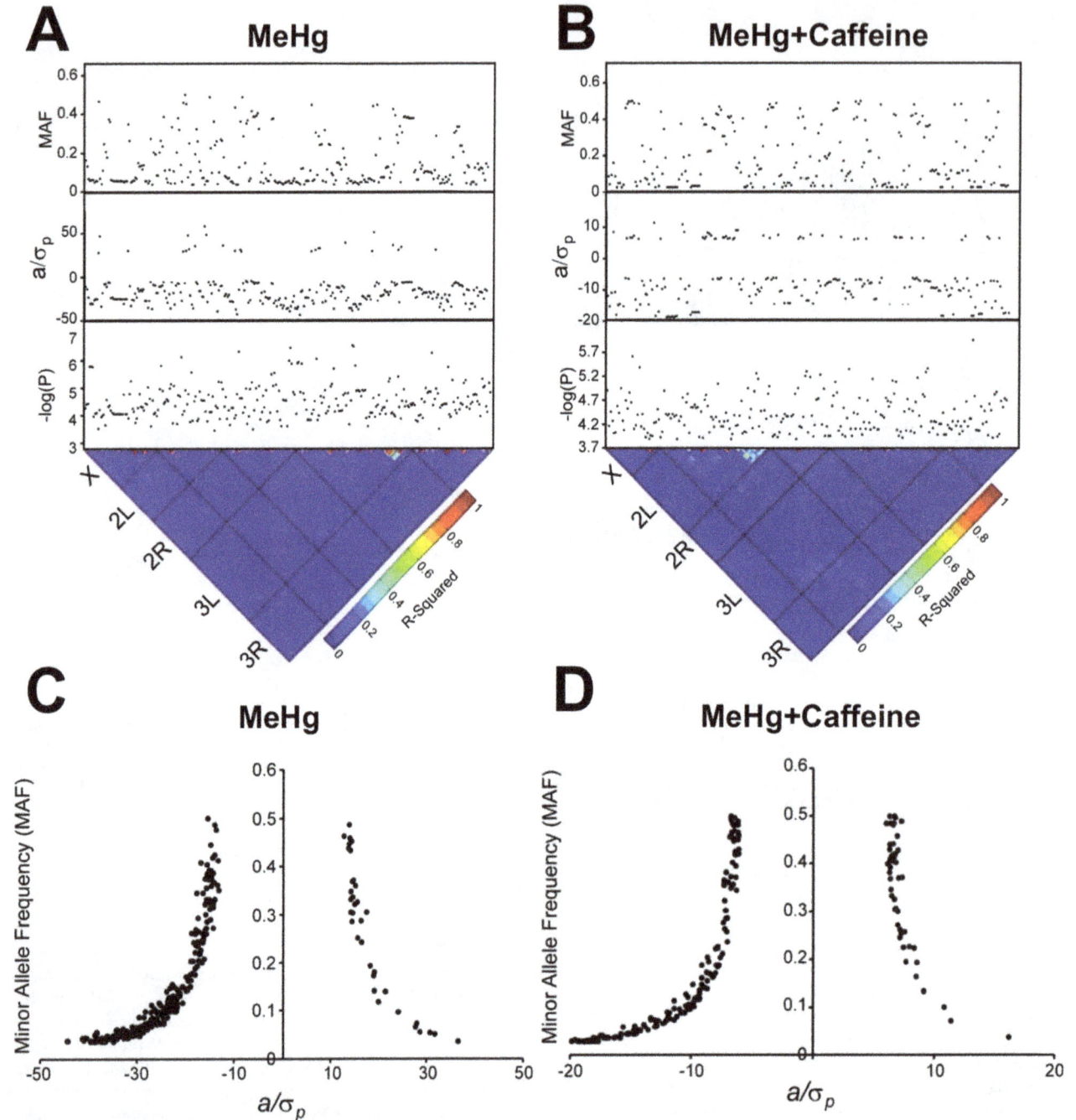

Figure 3. Genome-wide association analysis of eclosion on MeHg with and without caffeine supplementation. Single marker analyses using ANOVA of 2,180,555 (MeHg alone) and 2,357,353 (MeHg+caffeine) polymorphic alleles across 167 (MeHg alone) and 139 (MeHg+caffeine) DGRP lines, respectively, resolved (A) 350 and (B) 239 polymorphic markers ($p<10^{-4}$, MAF>3%). Depicted is a heat map for linkage disequilibrium (LD) based on r^2 values where the black bars represent the five major *Drosophila* chromosome arms. Red indicates high LD, while blue indicates low LD. The black dots represent polymorphic marker associations for eclosion on MeHg or MeHg+caffeine. p values ($\log_{10}(p)$), effect size (a/σ_p), and the minor allele frequency (MAF) are shown. (C,D) MAF *vs.* effect size. All 350 (C) or 239 (D) polymorphic markers associated with phenotypic variation for eclosion on MeHg or MeHg+caffeine, respectively, are depicted. a/σ_p indicates effect size ([mean of major allele class – mean of minor allele class]/2), where negative a/σ_p indicates the minor allele is associated with increased MeHg tolerance with respect to eclosion.

reaction (Bio-Rad; Hercules, CA). Twenty-five μL reactions containing 40 ng RNA template, iScript SYBR Green reaction mix (2x), iScript reverse transcriptase for one-step RT-PCR, and forward and reverse primers (10 μM final concentration) were used. The *Drosophila Rp49* gene was used for normalization of expression. Gene expression levels were determined by the comparative C_T method [63]. The following primers were used; *Rp49*: 5′-AGTATCTGATGCCCAACATCG-3′ and 5′-TTCC-GACCAGGTTACAAGAAC-3′, *GCLc*: 5′-ATGACGAGGAG-AATGAGCTG-3′, and 5′-CCATGGACTGCAAAATAGCTG-

MeHg   MeHg+Caffeine

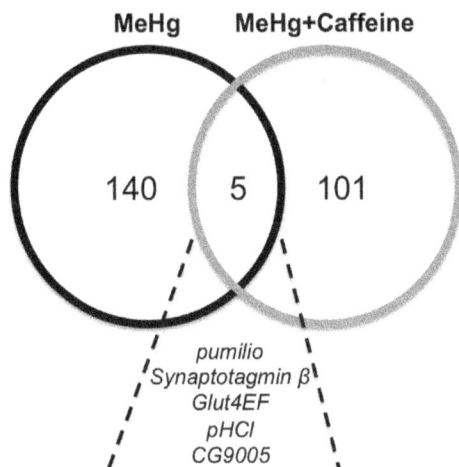

Figure 4. Overlap of genes identified by common polymorphic markers for variance in MeHg and MeHg+caffeine treatments. Candidate genes were identified from one or more associated polymorphisms in GWA analyses. A total of 145 and 106 genes were identified for MeHg and MeHg+caffeine, respectively. In common between the two treatments are 5 genes: *pumilio*, *Synaptotagmin* β, *Glut4EF*, *pHCl*, and *CG9005*.

3′, *kirre*: 5′-TGGACTGGCCATTAATCTTACC-3′ and 5′-AA-CGATCGCCACCGAAAT-3′.

Results

Quantitative genetic analysis of natural variation in tolerance to MeHg during development

To characterize natural genetic variation in tolerance and susceptibility to MeHg during development, we examined 176 DGRP lines in an eclosion assay on four concentrations ([MeHg] = 0, 5, 10, 15 µM, Table S3) of MeHg-containing food. We found substantial phenotypic variation in susceptibility to MeHg as measured by an eclosion index (see Materials and Methods for definition) across the DGRP lines, ranging from 0 to 244.6 (Fig.1). ANOVA for variation in MeHg tolerance across the four concentrations of MeHg showed significant variation for Line and the Dose × Line interaction term (Table 1). The broad sense heritability (H^2) was 0.86, indicating a strong genetic component to variation in MeHg tolerance, which provides a basis for GWA analysis (Table 1). To explore dietary modulation of MeHg toxicity, we examined the effects of caffeine co-administration (Fig. 2A, B, Table S3), which has previously been shown to attenuate MeHg developmental toxicity in a limited set of fly lines [26,27]. Addition of 2 mM caffeine to 10 µM MeHg supplemented food resulted in increased MeHg tolerance in the majority of the lines (Figure 2A, B), with only 12 lines exhibiting decreased MeHg tolerance with caffeine supplementation (Fig 2B). The variation in the modulating effect of caffeine can be seen clearly by subtraction of the eclosion rate on MeHg alone from that on MeHg+caffeine (Figure 2B). The modulating effect of caffeine on MeHg exposure varies significantly across the lines with $H^2 = 0.80$ (Table 2).

Polymorphic markers associated with MeHg tolerance

We performed single marker GWA analyses to identify polymorphic markers (single nucleotide polymorphisms (SNPs), insertions and deletions) associated with MeHg tolerance and the

modulatory effect of caffeine on MeHg toxicity. Line means for both traits were approximately normally distributed (Fig. S1); therefore, we did not transform the data. Quantile-quantile plots show a clear enrichment of variants with p-values $<10^{-4}$ for MeHg tolerance, but no evidence for enrichment for the modulatory effect of caffeine (Fig. S2). In the MeHg tolerance GWA, one SNP (in *pHCl*) nearly met a Bonferroni-adjusted significance level, and 14 SNPs (in *pHCl*, *eco*, *CG44245 CG33981* and *CG15221*) had false discovery rates (FDR) of FDR <0.2 (Table S1).

In *Drosophila*, we can use GWA analysis as an exploratory hypothesis-generating tool, and test hypotheses more rigorously in secondary screens using mutations or targeting RNAI for candidate genes implicated by the GWA analysis. Therefore, we used a lenient reporting threshold of $p<10^{-4}$ for both GWA analyses (Tables S1, S2). We identified 350 polymorphisms in or near 145 genes associated with variation in eclosion rates on MeHg food (Fig. 3A) and 239 polymorphisms in or near 106 genes (Fig. 3B) associated with the modulatory effect of caffeine on MeHg treatment. Most polymorphisms associated with the two analyses had MAF <0.15, and, as expected, we found inverse relationships of effect size and allele frequency (Figs. 3C, 3D). Rarer alleles were associated with greater tolerance to MeHg and modulation of the effect by caffeine with respect to larva-adult viability for a majority of the polymorphisms (Figs. 3C, 3D), suggesting these alleles have other deleterious effects on fitness. There was little linkage disequilibrium between the most significant polymorphisms (Figs. 3A, 3B).

Candidate genes associated with MeHg tolerance

We found five genes in common between the two GWA analyses for exposure to MeHg in the presence or absence of caffeine: *pumilio* (*pum*, *CG9755*), *Synaptotagmin* β (*Sytβ*, *CG42333*), *Glut4EF* (*CG34360*), *pHCl* (*CG44099*) and *CG9005* (Fig. 4). *pum* is an Armadillo repeat RNA binding protein involved in repression of translation and has two human homologs, PUM1 and PUM2. *pum* has functional implications in cellular and developmental processes including embryonic patterning, synaptic transmission, dendrite morphogenesis, pole cell migration, and learning and memory [28,29]. *Sytβ* is one of seven Synaptotagmin family members in *Drosophila* with predicted function in synaptic vesicle exocytosis [30]. *Sytβ* localizes to the developing CNS as well as to specific motor neuron termini [31]. *Glut4EF* is a zinc finger transcription factor and homolog of the human glucose transporter (GLUT4) enhancer factor [32]. In flies, *Glut4EF* influences wing positioning with mutants giving a stretched out wing phenotype [32]. Mammalian Glut4EF is important for glucose uptake in muscles and associates with the MEF2A transcription factor [33]. *pHCl* encodes a gamma-aminobutyric acid A receptor, a ligand gated chloride channel that is expressed in the embryonic nervous system [34]. *CG9005* encodes a protein of unknown function that has two human homologs, Hsap/KIAA1370 and Hsap/FAM214B. RNAi knockdown of *CG9005* in flies has effects on development of the notum [35]. Therefore, four of the five overlapping genes share functions in the domain of neuromuscular function and neural development, suggesting that these biological processes are likely associated with MeHg tolerance during development.

A particularly noteworthy gene identified by GWA with MeHg alone is the metal transcription factor (*MTF-1*) gene, which was represented with six SNPs (Table S1). *MTF-1* is a conserved transcriptional regulator and the primary responder to heavy metal toxicity insult [36]. While better known for conferring resistance to divalent metal ions and inorganic mercury [37], *MTF-1* regulates expression of metallothionein proteins and is

Table 3. GeneMANIA network analyses reveals both overlapping and unique biological functions among GWA genes from MeHg alone and MeHg+caffeine.

MeHg alone

Function	FDR	Network	Genome
muscle organ development	1.73E-11	24	247
muscle structure development	1.73E-11	26	294
muscle cell differentiation	3.38E-10	20	184
striated muscle cell differentiation	7.56E-09	18	170
muscle tissue development	7.58E-08	16	151
striated muscle tissue development	7.58E-08	16	150
skeletal muscle organ development	7.58E-08	17	173
visceral muscle development	1.83E-07	7	14
skeletal muscle tissue development	2.34E-07	15	141
urogenital system development	1.00E-06	11	71
renal system development	1.00E-06	11	71
cell migration	1.25E-06	17	217
cell motility	2.31E-06	17	227
localization of cell	3.63E-06	17	235
regulation of muscle tissue development	4.45E-06	10	65
regulation of striated muscle tissue development	4.45E-06	10	65
regulation of muscle organ development	8.81E-06	10	70
tube development	2.30E-05	13	149
tissue migration	3.42E-05	8	44
transmembrane receptor protein tyrosine kinase sig.	3.51E-05	12	129
enzyme linked receptor protein signaling pathway	4.00E-05	14	187
cell-cell adhesion	7.98E-05	10	90
striated muscle cell development	9.81E-05	12	144
renal filtration cell differentiation	9.81E-05	5	12

MeHg + Caffeine

Function	FDR	Network	Genome
transmembrane receptor protein tyrosine kinase signaling	3.40E-05	12	129
enzyme linked receptor protein signaling pathway	1.26E-04	13	187
muscle organ development	3.04E-04	14	247
muscle structure development	3.04E-04	15	294
regulation of cell differentiation	9.79E-04	14	283
tissue morphogenesis	1.05E-03	14	289
protein phosphorylation	1.53E-03	13	258
epithelium development	3.80E-03	13	286
adult behavior	3.80E-03	9	126
genital disc development	4.04E-03	6	44
response to organic substance	4.98E-03	12	261
embryonic morphogenesis	4.98E-03	11	213
regulation of nervous system development	4.98E-03	10	176
skeletal muscle organ development	4.98E-03	10	173
regulation of cell development	4.98E-03	11	217
neuromuscular junction development	5.23E-03	8	106
skeletal muscle fiber development	5.64E-03	8	108
morphogenesis of an epithelium	5.64E-03	12	269
muscle cell differentiation	5.76E-03	10	184
muscle tissue development	6.63E-03	9	151
striated muscle tissue development	6.63E-03	9	150
synapse organization	6.63E-03	9	149
hindgut morphogenesis	6.63E-03	6	55
muscle fiber development	6.80E-03	8	116

GeneMANIA derived biological network integration results from MeHg and MeHg+caffeine GWA genes. Overlapping networks with muscle related function (bold text)and receptor signaling/tyrosine kinase function (italicized bold text) are indicated. Also shown are False discovery rate (FDR) and coverage indicated by the number of the genes in the 'Network' with a given function relative to all the genes in the 'Genome' identified with that function.

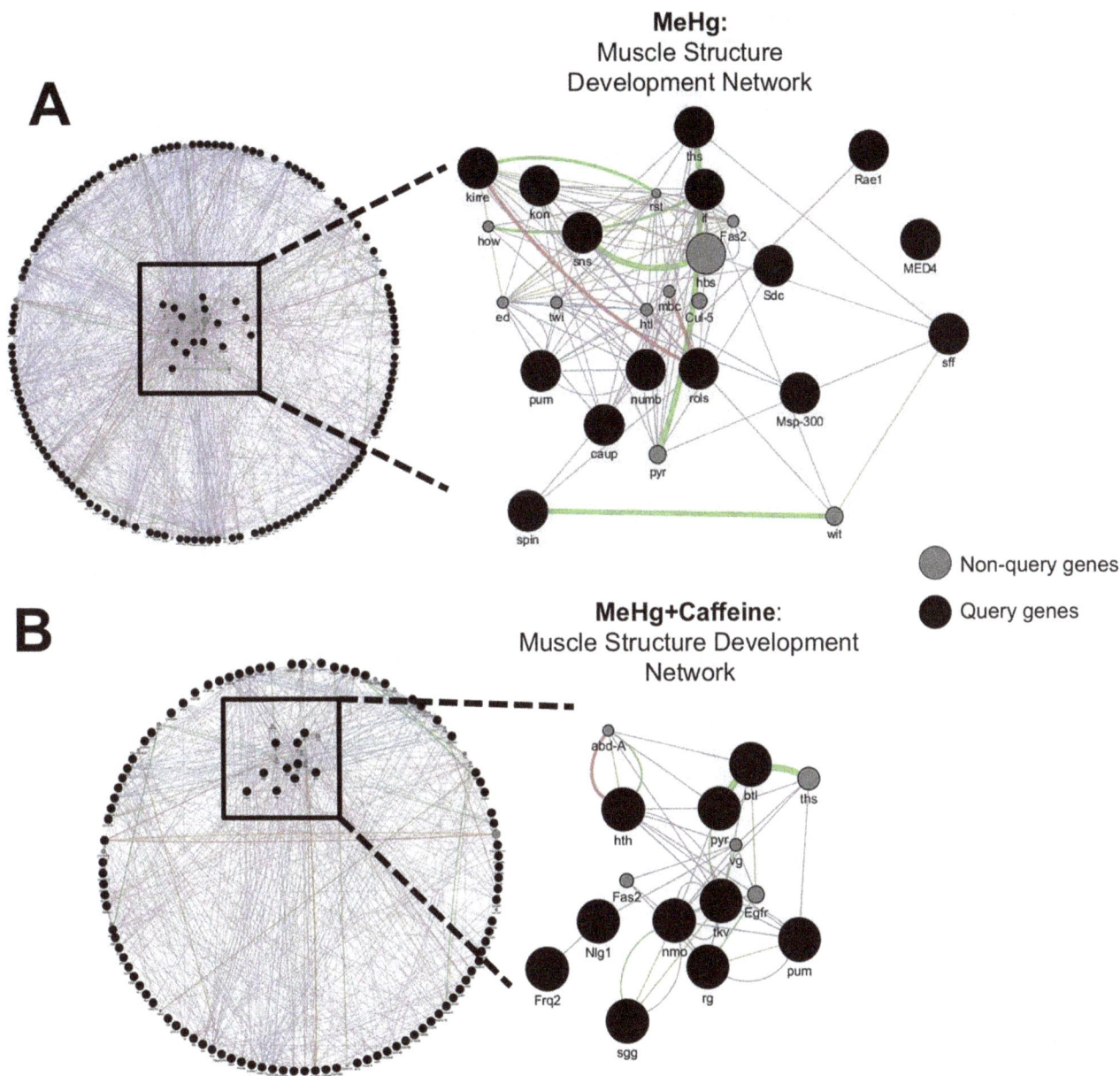

Figure 5. A shared functional network among non-overlapping genes identified in GWA of MeHg alone and MeHg+caffeine treatments. Network maps generated by GeneMANIA illustrate an example of a shared network for muscle structure development of genes identified in the GWA for MeHg alone (A) and MeHg+caffeine (B). Interactions are identified at the level of co-expression (purple lines), co-localization (blue lines), genetic interactions (green lines) and physical interactions (red lines). Query genes from the GWA analyses are represented by black circles and non-query genes (computationally recruited to complete network associations) are indicated by gray circles. The cluster of genes comprising the network is sorted to the inside of the circle, while the remaining query genes, not associated with the network, remain on the periphery.

effective in moderating oxidative stress [38]. MTF-1 responds robustly to inorganic mercury in *Drosophila*; however, its function in MeHg tolerance in flies remains to be investigated.

Functional gene networks related to MeHg tolerance

To assess to what extent genes associated with variation in MeHg tolerance can be assembled into genetic networks that represent biological pathways, we used the GeneMANIA [39] algorithm. The results from such analyses generate hypotheses regarding biological processes associated with phenotypic variation that can subsequently be tested by disrupting hub genes though

mutational analyses or targeted RNAi. Although only five candidate genes overlapped between the MeHg alone versus MeHg+caffeine treatments, we found substantial enrichment for genes associated with a number of muscle development functions for MeHg treatment both with and without caffeine and an additional overlap for the functional category of receptor protein kinase signaling (Table 3). One of the resulting GeneMANIA functional categories is assigned the "muscle structure development" network term and is shown schematically in Figure 5A and B for MeHg alone and MeHg+caffeine, respectively. Genes on the query list that are enriched for the "muscle structure develop-

Table 4. MeHg: Muscle structure development network query genes.

Gene Name	Fly Base ID	No. SNPs	Gene Function	Protein Function	CG #	Human Homolog
If	FBgn0001250	11	Axon guidance, muscle attachment, myofibril assembly	integrin alpha chain	CG9623	-
kirre	FBgn0028369	1	Myoblast fusion	immunoglobulin	CG3653	Kirrel/NEPH1
Sff	FBgn0036544	1	NMJ development	Protein serine/threonine kinase	CG6114	-
Rae1	FBgn0034646	1	Negative regulation of synaptic growth at NMJ	WD40 repeat	CG9862	RAE1
Rols	FBgn0041096	1	Myoblast fusion	zinc finger, ankyrin repeat	CG32096	-
Kon	FBgn0032683	1	Muscle attachment, muscle organ development, neurogenesis	laminin G domain, concanavalin A-like lectin	CG10275	CSPG4
pum	FBgn0003165	1	Regulation of synaptic growth at NMJ, synaptic transmission, dendrite morphogenesis, pole cell migration	RNA binding protein	CG9755	PUM1/PUM2
Msp-300	FBgn0261836	1	Skeletal muscle development, flight, locomotion	actin-binding, actinin-type, spectrin repeat	CG42768	
caup	FBgn0015919	1	Muscle cell fate commitment	Tale/IRO homebox family	CG10605	IRX3
spin	FBgn0086676	1	Regulation of synaptic growth at NMJ, locomotion, nervous system remodeling, glial migration	unknown	CG8428	SPNS1/2/3
ths	FBgn0033652	1	Myoblast migration, heart development	FGF receptor binding, growth factor	CG12443	-
sns	FBgn0024189	1	Myoblast fusion, nephrocyte diaphragm assembly	immunoglobulin	CG33141	NPHS1
Sdc	FBgn0010415	1	Regulation of synaptic growth at NMJ, motor neuron axon guidance	heparin sulfate proteoglycan, cytoskeletal binding	CG10497	-
numb	FBgn0002973	3	Muscle cell fate specification, neurogenesis	Notch binding	CG3779	NUMB/NUMBL
MED4	FBgn0035754	1	Muscle organ development	RNA Pol II cofactor activity	CG8609	MED4

Genes identified in the GWA for MeHg that constitute the "Muscle Structure Development" network are listed. The human homologs listed were identified in Flybase.

ment" category under analysis are listed in Tables 4 and 5. Five core genes in the network derived from the GWA analysis of the effects of MeHg alone act in muscle cell fusion and muscle attachment and development: *inflated* (*if*, *CG9623*), *kin of irre* (*kirre*, a.k.a. *dumbfounded* (*duf*) *CG3653*), *sticks and stones* (*sns*, *CG331441*), *kon-tiki* (*kon*, *CG10275*) and *rolling pebbles* (*rols*, *CG32096*) (Fig. 5A and Table 4). Also of note are five genes that play a role in neuromuscular junction development and integrity; *sugar free frosting* (*sff*, *CG6114*), *rae1* (*CG9862*), *pum* (*CG9755*), *spinster* (*spin*, *CG8428*) and *Syndecan* (*Sdc*, *CG10497*) (Fig. 5A and Table 4). Nine of these 15 genes have human homologs (Table 4). Among the 10 genes in the "muscle structure development" network with MeHg+caffeine treatments eight have human homologs (Table 5). It is of interest that eight of these 10 genes are transcriptionally responsive to caffeine (Table 5).

In summary, although the genes identified largely differ between the GWA analysis of effects of MeHg alone and MeHg+caffeine, the top GWA genes in both cases were functionally enriched for playing roles in muscle and neuromuscular junction development. This suggests that genes affecting muscle development contribute to the mechanisms of tolerance to MeHg toxicity during development.

Functional assessment of sensitivity of muscle development to MeHg toxicity

To functionally assay developing muscle tissue as a sensitive MeHg target, we examined flies that carry a gene known to affect MeHg tolerance expressed under the muscle specific enhancer *myocyte enhancing factor 2* (*Mef2*) [40]. We induced expression of glutamate-cysteine ligase (GCL), the highly conserved, rate-limiting enzyme for the synthesis of glutathione (GSH). GSH is a small molecule thiol compound present in all cells and is a first line of defense to MeHg toxicity, forming a conjugate that enhances MeHg excretion [41]. Elevated expression of GCL gives resistance to MeHg toxicity [42]. Using the *Gal4>UAS* system, we over-expressed the catalytic subunit of GCL (GCLc) using two independent lines carrying the *UAS-GCLc* construct (*GCLc5*, *GCLc6*) with the *Mef2-Gal4* driver (Fig. 6B). GCLc expression in muscle shows a robust enhancement of tolerance to MeHg in the eclosion assay (Fig. 6A), consistent with the notion that protection of muscle development is critical to overall development of the fly and completion of eclosion under MeHg toxicity stress.

We further examined flies carrying mutations in *kirre*, one of the core myogenic candidate genes identified in our GWA study. The *kirre*^G1566 mutation is a viable mutant of *kirre* carrying a *P*-element insert in exon 10 (Fig. 7A). The *kirre*^MI07148 and *kirre*^MI00678 mutations carry a *Minos* element insert in an intronic

Table 5. MeHg+caffeine: Muscle structure development network query genes.

Gene Name	Fly Base ID	No. SNPs	Gene Function	Protein Function	CG #	Human Homolog	Caff induced
Hth	FBgn0001235	3	Somatic muscle development, peripheral nervous system development, neuron differentiation	DNA binding protein	CG17117	MEIS1/2/3	low
Nmo	FBgn0011817	1	Positive regulation of synaptic growth at NMJ, eye and wing development	Protein kinase	CG7892	NLK	Y
pum	FBgn0003165	1	Regulation of synaptic growth at NMJ, synaptic transmission, dendrite morphogenesis, Pole cell migration	RNA binding protein	CG9755	PUM1/PUM2	Y
Tkv	FBgn0003716	1	Positive regulation of synaptic growth at NMJ, neuromuscular synaptic transmission, TGF-β signaling pathway	TGF beta receptor/Serine-threonine kinase	CG14026	BMPR1/ACVR1	Y
Pyr	FBgn0033649	1	Myoblast migration, cardioblast differentiation	FGF receptor binding, growth factor	CG13194	-	Y
Sgg	FBgn0003371	1	Negative regulation of synaptic growth at NMJ, circadian clock, epithelial cell morphogenesis	Serine threonine kinase	CG2621	GSK3A/B	Y
Nlg1	FBgn0051146	1	NMJ development, synaptic growth at NMJ	Carboxyesterase	CG31146	-	N
Frq2	FBgn0083228	1	NMJ development, synaptic transmission, regulation of neurotransmitter secretion	Ca++ binding guanylate cyclase activator	CG5744	KCNIP/NCALD	low
btl	FBgn0005592	1	Negative regulation of axon extension, tracheal outgrowth/open tracheal system	Tyrosine Kinase	CG32134	FGFR1/2/3/5, RET	N
rg	FBgn0266098	1	Neuromuscular junction development, mushroom body development, short-term memory, compound eye	Protein kinase binding	CG44835	LRBA, NBEA	low

Genes identified in the GWA for MeHg+caffeine that constitute the "Muscle Structure Development" network are listed. The human homologs listed were identified in Flybase. In addition, response to caffeine for each gene was adapted from the modENCODE treatment expression data in the Flybase record for each gene.

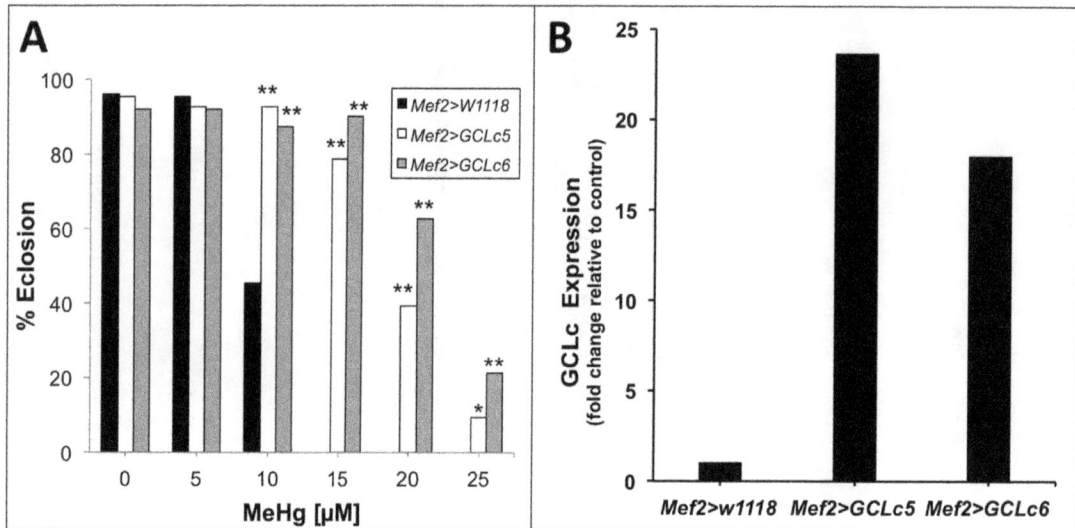

Figure 6. Muscle-specific expression of glutamate-cysteine ligase (GCL) rescues eclosion on MeHg. (A) Eclosion assays comparing flies expressing the catalytic subunit of GCL in two independent *UAS* responder lines (*GCLc5* and *GCLc6*) using the muscle-specific *Mef2-GAL4* driver (Mef2) with control flies (*Mef2>w^1118^*). Statistical analyses by z- test, n = 150 flies/bar. * p <0.001 and **p<0.0001 relative to *Mef2>w^1118^*. (B) Expression of *GCLc* mRNA in P12 pupae thoracic RNA extracts (pooled sample of n = 20 pupae) measured via qRT-PCR.

region of four of the seven transcripts (Fig. 7A). Analysis by qRT-PCR demonstrates that transcript levels are substantially reduced in the *kirre*^MI07148^ mutant and essentially absent in the *kirre*^G1566^ mutant (Fig. 7C), consistent with the predicted effect on transcripts encompassing the respective insertions. A moderate but significant reduction in MeHg tolerance is seen with both the *kirre*^MI07148^ and *kirre*^G1566^ mutants, specifically with exposure to 10 μM MeHg (Fig. 7B). Consistent with the corresponding expression levels, the *kirre*^G1566^ mutant showed the lowest MeHg tolerance with the *kirre*^MI07148^ mutant giving an intermediate tolerance relative to the *yw* control line at 10 μM exposures (Fig. 7B). Unexpectedly, the *kirre*^MI00678^ mutant shows a significant increase in MeHg tolerance at the 5, 10 and 15 μM exposures (Fig. 7B), and concomitantly, *kirre mRNA* levels are seen to be elevated relative to the y^1w^{67c23} control line (Fig. 7C). These findings are consistent with the notion that modulation in the levels of *kirre* expression affects sensitivity of muscle development to MeHg.

Identification of an indirect flight muscle phenotype in MeHg-exposed pupae

With MeHg exposure, eclosion commonly fails in late pupal stages resulting in the accumulation of dark pupae, particularly at 10–15 μM MeHg. Based on our GWA results we examined various stages of MeHg-exposed pupae for potential muscle phenotypes using a line of flies that expresses red fluorescent protein (RFP) constitutively in muscles (*Mef2>*RFP). *Mef2>*RFP larvae were exposed to various concentrations of MeHg and collected at various pupal stage endpoints. Fluorescent imaging of these pupae reveals the prominent pattern of the indirect flight muscle groups, notably the dorsal longitudinal muscles (DLMs) at their attachment sites under the notum. We observe an overall disruption of DLM morphogenesis with MeHg treatments, despite an apparent normal development of ectodermal structures as well as the specialized organs of the bristles and eyes (Fig. 8A, B). Following MeHg exposure, early pupae (P6) show a reduction in size of the DLM fiber bundles and an apparent displacement of forming DVM muscles (Fig. 8A, C upper panels, open arrows). At later stages (P12), in addition to reduced fiber size, a "ball" of

RFP-positive tissue is seen indicative of a failure of DLM myofiber maturation, elongation and attachment to tendon cells on the notum epithelium (Fig. 8B, C lower panels).

Discussion

Natural variation in tolerance to MeHg toxicity

We sought to elaborate a fuller spectrum of the molecular networks and the tissue targets that influence MeHg toxicity outcomes through an unbiased query of polymorphisms across the entire genome of a diverse panel of developing animals. We find that variation in MeHg tolerance is under significant genetic control. Furthermore, we demonstrate that variation in response to a dietary modifier of MeHg toxicity, caffeine, is also genetically variable. This latter finding adds an additional level of complexity to interpreting outcomes of MeHg exposure, namely, that dietary factors may influence MeHg toxicity, but their efficacy, in itself, is subject to genetic pre-disposition. Nonetheless, for more than two thirds of the fly lines assayed, a beneficial effect of caffeine on MeHg toxicity was observed. Translating this finding to mammalian models and to humans has the potential to identify a means of moderating MeHg exposure effects through the diet.

GWA identifies an association of MeHg tolerance with genes and networks in muscle development

We used a lenient statistical threshold of $p<10^{-4}$ for identification of candidate genes associated with phenotypic variation. This threshold does not reach genome-wide significance based on Bonferroni multiple testing correction or permutation, but nonetheless serves as a hypothesis generating mechanism which identifies the top polymorphisms in a population where all polymorphisms are known. Candidate genes identified at this nominal threshold can be further verified through the use of mutants or by asking to what extent candidate genes are members of a network with a probability that is significantly higher than expected by chance.

The top candidate genes from our GWA analyses of MeHg tolerance were enriched for functions in muscle development.

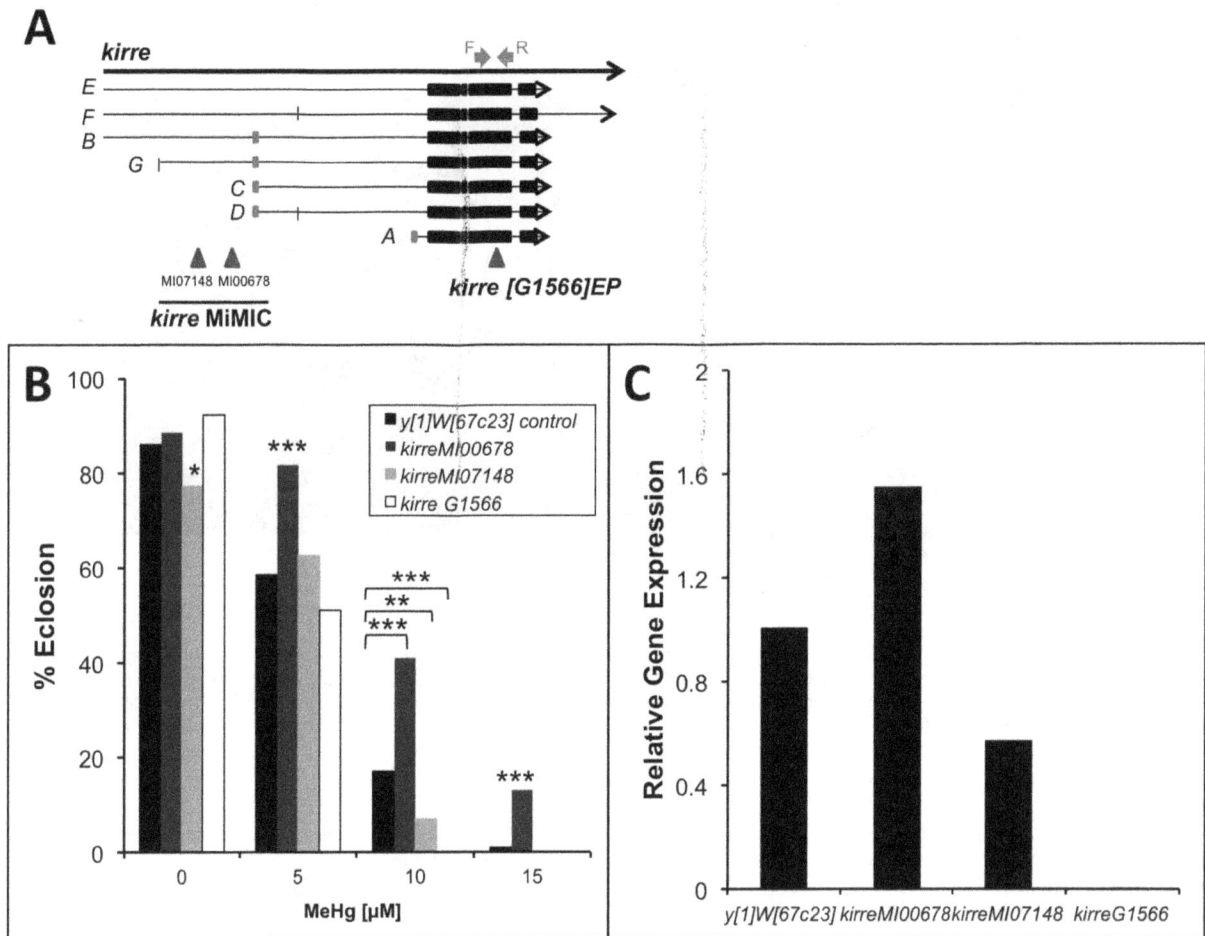

Figure 7. Altered MeHg tolerance in *kirre* mutant flies. (A) Schematic representation of mutant lines targeting *kirre*. *kirre*[M107148] is located at *X*: 2921499 targeting *kirre* transcripts E, F, B and G. *kirre*[MI00678] is located at *X*: 2952881 targeting E, F, B and G transcripts. *kirre*[G1566] is located at *X*: 3025507 in Exon 10 targeting all 7 splice variants of *kirre*. Gray arrows depict the location of the forward and reverse primers used for qRT-PCR analysis. (B) Eclosion assays of *kirre* mutant lines (MiMIC and EP) compared to y^1w^{67c23} control line. Statistical analyses done by z- test. $N = 300$ flies/bar. * $p < 0.01$, ** $p < 0.001$ and *** $p < 0.0001$ relative to y^1w^{67c23}. (C) Expression of *kirre* mRNA in P12 pupae thoracic RNA extracts (pooled sample of $n = 20$ pupae) measured via qRT-PCR.

kirre(*duf*), *sns*, *if*, *kon* and *rols* are central players in myoblast fusion, myotube elongation and attachment and myofibrillogenesis in both embryonic and adult muscle development in flies [43]. Adult indirect flight muscles (IFMs) are an excellent model to study muscle development [43]. IFMs are comprised of dorsal longitudinal muscles (DLMs) and dorso-ventral muscles (DVMs), which act antagonistically. DLMs form through a process whereby persistent larval oblique muscles serve as templates for the recruitment and fusion of myoblasts that migrate from the notal region of the imaginal wing disc [44]. Myoblast homing and fusion to growing myotubes is mediated by *kirre* and *sns*, these being Ig-domain proteins and cognate ligand partners that mediate cell adhesion and the formation of multinucleate syncytial cells [45]. In this process, *kirre* interacts directly with *rols*, a scaffolding protein, to facilitate myoblast fusion [46]. Subsequent splitting of the growing myotube occurs to generate three and then six distinct fiber bundles in each hemithorax. Concurrently the bundles elongate with the tips eventually anchoring to tendon cell attachment sites, a process that requires *kon* [43] [47]. The fact that polymorphisms in these five core pathway genes are among the top associations with variation in MeHg tolerance strongly supports the hypothesis that muscle development is a MeHg target.

MeHg appears to disrupt myotube growth, fiber bundle splitting and, apparently, anchoring at the myotendinous junction in DLMs at the pupal stage (Fig. 8), a phenotype entirely consistent with disruption of the function of one or more genes listed above. Furthermore, *kirre* mutants that reduce expression levels demonstrated enhanced susceptibility to MeHg toxicity (Fig. 7). This effect was moderate and likely due to a redundant function for *kirre* and *roughest* (*rst*) [48]. Nonetheless, a *kirre* mutant causing increased expression results in a corresponding increase in MeHg tolerance (Fig. 7) reinforcing the notion that *kirre* can moderate MeHg toxic effects. A concerted role for muscle development in MeHg toxicity is strongly supported by our finding that pan-muscular expression of the MeHg protective enzyme, GCLc, gives a robust rescue of the MeHg effect on eclosion. How MeHg interacts with myogenic genes and/or their products remains to be characterized. A potential role of these genes is consistent with the notion that failure to eclose, a behavior that requires concerted contractions of newly formed adult muscles, results from MeHg disrupting the integrity of forming muscles.

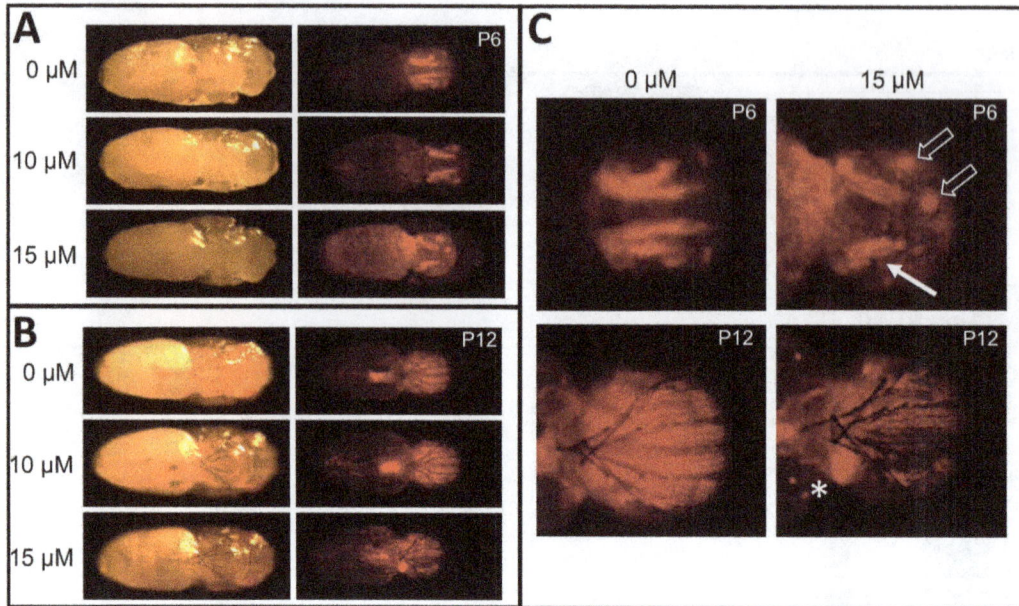

Figure 8. MeHg disruption of DLM muscle development. (A, B) *Mef2>RFP* pupae reared on indicated concentration of MeHg to stage P6 (A) or P12 (B) and imaged by bright light and red fluorescence to reveal DLM morphology. (C) Close-up image of selected panel from A and B. The solid arrow indicates reduced DLM bundle size and defects in DLM bundle splitting with MeHg. Open arrows indicate displacement of attachment sites of DVM bundles. Asterisks (*) indicates failure of extension and anchoring of the DLM resulting in myofibers coalesced in a ball. The development of eyes and bristles appear unaffected by MeHg treatment.

Genes with a function in formation of the neuromuscular junction (NMJ) were also enriched in our network analyses, notably, *sff, pum, spin, Sdc, nmo, tkv, sgg, nlg1* and *frq2*. NMJ-related genes are more highly represented in the MeHg+caffeine treatment. Six of the seven genes with NMJ function have human homologs and, intriguingly, are induced by caffeine as indicted by modENCODE treatment data in the FlyBase entry for each gene. Overall, these findings support the notion that clinical motor deficits may stem from effects of MeHg at the level of the motor unit. NMJ establishment and maintenance relies upon coordinated expression of factors that mediate targeting of the growing axon and connecting nerve terminals with muscle fibers, which are subsequently reinforced through a mechanism reliant upon vesicular trafficking [49]. It is therefore plausible that MeHg tolerance could arise from a favorable expression profile of the above genes that is influenced by natural polymorphic variation and can be modulated by caffeine. The effects of MeHg on the electrophysiological function of mature NMJ is well characterized [50], however, the extent to which MeHg alters NMJ formation in development is not clear.

Our findings present a paradigm shift in the hypothesis that the developing nervous system is preferentially targeted and exceptionally sensitive to MeHg. Myogenesis and neurogenesis share several conserved molecular pathways, and furthermore, the development of motor systems is intrinsically reliant on coordinated signals between developing muscle cells and neurons. It is not surprising that myogenesis could be an equally sensitive target for MeHg and that neurological deficits in animal models and humans observed thus far could include a yet unappreciated neuromuscular component.

We have previously characterized a MeHg-specific activation of the Notch receptor target gene *Enhancer of split mDelta* (*E(spl)mδ*) in *Drosophila* cells and embryos [51]. *E(spl)mδ* is unexpectedly expressed in embryonic muscle precursors as well as in differentiated larval muscles at late embryonic stages [51]. Furthermore, MeHg exposure during embryo development, as well as ectopic *E(spl)mδ* expression in muscle precursors, results in disrupted muscle patterning and concomitant defects in motor neuron outgrowth and branching [51,52]. While it remains to be seen if *E(spl)mδ* functions similarly in developing muscles at pupal stages, a recent study has identified a central role for Notch signals in maintaining migrating myoblast in a fusion incompetent state until encountering their myotube destination during IFM development [48]. Together, these data reinforce the notion that MeHg targets developing muscle in *Drosophila*, and further highlight the potential role for Notch signals, in addition to the gene candidates identified here, to mediate detrimental MeHg effects.

Multiple functions of candidate genes in development and MeHg toxicity

Several of the genes identified here have pleiotropic functions in development. Importantly, a number of genes have central functions in neurogenesis, neuronal differentiation and axon outgrowth, for example, *numb, if, kon, spin and Sdc*. While several steps in muscle development rely on autonomous signaling mechanisms and muscle-specific cues, in certain contexts a neural scaffold is required for appropriate muscle development [53]. Notably, denervation of the individual IFM fibers during early pupal stages affect subsequent myoblast proliferation contributing to reduced muscle bundle size in the DLM and DVM [53]. Thus, the phenotypes seen here may reflect an underlying MeHg-sensitive neural mechanism that remains to be characterized.

Alternatively, natural variation in MeHg tolerance may stem from effects on development of organs critical for dealing with toxic insult and excretion. *kirre* and *sns* are fundamental for morphogenesis of the Garland cell nephrocyte, a major site of waste removal and filtration of insect hemolymph [45,54]. *kirre/sns* function analogously to their vertebrate orthologs Neph1 and

Nephrin, which direct morphogenesis of the slit diaphragm of the podocyte in the mammalian glomerulus of the kidney [55,56]. In addition, *rols* functions in the normal morphogenesis of the Malpighian tubule, the renal organ of the fly [57]. Therefore, *kirre*, *sns* and *rols* may also influence MeHg tolerance by supporting development of essential excretory organs in addition to directing proper myogenesis during development.

Caffeine as a modulator of MeHg toxicity

We found that caffeine modulates MeHg toxicity, and in most cases shows an enhancement of tolerance to MeHg. Interestingly, caffeine has a positive impact on neurodegenerative disease, such as Parkinson's disease [58]. However, in some cases caffeine shows a negative effect on development with MeHg. There is a strong genetic component to the natural variation of the MeHg modulating effect of caffeine in flies, which parallels reports of genetic variation in caffeine effects in humans [59]. These findings emphasize the need to approach the issue of MeHg tolerance and susceptibility with a greater understanding of the role of individual genetic background, as well as dietary behaviors, particularly in investigations of fish-eating human population studies where MeHg exposures are most common.

Summary

We have identified muscle development as a prominent target for MeHg by associating genetic variation in the DGRP with MeHg toxicity. This finding expands the window of inquiry into mechanisms of MeHg toxicity. Candidate genes identified here set the stage for translational studies in vertebrates, and possibly in human populations, to assess to what extent muscle morphogenesis is compromised by this ubiquitous environmental toxin.

Supporting Information

Figure S1 Q-Q plots for eclosion train exhibited under MeHg alone and MeHg+caffeine treatments.

Figure S2 Q-Q plots for P-values under MeHg alone and MeHg+caffeine treatments.

Table S1 Genome Wide Association results for MeHg alone. Minor and major allele identities and counts are indicated for each polymorphism (SNP = single nucleotide polymorphism; INS = insertion; DEL = deletion; MNP = multiple nucleotide polymorphism). P-values and false discovery rates (FDR) are also indicated (See Materials and Methods). Gene identifications include the Flybase gene ID number (FB ID), gene name and location of the polymorphism.

Table S2 Genome Wide Association results for MeHg + Caffeine. (See legend for Table S2.)

Table S3 Eclosion assay raw data. Results of individual trials for eclosion assays are presented for each RAL line on food containing MeHg (concentration in μM) and Caffeine (2 mM) indicated in the column header. Results are expressed in percent (%) eclosion for 50 L1 larvae assayed in each trial (except for a few trials were 30 L1 larvae were assayed, indicated by italics).

Author Contributions

Conceived and designed the experiments: MDR SLM DV. Performed the experiments: SLM DV. Analyzed the data: SLM WH. Contributed reagents/materials/analysis tools: MDR RRHA TFCM. Wrote the paper: MDR SLM RRHA WH TFCM. Software development for analyses: WH.

References

1. Harada M (1995) Minamata disease: methylmercury poisoning in Japan caused by environmental pollution. Crit Rev Toxicol 25: 1–24.
2. Amin-Zaki L, Elhassani S, Majeed MA, Clarkson TW, Doherty RA, et al. (1974) Intra-uterine methylmercury poisoning in Iraq. Pediatrics 54: 587–595.
3. Harada M (1978) Congenital Minamata disease: intrauterine methylmercury poisoning. Teratology 18: 285–288.
4. Sabbagh K (1977) ECT and the media. Br Med J 2: 1215.
5. Choi BH, Lapham LW, Amin-Zaki L, Saleem T (1978) Abnormal neuronal migration, deranged cerebral cortical organization, and diffuse white matter astrocytosis of human fetal brain: a major effect of methylmercury poisoning in utero. J Neuropathol Exp Neurol 37: 719–733.
6. Al-saleem T (1976) Levels of mercury and pathological changes in patients with organomercury poisoning. Bull World Health Organ 53 Suppl: 99–104.
7. Debes F, Budtz-Jorgensen E, Weihe P, White RF, Grandjean P (2006) Impact of prenatal methylmercury exposure on neurobehavioral function at age 14 years. Neurotoxicol Teratol 28: 536–547.
8. Myers GJ, Davidson PW, Shamlaye CF, Axtell CD, Cernichiari E, et al. (1997) Effects of prenatal methylmercury exposure from a high fish diet on developmental milestones in the Seychelles Child Development Study. Neurotoxicology 18: 819–829.
9. Llop S, Engstrom K, Ballester F, Franforte E, Alhamdow A, et al. (2014) Polymorphisms in ABC transporter bgenes and concentrations of mercury in newborns - evidence from two Mediterranean birth cohorts. PLoS One 9: e97172.
10. Julvez J, Smith GD, Golding J, Ring S, Pourcain BS, et al. (2013) Prenatal methylmercury exposure and genetic predisposition to cognitive deficit at age 8 years. Epidemiology 24: 643–650.
11. Myers GJ, Davidson PW, Strain JJ (2007) Nutrient and methyl mercury exposure from consuming fish. J Nutr 137: 2805–2808.
12. Valera B, Dewailly E, Poirier P (2013) Association between methylmercury and cardiovascular risk factors in a native population of Quebec (Canada): a retrospective evaluation. Environ Res 120: 102–108.
13. Goodrich JM, Wang Y, Gillespie B, Werner R, Franzblau A, et al. (2013) Methylmercury and elemental mercury differentially associate with blood pressure among dental professionals. Int J Hyg Environ Health 216: 195–201.
14. Passos CJ, Mergler D (2008) Human mercury exposure and adverse health effects in the Amazon: a review. Cad Saude Publica 24 Suppl 4: s503–520.
15. Nyland JF, Fillion M, Barbosa F, Jr., Shirley DL, Chine C, et al. (2011) Biomarkers of methylmercury exposure immunotoxicity among fish consumers in Amazonian Brazil. Environ Health Perspect 119: 1733–1738.
16. Lee BE, Hong YC, Park H, Ha M, Koo BS, et al. (2010) Interaction between GSTM1/GSTT1 polymorphism and blood mercury on birth weight. Environ Health Perspect 118: 437–443.
17. Ramon R, Ballester F, Aguinagalde X, Amurrio A, Vioque J, et al. (2009) Fish consumption during pregnancy, prenatal mercury exposure, and anthropometric measures at birth in a prospective mother-infant cohort study in Spain. Am J Clin Nutr 90: 1047–1055.
18. Hughes WL (1957) A physicochemical rationale for the biological activity of mercury and its compounds. Ann N Y Acad Sci 65: 454–460.
19. Rabenstein DL, Evans CA (1978) The mobility of methylmercury in biological systems. Bioinorg Chem 8: 107–101, 104.
20. Morgan TJ, Mackay TFC (2006) Quantitative trait loci for thermotolerance phenotypes in *Drosophila melanogaster*. Heredity (Edinb) 96: 232–242.
21. Jordan KW, Craver KL, Magwire MM, Cubilla CE, Mackay TFC, et al. (2012) Genome-wide association for sensitivity to chronic oxidative stress in *Drosophila melanogaster*. PLoS One 7: e38722.
22. Weber AL, Khan GF, Magwire MM, Tabor CL, Mackay TFC, et al. (2012) Genome-wide association analysis of oxidative stress resistance in *Drosophila melanogaster*. PLoS One 7: e34745.
23. Morozova TV, Mackay TFC, Anholt RRH (2011) Transcriptional networks for alcohol sensitivity in *Drosophila melanogaster*. Genetics 187: 1193–1205.
24. Mackay TFC, Richards S, Stone EA, Barbadilla A, Ayroles JF, et al. (2012) The *Drosophila melanogaster* Genetic Reference Panel. Nature 482: 173–178.
25. Huang W, Massouras A, Inoue Y, Peiffer J, Ramia M, et al. (2014) Natural variation in genome architecture among 205 *Drosophila melanogaster* Genetic Reference Panel lines. Genome Res. 24: 1193–1208.
26. Rand MD, Lowe JA, Mahapatra CT (2012) Drosophila CYP6g1 and its human homolog CYP3A4 confer tolerance to methylmercury during development. Toxicology 300: 75–82.

27. Rand MD, Montgomery SL, Prince L, Vorojeikina D (2014) Developmental toxicity assays using the Drosophila model. Curr Protoc Toxicol 59: 1 12 11–11 12 20.

28. Baines RA (2005) Neuronal homeostasis through translational control. Mol Neurobiol 32: 113–121.

29. Dubnau J, Chiang AS, Grady L, Barditch J, Gossweiler S, et al. (2003) The staufen/pumilio pathway is involved in Drosophila long-term memory. Curr Biol 13: 286–296.

30. Littleton JT, Bai J, Vyas B, Desai R, Baltus AE, et al. (2001) synaptotagmin mutants reveal essential functions for the C2B domain in Ca2+-triggered fusion and recycling of synaptic vesicles in vivo. J Neurosci 21: 1421–1433.

31. Adolfsen B, Saraswati S, Yoshihara M, Littleton JT (2004) Synaptotagmins are trafficked to distinct subcellular domains including the postsynaptic compartment. J Cell Biol 166: 249–260.

32. Yazdani U, Huang Z, Terman JR (2008) The glucose transporter (GLUT4) enhancer factor is required for normal wing positioning in Drosophila. Genetics 178: 919–929.

33. Knight JB, Eyster CA, Griesel BA, Olson AL (2003) Regulation of the human GLUT4 gene promoter: interaction between a transcriptional activator and myocyte enhancer factor 2A. Proc Natl Acad Sci U S A 100: 14725–14730.

34. Schnizler K, Saeger B, Pfeffer C, Gerbaulet A, Ebbinghaus-Kintscher U, et al. (2005) A novel chloride channel in Drosophila melanogaster is inhibited by protons. J Biol Chem 280: 16254–16262.

35. Mummery-Widmer JL, Yamazaki M, Stoeger T, Novatchkova M, Bhalerao S, et al. (2009) Genome-wide analysis of Notch signalling in Drosophila by transgenic RNAi. Nature 458: 987–992.

36. Balamurugan K, Egli D, Selvaraj A, Zhang B, Georgiev O, et al. (2004) Metal-responsive transcription factor (MTF-1) and heavy metal stress response in Drosophila and mammalian cells: a functional comparison. Biol Chem 385: 597–603.

37. Egli D, Selvaraj A, Yepiskoposyan H, Zhang B, Hafen E, et al. (2003) Knockout of 'metal-responsive transcription factor' MTF-1 in Drosophila by homologous recombination reveals its central role in heavy metal homeostasis. EMBO J 22: 100–108.

38. Bahadorani S, Mukai S, Egli D, Hilliker AJ (2010) Overexpression of metal-responsive transcription factor (MTF-1) in Drosophila melanogaster ameliorates life-span reductions associated with oxidative stress and metal toxicity. Neurobiol Aging 31: 1215–1226.

39. Warde-Farley D, Donaldson SL, Comes O, Zuberi K, Badrawi R, et al. (2010) The GeneMANIA prediction server: biological network integration for gene prioritization and predicting gene function. Nucleic Acids Res 38: W214–220.

40. Olson EN, Perry M, Schulz RA (1995) Regulation of muscle differentiation by the MEF2 family of MADS box transcription factors. Dev Biol 172: 2–14.

41. Dutczak WJ, Ballatori N (1994) Transport of the glutathione-methylmercury complex across liver canalicular membranes on reduced glutathione carriers. J Biol Chem 269: 9746–9751.

42. Toyama T, Shinkai Y, Yasutake A, Uchida K, Yamamoto M, et al. (2011) Isothiocyanates reduce mercury accumulation via an Nrf2-dependent mechanism during exposure of mice to methylmercury. Environ Health Perspect 119: 1117–1122.

43. Weitkunat M, Schnorrer F (2014) A guide to study Drosophila muscle biology. Methods.

44. Fernandes JJ, Keshishian H (1996) Patterning the dorsal longitudinal flight muscles (DLM) of Drosophila: insights from the ablation of larval scaffolds. Development 122: 3755–3763.

45. Kesper DA, Stute C, Buttgereit D, Kreisköther N, Vishnu S, et al. (2007) Myoblast fusion in Drosophila melanogaster is mediated through a fusion-restricted myogenic-adhesive structure (FuRMAS). Dev Dyn 236: 404–415.

46. Bulchand S, Menon SD, George SE, Chia W (2010) The intracellular domain of Dumbfounded affects myoblast fusion efficiency and interacts with Rolling pebbles and Loner. PLoS One 5: e9374.

47. Devenport D, Bunch TA, Bloor JW, Brower DL, Brown NH (2007) Mutations in the Drosophila alphaPS2 integrin subunit uncover new features of adhesion site assembly. Dev Biol 308: 294–308.

48. Gildor B, Schejter ED, Shilo BZ (2012) Bidirectional Notch activation represses fusion competence in swarming adult Drosophila myoblasts. Development 139: 4040–4050.

49. Menon KP, Carrillo RA, Zinn K (2013) Development and plasticity of the Drosophila larval neuromuscular junction. Wiley Interdiscip Rev Dev Biol 2: 647–670.

50. Levesque PC, Atchison WD (1988) Effect of alteration of nerve terminal Ca2+ regulation on increased spontaneous quantal release of acetylcholine by methyl mercury. Toxicol Appl Pharmacol 94: 55–65.

51. Engel GL, Rand MD (2014) The Notch target E(spl)mdelta is a muscle-specific gene involved in methylmercury toxicity in motor neuron development. Neurotoxicol Teratol 43C: 11–18.

52. Engel GL, Delwig A, Rand MD (2012) The effects of methylmercury on Notch signaling during embryonic neural development in Drosophila melanogaster. Toxicol In Vitro 26: 485–492.

53. Fernandes JJ, Keshishian H (2005) Motoneurons regulate myoblast proliferation and patterning in Drosophila. Dev Biol 277: 493–505.

54. Zhuang S, Shao H, Guo F, Trimble R, Pearce E, et al. (2009) Sns and Kirre, the Drosophila orthologs of Nephrin and Neph1, direct adhesion, fusion and formation of a slit diaphragm-like structure in insect nephrocytes. Development 136: 2335–2344.

55. Srinivas BP, Woo J, Leong WY, Roy S (2007) A conserved molecular pathway mediates myoblast fusion in insects and vertebrates. Nat Genet 39: 781–786.

56. Sohn RL, Huang P, Kawahara G, Mitchell M, Guyon J, et al. (2009) A role for nephrin, a renal protein, in vertebrate skeletal muscle cell fusion. Proc Natl Acad Sci U S A 106: 9274–9279.

57. Putz M, Kesper DA, Buttgereit D, Renkawitz-Pohl R (2005) In Drosophila melanogaster, the rolling pebbles isoform 6 (Rols6) is essential for proper Malpighian tubule morphology. Mech Dev 122: 1206–1217.

58. Prediger RD (2010) Effects of caffeine in Parkinson's disease: from neuroprotection to the management of motor and non-motor symptoms. J Alzheimers Dis 20 Suppl 1: S205–220.

59. Yang A, Palmer AA, de Wit H (2010) Genetics of caffeine consumption and responses to caffeine. Psychopharmacology (Berl) 211: 245–257.

60. Orr WC, Radyuk SN, Prabhudesai L, Toroser D, Benes JJ, et al. (2005) Overexpression of glutamate-cysteine ligase extends life span in Drosophila melanogaster. J Biol Chem 280: 37331–37338.

61. Mahapatra CT, Bond J, Rand DM, Rand MD (2010) Identification of methylmercury tolerance gene candidates in Drosophila. Toxicol Sci 116: 225–238.

62. Bainbridge SP, Bownes M (1981) Staging the metamorphosis of Drosophila melanogaster. J Embryol Exp Morphol 66: 57–80.

63. Livak KJ, Schmittgen TD (2001) Analysis of relative gene expression data using real-time quantitative PCR and the 2(-Delta Delta C(T)) Method. Methods 25: 402–408.

6

Next-Generation Sequencing Analysis of MiRNA Expression in Control and FSHD Myogenesis

Veronica Colangelo[1][9], Stéphanie François[1][9], Giulia Soldà[2], Raffaella Picco[3], Francesca Roma[2], Enrico Ginelli[2], Raffaella Meneveri[1]*

1 Department of Health Sciences, University of Milano-Bicocca, Monza, Italy, 2 Department of Medical Biotechnology and Translational Medicine, University of Milan, Milan, Italy, 3 Department of Medical and Biological Sciences, University of Udine, Udine, Italy

Abstract

Emerging evidence has demonstrated that miRNA sequences can regulate skeletal myogenesis by controlling the process of myoblast proliferation and differentiation. However, at present a deep analysis of miRNA expression in control and FSHD myoblasts during differentiation has not yet been derived. To close this gap, we used a next-generation sequencing (NGS) approach applied to in vitro myogenesis. Furthermore, to minimize sample genetic heterogeneity and muscle-type specific patterns of gene expression, miRNA profiling from NGS data was filtered with FC≥4 ($\log_2$FC≥2) and p-value<0.05, and its validation was derived by qRT-PCR on myoblasts from seven muscle districts. In particular, control myogenesis showed the modulation of 38 miRNAs, the majority of which (34 out 38) were up-regulated, including myomiRs (miR-1, -133a, -133b and -206). Approximately one third of the modulated miRNAs were not previously reported to be involved in muscle differentiation, and interestingly some of these (i.e. miR-874, -1290, -95 and -146a) were previously shown to regulate cell proliferation and differentiation. FSHD myogenesis evidenced a reduced number of modulated miRNAs than healthy muscle cells. The two processes shared nine miRNAs, including myomiRs, although with FC values lower in FSHD than in control cells. In addition, FSHD cells showed the modulation of six miRNAs (miR-1268, -1268b, -1908, 4258, -4508- and -4516) not evidenced in control cells and that therefore could be considered FSHD-specific, likewise three novel miRNAs that seem to be specifically expressed in FSHD myotubes. These data further clarify the impact of miRNA regulation during control myogenesis and strongly suggest that a complex dysregulation of miRNA expression characterizes FSHD, impairing two important features of myogenesis: cell cycle and muscle development. The derived miRNA profiling could represent a novel molecular signature for FSHD that includes diagnostic biomarkers and possibly therapeutic targets.

Editor: Atsushi Asakura, University of Minnesota Medical School, United States of America

Funding: This work was supported by grants from the AFMTéléthon (reference Number 16547), http://www.afm-telethon.com, and the Italian Ministry of University and Research (PRIN 2008BEYKL8_002), www.prin.miur.it. The funders had no role in study design, data collection and analysis, decision to publish, or preparation of the manuscript.

Competing Interests: The authors have declared that no competing interests exist.

* Email: raffaella.meneveri@unimib.it

[9] These authors contributed equally to this work.

Introduction

Facioscapulohumeral muscular dystrophy (FSHD) is the third most common myopathy, with an incidence of 1 in 14.000 in the general population. Signs of FSHD become visible in an individual's 20's (men) or 30's (women) and include loss of muscle strength in the face, shoulders, and upper arms before eventually attaining the abdomen, legs and feet. FSHD is transmitted as an autosomal dominant trait and it is thought to be mainly associated to an epigenetic alteration leading to transcriptional imbalance of the responsible genes [1,2]. Almost all FSHD patients carry rearrangements reducing the copy number of a 3.3 kb tandemly repeated sequence (D4Z4) located at 4q35, and containing a conserved open reading frame for a double homeobox gene (DUX4). D4Z4 copy number is highly polymorphic in healthy individuals ranging between 11 and >100copies while FSHD patients carry fewer than 11 repeats [3]. Notably, although the number of D4Z4 repeats seems to be a critical determinant of the

age of onset and clinical severity of FSHD, patients without D4Z4 contraction (phenotypic FSHD or FSHD2) as well as healthy individuals with D4Z4 contraction (carrier) have been also identified [4,5]. All these observations strongly suggests that FSHD derives from the interplay of more complex genetic and epigenetic events than those already described; these additional events might take place at either 4q35 or elsewhere in the human genome.

Recently a unifying genetic model [6] that provides the expression of D4Z4 as a major cause of FSHD has been proposed. Another recent paper [7] defining the epigenetic regulation of 4q35 gene expression, demonstrated that D4Z4 deletion is associated to reduced epigenetic repression by Polycomb silencing in FSHD patients. Furthermore, *DBE-T*, a chromatin associated non-coding RNA is produced selectively in FSHD patients and it coordinates the de-repression of 4q35 genes. However, another study evaluating a large-scale population analysis of healthy and

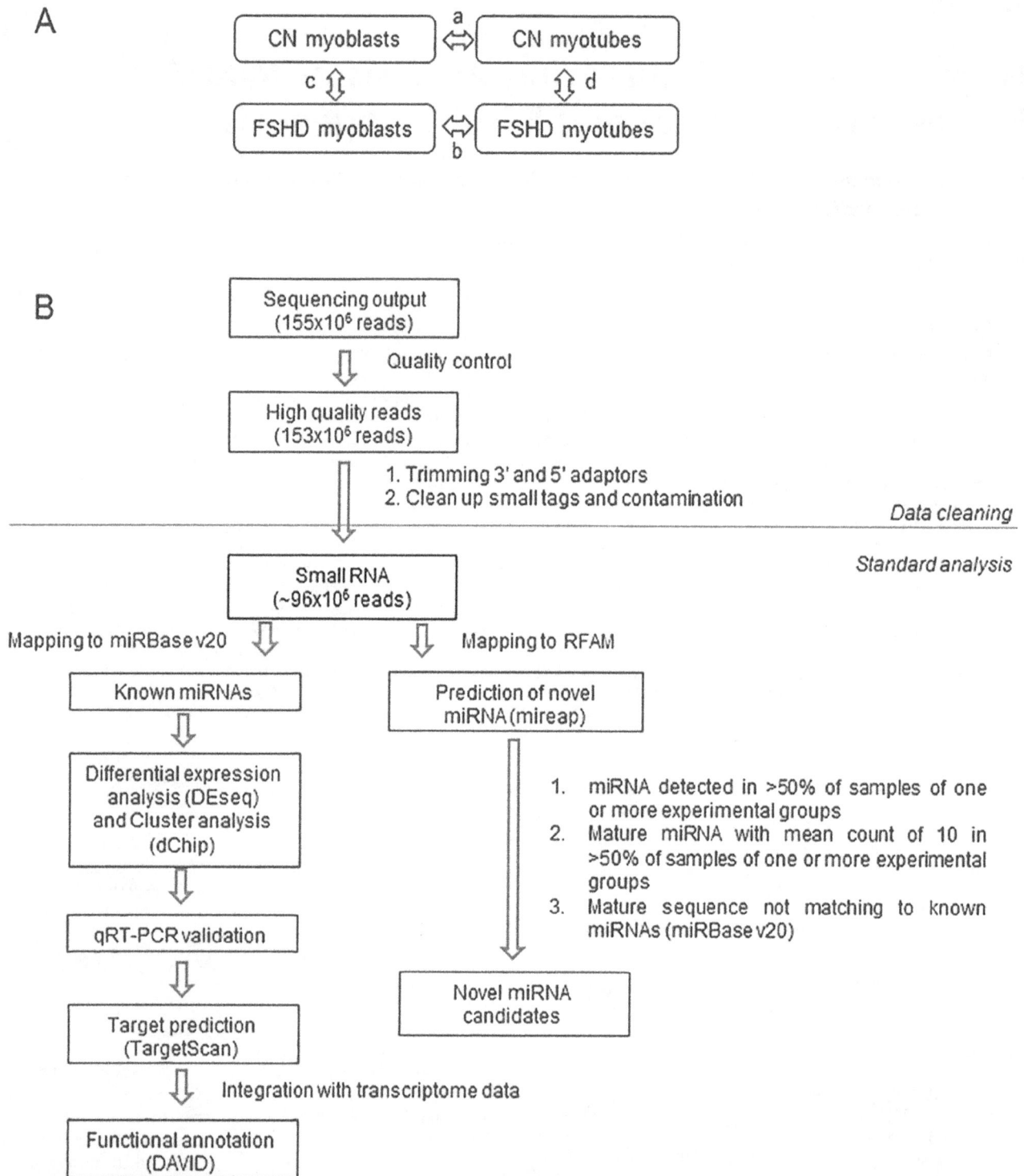

Figure 1. Study design and data analysis. A) Study design: Next-generation Sequencing (NGS) on three control and three FSHD myoblast cell lines before and after *in vitro* myogenic differentiation was used in order to derive miRNA modulation in: a) control myogenesis (CN myotubes vs CN myoblasts; arrow a); b) FSHD myogenesis (FSHD myotubes vs FSHD myoblasts; arrow b); c) FSHD myoblasts versus control myoblast (arrow c), and d) FSHD myotubes vs control myotubes (arrow d). B) Flow chart of filtering and analysis of NGS data. NGS generated a total of 153×10^6 high quality reads, that were filtered for rRNA, tRNA, snRNA, snoRNA, repeat associated RNAs and intron/exon. The filtered reads (approx. 99×10^6 reads, an average of 8×10^6/sample) were analyzed to derive known miRNAs (R/Bioconductor) and novel miRNAs (mireap). Differentially expressed miRNAs between samples were derived by $\log_2 FC \geq 2$ and p-value<0.05 parameters. The homogeneity of miRNA modulation among samples was evaluated by cluster analysis (dChip). miRNAs were then validated by qRT-PCR. Finally, target genes were predicted for modulated miRNAs and functionally annotated by DAVID.

unrelated FSHD patients reports that the genetic criteria in order to manifest FSHD (D4Z4 contraction associated with a specific chromosomal background 4A-161-p(A)- pathogenic haplotype) occur in 63.7% of the analyzed FSHD patients and in 1.3% of healthy subjects [8]. Although these data certainly represent a major advance toward the definition of the molecular basis of FSHD, many questions on the disease etiology remain unexplained. Also the reported high degree of variability of the disease, in term of onset, progression and severity strongly suggests that other mechanism(s) linked to the 4q subtelomere and/or to other regions of the human genome may play a role in the disease pathogenesis.

Various recent studies have demonstrated that both FSHD myoblasts and myotubes are characterized by an extensive gene expression dysregulation mainly affecting the myogenesis and including genes linked to cell cycle control, particularly G1/S and G2/M transitions, muscle structure, mitochondrial function, oxidative stress response, and cholesterol biosynthesis [9,10,11].

The deciphering of the molecular basis of FSHD has been further complicated by the finding that microRNAs (miRNAs) are involved in both control and pathological myogenesis [12,13,14]. MiRNAs are evolutionarily conserved short non-coding RNAs (~22 nts) that regulate the stability and/or the translational efficiency of target mRNAs. They have a very pervasive role since it is estimated that a single miRNA has the potential to regulate hundreds of target genes, and therefore, >90% of all human genes could be under miRNAs regulation [15]. MiRNAs are essential for normal mammalian development and are involved in fine-tuning of many biological processes, such as differentiation, proliferation and apoptosis [16,17]. Emerging evidence has demonstrated that miRNA sequences can regulate skeletal myogenesis by controlling the process of myoblast proliferation and differentiation, in particular, microRNA-1, -206 and -133a/b were defined as myomiRNAs to emphasize their crucial role in myogenesis [18,19]. More recently, a simultaneous microRNA/mRNA expression profiling of healthy myogenic cells during differentiation allowed to identify the involvement of miRNAs in the regulation of various biological processes such as cell cycle, transcription, transport, apoptosis and DNA damage [20]. Given these assumptions it was not surprising that miRNAs dysregulation was found to be involved in muscle dysfunctions [9,12,21].

To date, miRNA studies reported for FSHD were essentially based on the analysis of a restricted number of known miRNA sequences, thus not allowing the derivation of the full miRNA-based dysregulation network. To close this gap, here we report miRNAs expression analysis, derived by next-generation sequencing (NGS), in primary muscle cells from healthy and FSHD subjects during differentiation.

Results

Study design and NGS general results

In order to determine the entire small non coding RNAs (< 35 nts) transcriptome in control (CN) and FSHD primary myoblast cell lines, before and after *in vitro* myogenic differentiation, we used next-generation sequencing (NGS). Study design was organized to allow the comparison of small non-coding RNA expression profiles between FSHD and CN myoblasts and myotubes (Fig. 1A, arrows c and d respectively) and of the two differentiation processes (Fig. 1A, arrows a and b, respectively). In order to derive biological markers (i.e. miRNA dysregulation) commonly manifested by different affected muscle districts, we used two FSHD primary myoblasts cell lines deriving from rhomboid and one from ilio-psoas muscles, and three control

myoblasts from tensor fascia lata, quadriceps and vastus intermedius (Table S1).

As shown in the flow chart reported in Fig. 1B, small RNA sequencing generated a total of 153×10^6 high quality reads. Mature miRNAs make up the majority of sequences in the 18 to 25 nts size range (65% average), with a clear peak at 22 nts in all samples. The average of known miRNAs per sample was of 556, whereas un-annotated small RNAs (new miRNA candidates) per sample were 28.

The differential expression of known miRNAs was analyzed in the different stages of muscle differentiation by DEseq analysis. Furthermore, in order to assess the robustness of our approach, some of the miRNAs identified as differentially expressed were validated by qRT-PCR using specific TaqMan miRNA assays in primary FSHD and healthy myoblasts. For these experiments we employed the same cell lines used for NGS and additional ones from different muscles, including biceps and deltoid (Table S1). As reported in Materials and Methods, the nine control and the seven FSHD cell lines showed a highly comparable extent of Desmin-positive cells and of myogenic markers modulation upon differentiation (Fig. S1). Gene targets of differentially expressed miRNAs were predicted in both control and FSHD cellular systems by using the TargetScan algorithm. Derived gene targets were filtered on two independent transcriptome profiling experiments carried out on control and FSHD myogenesis [9,10], and shared targets were then functionally annotated by DAVID. Novel miRNAs were predicted by mireap and considered as novel candidates only if detected with a mean reads of ten in at least two out of three samples of one or more experimental groups (CN and FSHD myoblasts; CN and FSHD myotubes).

Modulation of miRNA expression during physiological and FSHD myoblast differentiation

We first analyzed the data regarding physiological myogenesis (control myotubes vs control myoblasts; Fig. 1A, arrow a). Filtered miRNA reads (mapping to miRBase v20) from the three control myoblasts samples and the corresponding myotubes were analyzed for differential expression by DEseq analysis, setting the $\log_2$ Fold Change ($\log_2$FC) at ≥ 2 and p-value<0.05. From this analysis we evidenced that during the control myogenesis 38 miRNAs showed a modulation in their expression, and that the great majority of them (34 out of 38) were up-regulated (Fig. 2A and B).

The hierarchical clustering analysis clearly separated proliferating from differentiated cells independently of the muscle district used (tensor fascia lata, quadriceps and vastus intermedius). As expected, the muscle specific miRNAs (myomiRs) hsa-miR-1, -133a, -133b and -206, were among the most up-regulated (Fig. 2B and Table S3). Twenty-six miRNAs were already reported to be involved in muscle differentiation either in human or in mouse cells, whereas 12 miRNAs, ten up-regulated (hsa-miR-95, -146a, -874, -1246, -1290, -3164, -4488, -208a, -944 and -3144) and two down-regulated (hsa-miR-3934 and -3165), were not previously known to be involved in muscle differentiation. The full list of the miRNAs modulated during control myoblasts differentiation with corresponding FC and p-value is reported in Table S3.

The same analysis was carried out on FSHD myogenesis (Fig. 1A, arrow b). As shown in Fig. 3A, the DEseq analysis evidenced the modulation of only 15 miRNAs during pathological muscle differentiation. Even in this case the hierarchical clustering analysis clearly separated proliferating from differentiated cells, independently of the muscle district (Fig. 3B). The majority of miRNAs was up-regulated (11 out of 15), including myomiR-1 and -206, although with a FC lower than that showed in control

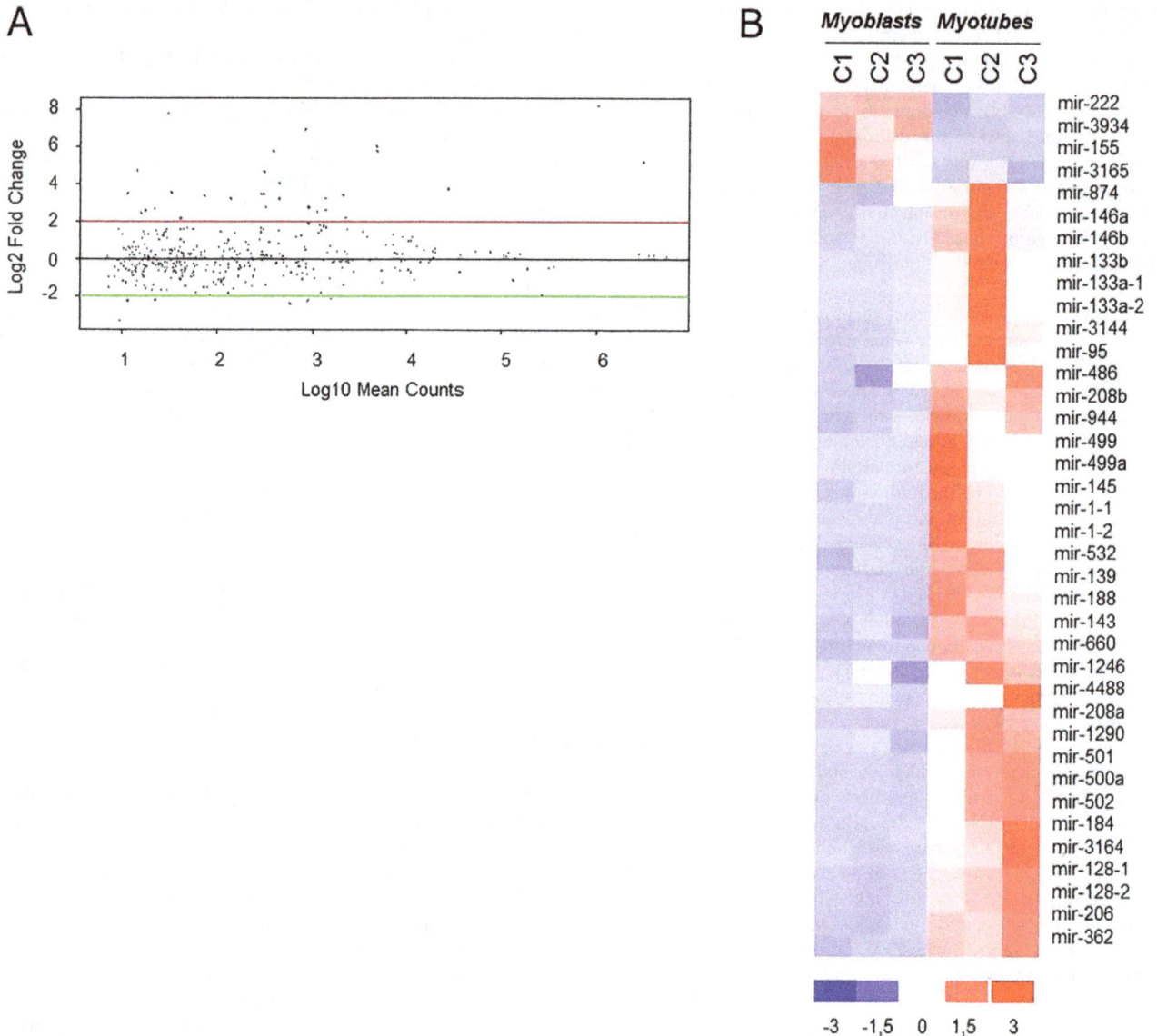

Figure 2. MiRNA modulation in control myogenesis. A) DEseq analysis of miRNAs differentially expressed in control myotubes vs control myoblasts (control differentiation). MiRNAs showing a modulation with $\log_2FC \geq 2$ and a p-value<0.05 are shown as red dots. B) Hierarchical clustering of the 38 modulated miRNAs (34 up-regulated and 4 down-regulated) in regard to the analyzed samples. C1:MX01010MBS; C2: MX03609MBS; C3: MX01110MBS, Control cell lines (see Table S1).

myogenesis (Table S4). MyomiR-133a and -133b showed up-regulation trend ($\log_2FC>5$) without reaching significance (p-value = 0.33). The full list of the miRNAs modulated in FSHD myogenesis, with corresponding FC and p-value, is reported in Table S4. Scatter plots of the reads of modulated miRNAs (for each control and FSHD proliferating and differentiated cell line) are reported in Fig. S2. To further support the results obtained by the sequencing approach, the same control and FSHD myoblast and myotube RNAs were used to analyze the expression of myomiRs (miR-1, miR-133a and miR-206) by qRT-PCR (Fig. S3). In both control and FSHD myotubes, we confirmed the general trend of myomiRs up-regulation derived by sequencing, with the pathological samples showing a lower extent of up-regulation than the normal ones.

Dysregulation of miRNA expression in FSHD myoblasts and myotubes

We next performed DEseq analysis of miRNAs differentially expressed in FSHD myoblasts and myotubes vs controls (Fig. 1, arrows c and d). No miRNAs were found significantly dysregulated ($\log_2FC \geq 2$ and p-value<0.05) in FSHD versus control myoblasts (Fig. 1, arrows c); this result was probably due to the high variability of miRNA expression observed in myoblasts. Conversely, 21 miRNAs were found dysregulated in FSHD myotubes (Table S5 and Fig. 4A), among these 12 miRNAs were up-regulated. The hierarchical clustering analysis clearly separated the pathological samples from the control ones and the three analyzed samples of each group resulted homogeneous in miRNAs dysregulation (Fig. 4B).

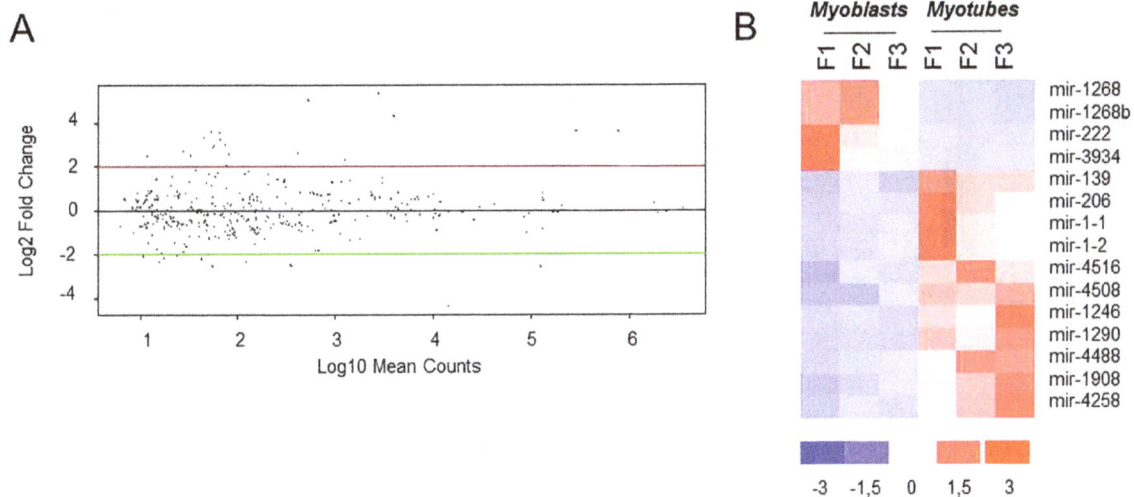

Figure 3. MiRNA modulation in FSHD myogenesis. A) DEseq analysis of miRNAs differentially expressed in FSHD myotubes vs FSHD myoblasts (FSHD differentiation). MiRNAs showing a modulation with $\log_2 FC \geq 2$ and a p-value<0.05 are shown as red dots. B) Hierarchical clustering of the 15 modulated miRNAs (11 up-regulated and 4 down-regulated) in regard to the analyzed samples. F1:MX00409MBS; F2: MX03010MBS; F3:MX04309MBS, FSHD cell lines (see Table S1).

qRT-PCR Validation

The effective validation of deep sequencing results was performed by the TaqMan miRNA assay on all the cell lines listed in Table S1, including those already used for the NGS experiment. Particularly, for myomiR-1, -133a and -206 the assay was carried out at different time points during myogenic differentiation (0, 3 and 7 days of differentiation) (Fig. 5A). In control cells, the myomiRs progressively increased their expression with the proceeding of time of differentiation, reaching the maximum of expression at seven days, with FC values ranging from approximately 350 folds (miR-1) to 28 folds (miR-206). In FSHD cells myomiRs showed an up-regulation significantly lower than that observed in controls, reaching at day seven an expression value similar to or slightly lower than that showed by control cells at day three. Comparable fusion indexes and expression values of myogenic markers in healthy and FSHD myoblasts and myotubes (see Fig. S1) support that the obtained results are not related to a

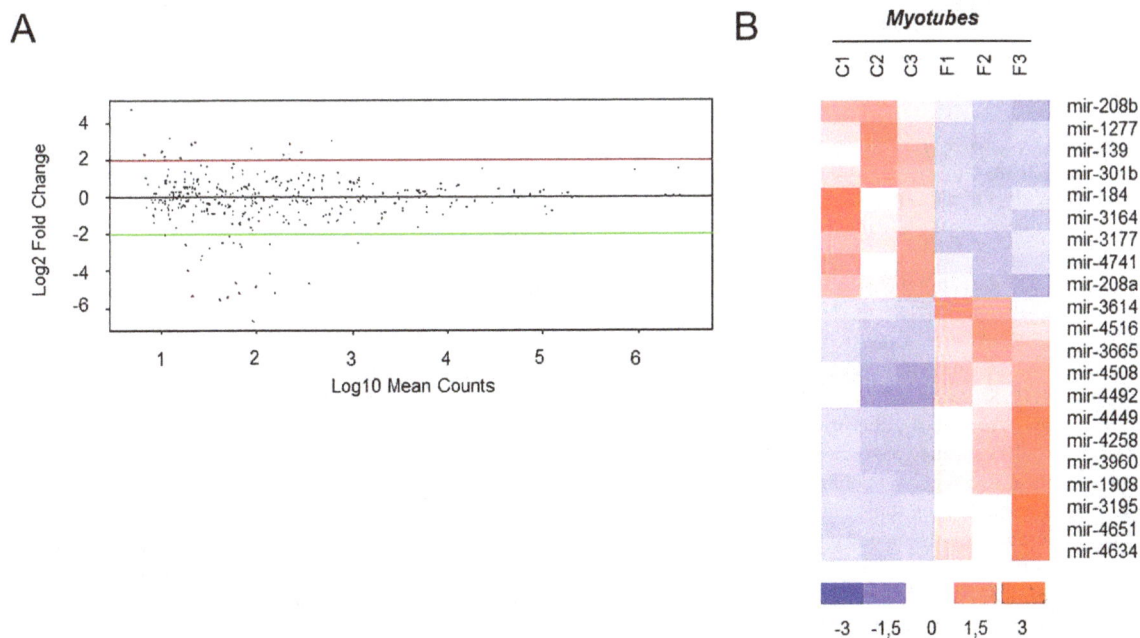

Figure 4. MiRNA dysregulation in FSHD myotubes. A) DEseq analysis of miRNAs differentially expressed in FSHD myotubes vs control myotubes. MiRNAs showing a differential expression of $\log_2 FC \geq 2$ and a p-value<0.05 are shown as red dots. B) Hierarchical clustering of the 21 modulated miRNAs (12 up-regulated and 9 down-regulated) in regard to the analyzed samples. C1:MX01010MBS; C2: MX03609MBS; C3: MX01110MBS, Control cell lines; F1:MX00409MBS; F2: MX03010MBS; F3:MX04309MBS, FSHD cell lines (see Table S1).

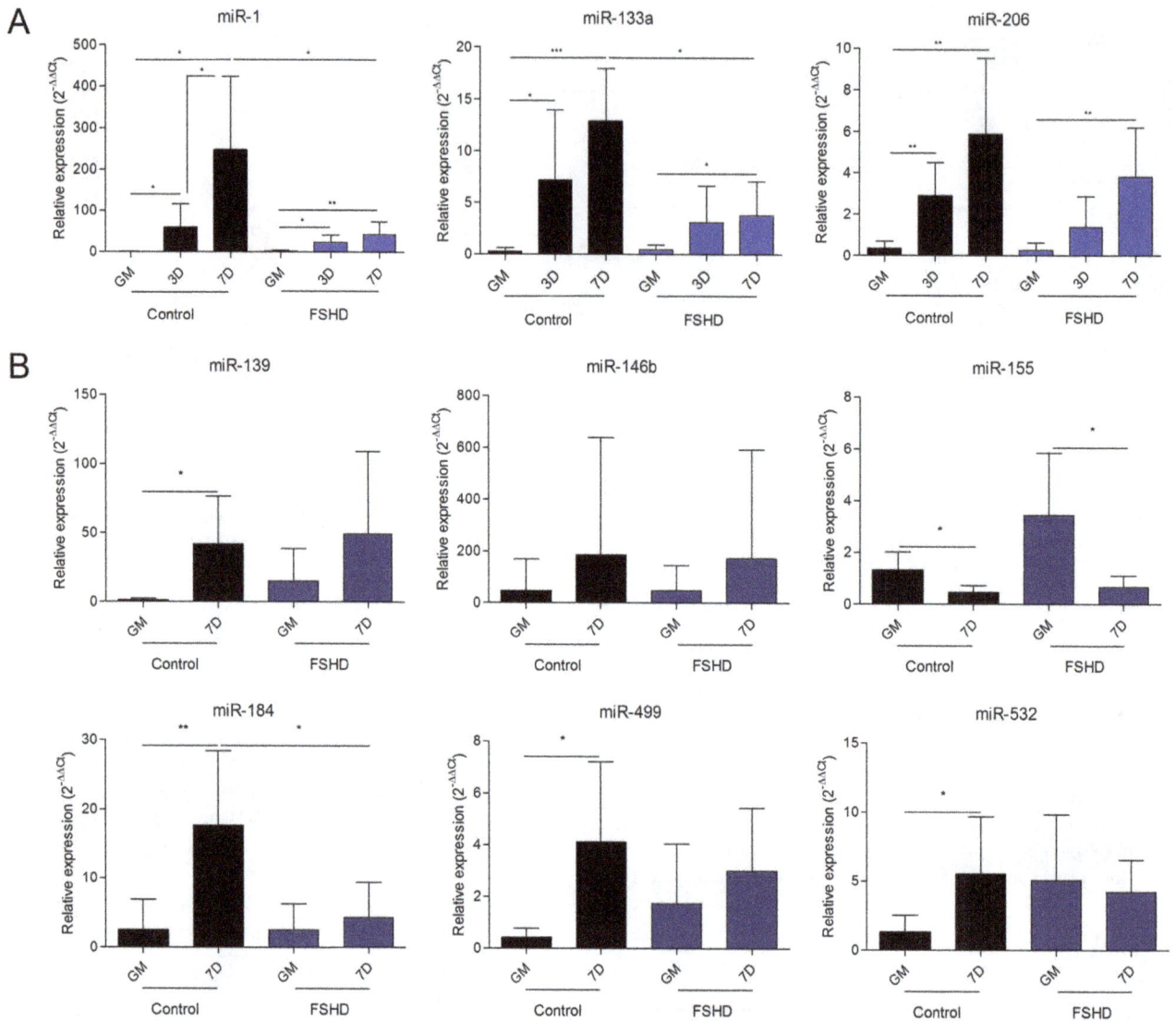

Figure 5. Validation of NGS data. A) qRT-PCR analysis of myomiRs (miR-1, miR-133a and miR-206) during control and FSHD myogenesis at 0, 3 and 7 days of differentiation. B) qRT-PCR analysis of six microRNAs modulated in control and/or FSHD myogenesis. GM: growth medium; 3D: 3 days of differentiation; 7D: 7 days of differentiation. * p-value<0.05; ** p-value<0.01; *** p-value<0.001.

different extent of differentiation between control and pathological samples.

Six additional miRNAs were evaluated for their expression by qRT-PCR (Fig. 5B). As shown in Fig. 5 and summarized in Table 1, the qRT-PCR assays validated about the 70% of the analyzed NGS data. Particularly, the up-regulation of hsa-miR-139 and hsa-miR-146b during, respectively, FSHD and control myogenesis, and the down-regulation of hsa-miR-206 in FSHD vs CN myotubes did not reach the statistical significance showed by NGS results, while maintaining the same trend. On the contrary, the up-regulation of miR-133a in FSHD myogenesis, the down-regulation of hsa-miR-1 and hsa-miR-133a in FSHD vs CN myotubes, and the down-regulation of hsa-miR-155 in FSHD myogenesis already observed in the NGS analysis became significant in the qRT-PCR analysis.

Comparison of FSHD and control myogenesis

The comparison of miRNA modulation between control and FSHD differentiation processes is reported in Fig. 6A, where black and striped bars identify the Fold Change of miRNAs up- and down-regulated, respectively, in control and FSHD myogenesis. From this comparison it was possible to derive that FSHD differentiation lacks the modulation of 29 miRNAs, the majority of which (27/29) was up-regulated in control differentiation (black bars in Fig. 6A, and Fig. 6B); while six miRNAs (4 up- and 2 down-regulated) were modulated only during the FSHD differentiation process (striped bars in Fig. 6A and Fig. 6B). Nine miRNAs showed the same trend in both processes (Fig. 6A and B), but with differences in Fold Change values. Among these, miRNAs pivotal for the myogenic process, such as hsa-miR-1, -206 and -222, were included. Thus, FSHD myogenesis differs from control myogenesis for the complete (35) or partial (9) dysregulation of a total of 44 miRNAs.

Table 1. qRT-PCR validation of NGS data.

miRNA	Control myogenesis				FSHD myogenesis				FSHD vs control myotubes			
	Deep seq		qRT-PCR		Deep seq.		qRT-PCR		Deep seq		qRT-PCR	
	FC	p-value	FC	p-value	FC	p-value	FC	p-value	FC	p-value	FC	p-value
miR-1	293.3	2E-05	352.2	0.007	12.4	3E-06	25.8	0.007	−2.8	0.510*	−5.7	0.033
miR-133a	64.5	2E-05	44.8	1E-04	40.7	0.335	7.9	0.03	−1.2	0.978*	−3.4	0.007
miR-139	24.7	4E-07	28.9	0.018	5.7	0.021	3.3	0.189*	−7.3	0.002	1.2	0.905*
miR-146b	5.5	0.016	3.9	0.419*	2.2	0.858	3.6	0.460	4	0.510	−1.1	0.955
miR-155	−5.6	0.003	−2.8	0.031	−1.4	0.869*	−5.2	0.019	2.7	0.263	1.4	0.413
miR-184	53.9	9E-11	7.1	0.003	7.4	0.145	1.7	0.493	−5.3	0.002	−4.1	0.035
miR-206	36.1	4E-17	28.7	0.009	12.03	1E-06	11.8	0.008	−2.9	0.002	−3.1	0.07*
miR-499a	122.1	0.005	9.7	0.031	5.9	0.143	1.7	0.359	−8.3	0.338	−1.4	0.525
miR-532	9.3	3E-06	4.3	0.037	3.9	0.131	1.2	0.699	−1.99	0.456	−1.3	0.511

Fold Change and p-value of nine miRNAs derived by deep sequencing and subsequently analyzed by qRT-PCR. Asterisked values refer to miRNAs that did not reach significance, although showing the same trend of variation in both analyses.

Prediction of miRNA target genes

To understand the functional impact of miRNA dysregulation during FSHD myogenesis we used TargetScan prediction software to derive potentially affected targets. In order to improve target prediction accuracy, a common approach is to combine the output of two or more prediction algorithms, however this strategy has been proved inefficient [20]. Therefore, we have used a single algorithm, TargetScan, which uses many parameters to predict target scoring without omitting miRNAs with multiple target sites [22]. Since the binding of a miRNA to the 3′ UTR of its mRNA target predominantly act to decrease target mRNA levels [23] we decide to essentially focalize our attention on mRNA targets showing an opposite expression value compared to the analyzed miRNA. Normally, this approach has been carried out on mRNA expression profile derived by using the same cells from which the miRNA expression profile has been derived [11,20,21]. However, the comparison of mRNA expression profiles derived by myoblast cell lines or biopsies from different FSHD patients and controls clearly evidenced a certain variability in the obtained results [5,9,10,11,24,25]. In addition, mRNA expression differences were also found by analyzing different muscles, such as biceps and deltoids [11]. To reduce sample variability, we filtered the predicted mRNA targets on two chip expression data (GSE26061 [9]; GSE26145 [10]), sharing *in vitro* myogenic differentiation protocol and platform although using primary FSHD and control cell lines different from those analyzed in this work. Functional classes corresponding to the filtered mRNAs were assigned by DAVID Gene Ontology Database (Table 2). As shown in Fig. 6, control myogenesis showed the modulation of 38 miRNAs (4 down- and 34 up-regulated), whereas FSHD myogenesis was characterized by 15 dysregulated miRNAs (4 down- and 11 up-regulated) and the lack of modulation of 29 miRNAs. Applying the rationale described above, we derived a total of 139 and 78 down- and up-modulated mRNAs in control myogenesis (potentially "validated" target, Table S6), and a total of 37 down- and 18 up-regulated transcripts in FSHD myogenesis (potentially "validated" target, Table S7). In control myogenic differentiation, the majority of down-regulated genes belonged to cell cycle (27 entries), DNA metabolic process (17 entries), cytoskeleton organization (11 entries), angiogenesis (8 entries) and signal transduction (19 entries); genes involved in cell adhesion (9 entries), regulation of cell migration (5 entries), muscle development (7 entries), lipid biosynthetic process (6 entries) and response to insulin (4 entries) were found up-regulated (Table 2). Conversely, in FSHD myogenesis genes belonging to muscle development (3 entries) and cell adhesion (5 entries) were down-regulated, whereas those involved in regulation of signal transduction (3 entries) were up-regulated. All the identified biological processes, except the down-regulation of cell adhesion in FSHD samples, showed a significant p-value ranging from 3.4E-10 to 3.2E-02 (see Table 2). It is noteworthy that target genes involved in two important biological processes of myogenesis (i.e. cell cycle and striated muscle development) subjected to miRNA control were, as expected, down- and up-regulated, respectively, in control cells. In FSHD myogenesis, on the contrary, the cell cycle was not down-regulated, and control of striated muscle development was down-regulated. It is important to notice that this analysis did not take into account the different FC showed by the nine miRNAs shared by control and FSHD myogenesis.

Identification of novel miRNAs

To identify novel potential miRNAs involved in human muscle system, the unclassified tags were further processed by mireap (http://sourceforge.net/projects/mireap). We considered only

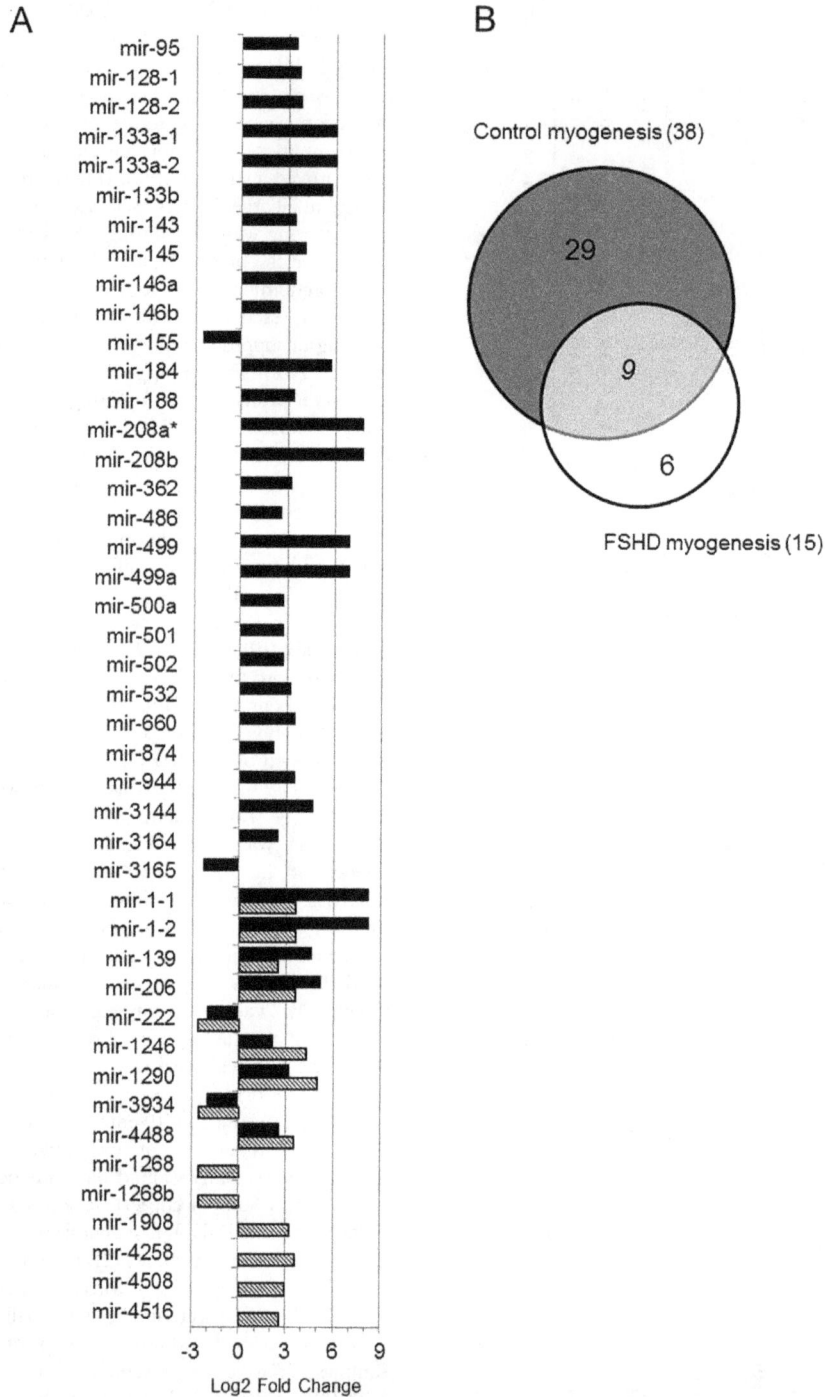

Figure 6. Comparison of miRNA modulation in control and FSHD myogenesis. A) Black and striped bars identify the Fold Change of miRNAs modulated respectively, in control and FSHD myogenesis. Bars on the left and on the right represent, respectively, down- and up-regulated miRNAs. *hsa-mir-208a showed infinite FC value (see Table S3). B) Venn diagram showing the number of miRNAs unique to FSHD (white) or control (grey), and shared (light grey) by FSHD and control differentiation processes.

tags meeting the default parameters, expressed in all experimental groups or preferentially expressed in one or more sample groups (i.e. proliferating vs differentiated cells, or FSHD vs control cells) and with mean read counts per group greater than ten. By using these criteria we identified a total of 13 novel candidate miRNA genes. In Table S8 are reported the main features of these novel miRNA genes, including chromosome location and genomic

organization, mfe (minimum free energy), sequence and structure of hairpin precursor, and sequence of 5p or 3p. A summary of these data is reported in Table 3: six miRNAs showed a preferential expression in myoblasts (both in FSHD and control) and four miRNAs seemed to be specific for myotubes. The remaining three miRNAs characterized all the considered groups (both control and FSHD myoblasts and myotubes). Among the 13

Table 2. Functional classification of predicted target genes in control and FSHD myogenesis.

Biological processes	CONTROL MYOGENESIS		FSHD MYOGENESIS	
	Down (p-value)	Up (p-value)	Down (p-value)	Up (p-value)
Cell cycle	+(3.4E-10)			
DNA metabolic process	+(2.8E-06)			
Cytoskeleton organization	+(2.7E-03)			
Angiogenesis	+(1.2E-03)			
Signal transduction	+(8.2E-03)			+(3.2E-02)
Cell migration		+(6.0E-03)		
Cell adhesion		+(1.2E-02)	+(7.4E-02)	
Striated muscle development		+(1.6E-02)	+(5.8E-03)	
Sterol biosynthetic process		+(1.0E-02)		+(5.0E-04)
Response to insulin		+(3.4E-03)		

Functional classification of predicted target genes of modulated miRNAs in control and FSHD myogenesis, filtered on GSE26061 [9] and GSE26145 [10]. For full lists of considered miRNAs and predicted target genes refer to Tables S3, S4 and S6, S7, respectively.

novel miRNAs, two miRNAs (namely hsa-miR-m1-3p and hsa-miR-m13-5p) had already been detected by analyzing prostate and breast tumor cells [26,27] and the mature hsa-mir-m9-3p showed 100% sequence similarity with hsa-mir-574 whose gene however differs in genomic location [28].

Furthermore, it is interesting to note that no sample showed reads generated from the D4Z4 region. This observation, derived either by the analysis of the filtered out repeats or by the re-mapping NGS raw data to specific D4Z4-bearing chromosome regions such as 4q and 10q, suggests that short transcribed sequences from D4Z4 array may have a length greater than 35 nts, the threshold used to build our libraries.

Discussion

The paper reports the first complete analysis of miRNA modulation during *in vitro* differentiation in both control and FSHD-derived myogenic cells. Myogenesis is a complex process that includes proliferation, differentiation, and formation of myotubes and myofibers. These molecular events are regulated by myogenic factors and miRNAs. MiRNAs specifically expressed in skeletal and cardiac muscles are called myomiRs, to imply their important roles in the regulation of muscle development and differentiation [13,19,29]. Recently miRNA dysregulation has been reported in FSHD [9,12,21]. However, due to the approaches used, these studies were limited for the number and type of miRNAs that could be simultaneously investigated; in addition they would not detect miRNAs expressed at low level and excluded discovery of novel miRNAs. Thus, to get the whole pattern of miRNA dysregulation in FSHD we used a next-generation sequencing (NGS) approach. Previous work aimed at identifying biomarkers in FSHD by the transcriptional profiling found muscle-type specific patterns of gene expression [11]. Similarly, DUX4-fl expression was found to vary between myotubes derived from different muscle groups [30]. Therefore, we tailored the experimental protocol to derive FSHD and control miRNA profiles common to different muscles. To this aim, due to inter-individual genetic heterogeneity, from deep sequencing data we considered only miRNA modulation with FC≥4 ($\log_2$FC≥2) and p-value<0.05. Then the derived miRNA expression in both FSHD and control myogenesis was validated by qRT-PCR in all the available FSHD and control cell lines.

Control myogenesis showed the modulation of 38 miRNAs, the majority of which (34 out 38) were up-regulated. The up-regulated miRNAs included those previously identified as key regulators of both proliferation and differentiation of myogenic cells and for this reason called myomiRs: hsa-miR-1, -133a, -133b and -206 [19,31,32]. The obtained results are in agreement but also expand what is known about miRNA modulation during *in vitro* human myogenic differentiation. Among the modulated miRNAs, 27 were in fact already reported to be involved in muscle differentiation either in human or in mouse cells [20,33]. Conversely, 12 miRNAs, ten up-regulated (hsa-miR-95, -146a, -874, -1246, -1290, -3164, -4488, -208a, -944 and -3144) and two down-regulated (hsa-miR-3934 and -3165), were not previously detected to be differentially expressed during control myogenesis. In comparison with a previous work [20], the reduced number of modulated miRNAs during control myogenesis that we derived is probably due to the choice of higher FC value (FC≥4). Furthermore, the observed up-regulation of myomiRs strongly supports the validity of used cell lines and differentiation protocol. Interestingly, some up-regulated miRNAs not previously reported to be involved in muscle differentiation, were previously shown to affect cell proliferation by targeting HDAC1 (hsa-miR-874), impairing cytokinesis (hsa-miR-1290), inhibiting cell growth (hsa-miR-95) and regulating differentiation of smooth muscle cells (hsa-miR-146a) [34,35,36,37].

Control myogenesis also showed the possible involvement of some of the novel miRNAs we derived by NGS. In this regard, six out of the 13 identified novel miRNAs (see Table 3) seem to characterize the proliferating status of muscle cells (myoblasts, miR-m2-3p, -m3-3p, -m4-5p, -m7-5p, -m12-3p, and -m13-5p) and one the differentiated status (myotubes, miR-m6-3p). Two, hsa-miR-m1-3p, and hsa-miR-m13-5p, have been previously identified by the NGS approach and validated in breast and prostate cancer cells (identified respectively as hsa-miR-B19 and hsa-novel-miR-08) [26,27]. Further experiments are thus necessary to validate and determine the possible involvement in muscle cells differentiation of these novel miRNAs.

The comparison of control and FSHD myogenesis clearly evidenced a reduced number of modulated miRNAs in FSHD than in control muscle cells, thus suggesting that a complex dysregulation of miRNA expression characterizes the dystrophy. In total, nine miRNAs were shared between the two processes and

Table 3. Novel miRNAs predicted by mireap.

Name	Chromosome location	Mature miRNA sequence	Length	Genomic context	Expression n.samples	Other evidence
hsa-miR-m1-3p	chr11:122022800–122022877	AAAAGGGGGCTGAGGTGGAGG	21	intronic	12/12 (higher expression in myoblasts)	PMID:21346806
hsa-miR-m2-3p	chr11:125757935–125758025	AGGGGCGCGGCCCAGGAGCTCAGA	24	intronic	5/6 myoblasts	no
hsa-miR-m3-3p	chr13:111102986–111103008	AGCTGGGGATGGAAGCTGAAGCC	23	intronic	4/6 myoblasts	no
hsa-miR-m4-5p	chr14:74998697–74998789	CTGCTCTGATGTCTGGCTGAGC	22	intronic	5/6 myoblasts	No
hsa-miR-m5-5p	chr15:41592311–41592403	ATCATTTGGCAGGGGGTAGAGTA	23	intergenic	3/3 FSHD myotubes	No
hsa-miR-m6-3p	chr15:45493361–45493452	TTGTGGAAACAATGGTACGGCA	22	overlaps repeat/tRNA	4/6 myotubes	No
hsa-miR-m7-5p	chr17:8042708–8042779	GAGTTAGCGGGGAGTGATATATT	23	overlaps repeat/tRNA	4/6 myoblasts	No
hsa-miR-m8-3p	chr:6:28918819–28918903	TCGGGCGGGAGTGGTGGCTTTT	22	overlaps repeat/tRNA	12/12	No
hsa-miR-m9-3p	chr8:79679467–79679541	TGAGTGTGTGTGTGTGAGTGTGA	23	intronic	9/12 (all groups)	mature miRNA identical to hsa-mir-574, different genomic location PMID:17604727
hsa-miR-m10-5p	chrX:18651329–18651427	AACTTTGGAATGTGGTAGGGTA	22	intronic	3/3 FSHD myotubes	No
hsa-miR-m11-5p	chrX:40478974–40479066	ATCATTTGGCAGGGGGTAGAGTA	23	intergenic	3/3 FSHD myotubes	No
hsa-miR-m12-3p	chr13:111102941–111103018	AGCTGGGGATGGAAGCTGAAGCC	23	intronic	4/6 myoblasts	No
hsa-miR-m13-5p	chr20:3194751–3194835	CAAAATGATGAGGTACCTGATA	22	Intronic	6/6 myoblasts	PMID:21152091

these included myomiR-1 and -206, with FC values of up-regulation during differentiation lower than those derived for control cells. Moreover, qRT-PCR analysis proved that in control cells the up-regulation of myomiRs is higher than in FSHD ones by a FC ranging from 2.4–5.5 × for hsa-miR-206 and 133a to 13× for hsa-miR-1. Furthermore, the kinetic of myomiRs up-regulation during FSHD myogenesis strongly suggests a defect in late stages of the differentiation process. Other differences between control and FSHD differentiation are represented by six miRNAs (i.e. hsa-miR-1268, -1268b, -1908, 4258, -4508- and -4516) not modulated in control cells and that therefore could be considered FSHD-specific, likewise three novel miRNAs (hsa-miR-m5-5p, hsa-miR-m10-5p and hsa-miR-m11-5p) that seem to be specifically expressed in FSHD myotubes (see Table 3). Of interest, hsa-miR-1268 exhibited a significant differential expression during the differentiation of pluripotent human embryonic stem cells into embryoid bodies [38]. In summary, FSHD myogenesis differed from control myogenesis by the loss of modulation of 29 miRNAs (black bars in Fig. 6A, and Fig. 6B) and the acquisition of modulation of six miRNAs, two down-regulated and four up-regulated (striped bars in Fig. 6A, and Fig. 6B). Among the nine miRNAs shared by the two differentiation processes (black and striped double bars in Fig. 6A), the myomiRs showed a significant deficit of expression in late phases of FSHD differentiation. Moreover, the comparison of miRNA expression between control and FSHD myoblasts or myotubes detected 21 dysregulated miRNAs only in myotubes (12 up-regulated and 9 down-regulated). The lack of differentially expressed miRNAs in FSHD myoblasts may be explained both by a high variance of miRNA expression showed by myoblasts and by the high FC used.

Some discrepancies between the data we derived and those recently reported in a similar cellular system [21] require several considerations. First, the methodological approach (NGS against transcriptome profiling), and consequently the cut-off used make the results obtained not comparable; second, both healthy and FSHD myoblast cell lines characterized by a high percentage of DES+ cells were induced to differentiate for three days [20] and for seven days (herein). Lastly, our study design was set up in order to derive a FSHD miRNA profiling possibly shared by different muscle districts and including all the microRNAs present in miRBase (release 20), as well as novel miRNAs. In this regard, it is noteworthy that if we had used the microRNA panel version 1.0 (a TaqMan low density array containing 365 miRNAs) instead of the NGS approach, we would have only detected the modulation of five miRNAs during differentiation of FSHD myoblasts (namely hsa-miR-1-1, 1-2, -206, -222 and -139), instead of the fifteen effectively found (see Table S4). Thus, as previously shown in other cellular systems [26,27,39,40] the deep sequencing approach allowed us to derive a more complete view of miRNA dysregulation in FSHD.

Our data strongly suggest that, in addition to the recently reported up-regulation in proliferating FSHD vs control cells, which however did not result in a complete down-regulation of the corresponding target genes [21], a defect of myomiRs expression also characterize late stages of FSHD differentiation. In fact, the extent of myomiRs expression in FSHD cells after seven days of differentiation was similar to or lower than that found at three days in control cells. Thus, besides the reported up-regulation of myomiRs in FSHD myoblasts due to the early euchromatization of their promoters [21] other defects could be responsible of their down-regulation during late stages of differentiation. In this regard it is possible to hypothesize a defect in FSHD myotubes at the myomiRs transcriptional or post-transcriptional levels, such as a decrease of myogenic differentiation factors controlling their transcription (i.e. *MEF2*) [41], or of factors controlling their processing. The latter hypothesis agrees with previous results showing that FSHD myotubes are characterized by the down-regulation of a gene (*Dicer1*) controlling the cytoplasmic maturation of pre-miRNAs [10].

Our data allowed us to confirm a few miRNAs previously found dysregulated by independent analysis of ten major skeletal muscle disorders, including FSHD [12,42]. Among the miRNAs we derived to be deregulated during FSHD muscle differentiation, four miRNAs (miR-146a, -146b, -155, -222) were consistently found up-regulated in six or more muscular disorders, including FSHD, whereas miR-501 was found dysregulated in five muscle diseases, but not in FSHD. Furthermore miRNA-486, a muscle enriched miRNA, previously found significantly reduced in patients with DMD [12], was found up-regulated in the present study. Interestingly overexpression of this miRNA in mouse primary myoblasts resulted in increased proliferation and thus in altered cell-cycle kinetics [43].

In order to understand the functional outcome of miRNA dysregulation in FSHD, the derived up- and down-regulated target genes were functionally clustered into biological processes. This approach when applied to healthy muscle differentiation evidenced two important features of myogenesis: cell cycle and muscle development. Effectively, as muscle differentiation proceeds, sustained by the up-regulation of myogenic markers (due to the down-regulation of the corresponding miRNA regulators), the cell proliferation program must slow down due to the up-regulation of miRNA controlling genes involved in this process. An opposite trend of the two biological processes was found to characterize FSHD myogenesis. In fact, down-regulated genes were essentially involved in the regulation of striated muscle tissue development, and no regulation of cell cycle was observed. Thus in FSHD cells miRNA dysregulation affects two important aspects of differentiation leading to a defect in myogenesis. These data are in agreement with previously reported studies [9,10,20].

By the NGS approach we derived that FSHD myogenesis is characterized by a profound dysregulation of miRNA expression showing the involvement of at least 38 known miRNAs, including the myomiRs and possibly three novel miRNAs, but excluding small RNAs previously reported to derive from the D4Z4 array [14]. This and previous works have clearly demonstrated that FSHD cells are characterized by a global dysregulation of mRNA, miRNA and protein expression essentially affecting the myogenic process [9,10,11,21,24,44].

The up-regulation of the last *DUX4* gene in individual showing a reduced numbers (≤8) of D4Z4 repeats at 4q35 combined with a specific molecular signature (4A(159/161/168) DUX4 polyadenylation signal (PA) haplotype) is supposed to underlie FSHD pathophysiology [6]. However, it has been recently reported that 1.3% of healthy individuals carry the same molecular signature and 19% of subjects affected by FSHD do not carry alleles with eight or fewer D4Z4 repeats [8]. Furthermore, a dysregulation of genes involved in myogenesis has been recently observed in FSHD fetuses; importantly, the DUX4-fl pathogenic transcript was detected in both FSHD and control samples [45], as well as in unaffected individuals, but not in all FSHD cases [8]. These data suggest that the molecular basis of FSHD might not be simply based on the overexpression of the single *DUX4* gene, but rather from a cascade of dysregulation mediated by the D4Z4 array contraction. This structural alteration, as previously shown, might induce conformational changes in the 4q35 region itself, and perhaps elsewhere in the human genome [46,47]. Furthermore, in the dysregulation cascade could also play a role lncRNAs, such as DBE-T [7].

Conclusions

By using the NGS approach, we derived the complete pattern of miRNAs regulating *in vitro* control and FSHD myogenesis. In addition to confirming previously reported FSHD-related miRNAs, we identified additional known and novel miRNAs that are differentially expressed between FSHD and control myogenesis and thus potentially contributing to the FSHD pathogenic mechanism. In general, the comparison of control and FSHD myogenesis reveals that the dystrophy is characterized by a complex alteration of miRNA expression, which also includes the significant down-regulation of myomiRs at late stages of differentiation, thus essentially affecting muscle differentiation and development.

Thus, the full range of molecular alteration(s) at the basis of FSHD is not yet fully deciphered and the miRNA profiling we derive could represent a novel molecular signature for FSHD that includes diagnostic biomarkers and possibly therapeutic targets.

Materials and Methods

Cell lines

Primary FSHD and control cell lines were obtained from Myobank-AFM (Institut de Myologie-Groupe Hospitalier Pitié-Salpetrière, Paris) and Boston Biomedical Research Institute (BBRI, Senator Paul D. Wellstone Muscular Dystrophy Cooperative, Research Center for FSHD). Six cell lines derived from biopsies of different healthy and FSHD muscles including vastus, tensor fascia lata, quadriceps femoris (controls) and ilio-psoas and rhomboid (FSHD) (Table S1) were used for deep small RNA sequencing. In addition, to these cell lines, five control and four FSHD cell lines from deltoid and biceps [48] (Table S1) were used to validate deep sequencing data by qRT-PCR. FSHD primary cell lines were derived from biopsies of mild or not affected muscles and showed a D4Z4 array contraction ranging from 5.9 to 28 kb as determined by Southern Blot after EcoRI/BnlI digestion. The results reported below were derived by the analysis of all the cell lines listed in Table S1, comprising nine controls and seven FSHD and thus including also the cells used for NGS. Control and FSHD myoblasts were at low population doubling (from 2 to 7) and highly comparable for the expression of the muscular marker Desmin (96–97%) and the proliferation marker Ki67 (62–65%), as determined by immunofluorescence (Fig. S1). Furthermore, control and FSHD cell lines showed a comparable extent of differentiation as demonstrated by the down-regulation of the proliferation marker Ki67 (by immunofluorescence) and of MYF5 (by qRT-PCR), and by the up-regulation of MYOG (by qRT-PCR), MYOD (by Western blot) and MHC (by qRT-PCR and Western blot), as well as a comparable extent of fusion index (40–45%) (Fig. S1). In addition, FSHD and control myoblasts and myotubes appeared similar when analyzed by immunofluorescence. The cell lines used for NGS originated results in the average comparable to those shown in Fig. S1. Cells were cultured as described in guidelines of BBRI and Cheli et al [9].

Immunofluorescence, image acquisition and analysis

Cell immunofluorescence was performed as described [49], with antibodies specific for Desmin (rAb, Sigma Aldrich), ki67 (rAb, Vector) and sarcomeric myosin MHC (MF20, from Developmental Studies Hybridoma Bank). Appropriate secondary antibodies conjugated with Alexa 488 (green, Cell Signalling) or Alexa 568 (red; Cell Signalling) were used for fluorescence detection, Nuclei were stained with Hoechst Stain Solution (H6024, SIGMA).

Fluorescent images were taken on confocal laser scanning microscope (Zeiss Lsm 01, Biorad mrc 600, Biorad 1024) using 12× magnification. Images showing double or triple fluorescence were separately acquired using appropriate filters, and the different layers were merged with ImageJ software.

For all control and FSHD cell lines used in this study, the quantification of Desmin and ki67 positive cells has been performed on myoblasts and myotubes. Furthermore, for all control and FSHD cell lines, the absolute fusion index has been calculated as the percentage of MHC-positive nuclei over total number of nuclei after 7 days in differentiation medium.

An average value was determined by counting cells (200–300 cells/field) in at least 5 microscopic fields per sample at 12× magnification.

RNA isolation and deep sequencing

Total RNA was isolated with the mirVana miRNA isolation kit (cat.# AM1560, Life Technologies) from myoblast cell lines derived from 3 FSHD patients and 3 control subjects, before and after *in vitro* differentiation. RNA was quantified by Nanodrop spectrophotometer (Thermo Scientific) and its integrity was evaluated on an Experion automated electrophoresis system (Bio-Rad); all samples had a RNA Quality Indicator (RQI) value ≥9.

20 micrograms of total RNA were used for PAGE purification of small RNA molecules shorter than 35 nucleotides, adaptor ligation, and small RNA library preparation. The obtained libraries were sequenced on a HiSeq 2000 platform (Illumina) at BGI, Hong Kong, giving approximately 12 million high quality reads per sample (submitted to SRA database under acc. number SRP034654).

Sequencing data analysis

MicroRNA differential expression analysis was performed using R/Bioconductor, by following the workflow implemented in the oneChannelGUI interface [50,51]. Briefly, adaptor sequences were trimmed from fastq files using a specific perl script, and then sequences were aligned to the reference human miRBase v.20 precursor dataset (www.mirbase.org) using bowtie 1.0.0. Data were filtered for count threshold (>8 reads in 50% of samples analyzed) and pairwise comparisons of differential miRNA expression were performed using DEseq ($\log_2$FC ≥ 2; p-value< 0.05). Hierarchical clustering of differentially expressed miRNA was performed with dChip (version 2010.01; https://sites.google.com/site/dchipsoft/).

Identification of novel miRNAs

After excluding all reads that matched known small RNA classes annotated in miRBase v.20 (known miRNAs) and Rfam (e.g. tRNA, snRNA, snoRNA), putative novel miRNAs were predicted using mireap (http://sourceforge.net/projects/mireap/). The program predicts novel miRNAs from deep sequenced small RNA libraries by taking into consideration miRNA biogenesis, sequencing depth, and structural features (hairpin structure and stability) to improve the sensitivity and specificity of miRNA identification. Among predicted novel miRNAs, we considered as plausible candidates those matching the following criteria: 1) the detection in several samples (at least 2 out of 3 samples of one or more experimental groups); 2) the mature miRNA had sufficient sequence support (at least a mean of 10 reads for each experimental group); 3) the sequence did not match to known miRNAs in miRBase v.20.

Quantitative Real-time PCR

Quantitative RT-PCR (qRT-PCR) analysis was performed on 7900 HT Fast Real-Time PCR System (Applied Biosystems) by TaqMan small RNA Assays to validate the miRNA sequencing data. The miRNA specific probes were from Applied Biosystems. 150 ng RNA was reverse transcribed by TaqMan MicroRNA Reverse Transcription Kit (cat.# 4366596; Applied Biosystems) at 16°C for 30 min, 42°C for 30 min and 85°C for 5 min. Each amplicon was analyzed in duplicate in 96-well plates. TaqMan small RNA Assays reactions were performed following manufacturer's protocol (cat.# 4440048; Applied Biosystems). RNU48 was used for normalization. Thermal cycling conditions for real time PCR were 2 min at 95°C, followed by 40 cycles at 95°C for 10 s and 60°C for 30 s. Results were analyzed using the comparative $2^{-\Delta\Delta Ct}$ method. qRT-PCR experiments for MYF5, MYOG, MHC and GAPDH gene expression analysis were performed as described [9]. The statistical analysis was performed using a two-tail unpaired t-test and the error bars on the graphs are referred to standard deviation. qRT-PCR probes and primers are listed in Table S2.

Derivation of target genes

The putative miRNAs target genes were predicted by TargetScan Human (http://www.targetscan.org/) [52]. The prediction tool is based on different parameters such as complementarity to the seed region, 3′ complementarity, local AU content, position contribution and conservation in different species [22]. Predicted target genes were then filtered on the basis of their inverse correlation with the expression of mRNAs of two different chip analysis on Affymetrix human exon 1.0 ST array [9,10], using a FC ≥ 1.5 and a p-value < 0.05.

Pathway and functional annotation analysis

The derived predicted target genes, inversely correlated to the miRNAs expression, were subjected to the analysis of Gene Ontology terms (biological processes) by DAVID (Database for Annotation, Visualization and Integrated Discovery, v6.7) [53,54]. The target genes were mapped to the GO annotation dataset, and the enriched biological processes were extracted using the EASE score, a modified Fisher exact p-value.

Protein extracts and Immunoblot analysis

Cells were collected in RIPA Buffer (50 mM TrisHCl pH = 7,4, 150 mM NaCl, 0,1% SDS, 0,5% Deoxycholate Sodium, 1% NP-40 and protease inhibitor cocktail 1X-cat.# P2714-1BTL, Sigma MO, USA), and centrifuged 15 minutes at 13000 rpm at 4°C to discard cellular debris. Sample preparation and Western blot analyses were performed as described in Pisconti et al [55]. After electrophoresis, polypeptides were electrophoretically transferred to nitrocellulose filters (Thermo Scientific) and antigens revealed by the respective primary Abs and the appropriate secondary Abs, through autoradiography using enhanced chemiluminescence (LiteAblot Plus, cat.# EMP011005, Euroclone). In Western blot analyses, primary antibodies against MHC (MF20, from Developmental Studies Hybridoma Bank), MYOD (cat.# sc-31942, Santa Cruz) and housekeeping gene GAPDH (cat.# G8795; Sigma) were used.

Supporting Information

Figure S1 Characterization of control and FSHD myoblasts cell lines. A) Example of immunostaining experiment on proliferating and differentiated primary myoblasts (control: MX01010MBS; FSHD: MX04309MBS). Images have been taken at confocal laser scanning microscope at 12× magnification. Nuclei were stained with Hoescht (blue). Panels I–IV show localization of Desmin and Ki67 in proliferating myoblasts; panels I–II show immunostaining experiment using the polyclonal anti-Desmin (red); panels II–IV show immunostaining experiment using the polyclonal anti-Ki67 (red). Panels V–VIII show co-localization of Desmin or Ki67 and MHC on differentiated primary myoblasts: panels V–VI show immunostaining with polyclonal anti-Desmin and monoclonal anti-MHC (Ab-Desmin-red and Ab-MHC-green); panels VII–VIII show immunostaining with polyclonal anti-Ki67 and monoclonal anti-MHC (Ab-Ki67-red and Ab-MHC-green). Scale bar = 100 μm. B) Percentage of Desmin and Ki67 positive cells in myoblasts and myotubes after 7 days of differentiation derived from immunostaining with appropriate antibodies (Ab-Desmin and Ab-Ki67). Results are expressed as mean±SD of independent experiments performed on all cell lines described in Table S1. C) Absolute fusion index was determined at day 7 of differentiation (D7), counting the percentage of MHC- positive nuclei over the total number of nuclei. An average value was determined by counting cells in at least 5 microscopic fields (200–300 cells/field). Results are expressed as mean±SD of independent experiments performed on all cell lines (see Table S1). *p<0.05. D) Myogenic differentiation was evaluated by qRT-PCR analysis for MYF5, MYOG, MHC expression. All data points were calculated in triplicate as gene expression relative to endogenous GAPDH expression. Data are represented as the mean±SD of independent experiments performed on all cell lines described in Table S1. GM: growth medium; 7D: seven days of differentiation. *p<0.05, **p<0.01. E) Example of Western blot analysis with specific antibodies against MYOD and MHC in control and FSHD myoblasts at different time points during myogenic differentiation (GM: growth medium; 3D: three days of differentiation; 7D: seven days of differentiation). GAPDH protein level was used as an internal loading control. Graphs show mean values ±SD obtained from the ratio of densitometric values of protein/GAPDH bands. Data are representative of independent experiments performed on all cell lines described in Table S1. The Western blot in E shows a representative experiment (control: MX01010MBS; FSHD: MX04309MBS). *p<0.05, **p<0.01.

Figure S2 Scatter plots of the reads of miRNAs modulated in control and FSHD myogenesis. C1: MX01010MBS; C2: MX03609MBS; C3: MX01110MBS, Control cell lines; F1:MX00409MBS; F2: MX03010MBS; F3:MX04309MBS, FSHD cell lines (see Table S1).

Figure S3 Authentication of NGS data by qRT-PCR. qRT-PCR analysis of myomiRs (miR-1, miR-133a and miR-206)

during control and FSHD myogenesis at 0 and 7 days of differentiation on the three control and three FSHD cell lines used in the NGS experiment (MX01010MBS; MX03609MBS; MX01110MBS; MX00409MBS; MX03010MBS; MX04309MBS). GM: growth medium; 7D: seven days of differentiation. *p<0.05; **p<0.01.

Table S1 Primary myoblasts cell lines used in this study. Cell lines have been obtained from Myobank-AFM Istitut de Myologie (Paris)*and Boston Biomedical Research Institute (BBRI, Boston).

Table S2 Taqman probes and primers used in qRT-PCR experiments.

Table S3 List of microRNAs modulated in control myogenesis resulting by DEseq analysis.

Table S4 List of microRNAs modulated in FSHD myogenesis resulting by DEseq analysis.

Table S5 List of microRNAs modulated in FSHD vs control myotubes resulting by DEseq analysis.

Table S6 Potentially "validated" targets. List of predicted target genes of miRNAs modulated in control myogenesis, filtered on GSE26061 [9] and GSE26145 [10].

Table S7 Potentially "validated" targets. List of predicted target genes of miRNAs modulated in FSHD myogenesis, filtered on GSE26061 [9] and GSE26145 [10].

Table S8 Novel miRNAs predicted by mireap.

Acknowledgments

We would like to thank Dr. J. Chen for the cell lines used in the study. We are grateful to Cristina D'Orlando for her technical support and to Donatella Barisani for critical reading of the manuscript.

Author Contributions

Conceived and designed the experiments: RM EG VC SF. Performed the experiments: VC SF GMS RP FR. Analyzed the data: VC SF GMS RP. Contributed reagents/materials/analysis tools: FR. Wrote the paper: RM EG.

References

1. Padberg GW, Lunt PW, Koch M, Fardeau M (1992) Diagnostic criteria for facioscapulohumeral muscular dystrophy. Neuromuscl Disord 1:231–4
2. Tupler R, Gabellini D (2004) Molecular basis of facioscapulohumeral muscular dystrophy. Cell Mol Life Sci 61:557–66
3. Wijmenga C, Hewitt JE, Sandkuijl LA, Clark LN, Wright TJ, et al. (1992) Chromosome 4q DNA rearrangements associated with facioscapulohumeral muscular dystrophy. Nature Genetics 2:26–30
4. Yamanaka G, Goto K, Ishihara T, Oya Y, Miyajima T, et al. (2004) FSHD-like patients without 4q35 deletion. J Neurol Sci 219:89–93
5. Arashiro P, Eisenberg I, Kho AT, Cerqueira AM, Canovas M, et al. (2009) Transcriptional regulation differs in affected facioscapulohumeral muscular dystrophy patients compared to asymptomatic related carriers. Proc Natl Acad Sci USA 106:6220–5
6. Lemmers RJ, van der Vliet PJ, Klooster R, Sacconi S, Camaño P, et al. (2010) A unifying genetic model for facioscapulohumeral muscular dystrophy. Science 329:1650–3

7. Cabianca DS, Casa V, Bodega B, Xynos A, Ginelli E, et al. (2012) A long ncRNA links copy number variation to a polycomb/trithorax epigenetic switch in FSHD muscular dystrophy. Cell 149:819–31
8. Scionti I, Greco F, Ricci G, Govi M, Arashiro P, et al. (2012) Large-scale population analysis challenges the current criteria for the molecular diagnosis of facioscapulohumeral muscular dystrophy. Am J Hum Genet 90:628:35
9. Cheli S, François S, Bodega B, Ferrari F, Tenedini E, et al. (2011) Expression profiling of FSHD-1 and FSHD-2 cells during myogenic differentiation evidences common and distinctive gene dysregulation patterns. PLoS One 6(6)
10. Tsumagari K, Chang SC, Lacey M, Baribault C, Chittur SV, et al. (2011) Gene expression during normal and FSHD myogenesis. BMC Med Genomics 4:67
11. Rahimov F, King OD, Leung DG, Bibat GM, Emerson CP Jr, et al. (2012) Transcriptional profiling in facioscapulohumeral muscular dystrophy to identify candidate biomarkers. Proc Natl Acad Sci USA 109:16234–9

12. Eisenberg I, Eran A, Nishino I, Moggio M, Lamperti C, et al. (2007) Distinctive patterns of microRNA expression in primary muscular disorders. Proc Natl Acad Sci USA 104:17016–21

13. Ge Y, Chen J (2011) MicroRNAs in skeletal myogenesis. Cell Cycle 10:441–8

14. Snider L, Asawachaicharn A, Tyler AE, Geng LN, Petek LM, et al. (2009) RNA transcripts, miRNA-sized fragments and proteins produced from D4Z4 units: new candidates for the pathophysiology of facioscapulohumeral dystrophy. Hum Mol Genet 18:2414–30

15. Bartel DP (2009) MicroRNAs: target recognition and regulatory functions. Cell 136:215–33

16. Kloosterman WP, Plasterk RH (2006) The diverse functions of microRNAs in animal development and disease. Dev Cell 11:441–50

17. Mendell JT (2005) MicroRNAs: critical regulators of developmment, cellular physiology and malignancy. Cell Cycle 4:1179–84

18. Zhao Y, Samal E, Srivastava D (2005) Serum response factor regulates a muscle-specif microRNA that targets Hand2 during cardiogenesis. Nature 436:214–20

19. Chen JF, Mandel EM, Thomson JM, Wu Q, Callis TE, et al. (2006) The role of microRNA-1 and microRNA-133 in skeletal muscle proliferation and differentiation. Nat Genet 38:228–33

20. Dmitriev P, Barat A, Polesskaya A, O'Connell MJ, Robert T, et al. (2013) Simultaneous miRNA and mRNA transcriptome profiling of human myoblasts reveals a novel set of myogenic differentiation-associated miRNAs and their target genes. BMC Genomics 14:265

21. Dmitriev P, Stankevicins L, Ansseau E, Petrov A, Barat A, et al. (2013) Defective regulation of microRNA target genes in myoblasts from facioscapulohumeral dystrophy patients. J Biol Chem 288:34989–5002

22. Witkos TM, Koscianska E, Krzyzosiak WJ (2011) Practical Aspects of microRNA Target Prediction. Current Molecular Medicine 11:93–109

23. Guo H, Ingolia NT, Weissman JS, Bartel DP (2010) Mammalian microRNAs predominantly act to decrease target mRNA levels. Nature 466:835–40

24. Winokur ST, Chen YW, Masny PS, Martin JH, Ehmsen JT, et al. (2003) Expression profiling of FSHD muscle supports a defect in specific stages of myogenic differentiation. Hum Mol Genet 12:2895–907

25. Osborne RJ, Welle S, Venance SL, Thornton CA, Tawil R (2007) Expression profile of FSHD supports a link between retinal vasculopathy and muscular dystrophy. Neurology 68:569–77

26. Ryu S, Joshi N, McDonnell K, Woo J, Choi H, et al. (2011) Discovery of novel human breast cancer microRNAs from deep sequencing data by analysis of pri-microRNA secondary structures. PLoS One 6:e16403

27. Xu G, Wu J, Zhou L, Chen B, Sun Z, et al. (2010) Characterization of the small RNA transcriptomes of androgen dependent and independent prostate cancer cell line by deep sequencing. PLoS One 5:e15519

28. Landgraf P, Rusu M, Sheridan R, Sewer A, Iovino N, et al. (2007) A mammalian microRNA expression atlas based on small RNA library sequencing. Cell 129:1401–14

29. Luo W, Nie Q, Zhang X (2013) MicroRNAs involved in skeletal muscle differentiation. J Genet Genomics 40:107–16

30. Ferreboeuf M, Mariot V, Bessières B, Vasiljevic A, Attié-Bitach T, et al. (2014) DUX4 and DUX4 downstream target genes are expressed in fetal FSHD muscles. Hum Mol Genet 23:171–81

31. McCarthy JJ (2008) MicroRNA-206: the skeletal muscle-specific myomiR. Biochim Biophys Acta 1779:682–91

32. Townley-Tilson WH, Callis TE, Wang D (2010) MicroRNAs 1, 133, and 206: critical factors of skeletal and cardiac muscle development, function, and disease. Int J Biochem Cell Biol 42:1252–5

33. Callis TE, Deng Z, Chen JF, Wang DZ (2008) Muscling through the microRNA world. Experimental biology and medicine 233:131–138

34. Wu J, Ji X, Zhu L, Jiang Q, Wen Z, et al. (2013) Up-regulation of microRNA-1290 impairs cytokinesis and affects the reprogramming of colon cancer cells. Cancer Lett 329:155–63

35. Nohata N, Hanazawa T, Kinoshita T, Inamine A, Kikkawa N, et al. (2013) Tumor-suppressive microRNA-874 contributes to cell proliferation through targeting of histone deacetylase 1 in head and neck squamous cell carcinoma. Br J Cancer 108:1648–58

36. Cheng AM, Byrom MW, Shelton J, Ford LP (2005) Antisense inhibition of human miRNAs and indications for an involvement of miRNA in cell growth and apoptosis. Nucleic Acids Res 33:1290–7

37. Dong S, Xiong W, Yuan J, Li J, Liu J, et al. (2013) MiRNA-146a regulates the maturation and differentiation of vascular smooth muscle cells by targeting NF-kB expression. Mol Med Rep 8:407–12

38. Morin RD, O'Connor MD, Griffith M, Kuchenbauer F, Delaney A, et al. (2008) Application of massively parallel sequencing to microRNA profiling and discovery in human embryonic stem cells. Genome Res 18:610:21

39. Yang Q, Hua J, Wang L, Xu B, Zhang H, et al. (2013) MicroRNA and piRNA profiles in normal human testis detected by next generation sequencing. PLoS One 8:e66809

40. Schee K, Lorenz S, Worren MM, Günther CC, Holden M, et al. (2013) Deep sequencing the microRNA transcriptome in colorectal cancer. PLoS One 8:e66165

41. Rao PK, Kumar RM, Farkhondeh M, Baskerville S, Lodish HF (2006) Myogenic factors that regulate expression of muscle-specific microRNAs. Proc Natl Acad Sci USA 103:8721–6

42. Goljanek-Whysall K, Sweetman D, Münsterberg AE (2012) microRNAs in skeletal muscle differentiation and disease. Clin Sci (Lond) 123:611–25

43. Alexander MS, Casar JC, Motohashi N, Myers JA, Eisenberg I, et al. (2011) Regulation of DMD pathology by an ankyrin-encoded miRNA. Skelet Muscle 1:27.

44. Celegato B, Capitanio D, Pescatori M, Romualdi C, Pacchioni B, et al. (2006) Parallel protein and transcript profiles of FSHD patient muscles correlate to the D4Z4 arrangement and reveal a common impairment of slow to fast fibre differentiation and a general deregulation of MyoD-dependent genes. Proteomics 6:5303–21

45. Broucqsault N, Morere J, Gaillard MC, Dumonceaux J, Torrents J, et al. (2013) Dysregulation of 4q35- and muscle-specific genes in fetuses with a short D4Z4 array linked to facio-scapulo-humeral dystrophy. Hum Mol Genet 22:4206–14

46. Ottaviani A, Schluth-Bolard C, Rival-Gervier S, Boussouar A, Rondier D, et al. (2009) Identification of a perinuclear positioning element in human subtelomeres that requires A-type lamins and CTCF. EMBO J 28:2428–36

47. Bodega B, Ramirez GD, Grasser F, Cheli S, Brunelli S, et al. (2009) Remodeling of the chromatin structure of the facioscapulohumeral muscular dystrophy (FSHD) locus and upregulation of FSHD-related gene 1 (FRG1) expression during human myogenic differentiation. BMC Biol 7:41

48. Homma S, Chen JC, Rahimov F, Beermann ML, Hanger K, et al. (2012) A unique library of myogenic cells from facioscapulohumeral muscular dystrophy subjects and unaffected relatives: family disease and cell function. Eur J Hum Genet 20:404–10

49. Brunelli S, Tagliafico E, De Angelis FG, Tonlorenzi R, Baesso S, et al. (2004) Msx2 and necdin combined activities are required for smooth muscle differentiation in mesoangioblast stem cells. Circ Res 94: 1571–8

50. Sanges R, Cordero F, Calogero RA (2007) oneChannelGUI: a graphical interface to Bioconductor tools, designed for life scientists who are not familiar with R language. Bioinformatics 23:3406–8

51. Cordero F, Beccuti M, Arigoni M, Donatelli S, Calogero RA (2012) Optimizing a massive parallel sequencing workflow for quantitative miRNA expression analysis. PLoS One 7:e31630

52. Lewis BP, Shih IH, Jones-Rhoades MW, Bartel DP, Burge CB (2003) Prediction of mammalian microRNA targets. Cell 115:787–798

53. da Huang W, Sherman BT, Lempicki RA (2009) Systematic and integrative analysis of large gene lists using DAVID bioinformatics resources. Nat Protoc 4:44–57

54. da Huang W, Sherman BT, Lempicki RA (2009) Bioinformatics enrichment tools: paths toward the comprehensive functional analysis of large gene lists. Nucleic Acids Res 37:1–13

55. Pisconti A, Brunelli S, Di Padova M, De Palma C, Deponti D, et al. (2006) Follistatin induction by nitric oxide through cyclic GMP: a tightly regulated signaling pathway that controls myoblast fusion. J Cell Biol 172:233–44

7

Maternal Socioeconomic Status and the Risk of Congenital Heart Defects in Offspring

Di Yu[℧], **Yu Feng**[℧], **Lei Yang, Min Da, Changfeng Fan, Song Wang, Xuming Mo***

Department of Cardiothoracic Surgery, Nanjing Children's Hospital, Nanjing Medical University, Nanjing, China

Abstract

Background: We conducted this meta-analysis to address the open question of a possible association between maternal socioeconomic status and congenital heart defects (CHDs).

Methods: We searched MEDLINE and EMBASE from their inception to January 1, 2014 for case-control and cohort studies that assessed the association between maternal socioeconomic status and the risk of CHDs. Study-specific relative risk estimates were polled according to random-effect or fixed-effect models.

Results: From 3343 references, a total of 31 case-control studies and 2 cohort studies were enrolled in this meta-analysis, including more than 50,000 cases. We observed that maternal educational attainment, family income and maternal occupation were negatively associated with an 11% (pooled RR = 1.11, 95% CI: 1.03, 1.21), 5% (pooled RR = 1.05, 95% CI: 1.01, 1.09) and 51% (pooled RR = 1.51, 95% CI: 1.02, 2.24) increased risk of CHDs, respectively. In a subgroup analysis by geographic region, the results were inconsistent for the European region (RR = 1.29, 95% CI: 0.99–1.69) and USA/Canada region (RR = 1.06, 95% CI: 0.97, 1.16) in maternal educational attainment.

Conclusion: In summary, this meta-analysis suggests that a lower degree of maternal socioeconomic status is modestly associated with an increased risk of CHDs. However, further investigations are needed to confirm the association.

Editor: Graham R. Wallace, University of Birmingham, United Kingdom

Funding: The authors have no support or funding to report.

Competing Interests: The authors have declared that no competing interests exist.

* Email: mohsuming15@sina.com

℧ These authors contributed equally to this work.

Introduction

Congenital heart defects (CHDs) are the most common group of congenital malformations, affecting almost 1% of live births throughout the world [1]. CHDs represent approximately one-third of all congenital anomalies and are the leading cause of perinatal mortality [2]. Although cardiovascular diagnostics and cardiothoracic surgery have achieved massive breakthroughs over the past century, leading to an increased survival of newborns with CHDs, the etiology of most congenital heart defects is still unknown. Several chromosomal anomalies, certain maternal illnesses, and prenatal exposures to specific therapeutic drugs are recognized risk factors. It is difficult to establish the role of a single factor because the cause of a defect is believed to be multifactorial in many cases, including the combination of environmental teratogens with genetic and chromosomal conditions [3]. A review published in 2007 provided a summary of currently available literature on noninherited risk factors that might alter the risk of CHDs [4]. Moreover, CHDs include several distinct subtypes (e.g., conotruncal defects, left ventricular outflow track defects, and septal defects), and there is a potential for etiologic heterogeneity.

Therefore, it is not surprising that studies for categories of CHDs report different or opposite results.

Various approaches to the conceptualization and measurement of socioeconomic status (SES) have been taken, reflecting both different theoretical orientations and the exigencies of conducting studies. In our study, we used the most common measures, indexes, and ecological measures of SES, which is typically characterized by educational attainment, family income level and occupational prestige. According to the International Standard Classification of Occupations (ISCO-08) [5], skill level is used as the criterion for dividing occupations into groups and can be defined as a function of the complexity and range of tasks and duties to be performed in an occupation. This ranking of occupations consists of 4 groups ranging from a low to a high level. Lower SES often has connections with health-damaging lifestyles that result in the development of poor dietary habits and show the influence of behaviors related to physical activity and smoking [6–10]. Previous studies have reported that lower SES increases the risk of diabetes mellitus and cardiovascular disease [6,11–14]. Recently, there has been a steep increase in the number of maternal SES studies with CHDs as the primary health

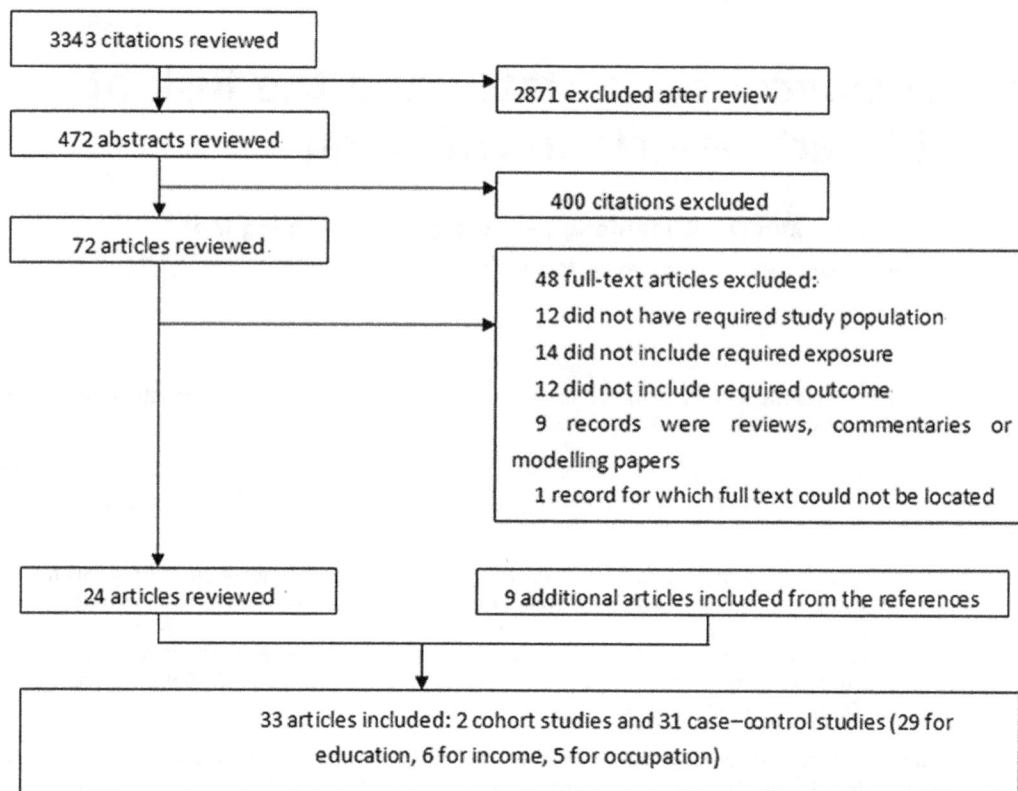

Figure 1. References searched and selection of studies in the meta-analysis.

outcome, with several studies showing positive associations and others providing null results.

An increasing number of studies to date have focused on the association between maternal SES and CHDs; however, the results have been ambiguous, perhaps because of inadequate sample sizes. Therefore, we conducted a meta-analysis to quantitatively assess the effect of maternal SES on CHDs.

Materials and Methods

Literature search

A computerized literature search was conducted by two independent investigators (Yu and Feng) in MEDLINE and EMBASE from their inception to January 1, 2014. We searched relevant studies using the following strategy: ("Socioeconomic Status" OR "Social Class" OR "Middle Class Population" OR "Caste" OR "education" OR "occupation" OR "income") AND ("abnormalities" OR "birth defects" OR "congenital anomaly" OR "malformations" OR "congenital malformations" OR "congenital heart defect" OR "Heart Abnormality" OR "Malformation of heart" or "CHD") AND ("maternal" OR "mother" OR "periconceptional" OR "pregnant" OR "gestation"). In addition, we conducted a search with a broader range on environmental teratogens and CHDs and checked the references in relevant retrieved and review articles. In this way, we identified information about other related studies.

Eligibility Criteria

We selected articles that (1) were original epidemiologic studies (i.e., case–control and cohort), (2) examined the association between maternal SES and CHDs overall or any one of the

CHD subtypes in infants, (3) were published in the English language, (4) reported RRs (i.e., risk ratios or odds ratios) and associated 95% confidence intervals (CIs) or had raw data available, and (5) defined CHDs or one of the CHD subtypes as an outcome. Articles that reported results from more than one population were considered separate studies. Non-peer-reviewed articles, experimental animal studies, ecological assessments, correlational studies and mechanistic studies were excluded.

Data extraction

Data extraction was conducted separately by two reviewers (Yu and Feng) working independently. If differences of opinion arose, these were resolved by discussion between the two. The studies meeting the inclusion criteria were reviewed to retrieve information of interest including study characteristics (i.e., authors, year of publication, geographic region, periods of data collection, study design, sample size, case classification, source of exposure data, and maternal SES (family income, occupational prestige, and education attainment)) and to record reported effect estimates and the associated 95% confidence interval (CI). If effect estimates were not available, raw data were extracted. For original studies that reported risk estimates in association with SES according to more than one measure, each estimate was extracted and its own association with the specific SES then analyzed. The information on the country where the study was conducted was then classified according to the geographical area (USA/Canada, Europe, Asia and Africa) and the country's income level (high-income and middle or low income). The education attainment and family income were evaluated by the lowest vs. the highest reported in the enrolled studies. Information on occupational social class was collected in the evaluated studies and coded according to the

Table 1. Overall characteristics of included studies.

First Author	Year of Publication	Country	Study Design	No. of Cases [a]	Study Period	Maternal SES Measure
Stoll	1989	Europe	CC	801	1979–1986	Education
Tikkanen	1992	Europe	CC	408	1982–1983	Education
Pradat	1993	Europe	CC	1108	1982–1986	Occupation
Wasserman	1996	USA/Canada	CC	207	1987–1988	Education
Fixler	1998	USA/Canada	CC	89	-	Education; Income
Torfs	1999	USA/Canada	CC	385	1991–1993	Education; Income
Botto	2000	USA/Canada	CC	957	1968–1980	Education
Bassili	2000	Africa	CC	894	1995–1997	Education
Carmichael	2003	USA/Canada	CC	131	1987–1988	Education
Williams	2004	USA/Canada	CC	122	1968–1980	Education
McBride	2005	USA/Canada	CC	476	1991–2001	Education
Batra	2007	USA/Canada	CC	3489	1987–2003	Education
Yang	2008	USA/Canada	CC	397	1997–2000	Education; Income
Grewal	2008	USA/Canada	CC	323	1999–2003	Education
Malik	2008	USA/Canada	CC	3067	1997–2002	Education
Van Driel	2008	Europe	CC	292	2003–2008	Education
Liu	2009	Asia	CC	164	2004–2005	Education
Smedts	2009	Europe	CC	276	2003–2006	Education
Hobbs	2010	USA/Canada	CC	572	1998–2007	Education; Income
Kuciene	2010	Europe	CC	261	1995–2005	Education; Occupation
Long	2010	USA/Canada	CC	1576	1999–2004	Education
Van Beynum	2010	Europe	CC	611	1996–2005	Education
Agha	2011	USA/Canada	Cohort	28302	1994–2007	Education; Income
Alverson	2011	USA/Canada	CC	2525	1981–1989	Education
Karatza	2011	Europe	CC	157	2006–2009	Education
Materna-Kiryluk	2011	Europe	CC	1673	2005–2006	Education
Agopian	2012	USA/Canada	Cohort	563	1999–2008	Education
Lupo	2012	USA/Canada	CC	1907	1997–2002	Education; Occupation
Mateja	2012	USA/Canada	CC	237	1996–2005	Education
Patel	2012	USA/Canada	CC	187	1997–2005	Income
Vereczkey	2012	Europe	CC	302	1980–1996	Occupation
Padula	2013	USA/Canada	CC	822	1997–2006	Education
Vereczkey	2013	Europe	CC	77	1980–1996	Occupation

CC: case–control study; SES: socio-economic status.
[a]Reported number of cases and control subjects with available exposure information.

International Standard Classification of Occupations (ISCO-08) (unskilled (I), skilled (II), intermediate (III) and professional (IV)).

RR was used as the measure for the summary statistics of associations of maternal SES with CHD risk. To simplify the procedure, RR was used to represent all reported study-specific results from cohort studies and OR from case-control studies. RR estimates and 95% CIs were extracted from each study for CHDs overall and CHD subtypes. To augment the comparability of different SES categories in the studies, the lowest and highest SES categories were compared. We back-calculated the point estimate and 95% CI if the original study did not report the risk estimates in this order. If the original study did not report estimates in the form of RR or OR, we used standard equations to recalculate the risk estimates and 95% CI from the raw data presented in the study.

Statistical analysis

The strength of the association of maternal SES with CHD risk was evaluated by RR with a 95% CI. We calculated pooled RR and accompanying 95% CI for the lowest vs. the highest categories of both income and education. Occupation included 4 groups, from low to high level, and pooled RRs with 95% CIs were calculated for the first vs. the fourth level, the second vs. the fourth, and the third vs. the fourth.

Cochran Q and I^2 statistics were used to test for heterogeneity across studies [16]. If there was evidence of heterogeneity ($P<0.05$ or $I^2 \geq 50\%$), the random-effects model was used. This model provided a more appropriate summary effect estimate among heterogeneous study-specific estimates. If the study showed no evidence of heterogeneity, a fixed-effects analysis was used,

Study ID		ES (95% CI)	% Weight
Stoll (1989)		0.74 (0.46, 1.20)	2.05
Tikkanen (1992)		1.09 (0.37, 3.18)	0.52
Wasserman (1996)		1.29 (0.75, 2.20)	1.72
Fixed (1998)		2.07 (0.98, 4.40)	0.99
Trofs (1999)		1.20 (0.89, 1.63)	3.67
Bassili (2000)		1.40 (1.13, 1.73)	4.97
Botto (2000)		1.04 (0.82, 1.32)	4.57
Cavmichael (2003 TOF)		0.50 (0.20, 1.20)	0.73
Cavmichael (2003 TGA)		2.00 (0.80, 4.80)	0.73
Williams (2004)		0.93 (0.51, 1.69)	1.45
Mcbride (2005 AVS)		0.52 (0.30, 0.88)	1.72
Mcbride (2005 COA)		0.83 (0.61, 1.13)	3.59
Mcbride (2005 HLHS)		1.15 (0.78, 1.69)	2.76
Yang (2007 TOF)		1.20 (0.80, 1.70)	2.85
Yang (2007 TGA)		1.30 (0.80, 2.10)	2.03
Batra (2007)		0.99 (0.87, 1.12)	6.43
Grewal (2008)		0.72 (0.50, 1.02)	3.05
Malik (2008)		1.06 (0.91, 1.22)	6.09
van Driel (2008)		1.28 (0.81, 2.03)	2.18
Kuciene (2009)		4.02 (2.13, 7.62)	1.31
Liu (2009)		3.61 (1.04, 12.52)	0.40
Smedts (2009)		1.08 (0.69, 1.69)	2.26
Hobbs (2010)		1.19 (0.76, 1.85)	2.28
Long (2010 TA)		0.82 (0.46, 1.47)	1.53
Long (2010 TGA)		0.96 (0.76, 1.22)	4.59
van Beynum (2010)		1.24 (0.93, 1.66)	3.83
Agha (2011)		1.27 (1.21, 1.31)	7.50
Alverson (2011)		1.03 (0.89, 1.18)	6.18
Karatza (2011)		1.46 (0.69, 3.09)	1.00
Materna-kiryluk (2011)		1.25 (0.96, 1.62)	4.22
Agopian (2012)		1.10 (0.90, 1.30)	5.46
Lupo (2012)		1.17 (0.95, 1.43)	5.11
Mateja (2012)		1.29 (0.60, 2.76)	0.97
Padula (2013)		1.22 (0.63, 2.35)	1.25
Overall (I-squared = 61.3%, p = 0.000)		1.11 (1.03, 1.21)	100.00

NOTE: Weights are from random effects analysis

.0799   1   12.5

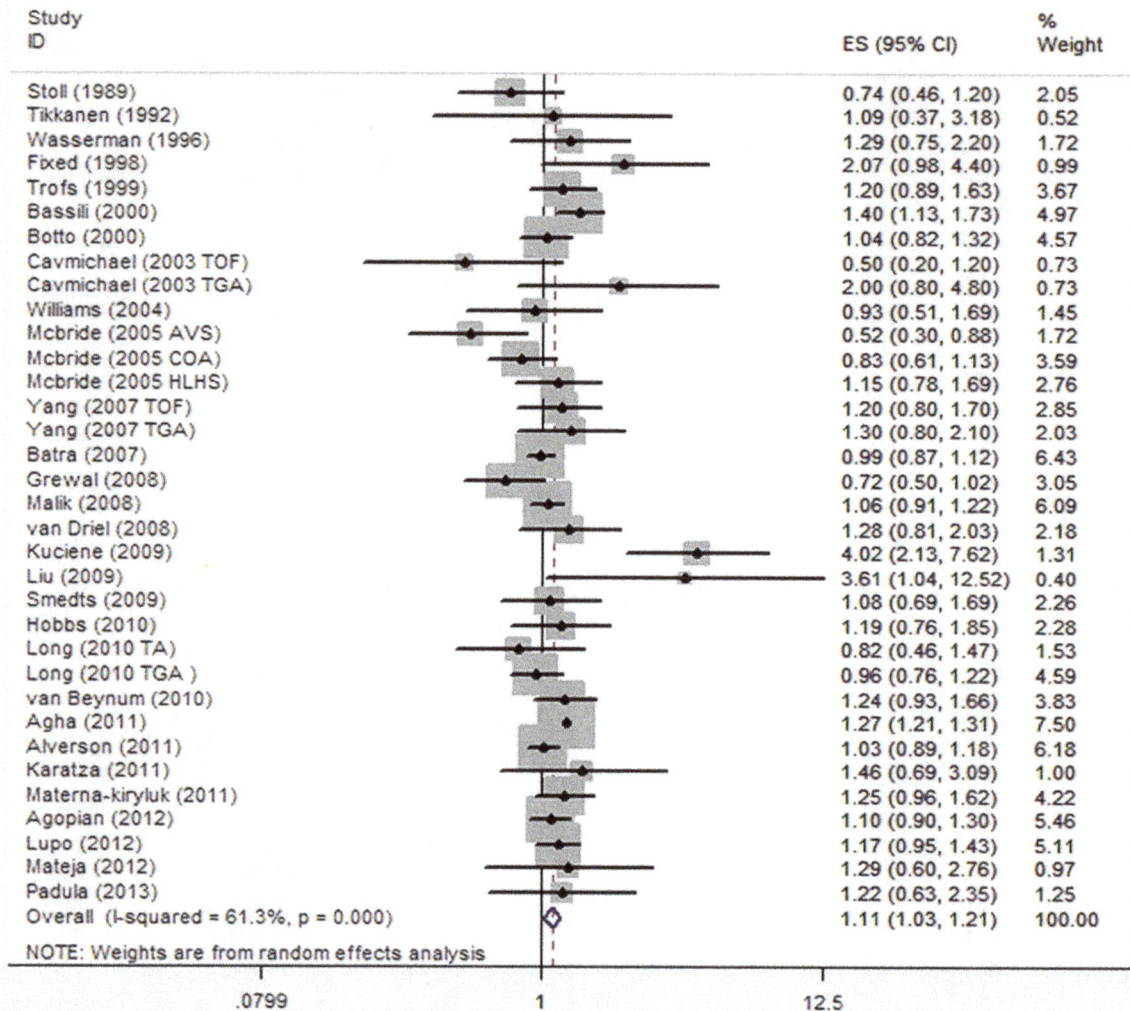

Figure 2. Estimates (95% CIs) of maternal educational attainment (lowest vs. highest category) and congenital heart defect (CHD) risk.

applying inverse variance weighting to calculate summary RR estimates [17].

Publication bias was assessed by visual inspection of a funnel plot with asymmetry, using both Egger's linear regression [18] and Begg's rank correlation [19] methods. Significant statistical publication bias was defined as a P value of <0.05 for the two above-mentioned tests. All statistical analyses were performed with STATA (version 11.0; StataCorp, College Station, Texas, USA).

Results

Study characteristics

The search strategy generated 3343 citations, among which 33 were identified in the final analysis for 53,358 incident cases (Figure 1). All of the studies were published from 1989 to 2013. There were 31 case–control studies [15,20–37,39–41,43–48,52–54] and 2 cohort studies [38,42]. The main characteristics of the included studies are shown in Table 1. In all, 20 studies [16,23–25,27–33,35,38,39,42–44,46,52,54] were conducted in the United States/Canada, 11 in Europe [20–22,34,36,37,40,41,45,47,53], and 2 in other regions (1 in China and 1 in Egypt) [26,35]; 29 studies were conducted in high-income countries and 4 in middle or low-income countries. Of the studies examined, 29 investigated the association of educational attainment with CHD risk [16,20,21,23–45,47,52,53], 6 examined the association of family income level with CHD risk [23,24,29,31,39,54], and 5 examined the association of occupational categories with CHD risk [22,34,44,46,48].

Overall results

The overall results of this meta-analysis provided evidence for a significant increase in the risk of CHDs among the lowest socioeconomic categories for all 3 socioeconomic indicators (figure 2–4). Heterogeneity was observed for education and occupation (p<0.01) (Table 2).

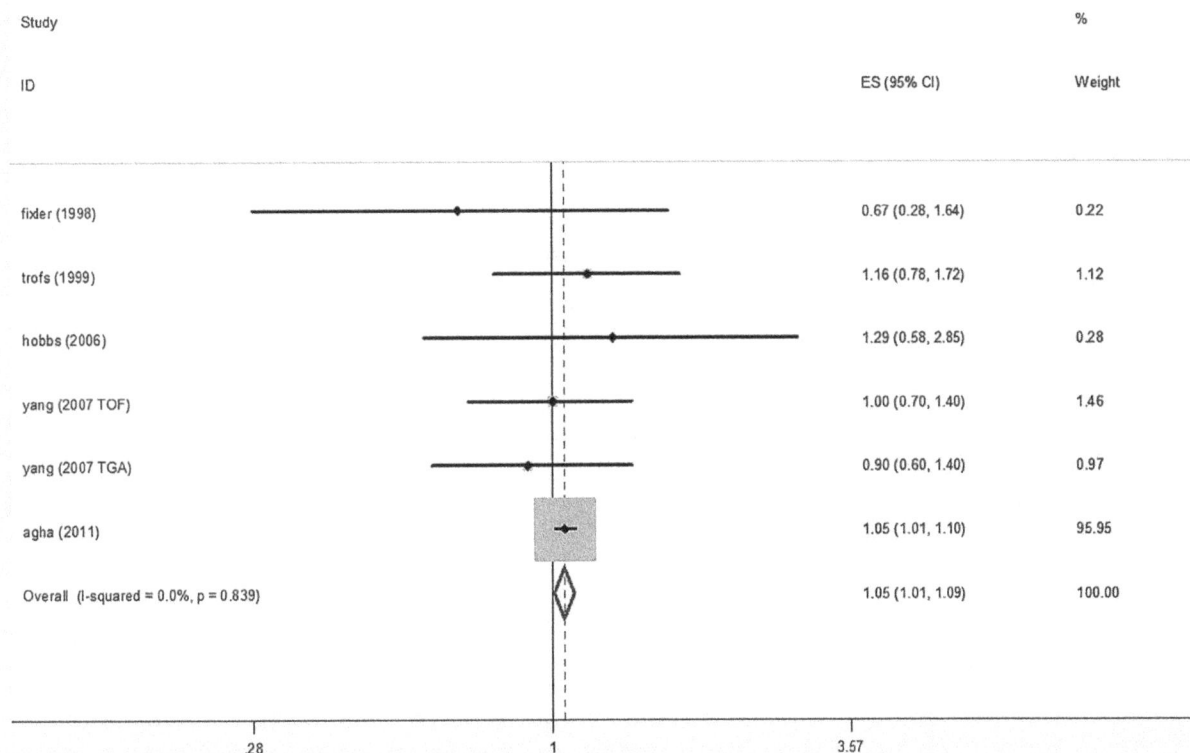

Figure 3. Estimates (95% CIs) of income level (lowest vs. highest category) and congenital heart defect (CHD) risk.

Association of SES categories with CHD risk

A total of 29 studies evaluated the association between maternal educational attainment and CHDs as a group (Table 2). We found that decreasing maternal educational attainment was associated with an 11% increase in the risk of CHDs (RR = 1.11, 95% CI: 1.03, 1.21) (Figure 2). Statistically significant heterogeneity was detected (Q = 85.32, $P<0.001$, $I^2 = 61.3\%$), with no publication bias (Begg's test: P = 0.68, Egger's test: P = 0.14) (Figure 5). After stratification by the countries' income group, a 10% increment was found among high-income countries (RR = 1.10, 95% CI: 1.01, 1.19); additionally, a significant increment (44%) was found among middle- or low-income countries in education (RR = 1.44, 95% CI: 1.17, 1.77). Furthermore, in a subgroup analysis by geographic region, null results were found for European (RR = 1.29, 95% CI: 0.99–1.69) and North American studies (RR = 1.06, 95% CI: 0.97, 1.16). If the analysis was limited to cohort studies (RR = 1.26, 95% CI: 1.21, 1.31) and sample sizes under 1000 (RR = 1.26, 95% CI: 1.05, 1.52), the results were generally consistent with the overall summary measure. Moreover, a significant association was observed in case-control studies (RR = 1.10, 95% CI: 1.01, 1.20), whereas no significant association was observed for sample sizes greater than 1000 (RR = 1.09, 95% CI: 0.99, 1.19). An increased incidence of CHD was observed if studies on family income were pooled (RR = 1.05, 95% CI: 1.01, 1.09) (Table 2), with no heterogeneity (Q = 2.49, P = 0.87, $I^2 = 0.0\%$) or publication bias (Egger's test: $P = 0.475$) (Figure 3). For the influence of occupation, Figure 4 shows the relationship between SES, categorized by occupation in classes one to four, and CHDs. In most of the studies reviewed, the risk of CHDs was higher in the lowest classes and affected the entire SES spectrum: the first vs. the fourth (RR = 1.51, 95% CI: 1.09–2.24), the second vs. the fourth (RR = 1.12, 95% CI: 1.00, 1.26) and the third vs. the fourth (RR = 1.18, 95% CI: 1.00, 1.39). In high-income countries, lower maternal occupational prestige was associated with a 7% increased risk of CHDs (for the first level vs. the fourth level, RR = 1.07, 95% CI: 0.92, 1.24); additionally, a significant increment (155%) was found for middle- or low-income countries (RR = 2.55, 95% CI: 1.76, 3.70).

Discussion

The purpose of the current study was to investigate the association between SES and CHDs. SES is customarily determined by educational achievement, family income and occupational prestige. To examine this association, we conducted a meta-analysis including 29, 6 and 5 studies of maternal education, family income and maternal occupation, respectively. Our findings indicated an increased incidence of CHDs among the lowest SES classifications in maternal education (11%), family income (5%) and maternal occupation (51%) compared with the highest classification of the corresponding SES. Moreover, for education stratified by the country's income group, a 10% increment was found among high-income countries. Furthermore, a significant increment (44%) was found among middle- or low-income

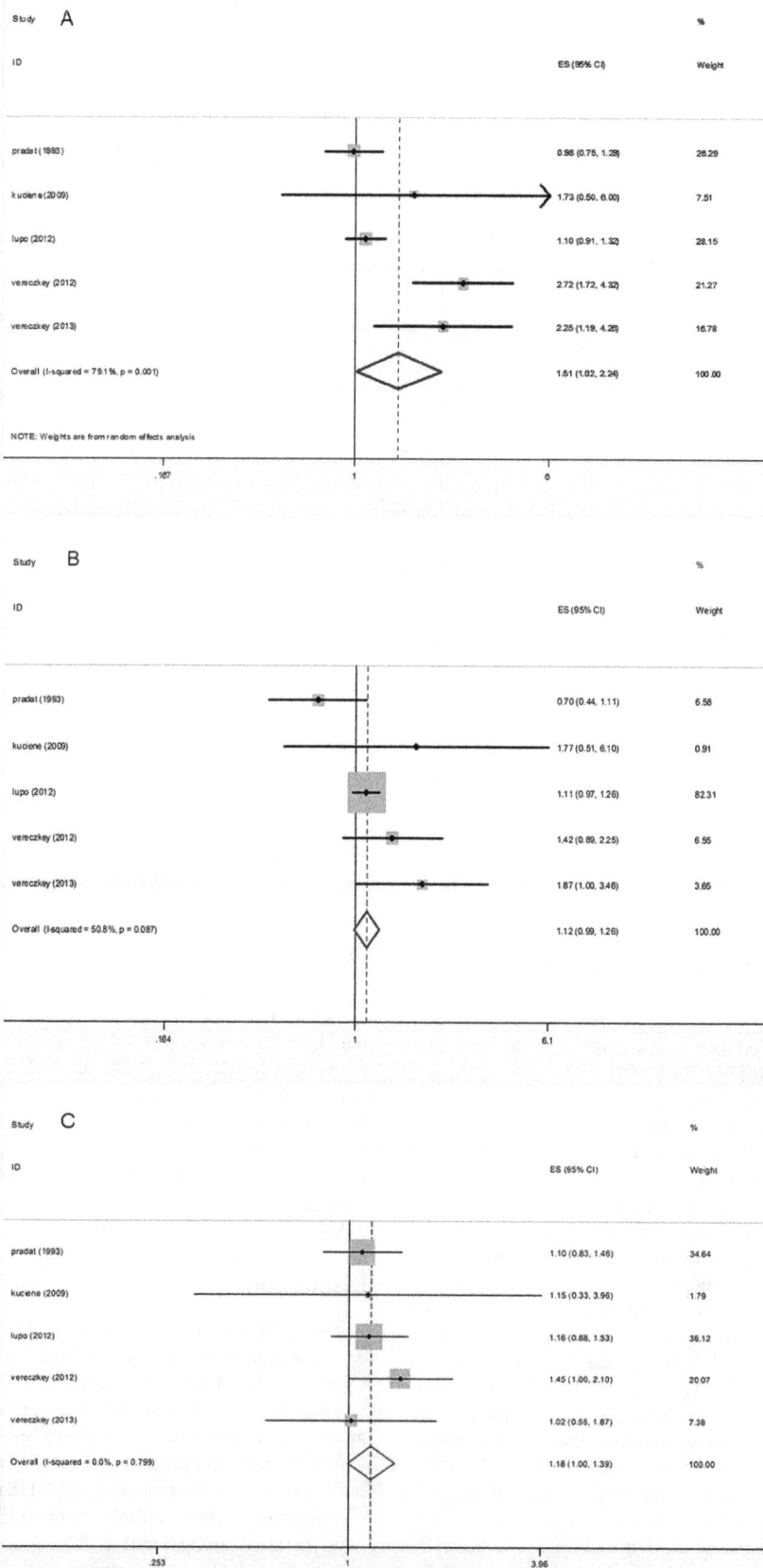

Figure 4. Estimates (95% CIs) of maternal occupational prestige and congenital heart defect (CHD) risk. A: Level I vs. level IV occupation; B: Level II vs. level IV occupation; C: Level III vs. level IV occupation.

Table 2. Pooled estimates for socioeconomic category and incidence of CHDs in series of subgroup analyses.

Subgroup analysis	No. of studies	No. of cases	Education	P [a]	I²(%)	No. of studies	No. of cases	Income	P [a]	I²(%)	No. of studies	No. of cases	Occupation	P [a]	I²(%)
Summary pooled estimate	29	51684	1.11(1.03–1.21)	<0.01	61.3	6	29932	1.05(1.01–1.09)	0.87	0.0	5	3655	1.51(1.02–2.24)	<0.01	79.1
Country's income group															
High	27	50626	1.10(1.01–1.19)	<0.01	61.3	6	29932	1.05(1.01–1.09)	0.87	0.0	3	3276	1.07(0.92–1.24)	0.59	0.0
Middle or low	2	1058	1.44(1.17–1.77)	0.14	53.8	0	-	-	-	-	2	379	2.55(1.76–3.70)	0.64	0.0
Geographical region															
USA/Canada	19	46147	1.06(0.97–1.16)	<0.01	62.6	6	29932	1.05(1.01–1.09)	0.87	0.0	1	1907	1.10(0.91–1.32)	-	-
Europe	8	4479	1.29(0.99–1.69)	0.01	61.4	0	-	-	-	-	4	1748	1.76(0.93–3.32)	<0.01	82.2
Asia	1	164	3.61(1.04–12.52)	-	-	0	-	-	-	-	0	-	-	-	-
Africa	1	894	1.40(1.13–1.73)	-	-	0	-	-	-	-	0	-	-	-	-
Sample size															
≤1000	9	2273	1.26(1.05–1.52)	0.33	12.7	3	1046	1.12(0.86–1.48)	0.81	0.0	1	302	2.73(1.72–4.32)	-	-
>1000	20	49411	1.09(0.99–1.19)	<0.01	69.2	3	28886	1.05(1.00–1.09)	0.86	0.0	4	3353	1.11(0.96–1.29)	0.11	50.2
Publication period															
Before 2004	8	3872	1.19(1.05–1.35)	0.07	44.2	2	89	1.06(0.74–1.52)	0.27	19.0	1	1108	0.98(0.75–1.29)	-	-
2004 or after	21	47812	1.10(1.00–1.20)	<0.01	66.2	3	29843	1.05(1.01–1.09)	0.87	0.0	4	2547	1.80(1.01–3.24)	<0.01	81.3
Design															
Case-control	27	22819	1.10(1.01–1.20)	<0.01	47.8	5	1630	1.02(0.86–1.22)	0.79	0.0	5	3655	1.51(1.02–2.24)	<0.01	79.1
Cohort	2	28865	1.26(1.21–1.31)	0.13	55.4	1	28302	1.05(1.01–1.10)	-	-	0	-	-	-	-

[a]P value for heterogeneity.

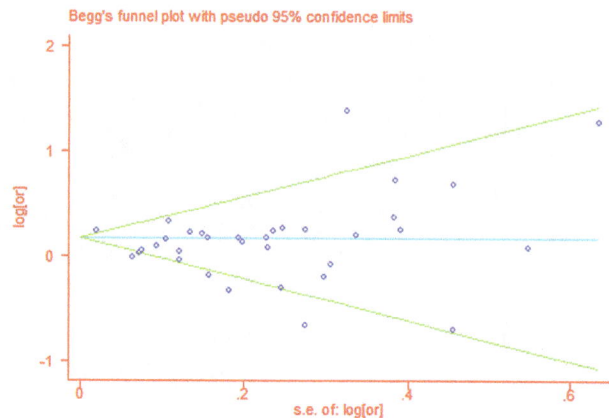

Figure 5. Begg's test of studies for maternal educational attainment and congenital heart defects (CHD).

countries. The results were inconsistent with the findings of the subgroup analysis of geographic regions, most likely indicating that in addition to individual SES, regional SES differences, especially between the developed countries and developing countries, may also have an influence on the risk of CHDs. Many factors could explain this finding. First, poor education is always accompanied by smoking [49] and diabetes mellitus [50], which have both been shown to be associated with an increased risk of CHDs [10,51]. Moreover, education has attracted much attention in developed countries, where everyone can receive a good education. Thus, there are no significant differences among individuals, and this uniformity may lead directly to the results. However, the opposite pattern occurs in developing countries. The unequal distribution of educational resources contributes to the obvious individual differences. However, the number of studies including a subgroup analysis of middle- or low-income countries is limited; for this reason, more studies need to be included in the meta-analysis to further confirm our findings. In regard to family income and occupational prestige, our results showed that the lowest income category and occupation levels increased the risk of CHDs compared with the highest. This result was consistent with the conclusions for education. Moreover, the percentage increment in risk decreased as the occupational level increased (Level I vs. Level IV RR = 1.51, 95% CI: 1.09–2.24; Level II vs. Level IV RR = 1.12, 95% CI: 1.00, 1.26; Level III vs. Level IV RR = 1.18, 95% CI: 1.00, 1.39). However, as a limited number of studies were included, the results need further confirmation. Moreover, note that case-control studies are always accompanied by selection and information biases. The studies that we examined are almost all case-control studies. Accordingly, the estimates of the relationship of SES in education to CHD risk based on case-control studies (RR = 1.10, 95% CI: 1.01, 1.20) are consistent with the conclusion drawn from the pooled analysis.

Several limitations of our study should be considered. First, a total of 31 case-control studies and 2 cohort studies were recruited in our meta-analysis, and we extracted our raw data principally from case-control studies, which are susceptible to selection and information biases. Therefore, our results cannot be viewed as an inevitable relationship, and further investigations including more high-quality studies are needed. Second, our meta-analysis was limited to studies published in English, but no evidence of publication bias was found, whereas heterogeneity exists in the component studies. This finding may reflect the differences among study designs and study populations as well as other unknown factors associated with the included studies. Moreover, as our study was limited to English-language publications, our results may have been affected by the lack of data from studies published in other languages, especially in the middle- or low-income countries, where SES may be associated with CHDs. Therefore, any general conclusions must be considered carefully. The third limitation of our meta-analysis is the possible differences in the classification and definition of SES among the examined studies. Countries' overall economic and educational levels have a significant impact on the SES categories of maternal educational achievement, family income and maternal occupational prestige. Meanwhile, family income cannot strictly be reported as a solely maternal characteristic and we used it to instead of maternal income, which would bring confounding factors. Finally, our small sample size for family income and maternal occupation levels may have been underpowered to detect any influence of SES on the risk of CHD. Additionally, lacking a large set of data, we did not conduct a subgroup analysis of CHD subtypes.

However, our study offers several important strengths. Due to the difficulty of evaluating the socioeconomic level and the lack of sufficient available literature, no meta-analysis had previously been performed to investigate the association of maternal SES with the risk of CHDs. Therefore, to our knowledge, this is the first meta-analysis to report an association between maternal SES and CHDs, including more than 50,000 cases. Moreover, our literature search was conducted on multiple databases, and the references in the relevant retrieved and review articles were fully scrutinized to obtain the missing data. Moreover, both Egger's linear regression and Begg's rank correlation tests showed no significant publication bias.

In summary, this study provides evidence of an association between low SES and an increased risk of CHDs, including maternal educational attainment, family income and maternal occupational prestige, whereas no clear relationship was found between socioeconomic status and CHDs in developed countries. Our findings could make public health policy focus more strongly on at-risk populations and could be used in the development of population-based prevention strategies to reduce the incidence and burden of CHDs, particularly for the regions with a lower level of economic development. Moreover, as previous studies have found a correlation between educational level and CHD risk factors [50,51], maternal education attainment appears to be the main target for preventing the development of CHDs.

Supporting Information

Checklist S1   PRISMA checklist.

Acknowledgments

We sincerely thank Dr. Hongcheng Zhu and Dr. Xi Yang for their help in improving this manuscript.

Author Contributions

Conceived and designed the experiments: XMM. Performed the experiments: DY YF XMM. Analyzed the data: YF DY MD LY. Contributed reagents/materials/analysis tools: LY MD CFF SW. Wrote the paper: DY YF.

References

1. Hoffman JI, Kaplan S (2002) The incidence of congenital heart disease. J Am Coll Cardiol 39: 1890–1900.
2. Boneva RS, Botto LD, Moore CA, Yang Q, Correa A, et al. (2001) Mortality associated with congenital heart defects in the United States: trends and racial disparities, 1979–1997. Circulation 103: 2376–2381.
3. Brent RL (2004) Environmental causes of human congenital malformations: the pediatrician's role in dealing with these complex clinical problems caused by a multiplicity of environmental and genetic factors. Pediatrics 113: 957–968.
4. Jenkins KJ, Correa A, Feinstein JA, Botto L, Britt AE, et al. (2007) Noninherited risk factors and congenital cardiovascular defects: current knowledge: a scientific statement from the American Heart Association Council on Cardiovascular Disease in the Young: endorsed by the American Academy of Pediatrics. Circulation 115: 2995–3014.
5. Ariza-Montes A, Muniz NM, Montero-Simo MJ, Araque-Padilla RA (2013) Workplace bullying among healthcare workers. Int J Environ Res Public Health 10: 3121–3139.
6. Kaplan GA, Keil JE (1993) Socioeconomic factors and cardiovascular disease: a review of the literature. Circulation 88: 1973–1998.
7. Gonzalez MA, Rodriguez Artalejo F, Calero JR (1998) Relationship between socioeconomic status and ischaemic heart disease in cohort and case-control studies: 1960–1993. Int J Epidemiol 27: 350–358.
8. Tusting LS, Willey B, Lucas H, Thompson J, Kafy HT, et al. (2013) Socioeconomic development as an intervention against malaria: a systematic review and meta-analysis. Lancet 382: 963–972.
9. Hackshaw A, Rodeck C, Boniface S (2011) Maternal smoking in pregnancy and birth defects: a systematic review based on 173 687 malformed cases and 11.7 million controls. Hum Reprod Update 17: 589–604.
10. Lee LJ, Lupo PJ (2013) Maternal smoking during pregnancy and the risk of congenital heart defects in offspring: a systematic review and metaanalysis. Pediatr Cardiol 34: 398–407.
11. Adler NE, Boyce T, Chesney MA, Cohen S, Folkman S, et al. (1994) Socioeconomic status and health. The challenge of the gradient. Am Psychol 49: 15–24.
12. Strike PC, Steptoe A (2004) Psychosocial factors in the development of coronary artery disease. Prog Cardiovasc Dis 46: 337–347.
13. Tamayo T, Christian H, Rathmann W (2010) Impact of early psychosocial factors (childhood socioeconomic factors and adversities) on future risk of type 2 diabetes, metabolic disturbances and obesity: a systematic review. BMC Public Health 10: 525.
14. Manrique-Garcia E, Sidorchuk A, Hallqvist J, Moradi T (2011) Socioeconomic position and incidence of acute myocardial infarction: a meta-analysis. J Epidemiol Community Health 65: 301–309.
15. Carmichael SL, Nelson V, Shaw GM, Wasserman CR, Croen LA (2003) Socioeconomic status and risk of conotruncal heart defects and orofacial clefts. Paediatr Perinat Epidemiol 17: 264–271.
16. Higgins JP, Thompson SG, Deeks JJ, Altman DG (2003) Measuring inconsistency in meta-analyses. BMJ 327: 557–560.
17. Woolf B (1955) On estimating the relation between blood group and disease. Ann Hum Genet 19: 251–253.
18. Egger M, Davey Smith G, Schneider M, Minder C (1997) Bias in meta-analysis detected by a simple, graphical test. BMJ 315: 629–634.
19. Begg CB, Mazumdar M (1994) Operating characteristics of a rank correlation test for publication bias. Biometrics 50: 1088–1101.
20. Stoll C, Alembik Y, Roth MP, Dott B, De Geeter B (1989) Risk factors in congenital heart disease. Eur J Epidemiol 5: 382–391.
21. Tikkanen J, Heinonen OP (1992) Occupational risk factors for congenital heart disease. Int Arch Occup Environ Health 64: 59–64.
22. Pradat P (1993) Maternal occupation and congenital heart defects: a case-control study. Int Arch Occup Environ Health 65: 13–18.
23. Fixler DE, Threlkeld N (1998) Prenatal exposures and congenital heart defects in Down syndrome infants. Teratology 58: 6–12.
24. Torfs CP, Christianson RE (1999) Maternal risk factors and major associated defects in infants with Down syndrome. Epidemiology 10: 264–270.
25. Botto LD, Mulinare J, Erickson JD (2000) Occurrence of congenital heart defects in relation to maternal mulitivitamin use. Am J Epidemiol 151: 878–884.
26. Bassili A, Mokhtar SA, Dabous NI, Zaher SR, Mokhtar MM, et al. (2000) Risk factors for congenital heart diseases in Alexandria, Egypt. Eur J Epidemiol 16: 805–814.
27. Williams IJ, Correa A, Rasmussen S (2004) Maternal lifestyle factors and risk for ventricular septal defects. Birth Defects Res A Clin Mol Teratol 70: 59–64.
28. McBride KL, Marengo L, Canfield M, Langlois P, Fixler D, et al. (2005) Epidemiology of noncomplex left ventricular outflow tract obstruction malformations (aortic valve stenosis, coarctation of the aorta, hypoplastic left heart syndrome) in Texas, 1999–2001. Birth Defects Res A Clin Mol Teratol 73: 555–561.
29. Hobbs CA, Cleves MA, Karim MA, Zhao W, MacLeod SL (2010) Maternal folate-related gene environment interactions and congenital heart defects. Obstet Gynecol 116: 316–322.
30. Batra M, Heike CL, Phillips RC, Weiss NS (2007) Geographic and occupational risk factors for ventricular septal defects: Washington State, 1987–2003. Arch Pediatr Adolesc Med 161: 89–95.
31. Yang J, Carmichael SL, Canfield M, Song J, Shaw GM, et al. (2008) Socioeconomic status in relation to selected birth defects in a large multicentered US case-control study. Am J Epidemiol 167: 145–154.
32. Grewal J, Carmichael SL, Ma C, Lammer EJ, Shaw GM (2008) Maternal periconceptional smoking and alcohol consumption and risk for select congenital anomalies. Birth Defects Res A Clin Mol Teratol 82: 519–526.
33. Malik S, Cleves MA, Honein MA, Romitti PA, Botto LD, et al. (2008) Maternal smoking and congenital heart defects. Pediatrics 121: e810–816.
34. Kuciene R, Dulskiene V (2010) Parental cigarette smoking and the risk of congenital heart septal defects. Medicina (Kaunas) 46: 635–641.
35. Liu S, Liu J, Tang J, Ji J, Chen J, et al. (2009) Environmental risk factors for congenital heart disease in the Shandong Peninsula, China: a hospital-based case-control study. J Epidemiol 19: 122–130.
36. Smedts HP, de Vries JH, Rakhshandehroo M, Wildhagen MF, Verkleij-Hagoort AC, et al. (2009) High maternal vitamin E intake by diet or supplements is associated with congenital heart defects in the offspring. BJOG 116: 416–423.
37. Long J, Ramadhani T, Mitchell LE (2010) Epidemiology of nonsyndromic conotruncal heart defects in Texas, 1999–2004. Birth Defects Res A Clin Mol Teratol 88: 971–979.
38. van Beynum IM, Kapusta L, Bakker MK, den Heijer M, Blom HJ, et al. (2010) Protective effect of periconceptional folic acid supplements on the risk of congenital heart defects: a registry-based case-control study in the northern Netherlands. Eur Heart J 31: 464–471.
39. Agha MM, Glazier RH, Moineddin R, Moore AM, Guttmann A (2011) Socioeconomic status and prevalence of congenital heart defects: does universal access to health care system eliminate the gap? Birth Defects Res A Clin Mol Teratol 91: 1011–1018.
40. Alverson CJ, Strickland MJ, Gilboa SM, Correa A (2011) Maternal smoking and congenital heart defects in the Baltimore-Washington Infant Study. Pediatrics 127: e647–653.
41. Karatza AA, Giannakopoulos I, Dassios TG, Belavgenis G, Mantagos SP, et al. (2011) Periconceptional tobacco smoking and isolated congenital heart defects in the neonatal period. Int J Cardiol 148: 295–299.
42. Materna-Kiryluk A, Wieckowska B, Wisniewska K, Borszewska-Kornacka MK, Godula-Stuglik U, et al. (2011) Maternal reproductive history and the risk of isolated congenital malformations. Paediatr Perinat Epidemiol 25: 135–143.
43. Agopian AJ, Moulik M, Gupta-Malhotra M, Marengo LK, Mitchell LE (2012) Descriptive epidemiology of non-syndromic complete atrioventricular canal defects. Paediatr Perinat Epidemiol 26: 515–524.
44. Lupo PJ, Symanski E, Langlois PH, Lawson CC, Malik S, et al. (2012) Maternal occupational exposure to polycyclic aromatic hydrocarbons and congenital heart defects among offspring in the national birth defects prevention study. Birth Defects Res A Clin Mol Teratol 94: 875–881.
45. Mateja WA, Nelson DB, Kroelinger CD, Ruzek S, Segal J (2012) The association between maternal alcohol use and smoking in early pregnancy and congenital cardiac defects. J Womens Health (Larchmt) 21: 26–34.
46. Vereczkey A, Kosa Z, Csaky-Szunyogh M, Urban R, Czeizel AE (2012) Birth outcomes of cases with left-sided obstructive defects of the heart in the function of maternal socio-demographic factors: a population-based case-control study. J Matern Fetal Neonatal Med 25: 2536–2541.
47. Padula AM, Tager IB, Carmichael SL, Hammond SK, Yang W, et al. (2013) Ambient air pollution and traffic exposures and congenital heart defects in the San Joaquin Valley of California. Paediatr Perinat Epidemiol 27: 329–339.
48. Vereczkey A, Kosa Z, Csaky-Szunyogh M, Czeizel AE (2013) Isolated atrioventricular canal defects: birth outcomes and risk factors: a population-based Hungarian case-control study, 1980-1996. Birth Defects Res A Clin Mol Teratol 97: 217–224.
49. Huisman M, Van Lenthe FJ, Giskes K, Kamphuis CB, Brug J, et al. (2012) Explaining socio-economic inequalities in daily smoking: a social-ecological approach. Eur J Public Health 22: 238–243.
50. Kavanagh A, Bentley RJ, Turrell G, Shaw J, Dunstan D, et al. (2010) Socioeconomic position, gender, health behaviours and biomarkers of cardiovascular disease and diabetes. Soc Sci Med 71: 1150–1160.
51. Lisowski LA, Verheijen PM, Copel JA, Kleinman CS, Wassink S, et al. (2010) Congenital heart disease in pregnancies complicated by maternal diabetes mellitus. An international clinical collaboration, literature review, and meta-analysis. Herz 35: 19–26.
52. Wasserman CR, Shaw GM, O'Malley CD, Tolarova MM, Lammer EJ (1996) Parental cigarette smoking and risk for congenital anomalies of the heart, neural tube, or limb. Teratology 53: 261–267.
53. van Driel LM, Smedts HP, Helbing WA, Isaacs A, Lindemans J, et al. (2008) Eight-fold increased risk for congenital heart defects in children carrying the nicotinamide N-methyltransferase polymorphism and exposed to medicines and low nicotinamide. Eur Heart J 29: 1424–1431.
54. Patel SS, Burns TL, Botto LD, Riehle-Colarusso TJ, Lin AE, et al. (2012) Analysis of selected maternal exposures and non-syndromic atrioventricular septal defects in the National Birth Defects Prevention Study, 1997–2005. Am J Med Genet A 158A: 2447–2455.

8

Characterization of Dystrophin Deficient Rats: A New Model for Duchenne Muscular Dystrophy

Thibaut Larcher[1,9,¶], Aude Lafoux[2,9,¶], Laurent Tesson[3], Séverine Remy[3], Virginie Thepenier[3], Virginie François[4], Caroline Le Guiner[4,5], Helicia Goubin[1], Maéva Dutilleul[1], Lydie Guigand[1], Gilles Toumaniantz[2], Anne De Cian[6], Charlotte Boix[6], Jean-Baptiste Renaud[6], Yan Cherel[1,†], Carine Giovannangeli[6], Jean-Paul Concordet[6], Ignacio Anegon[3*], Corinne Huchet[2*]

1 INRA, UMR703 APEX, Oniris, Atlantic Gene Therapies, Université de Nantes, Oniris, École nationale vétérinaire, agro-alimentaire et de l'alimentation, Nantes, France, 2 INSERM, UMR 1087/CNRS 6291 Institut du Thorax, Université de Nantes, Faculté des Sciences et des Techniques, Nantes, France, 3 INSERM, UMR 1064-Center for Research in Transplantation and Immunology, ITUN, CHU Nantes, Université de Nantes, Faculté de Médecine, Nantes, France, 4 INSERM, UMR 1089, Atlantic Gene Therapies, Thérapie génique pour les maladies de la rétine et les maladies neuromusculaires, Université de Nantes, Faculté de Médecine, Nantes, France, 5 Genethon, Evry, France, 6 INSERM, U1154, CNRS, UMR 7196, Muséum National d'Histoire Naturelle, Paris, France

Abstract

A few animal models of Duchenne muscular dystrophy (DMD) are available, large ones such as pigs or dogs being expensive and difficult to handle. *Mdx* (*X-linked muscular dystrophy*) mice only partially mimic the human disease, with limited chronic muscular lesions and muscle weakness. Their small size also imposes limitations on analyses. A rat model could represent a useful alternative since rats are small animals but 10 times bigger than mice and could better reflect the lesions and functional abnormalities observed in DMD patients. Two lines of *Dmd* mutated-rats (*Dmd*[mdx]) were generated using TALENs targeting exon 23. Muscles of animals of both lines showed undetectable levels of dystrophin by western blot and less than 5% of dystrophin positive fibers by immunohistochemistry. At 3 months, limb and diaphragm muscles from *Dmd*[mdx] rats displayed severe necrosis and regeneration. At 7 months, these muscles also showed severe fibrosis and some adipose tissue infiltration. *Dmd*[mdx] rats showed significant reduction in muscle strength and a decrease in spontaneous motor activity. Furthermore, heart morphology was indicative of dilated cardiomyopathy associated histologically with necrotic and fibrotic changes. Echocardiography showed significant concentric remodeling and alteration of diastolic function. In conclusion, *Dmd*[mdx] rats represent a new faithful small animal model of DMD.

Editor: Atsushi Asakura, University of Minnesota Medical School, United States of America

Funding: Funding was provided by Région Pays de la Loire through Biogenouest, IBiSA program, Fondation Progreffe and TEFOR (Infrastructures d'Avenir of the French goverment), and the integrative genomic facility of Nantes for sequencing experiments. The authors thank the Wolfson Centre for Inherited Neuromuscular Disease for kindly supplying dystrophin monoclonal antibody. The funders had no role in study design, data collection and analysis, decision to publish or preparation of the manuscript.

Competing Interests: The authors have declared that no competing interests exist.

* Email: ianegon@inserm.fr (IA); corinne.huchet2@univ-nantes.fr (CH)

9 These authors contributed equally to this work.

¶ These authors are first authors on this work.

† Deceased.

Introduction

DMD is a severe X-linked muscular dystrophy due to mutations of the *DMD* gene. It affects all voluntary muscles as well as the heart and breathing muscles in later stages. Despite recent promising new treatments the average life expectancy is severely reduced in DMD patients and a better understanding of the disease and faster evaluation of new treatments are needed [1].

Both large and small animal species deficient for dystrophin have been described and have been extremely useful for pre-clinical studies of DMD. Although they display more features of the human clinical phenotype than *mdx* mice, large dystrophin-deficient animals such as dogs [2] and pigs [3,4], suffer from individual variability and are costly and time consuming. *Mdx*

mice [5,6] have the advantage of low maintenance costs. Sufficient numbers of animals can also be easily characterized to reach high statistical power. On the other hand, *mdx* mice exhibit only minor clinical dysfunction [6] and their small size imposes limitations in the analysis of several aspects of the disease. Although each animal model has its own limitations, they have all been essential for the development of treatment strategies that target dystrophin absence, disease progression or muscle regeneration [7]. Nevertheless, new animal models are needed to help pre-clinical research on DMD.

We hypothesized that the rat could represent a useful model of DMD. One of its advantages of over mice is that its behavior is much better characterized. Rats have finer and more accurate motor coordination than mice and exhibit a richer behavioral display, including more complex social traits [8]. Rats have a convenient size since they are 10 times larger than mouse but are still a small laboratory animal model and allow studies with high statistical power. Until recently, the rat model lacked genetic engineering tools for introducing targeted genetic mutations. But in the last years, we and others have used in rats sequence-specific nucleases, such as meganucleases, zinc-finger nucleases, TALENs and CRISPRs/Cas9, to efficiently generate precise gene mutations [9–13].

To generate dystrophin-deficient rats, we generated TALENs for *Dmd* that were microinjected in rat zygotes allowing generation of two *Dmd^{mdx}* rat lines. The muscles of both lines displayed undetectable levels of dystrophin as evaluated by western blot analysis and less than 5% of dystrophin positive fibers by immunohistochemistry. At 3 months of age, forelimb, hindlimb and diaphragm muscles showed severe fiber necrosis and a strong regeneration activity. At 7 months of age, regeneration activity was decreased and muscle showed abundant peri- and endomysial fibrosis with some adipose tissue infiltration. Muscle strength and spontaneous activity were decreased and fatigue was a prominent finding of muscle function analysis. Cardiac muscle was also affected with necrosis and fibrosis and showed signs of progressive dilated cardiomyopathy. Echocardiography showed significant concentric remodeling and alteration of diastolic function. These lesions in skeletal muscle and heart closely mimic those observed in DMD patients. These results indicate that *Dmd^{mdx}* rats represent a new invaluable small animal model for pre-clinical research on DMD.

Materials and Methods

Animals

This study was approved by the Ethics Committee on Animal Experimentation of the Pays de la Loire Region, France, in accordance with the guidelines from the French National Research Council for the Care and Use of Laboratory Animals (Permit Numbers: CEEA-PdL-2011-45 and CEEA-PdL-01579.01). All efforts were made to minimize suffering. Sprague-Dawley (SD/Crl) rats were obtained from Charles River (L'Arbresle, France). The rats were housed in a controlled environment (temperature $21 \pm 1°C$, 12-h light/dark cycle). Before blood collection, animals were anesthetized with a mixture of ketamine (100 mg/kg, Imalgene, Merial, Lyon, France) and xylazine (10 mg/kg, Rompun, Bayer, Leverkusen, Germany). Rats were then sacrificed by intravenous administration of sodium pentobarbital (300 mg, Dolethal, Vetoquinol UK Ltd, Buckingham, UK). Just after sacrifice, the body weight (g) and the body length (cm) of each rat were determined to define the body mass index calculated as body weight/(body length)^2 (g/cm^2).

Design and production of TALE nucleases

TALE nucleases targeting *Dmd* exon 23 were designed to recognize the following sites on Rat genome Assembly (Rnor_5.0 - GCA_000001895.3/chrX: 51,878,333-52,510,293).

TCTGCAAAGCTCTTTGAAAGAGCAACAAAATGGCTT
CAACTATCTGAATGCCA.

TALE nucleases were produced, as previously described, by unit assembly method adapted from Huang et al. 2011 [14–16]. For each TALE nuclease subunit, the fragment containing the 16 RVD segment was obtained from single unit vectors: A (NI), T (NG), G (NN) and C (HD), derived from plasmids kindly provided by the laboratory of Dr B. Zhang (Peking University, China). The assembled TALE RVD (DMD-L: TCTGCAAAGCTCTTT-GAAA; DMD-R: TGGCATTCAGATAGTTGA) were subcloned in the pVax vector containing the 17th half RVD, the Δ152/+63 N- and C-terminal truncation points as described by Miller et al. 2011 [17] and contained the wt Fok I catalytic domains or FokI heterodimer-forcing mutations (ELD or KKR) [18] that were constructed by site-directed mutagenesis starting from Addgene plasmids 21872 et 21873 kindly made available by the Joung lab. Assembled TALE nucleases DNAs were cloned into the pVAX vector for mRNA production and *in vitro* cell transfection.

In vitro assay of TALE nucleases

Each subunit of TALE nuclease (0.75 and 1.5 µg each) was nucleofected into 4.10^5 C6 cells (Sigma) in 20 µL of V solution using AMAXA FF137 program. Cells were grown into 12 well dishes and collected 48 h after nucleofection. Genomic DNA was extracted with E.Z.N.A. Tissue DNA Kit (OMEGA Biotek). The genomic region encompassing the TALE nuclease target sites was amplified with the primers listed below, typically using 50 ng of genomic DNA, and phusion polymerase (New England Biolabs) following manufacturer instructions. Unpurified PCR product (10 µL out of 25 µL of PCR reaction) was supplemented with 10 µL of NEBuffer 2 (2X), melted, and annealed (5 min at 95°C, 95°C to 25°C at −0.5°C/30 sec, and 15 min at 4°C) to form heteroduplex DNA. The annealed DNA was treated with 1.5 units of T7 endonuclease I (New England BioLabs) for 20 min at 37°C and run on a 2.4% agarose gel.

Off-target mutation analysis

Potential off-target sites were identified by bioinformatics using the PROGNOS program [19] with parameters set for heterodimer-forcing TALEN subunits and looking for pairs of potential TALEN binding sequences with up to 6 mismatches to each cognate target sequences of the TALEN pair separated by spacer of 10 to 30 bp. Candidate off-target sites that were identified had at least a total of 8 mismatches to the TALEN target sequence in the DMD gene. We chose to further look for mutations at the 2

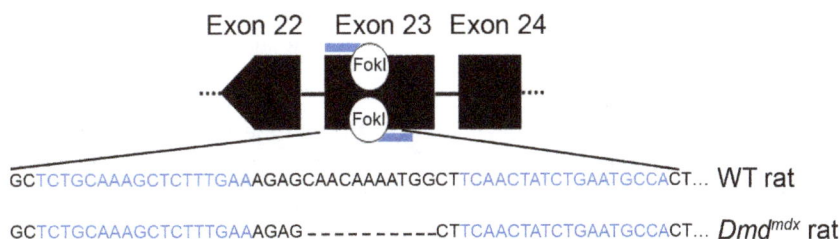

Figure 1. Targeted deletion in exon 23 of the rat *Dmd* gene via TALE nucleases. DNA cleavage takes place in the spacer sequence located between TALEN DNA binding sequences indicated in blue. *Dmd^{mdx}* rats carries a deletion of 11 nucleotides in exon 23 of the *Dmd* gene.

Table 1. Rat weight and size as well as isolated muscles weight and heart characteristics of Dmd^{mdx} male rats.

Age	3 month-old		7 month-old	
	WT	Dmd^{mdx}	WT	Dmd^{mdx}
Rat characteristics				
Body weight (g)	456.0±21.2	409.4±18.9	749.7±45.9	501.6±30.1*
Body length (cm)	25.9±0.4	25.5±0.3	28.9±0.4	27.1±0.4*
Body mass index (g/cm^2)	0.68±0.02	0.63±0.02	0.89±0.03	0.68±0.03*
Muscle weight				
Tibialis Anterior (mg)	751.8±41.2	822.9±50.2	1009.2±57.1	773.1±58.1*
Tibialis Anterior (mg/g)	1.65±0.02	2.00±0.07	1.35±0.07	1. 54±0.06*
Extensor Digitorum Longus (mg)	200.6±10.4	205.0±10.5	257.2±14.0	193.2±15.9
Extensor Digitorum Longus (mg/g)	0.44±0.01	0.50±0.01	0.34±0.02	0.38±0.01
Soleus (mg)	184.9±6.5	199.9±10.8	206.2±8.2	189.0±13.7
Soleus (mg/g)	0.41±0.01	0.49±0.02	0.28±0.01	0.38±0.02
Heart characteristics				
Weight (mg)	1379.5±84.7	1732.2±271.1	2049.8±72.6	1766.3±84.4*
Weight (mg/g)	3.02±0.13	4.21±0.58	2.75±0.11	3.55±0.16*
p/LV (%)	56.0±0.1	44.9±0.03 §	43.1±1.8	35.8±3.9 §
n	5	5	4	5

body mass index calculated as body weight/(body length)^2; (g/cm^2).
Values are means ± SEM; *: p<0.05 *vs* wild-type controls (WT); §: p<0.0001 *vs* WT. n, numbers of rats.

candidate sites lying on the X-chromosome because they could potentially cosegregate with the DMD mutations and affect the phenotype of the rats investigated. The candidate sites found on X-chromosomes were TttCATTCAGcTAtTTGAAATGGGAA-GACAGCACACTGATCCATTTttAAGAGgTTTGCAGc and TCTatAAcGCcCTTTaAAAATGGAATAAGATCCTTTGCAA GTGATTtAcCTATaTGAATGtgA with potential TALEN binding sites underlined and lowercase indicating mismatches to the TALEN DMD target sequence. The presence of mutations was tested using the T7 endonuclease E1 assay as detailed above using specific primers for PCR listed below. No mutations could be detected at candidate off-target sites from the X-chromosome in tail DNA from 6 F0 (including the founder rats #61 and #71) and 8 F2 rats derived from founder rat #61 (including 4 DMD mutant animals) using the T7 endonuclease 1 assay (data not shown).

Primer sequences:

DMDs CAAGTATGCATCGGTTAGTGTA
DMDas GCATCAATAACTTTGAGGGACT
XoffN1s CGAAATAGAGCTAGAATCCCCAGG
XoffN1as GTT TCC AAG GGA CAG ACA ACA CAG
XoffN2s CAG TCC TCT TTG CTC ATG TGC TAG
XoffN2as CCC TTG TGT GTG TGT GTG TGT ATG

In vitro transcription of TALE nucleases mRNA

As previously described, TALE nuclease plasmids were *in vitro* transcribed to mRNA and polyadenylated using the mMessage mMachine T7 Ultra kit (Ambion, Austin, TX) following the manufacturer protocol and purified using the MegaClear Kit (Ambion, Austin, TX), quantitated using a NanoDrop-1000 (Thermo Scientific) and stored at −80°C until use [12]. mRNAs encoding each monomer of TALE nucleases for each target sequence were mixed in TE 5/0.1 (5 mM Tris-Cl pH 7.5,

0.1 mM EDTA in RNase DNase free water) and stored at −80°C until use. mRNAs were diluted to the working concentration and kept on ice during one day micro-injection procedures and then discarded.

Microinjection of rat one-cell embryos

Prepubescent females (4–5 weeks old) were super-ovulated with pregnant mare serum gonadotropin (30 IU; Intervet, France) and followed 48 hours later with human chorionic gonadotropin (20 IU; Intervet, France) before breeding. Zygotes were collected for subsequent microinjection using a previously published procedure [12,20]. Briefly, a mixture of TALE nucleases mRNA was microinjected into the cytoplasm of fertilized one-cell stage embryos. Microinjected embryos were maintained under 5% CO_2 at 37°C until reimplantation. Surviving embryos were then implanted immediately in the oviduct of pseudo-pregnant females (0.5 dpc) and allowed to develop until term.

Analysis of Dmd mutation events

DNA from embryos or neonates was extracted from tail biopsy following treatment with Proteinase K as previously described [12]. To analyze mutations, we first used PCR followed by the T7 endonuclease I assay (as described above in the *in vitro* analysis of mutations) to identify rats carrying mutations and PCR products were sequenced.

Dystrophin messenger analysis

Total RNA was extracted from muscles with TRIzol reagent (Invitrogen) and 500 ng of this RNA was reverse transcribed using random primers (Invitrogen) and M-MLV reverse transcriptase (Invitrogen). Detection of the Dystrophin mRNA was done by 25 cycles of PCR1 followed by 30 cycles of PCR2 using 1 μl of ADNc or 1 μl of PCR1, 2,5 U of GoTaq DNA polymerase (Promega), 1,5 mM of MgCl2 and 0,2 μM of each following primers (PCR1:

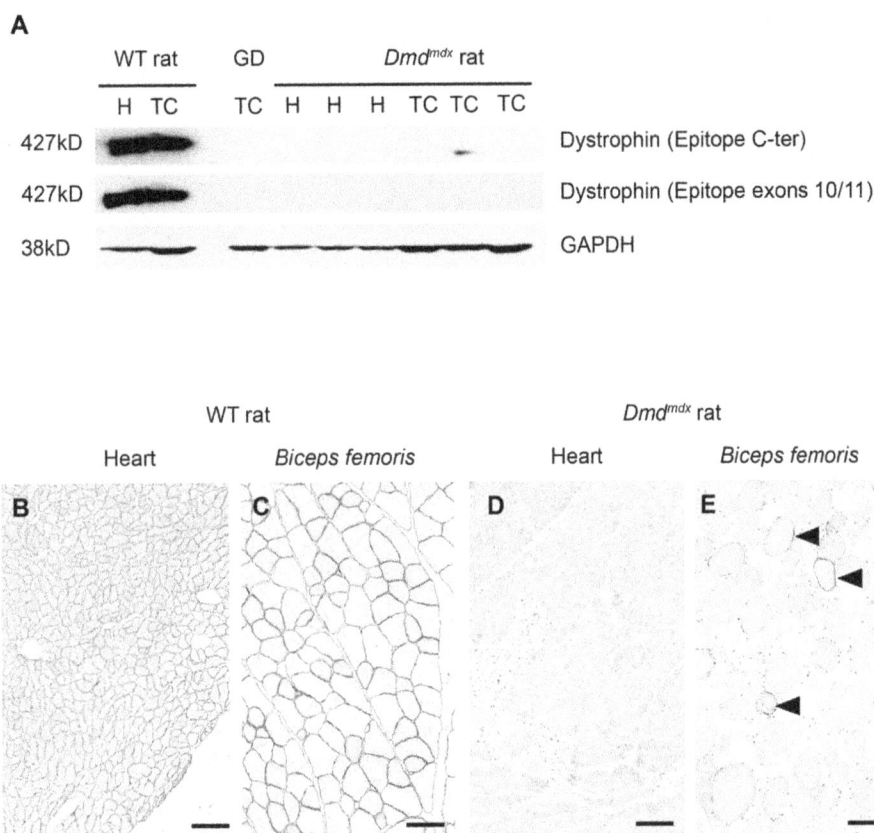

Figure 2. No dystrophin expression was detected in cardiac and skeletal muscles of *Dmd^mdx* rats. (*A*) Male 7 month-old rats of line 61, wild-type littermate controls (WT) and *Dmd^mdx* were sacrificed and biopsies from *tibialis cranialis* muscles (TC) and hearts (H) were harvested. Western-blot of total proteins (50 µg) was incubated with NCL-DYS2 and Manex1011C monoclonal antibodies (C-terminal and exons 10/11 epitopes, respectively). This revealed undetectable levels of the 427 kDa dystrophin band in line 61 *Dmd^mdx* rats. Muscle from a GRMD dog (GD) was used as negative control and samples from WT rats were used as positive controls. Staining with an anti-GAPDH polyclonal antibody validated equal protein loadings. (*B–E*) Heart and *biceps femoris* muscles from the same wild-type (*B* and *C*) and *Dmd^mdx* rats (*D* and *E*) were assessed for dystrophin expression using immunohistochemistry with Mandys110 monoclonal antibody (against exons 38–39 epitope). Compared to the subsarcolemmal expression of dystrophin in wild-type muscles, no dystrophin was detected in *Dmd^mdx* rats except for the presence in skeletal muscle of only rare scattered revertant positive fibers (arrowheads). Immunolabelling of dystrophin (*B–E*) Bar = 100 µm.

ex20ratDyst-F TCA GAC AAG CCT CAG AAC AA and ex26ratDyst-R AGTTTTATCCAAACCAGCCT annealing 50°C; PCR2:ex22ratDyst-F AAT GCG CTA TCA AGA GAC AA and ex24ratDyst-R TCT GCA CTG TTT GAG CTG TT annealing 55°C). Final PCR products were migrated on an 1.5% agarose gel and revealed with ethidium bromide staining (**Fig. S6**).

Western blot analysis

Total proteins from muscle were extracted using 400 µL of RIPA extraction buffer containing protease inhibitors (Roche) and ground with TissueLyser II (Qiagen). 50 µg of protein extracts were loaded on a 3–8% Tris-Acetate Precast polyacrylamide gel of NuPAGE Large Protein blotting kit (Invitrogen). After Red Ponceau staining, membranes were incubated with two different mouse anti-Dystrophin antibodies: NCL-DYS2 (1:100, Novocastra), MANEX 1011C (1:100, MDA Monoclonal Antibody Resource) [21] and with an anti-GAPDH antibody (1:10000, Imgenex). Detection was performed using a secondary anti-mouse IgG HRP-conjugated antibody (1:2000, Dako) or secondary anti-goat IgG HRP-conjugated antibody (1:2000, Dako). Immunoblots were revealed with ECL Western blotting substrate (Pierce) and exposed to ECL-Hyperfilm (Amersham).

Histopathological evaluation

After gross examination, *tibialis cranialis, extensor digitorum longus, biceps femoris* muscles of the hindlimb, *soleus* and *biceps brachii* muscles of the forelimb, diaphragm and heart were sampled and divided in two parts. Furthermore, *tibialis cranialis, extensor digitorum longus, soleus* and heart were weighted. The first part was frozen and 8-µm-thick serial sections were cut for immunohistology and histoenzymology assays, the second part was fixed in 10% neutral buffered formalin, embedded in paraffin wax, and 5-µm-thick sections were cut and routinely stained with hematoxylin eosin saffron for histopathological evaluation. Additional sections were stained with Picrosirius red stain for collagen, Gomori's trichrome and Alizarin red staining for calcium. All lesions were reported by a skilled pathologist certified by the European College of Veterinary Pathology.

Immunohistochemistry and histoenzymology assays

Immunohistochemical analysis involved the use of immunoperoxidase techniques on frozen sections. MANDYS110 clone 3H10, provided by the Wolfson Centre for Inherited Neuromuscular Disease, NCL-DYSB and NCL-DYS2 (Novocastra Laboratories, Newcastle on Tyne, UK) mouse monoclonal antibodies were used for dystrophin protein detection (1:50 for each ones). Other

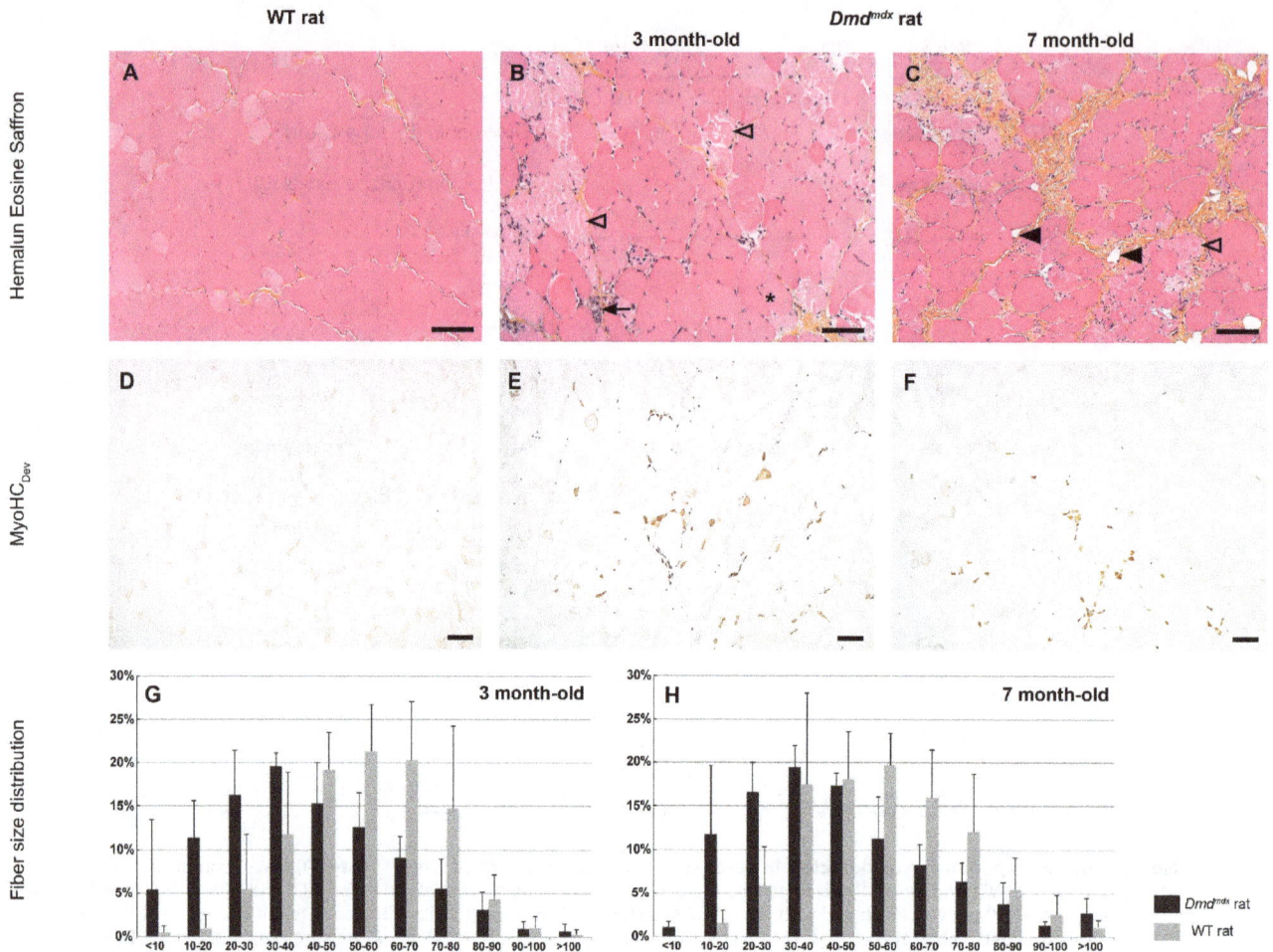

Figure 3. Severe and progressive muscle changes in *Dmd^mdx* rats. *Biceps femoris* muscle was sampled at 3 and 7 month-old from wild-type littermate controls (WT) and *Dmd^mdx* rats. Compared to controls (*A*), 3 month-old *Dmd^mdx* rat skeletal muscles (*B*) displayed individual fiber necrosis (open arrowhead) associated with foci of small newly regenerating centronucleated fibers (arrow). In addition, 7 month-old muscles displayed a progessive replacement of fibers by fibrosis (*) and fat (black arrowhead) tissue (*C*). The regenerative activity was assessed using a specific antibody against the Myosin Heavy Chain developmental isoform (MyoHC$_{Dev}$). No regenerative activity was observed in control rats (*D*) whereas newly regenerating fibers were numerous in *Dmd^mdx* rats (*E*) with a decrease in their number with age (*F*). Histological changes were quantified. Fiber minimal Ferret diameter was measured. A global switch towards lower sized fibers was observed in the fiber size distribution in mutated rats compared to controls at 3 and 7 month of age (*G–H*). Hemalun eosin saffron staining (*A–C*), immunolabelling of Myosin Heavy Chain developmental isoform (MyoHC$_{Dev}$) (*D–F*). Bar = 100 μm (*A–C*) and 200 μm (*D–F*).

primary antibodies used were IVD3₁A9 (1:50, Developmental Studies Hybridoma Bank, Iowa City, IA), NCL-g-Sarc (1:10, Novocastra Laboratories) and NCL-DRP2 (1:60, Novocastra Laboratories) for alpha-sarcoglycan, gamma-sarcoglycan, and utrophin detection respectively, NCL-MHCd (1:20, Novocastra Laboratories) for developmental myosin heavy chain isoform, anti-complement 5b-9 (1:250, Calbiochem, Strasbourg, France) and anti-CD3 (1:100, Dako, Glostrup, Denmark) for lymphocytes. Briefly, transverse cryosections were incubated in PBS with 5% normal goat serum (Dako) for 1 hour at room temperature. They were then incubated with primary antibody in 5% rat serum overnight at 4°C and with biotinylated secondary antibodies (E433, 1:300, Dako) in PBS with 5% rat serum for 1 hour. Bound antibodies were detected either with streptavidin (P397; Dako) and DAB Liquid Substrate (Dako) for immunoperoxidase. Fiber type was determined using histochemical myosin-ATPase reaction after preincubation at pH 4.2, 4.35, and 10.4 as previously described [22].

Histomorphometry

Morphometric analysis was done using a digital camera (Nikon DXM 1200; Nikon Instruments, Badhoevedorp, the Netherlands) combined with image-analysis software (NIS; Nikon). To determine the proportion of dystrophin-positive fibers in the *biceps femoris* muscle of 3 and 7 month-old *Dmd^mdx* rats (5 individuals in each group), a total of 1,000 fibers were counted in sections immunolabelled with Mandys110 antibody and the percentage of fibers expressing dystrophin was determined. To determine the percentage of MyoHC$_{Dev}$ fibers, at least 1,000 fibers were numbered on randomly selected microscopic fields. Fiber size was documented measuring minimal Ferret diameter on at least 250 fibers (292±61 per sample), on randomly selected microscopic fields in picrosirius red-stained sections. Fibrosis was determined as the ratio of areas rich in collagen using specific picrosirius staining on the total muscle area in an overall cross section. Repeatability was tested by the same operator measuring five times the same

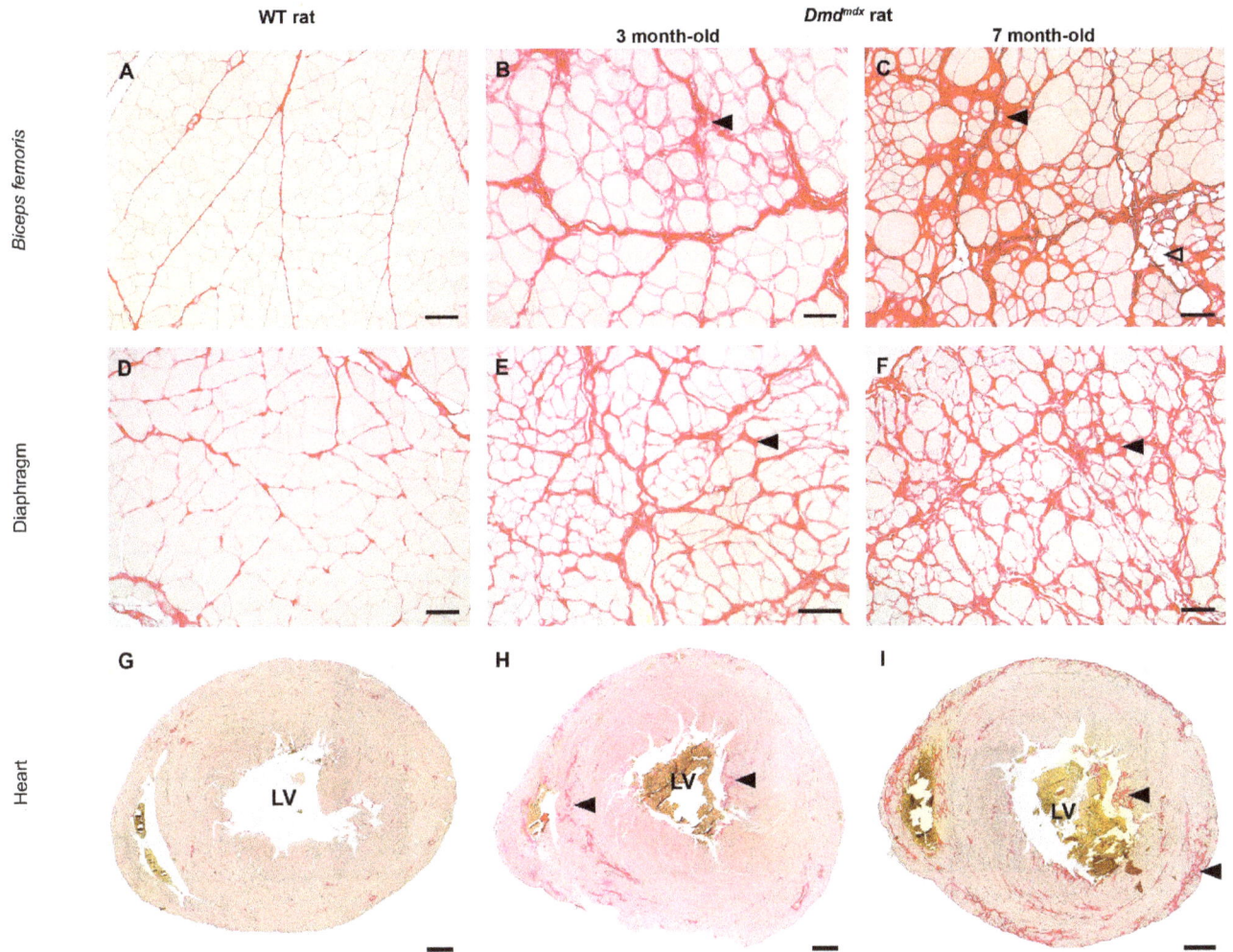

Figure 4. Progressive myofiber replacement by fibrotic and fat tissue in *Dmd^mdx* rats. A picrosirius red staining specific for fibrosis was performed on *biceps femoris* (A–C), respiratory (D–F) and heart muscle (G–I) samples obtained from wild-type littermate controls (WT) and 3 month-old and 7 month-old *Dmd^mdx* rats. Compared to control rats (left panel), a progressive increase in the amount of fibrotic tissue (black arrowhead) was noticed in 3 (mid panel) and 7 month-old *Dmd^mdx* rats (right panel). Note the focal presence of fat tissue infiltration (open arrowhead). In the heart, fibrosis was most marked in papillary muscle of the left ventricle (LV), in the septum and in the ventricular subepicardic area. Picrosirius red staining. Bar = 100 μm (A–F) and 1 mm (G–I).

sample. The intra-assay variations coefficients were always lower than 5%.

Serum creatine phosphokinase levels

Blood was collected just before sacrifice under anesthesia. Creatine phosphokinase (CK) activity was determined at the Central Laboratory of the National Veterinary School (E.N.V) at Nantes using a commercial kit (LaRoche Diagnostic). Results were expressed as CK levels [23].

Skeletal muscle function

The *in vivo* tests were performed in the same sequence for each rat, with equivalent time of rests between the tests [23].

Grip test. Rats were placed with their forepaws or four paws on a grid and were gently pulled backward until they released their grip [24]. A grip meter (Bio-GT3, BIOSEB, France), attached to a force transducer, measured the peak force generated. Five tests were performed in sequence with a short latency between each test, and the reduction in strength between the first and the last

determination was taken as an index of fatigue [24]. Results are expressed in grams (g) and normalized to the body weight (g/g).

Actimeter. The motor behavior was examined with an open field actimeter [25]. For this analysis, rats were individually placed in an automated photocell activity chamber (Letica model LE 8811, Bioseb, France) which consists of a plexiglass chamber (45 cm×45 cm×50 cm) surrounded by two rows of infrared photobeams. The first row of sensors was raised at a height of 3 cm for measuring horizontal activity and the second row placed above the animal for vertical activity. The spontaneous motor activity was measured for 5 min using a movement analysis system (Bioseb, France), which dissociates resting and ambulatory time (s), distance traveled (cm), stereotyped, and rearing movements.

Echocardiography

Two-dimensional (2-D) echocardiography was performed on rats using a MyLab70 Family - Ultrasound Systems – Esaote with a CA129 transducer [26]. In order to look for possible structural remodelling, the left ventricular end-diastolic diameter and the

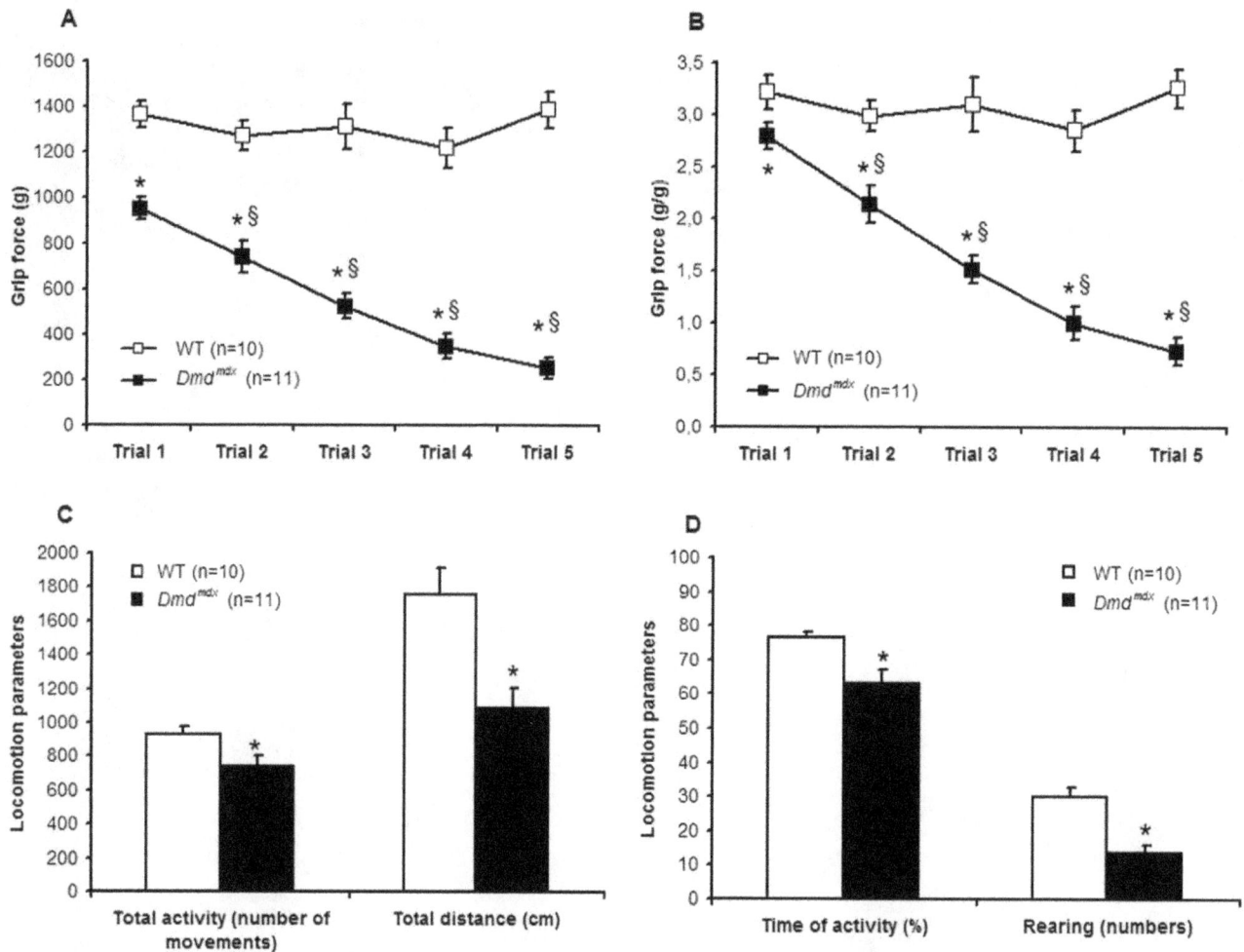

Figure 5. Dmd^mdx rats are characterized by muscle weakness and by a decrease in a spontaneous activity. Forelimb grip test and locomotor activity of Dmd^mdx rats were analyzed at the age of 3 months. Compared to wild-type littermate controls (WT), Dmd^mdx rats were characterized by muscle weakness and showed a lower strength in absolute values (A) and when normalized to body weight (B). Dmd^mdx rats showed fatigue with a decreased force grip across the five pulls, and significant values are obtained for the two last trials that were reduced to ~70% when compared to Trial 1. At 3 months, when compared to WT, Dmd^mdx rats were less active and the number of movements (C), the total distance travelled (C), the time of activity (D) and the number of rearing (D) were significantly lower. Values are means ± SEM; *p<0.05 vs WT; §: p<0.05 vs Trial 1.

free wall end-diastolic thickness were measured during the diastole from long and short-axis images obtained by M-mode echocardiography. Furthermore, the systolic function was assessed by the ejection fraction, while transmitral flow measurements of ventricular filling velocity were obtained using pulsed Doppler, with an apical four-chamber orientation. Thus, Doppler–derived mitral deceleration time, the early diastolic (E), the late diastolic (A) and the ratio E/A were obtained to assess diastolic dysfunction associated to an evaluation of the isovolumetric relaxation time. To avoid bias in the analysis, experiments were done in a blind fashion.

Statistics

Differences between means were analyzed by unpaired Student's t-test. When the Student test was not applicable a Mann Whitney test was used. All analyses were performed using SigmaStat 3.1 software (Logi Labo, Paris, France) and Xlstat (Addinsoft, Paris, France). The data are presented as mean ± S.E.M. with the significance level set at $p<0.05$.

Results

Generation of Dmd mutated rats

We generated a pair of TALENs targeting exon 23 of the rat Dmd gene in order to generate rats with Dmd mutations comparable to that of the mdx mouse (**Fig. S1**) [11,14]. The TALE DNA-binding domains were fused either to the wild-type FokI nuclease domain that can form homodimers or to mutated forms (so called ELD or KKR mutants for each monomer) that can only form heterodimers and thus will have less potential off-target effects. Plasmids encoding wild-type or ELD/KKR TALENs were transfected into rat C6 cells at different doses and were found to induce mutations in dose dependent fashion as evidenced by the PCR-T7 endonuclease I assay (**Fig. S1**). Even if that the wild-type pair was more effective at inducing mutations, we chose to use the heterodimeric ELD/KKR TALEN pair for the sake of improved specificity.

The mRNAs of each ELD/KKR TALEN monomer were microinjected into the cytoplasm of 387 rat zygotes, 320 were viable and 294 were transferred into pseudopregnant females, as previously described (**Table S1**) [11,12]. This resulted in the 88

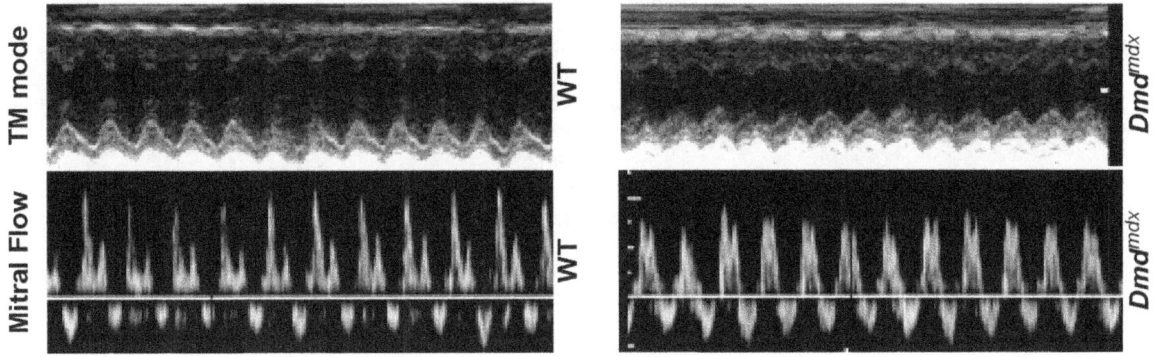

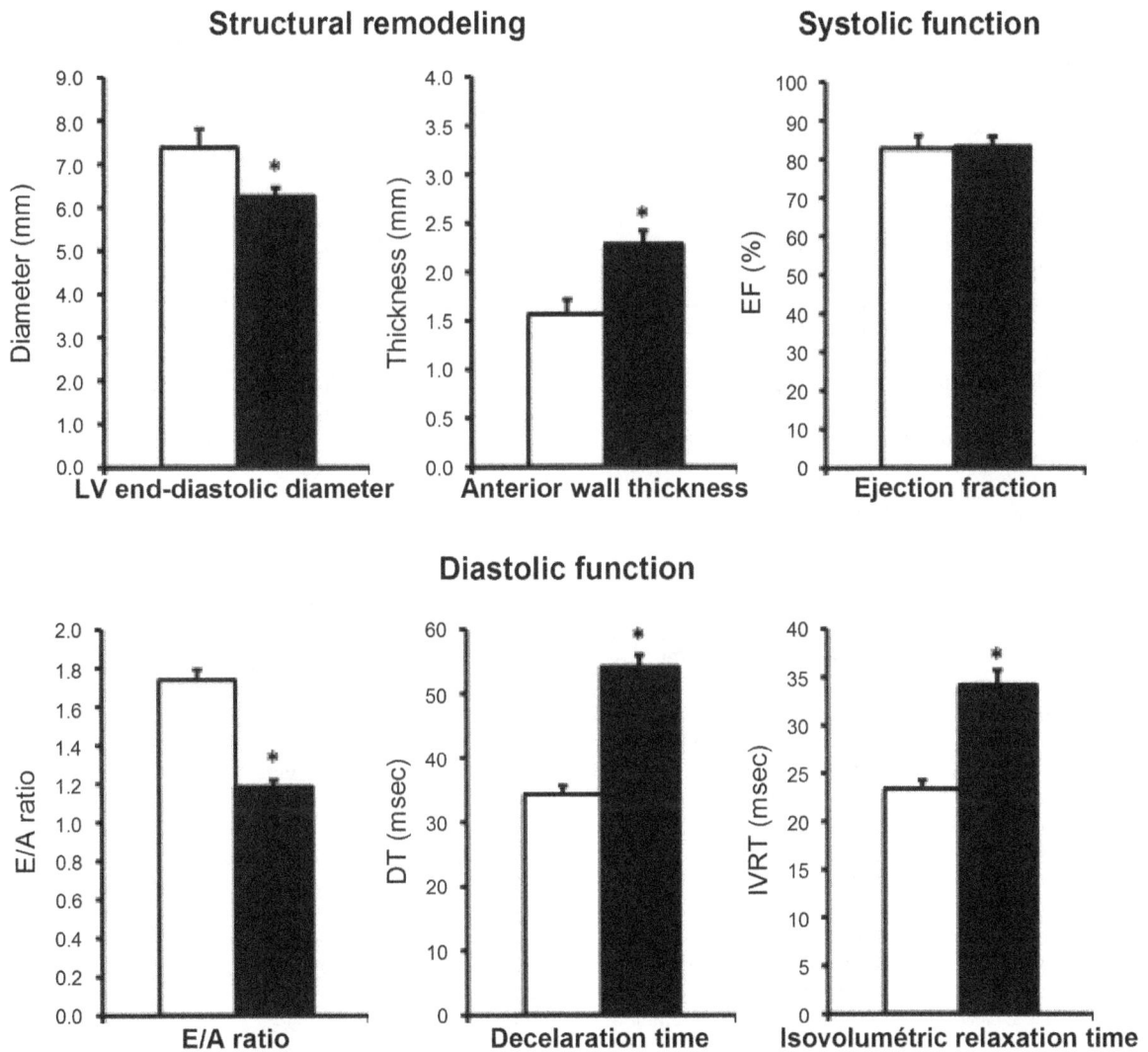

Figure 6. Structural remodelling and altered diastolic function in hearts of Dmd^{mdx} rats. Echocardiography was performed on 3 month-old Dmd^{mdx} rats. The structural remodeling and the systolic and diastolic functions were assessed respectively by Two-dimensional (2-D) echocardiography and pulsed Doppler (n = 6 for each condition); Values are mean ± SEM; *p<0.05). (TM mode: Time movement mode; LV: Left ventricular; E/A ratio: early diastolic (E)/late diastolic (A) ratio.

newborns and among them 11 showed mutations (3.74% of the transferred zygotes and 12.5% of newborns). These results are similar to our previous ones injecting other engineered sequence-specific nucleases and thus Dmd TALENs did not trigger embryo toxicity and allowed to induce Dmd gene mutations as expected [9–12].

The Dmd mutations observed in newborn rats were deletions ranging from 1 to 24 bp and insertions of a few nucleotides

Table 2. Comparison of pathological and functional characteristics between patients and animal models of Duchenne disease.

	DMD patient	GRMD dog	DMD pig	HFMD cat	Mdx mouse	Dmd^{mdx} rat
Muscle histopathology						
Revertant fibers	1 to 3%	<1%	ND	ND	5%	**5%**
Necrotic fibers	0.5 to 3.5%	2%	absent	present	5%	**>10%**
Regeneration	ND	15%	ND	present	10%	**10%**
Calcification	mild	mild to marked	ND	severe	mild	**absent**
Fibrosis	marked	present	present	diaphragm mainly	late & mainly diaphragm	**marked**
Lipomatosis	severe	absent	absent	absent	absent	**mild**
Cardiomyopathy	marked, major cause of death	mild	absent	present	absent or late	**marked**
Muscle function						
Strength reduction	marked	marked	ND	ND	mild	**marked**
Locomotion	severely impaired	impaired	impaired	ND	normal	**impaired**

(**Table S2**). Two of the newborns presented mutations that did not disrupt the open reading frame and were not further analyzed. In the other newborns, the mutations disrupted the open reading frame and resulted in premature stop codons. Some animals with *Dmd* mutations detected in tail DNA did not transmit the mutation to their offspring, most likely because it was not present in the germ line due to late activity of TALENs after injection, while several of them transmitted the mutations to the offspring in a Mendelian manner: two of these rat lines were further analyzed (lines 61 and 71). Both lines displayed the same muscle anatomopathological and functional characteristics and therefore only line 61, hereafter called Dmd^{mdx} is presented in this report. Dmd^{mdx} mutation is a deletion of 11 bp in exon 23 leading to a +1 frame shift and premature stop codon 81 bp after the mutation (**Fig. 1** and **Table S2**).

Female offspring of mutant founders that carried the mutation did not show premature death and were indistinguishable from wild-type littermate controls (**Fig. S2 A**). X-linked DMD mainly affects boys and we have therefore only characterized male Dmd^{mdx} rats in this study. When compared to wild-type littermate controls of the same ages, Dmd^{mdx} rats are smaller and less heavy (**Table 1** and **Fig. S2 B**). As illustrated by the increase in weight with age, the growth of Dmd^{mdx} rats is significantly altered as soon as 4 weeks of development (**Fig. S2 C**).

Founder 61 and its offspring were analyzed for potential off target mutations at other loci. The closest homologous sequences to the ones recognized by *Dmd* TALENs showed eight mismatches. We chose to further look for mutations at the 2 candidate sites located on the X-chromosome because they could potentially cosegregate with the *Dmd* mutations and affect the mutant rat phenotype (**Table S2**). They were analyzed no mutations could be detected by T7 endonuclease I assays (data not shown). Altogether, these results indicate that rat lines with a specific disruption of the *Dmd* coding frame by premature stop codons were successfully generated.

Analysis of dystrophin expression

Western blot analysis of dystrophin expression in *tibialis cranialis* and cardiac muscles was performed using two antibodies directed against epitopes located at the C terminus and within exons 10–11. No expression of dystrophin or of a truncated dystrophin could be detected in muscle tissues analyzed from *Dmd* mutated animals (**Fig. 2A**).

Immunohistological analysis of *biceps femoris* and cardiac muscles using mouse monoclonal antibodies directed against three different epitopes (located within exons 11–12, 38–39 and exon 78) revealed rare scattered dystrophin positive fibers in all *Dmd* mutated animals (**Fig. 2B–E** for epitope within exons 38–39 and data not shown). Dystrophin positive fibers in *biceps femoris* of mutated animal of line 61 after immunolabelling represented $4.9 \pm 0.8\%$ (n = 5) of total number of fibers. Such dystrophin positive fibers are frequently observed in animal models of dystrophin deficiency [26]. The exact mechanism involved in the generation of the so-called revertant myofibers remains poorly understood.

The dystrophin-associated protein complex is severely affected in skeletal muscle of DMD patients. We therefore examined the expression of alpha- and gamma- sarcoglycan, two proteins of the dystrophin-associated complex. The expression of alpha-sarcoglycan (**Fig. S3 A, B**) and gamma (data not shown) was weak and scarce compared to the systematic expression in myofibers of wild-type rats [27]. Altogether, these results indicate that Dmd^{mdx} rats are *bona fide* dystrophin deficient animals.

Histopathological evaluation of muscles

Before histopathological evaluation, skeletal muscle and heart weights were measured in 3 and 7 month-old rats. Fast muscles *tibialis anterior* and *extensor digitorum longus* muscles reached significantly lower weights, of around 25% less, in 7 month-old Dmd^{mdx} rats compared to wild-type controls (**Table 1**). In contrast, the slow *soleus* skeletal muscle was not significantly affected. The loss of skeletal muscle mass that was detected is consistent with the lower body weight of Dmd^{mdx} rats (**Fig. S2 C**) and indicates that muscular atrophy takes place in Dmd^{mdx} rats, preferentially in fast muscles. The heart weights were slightly higher at 3 months and significantly lower at 7 months in Dmd^{mdx} rats compared to littermates (**Table 1**). At 3 months, hearts of Dmd^{mdx} rats were markedly dilated with left ventricular wall thinning and increased diameter of the ventricular cavity. Ratio between left ventricle wall thickness and ventricular cavity diameter was significantly decreased compared to age-related wild-type littermate controls at 3 and 7 months (**Table 1**). All together these changes in heart morphology were indicative of a progressive dilated cardiomyopathy in Dmd^{mdx} rats.

Histopathological evaluation of *tibialis cranialis, extensor digitorum longus, biceps femoris, biceps brachii, soleus* and diaphragm skeletal muscles and heart was performed at 3 and 7 month-old. Similar severe lesions were present in all examined striated skeletal muscles and were characterized by excessive fiber size variation with some large rounded hypertrophic fibers and clusters of small centronucleated regenerating fibers (**Fig. 3** and **Fig. S4**). Numerous individual necrotic fibers appeared fragmented or contained invading inflammatory cells (**Fig. 3** and **Fig. S4**), some being CD3-positive lymphocytes (data not shown). No calcium deposit was noticed after HES or Alizarin red stainings. ATPase staining revealed a predominance of type 1 fibers with perturbation of the normal random distribution of muscle fiber types, abnormal fiber grouping and presence of some type 2C fibers, described to be observed in case of a regenerating process after fiber necrosis (**Fig. S5**). These lesions were accompanied by a patchy increase in endomysial connective tissue at 3 months and a marked endomysial and perimysial fibrosis at 7 months (**Fig. 3 A–C**). The main lesions observed in striated skeletal muscles were quantified by histomorphometry on the *biceps femoris* muscle. The mean minimal Ferret diameter was 41.1±20.9 μm *vs* 54.7±16.7 μm and 45.0±22.5 μm *vs* 54.7±18.5 μm, respectively at 3 and 7 months in Dmd^{mdx} rat *versus* age-related controls (n = 5 in each group) showing a significant decrease in individual fiber size in Dmd^{mdx} rats (p<0.0001) (**Fig. 3 G, H**). This decreased diameter was illustrated by the modal value that was 30 to 40 μm in 3 and 7 month-old Dmd^{mdx} rats (19.5±1.6% and 19.5±2.6%, respectively), whereas it corresponded to 50 to 60 μm in wild-type littermate controls (19.5±1.6% and 19.7±3.8%, respectively). The smallest fibers (with diameter less than 10 μm) represented 5.4±8.0% and 1.1±0.6% in 3 and 7 month-old Dmd^{mdx} rats compared to the lower percentage in age-related control rats (0.5±0.7% and 0.0±0.0%, respectively). The strong regenerative activity suggested by this last result was assessed by examining expression of developmental myosin heavy chain isoform ($MyoHC_{Dev}$), a marker specific to developing and regenerating myofibers. Abundant $MyoHC_{Dev}$ positive fibers were observed in *biceps femoris* muscles of 3 and 7 month-old Dmd^{mdx} rats *vs* control rats (13.1±6.7% and 10.2±4.4%, respectively, illustrating a slight decrease with time) and were absent in muscles of wild-type littermate controls (**Fig. 3 D–F**). In addition, utrophin expression was found in fiber clusters with continuous membrane immunolabelling, also suggestive of regenerative processes as previously described in other models [28] (data not shown).

The accumulation of connective tissue in *biceps femoris*, diaphragm and cardiac muscles was confirmed using a specific picrosirius staining in 7 month-old Dmd^{mdx} rats and was associated mostly with muscle fiber focal replacement by adipose tissue (**Fig. 4**). In the heart, fibrosis was most prominent in papillary muscle of the left ventricle (**Fig. 4 G–I**), in the septum and in the ventricular subepicardic area.

Total fibrosis was assessed by measuring the area occupied by connective tissue in transverse sections of *biceps femoris*. Connective tissue represented 16.9±4.7% and 24.3±8.7% in Dmd^{mdx} rats compared to 6.7±3.2% and 6.4±2.3% in control rats at 3 and 7 months of age, respectively, demonstrating a significant effect of the mutation on total fibrosis (p<0.0001).

In the cardiac muscle, individual fiber necrosis associated with inflammatory cell infiltration and fibrosis predominated at 3 and 7 months in Dmd^{mdx} rats (**Fig. S4 C**). Fibrous connective tissue represented 8.0±4.9% and 15.7±7.9% of the total cardiac muscle area in Dmd^{mdx} rats compared to 4.3±3.1% and 3.4±1.4% in

wild-type littermate controls at 3 and 7 months, respectively. The increase in the fibrotic area from 3 to 7 months in Dmd^{mdx} rats was significant (p<0.0001).

Dmd^{mdx} rats have a high level of serum creatine kinase

In DMD patients, muscle damage is characterized by an increase of creatine kinase (CK) activity in the serum. In Dmd^{mdx} rats, CK levels showed a 10-fold augmentation when compared to wild-type littermate controls (CK (U.L^{-1}). At 3 month-old, CK levels were 1945±620 in Dmd^{mdx} rats (n = 5) and 185±17 in wild-type rats (n = 5) (p<0.05); At 7 month, these levels were 3965±817 in Dmd^{mdx} rats (n = 5) and 588±397 in wild-type rats (n = 5) (p<0.05). These results suggest an increased fragility of muscle membrane and confirmed that Dmd^{mdx} rats develop a dystrophic pathology.

Significant muscle weakness in Dmd^{mdx} rats

To examine whether the absence of dystrophin affected muscle function, we performed a grip test on 3 month-old Dmd^{mdx} rats. A significant decrease in forelimb grip strength indicated a generalized alteration in the whole body muscular performance. As illustrated in Fig. 5 (**A, B**), at the first trial to the grip a 30% weaker force was exerted by Dmd^{mdx} rats compared to wild-type littermates. Furthermore, while control rats maintain the same force over five successive pulls, the grip force of Dmd^{mdx} rats decreased significantly and became 70% weaker than that of control rats. Therefore, Dmd^{mdx} rats presented a decreased grip strength test demonstrating muscle weakness, a typical sign observed in DMD patients [24,29].

Decrease in the spontaneous activity in Dmd^{mdx} rats

Behavioral and locomotor measurements are important parameters that help to define the phenotypes of animal models with neuromuscular disorders such as muscular dystrophies [25]. These assessments were done in an open field plexiglass chamber equipped with multiple photocell receptors and emitters. At the age of 3 months, Dmd^{mdx} rats were less active than control littermates and the number of movements, the total distance travelled and the time of activity were significantly lower respectively 21±7%; 38±7%; 17±5% lower in Dmd^{mdx} than in wild-type littermate controls (p<0.05) (**Fig. 5C, D**). Furthermore, the horizontal activity as rearing was also significantly impaired and reduced by 55±8% (p<0.05) in Dmd^{mdx} rats compared to wild-type littermate controls (**Fig. 5D**). These data demonstrate that motor activity and locomotor behavior are severely affected in Dmd^{mdx} rats.

Modification of the heart function in Dmd^{mdx} rats

At 3 months, Dmd^{mdx} rat and wild-type control littermates showed the same heart rate values under anaesthesia ($HR_{WT} = 221±13$ bpm and $HR_{Dmd}{}^{mdx} = 219±12$ bpm) (**Fig. 6**). The echocardiographic investigations revealed structural remodeling with a significant decrease in the LV end-diastolic diameter ($EDD_{WT} = 7.4±0.4$ mm and $EDD_{Dmd}{}^{mdx} = 6.3±0.2$ mm) associated with thickening of the LV end-diastolic anterior wall in Dmd^{mdx} rats ($EDAW_{WT} = 1.6±0.1$ mm and $EDAW_{Dmd}{}^{mdx} = 2.3±0.1$ mm). These observations suggest the presence of a concentric remodeling in Dmd^{mdx} rats heart at the age of 3 months. We did not observe any change in systolic function as ejection fraction ($EF_{WT} = 82.3±3.3\%$ and $EF_{Dmd}{}^{mdx} = 83.3±2.6\%$) and shortening fractions ($SF_{WT} = 44.3±4.4\%$ and $SF_{Dmd}{}^{mdx} = 47.8±2.8\%$; data not shown). However, diastolic dysfunction markers were observed. We observed changes in E/A ratio (E/A$_{WT}$ = 1.7±0.1 and E/A$_{Dmd}^{mdx}$ = 1.2±0.1), in Doppler–derived mitral deceleration time (DT$_{WT}$ = 34.3±1.3 msec and DT$_{Dmd}^{mdx}$ = 54.2±1.8 msec) and in isovolumetric relaxation time (IRT$_{WT}$ = 23.3±0.9 msec and IRT$_{Dmd}^{mdx}$ = 34.2±1.5 msec) (**Fig. 6**). These data demonstrate structural remodeling and altered diastolic function in hearts of Dmd^{mdx} rats.

Discussion

In this study, we describe the generation and the full characterization of a novel dystrophin deficient rat model (Dmd^{mdx} rats) with phenotypic properties very close to the human DMD pathology.

DMD is the most common neuromuscular disorder, accounting for approximately 30% of muscular dystrophy patients [30–32]. The responsible gene is located in the short arm of the X chromosome, Xp21, and contains 14 kbp of dystrophin protein coding sequence split in 79 exons. The primary molecular characteristic of DMD patients is the absence of dystrophin protein in skeletal and cardiac muscles. The present results show that dystrophin was undetectable by western blot in skeletal muscles of Dmd^{mdx} rats. In agreement, immunohistological analysis demonstrated that dystrophin is undetectable in 95% of fibers in skeletal muscles and only some revertant fibers were visible, as in DMD patients and other animal models of DMD [27]. Immunostaining and western blot analyses demonstrated that dystrophin was also undetectable in hearts of Dmd^{mdx} rats. The absence of dystrophin, as encountered in DMD patients, leads to absence of the dystro-associated protein complex, instability of the muscle cell membrane and uncontrolled influx of calcium. This initiates a set of downstream pathological processes that ultimately lead to loss of muscle cell proteins, myofiber damage and progressive muscle wasting [33]. Similar to muscle damages in DMD patients, Dmd^{mdx} rats are characterized by a loss of skeletal muscle mass, muscle fiber necrosis, increased variation in fiber size due to the simultaneous presence of hypertrophic fibers and small centronucleated regenerating fibers, infiltration by inflammatory cells, regenerative activity and interstitial fibrosis in skeletal muscles. All these signs of tissue changes observed in limb and forelimb skeletal muscles are correlated with a dramatic increase of serum creatine kinase activity in 3 and 7 month-old Dmd^{mdx} rats. As observed in DMD patients but not in mdx mice [34], fibrosis was severe in all skeletal muscle examined as well as in cardiac muscle of Dmd^{mdx} rats (**Fig. 4**). It is well described in dystrophin deficient skeletal myofibers that Ca^{2+} homeostasis alteration is pivotal in the initiation of the fiber damages [35,36]. In skeletal muscle from Dmd^{mdx} rats at 3 and 7 months, while severe lesions were present no calcification was observed. Thus it will be interesting to further investigate calcium regulation in Dmd^{mdx} rat skeletal muscles.

Consistent with histomorphological muscle damages, at 3 and 7 months of age, Dmd^{mdx} rats exhibit muscle dysfunction with decrease in muscle force and reduced locomotor activity. The clinical manifestations of Dmd^{mdx} rats evaluated at 3 and 7 months of age were progressive with gradual loss of body weight and muscle mass (mainly of fast twitch muscles such as *tibialis anterior*) as well as in necrosis and degeneration of skeletal muscle fibers including fundamental muscles such as the diaphragm. Additionally, Dmd^{mdx} rats showed a progressive dilated cardiomyopathy as indicated by morphological analyses, concentric structural remodeling and diastolic dysfunction. Since the majority of DMD patients die from cardiac failure [37] the cardiac dysfunction described in Dmd^{mdx} rats is an important feature of this new DMD model.

All skeletal and cardiac lesions as well as functional deficits are important features of the rat Dmd^{mdx} model since they are similar to those of patients. It will be interesting to study the disease onset in this new model as was previously studied in the mdx mouse, in particular to identify the early prenecrotic stages. Importantly, the changes observed between the two time points of this study i.e 3 and 7 month-old showed a progressive evolution of the pathology in this rat model, in the same manner than those described in DMD patients. Therefore, Dmd^{mdx} rats will be useful for preclinical assessment of new therapies. In particular, as analyses of dystrophin mRNA in the muscles of Dmd^{mdx} rats showed that the mutated exon 23 was transcribed (**Fig. S6**). Dmd^{mdx} rats could therefore be used for the evaluation of DMD treatment using an exon skipping approaches.

The use of TALENs targeting the rat Dmd gene was efficient at introducing Dmd mutations. Other types of gene-specific nucleases are available, such as ZFNs and more recently CRISPRs/Cas9, that have also been used successfully in rats [9–13,38]. Therefore, the generation of new rat models of neuromuscular disease using artificial sequence-specific nucleases now appears very promising. While this manuscript was in preparation, dystrophin-deficient rats generated with CRISPR/Cas9 targeting exon 2 and exon 16 were reported [38]. The phenotype of the corresponding F0 rats showed characteristics similar to the Dmd^{mdx} rats in skeletal muscle although progression of the disease was not documented. In the heart, in contrast, even though morphological changes were reported in some 3-month old F0 animals, they were not found statistically significant. The difference between the two models could be due to differences in genetic backgrounds (Wistar-Imamichi inbred *vs* Sprague-Dawley outbred strain for Dmd^{mdx}) or to mosaicism of Dmd mutations in F0 animals studied in the latter report.

The genetic heterogeneity and body size of the Golden Retriever Muscular Dystrophy dog model (GRMD) is closer to humans than the mice (**Table 2**). Therefore, GRMD is an important animal model for DMD. However, one major disadvantage is the large heterogeneity in the clinical phenotypes of GRMD dogs that makes the functional evaluation of potential therapies difficult. The mdx mouse has been important as a DMD model for development of gene therapy approaches. Murine models as mdx, mdx^{5cv}, $mdx52$ will continue to provide important findings for the basic study of pathogenesis and development of therapies but they do not show significant muscle weakness or cardiac alterations nor any of the skeletal or cardiac muscle lesions developing at late time points in DMD patients, such as fibrosis and muscular atrophy. In contrast, it was shown that mdx/mTR^{KO} mice, combining the mdx mutation with deletion of the gene for the RNA component of telomerase, show skeletal muscle dysfunction and progressive dilated cardiomyopathy as a result of shortened telomers and accelerated depletion of stem cell reservoirs [39,40]. Given the much aggravated skeletal and cardiac muscle phenotype of Dmd^{mdx} rats compared to mdx mice, it will be interesting to look for any difference in telomere dynamics or stem cell biology between these two closely related species that could be involved.

In summary, our study shows that Dmd^{mdx} rats represent a promising model to further elucidate mechanisms of DMD and to test novel therapies, in particular those aiming to curtail heart alterations and skeletal muscle disease progression.

Supporting Information

Figure S1 Mutation rates of TALE nucleases targeting exon 23 of the rat Dmd gene. The mutation rates induced by

TALE nuclease were determined using T7 endonuclease assay in C6 cells transfected with indicated amounts of rat *Dmd* TALE nucleases expression vectors. Homodimeric (WT, 0.75 μg each) and heterodimeric (EDL/KKR; 0.75 μg and 1.5 μg each) TALEN pairs (DMD-L+DMD-R) have been evaluated as well as left and right part of the TALEN pair (1.5 μg). The expected sizes of digested fragments are indicated near the gel. The rates of insertion and deletion mutations (indels) detected are indicated below each lane. The TALEN binding sites in exon 23 are indicated in blue.
(TIFF)

Figure S2 Growth of *Dmd^mdx* rats. Newborns at day 2 (*A*). While wild-type littermate controls (WT) and *Dmd^mdx* rats were indistinguishable after birth, *Dmd^mdx* rats were noticeably smaller at the age of 16 weeks (*B*). Body weight values from WT and *Dmd^mdx* rats from 4 to 12 weeks postnatal (*C*). Values are mean ± SEM; *p<0.05 *Dmd^mdx* vs WT, analyzed by unpaired *t* test.
(TIF)

Figure S3 The absence of dystrophin in skeletal muscle fibers is associated with a strong reduction in alpha-sarcoglycan expression in *Dmd^mdx* rats. *Biceps femoris* muscle samples were assessed for the expression of proteins of the dystrophin associated protein complex by immunohistochemistry. Compared to the generalized subsarcolemmal expression of alpha-sarcoglycan in wild-type littermate control (WT) rats (*A*), only some fibers expressed the protein in 7 month-old *Dmd^mdx* rats (*B*). Bar = 100 μm.
(TIF)

Figure S4 Muscle changes were present in all striated muscle including skeletal, respiratory and cardiac ones in *Dmd^mdx* rats at 7 month-old. In *extensor digitorum longus* (*A*) and diaphragm (*B*) muscles, hypercontracted hyalin giant fibers (°) and individual necrotic fibers (open arrowhead), sometimes surrounded by some inflammatory cells (#), were associated with foci of small regenerative centro-nucleated fibers (arrow). Some muscle fibers were replaced by fibrotic (*) and fat (black arrowhead) tissue. In cardiac muscle (*C*), individual fiber necrosis

(open arrowhead) elicited some inflammatory cell infiltration (#). Focal thick bundles of fibrotic tissue surround cardiomyocytes (*). Hemalun eosin saffron staining (*A–C*). Bar = 100 μm.

Figure S5 Abnormal fiber type pattern in *Dmd^mdx* rats. Compared to wild-type littermate control (WT) rats, *biceps femoris* muscles from 7-month-old *Dmd^mdx* rats were characterized by type 1 predominance and grouping (type 1 fibers in black) and abnormal presence of type 2C (intermediate grey staining, open arrowhead). ATPase staining, pH 4.35. Bar = 100 μm.

Figure S6 Dystrophin messenger RNA expression in skeletal muscles and hearts of *Dmd^mdx* rats. Dystrophin mRNA was detected by nested RT-PCR from exon 22 to 24 in *tibialis cranialis* muscles and in hearts sampled in 7 month-old *Dmd^mdx* rats and wild-type control littermates. No skipping of the mutated exon 23 was detected in muscles from *Dmd^mdx* rats.

Table S1 Efficacy of generation of *Dmd* mutants by TALE nuclease microinjection.

Table S2 *Dmd* mutations and potential off target sequences analyzed in founder animals.

Acknowledgments

Yan Cherel died during the course of the study. This manuscript is dedicated to his memory.

Author Contributions

Conceived and designed the experiments: YC CG JPC IA CH. Performed the experiments: TL AL LT SR VT VF CLG HG MD LG GT ADC CB JBR YC CG JPC IA CH. Analyzed the data: TL AL LT SR VT VF CLG HG MD LG GT ADC CB JBR YC CG JPC IA CH. Contributed to the writing of the manuscript: IA CH.

References

1. Jarmin S, Kymalainen H, Popplewell L, Dickson G (2014) New developments in the use of gene therapy to treat Duchenne muscular dystrophy. Expert Opin Biol Ther 14: 209–230.2.
2. Sharp NJ, Kornegay JN, Van Camp SD, Herbstreith MH, Secore SL, et al. (1992) An error in dystrophin mRNA processing in golden retriever muscular dystrophy, an animal homologue of Duchenne muscular dystrophy. Genomics 13: 115–121.
3. Klymiuk N, Blutke A, Graf A, Krause S, Burkhardt K, et al. (2013) Dystrophin-deficient pigs provide new insights into the hierarchy of physiological derangements of dystrophic muscle. Hum Mol Genet 22: 4368–4382.
4. Hollinger K, Yang CX, Montz RE, Nonneman D, Ross JW, et al. (2014) Dystrophin insufficiency causes selective muscle histopathology and loss of dystrophin-glycoprotein complex assembly in pig skeletal muscle. FASEB J 28: 1600–1609.
5. Bulfield G, Siller WG, Wight PA, Moore KJ (1984) X chromosome-linked muscular dystrophy (mdx) in the mouse. Proc Natl Acad Sci U S A 81: 1189–1192.
6. Partridge TA (2013) The mdx mouse model as a surrogate for Duchenne muscular dystrophy. FEBS J 280: 4177–4186.
7. Nakamura A, Takeda S (2011) Mammalian models of Duchenne Muscular Dystrophy: pathological characteristics and therapeutic applications. J Biomed Biotechnol. 2011: 184393.
8. Jacob HJ, Kwitek AE (2002) rat genetics: attaching physiology and pharmacology to the genome. Nat Rev Genet 3: 33–42.
9. Ménoret S, Fontanière S, Jantz D, Tesson L, Thinard R, et al. (2013) Generation of Rag1-knockout immunodeficient rats and mice using engineered meganucleases. FASEB J 27: 703–711.
10. Geurts AM, Cost GJ, Freyvert Y, Zeitler B, Miller JC, et al. (2009) Knockout rats via embryo microinjection of zinc-finger nucleases. Science 325: 433.
11. Remy S, Tesson L, Menoret S, Usal C, De Cian A, et al. (2014) Efficient gene targeting by homology-directed repair in rat zygotes using TALE nucleases. Genome Res, in press.
12. Tesson L, Usal C, Ménoret S, Leung E, Niles BJ, et al. (2011) Knockout rats generated by embryo microinjection of TALENs. Nat Biotechnol 29: 695–696.
13. Li D, Qiu Z, Shao Y, Chen Y, Guan Y, et al. (2013) Heritable gene targeting in the mouse and rat using a CRISPR-Cas system. Nat Biotechnol 31: 681–683.
14. Auer TO, Duroure K, De Cian A, Concordet JP, Del Bene F (2014) Highly efficient CRISPR/Cas9-mediated knock-in in zebrafish by homology-independent DNA repair. Genome Res 24: 142–153.
15. Piganeau M, Ghezraoui H, De Cian A, Guittat L, Tomishima M, et al. (2013) Cancer translocations in human cells induced by zinc finger and TALE nucleases. Genome Res 23: 1182–1193.
16. Huang P, Xiao A, Zhou M, Zhu Z, Lin S, et al. (2011) Heritable gene targeting in zebrafish using customized TALENs. Nat Biotechnol 29: 699–700.
17. Miller JC, Tan S, Qiao G, Barlow KA, Wang J, et al. (2011) A TALE nuclease architecture for efficient genome editing. Nat Biotechnol 29: 143–148.
18. Doyon Y, Vo TD, Mendel MC, Greenberg SG, Wang J, et al. (2011) Enhancing zinc-finger-nuclease activity with improved obligate heterodimeric architectures. Nat Methods 8: 74–79.
19. Fine EJ, Cradick TJ, Zhao CL, Lin Y, Bao G (2014) An online bioinformatics tool predicts zinc finger and TALE nuclease off-target cleavage. Nucleic Acids Res42: e42.
20. Geurts AM, Cost GJ, Rémy S, Cui X, Tesson L, et al. (2010) Generation of gene-specific mutated rats using zinc-finger nucleases. Methods Mol Biol 597: 211–225.
21. Bartlett RJ, Stockinger S, Denis MM, Bartlett WT, Inverardi L, et al. (2000) In vivo targeted repair of a point mutation in the canine dystrophin gene by a chimeric RNA/DNA oligonucleotide. Nat Biotechnol 18: 15–622.

22. Dubowitz V, Brooke MH (1985) Muscle Biopsy: a Modern Approach. Saunders: London: 335–356.

23. Carre-Pierrat M, Lafoux A, Tanniou G, Chambonnier L, Divet A, et al. (2011) Pre-clinical study of 21 approved drugs in the mdx mouse. Neuromuscul Disord 21: 313–327.

24. De Luca AM (2008) Use of grip strength meter to assess the limb strength of mdx mice. SOP DMD_M.2.2.001.

25. Kanneboyinna N (2008) Behavioural and Locomotor Measurements Using Open Field Animal Activity Monitoring System. SOP DMD_M.2.1.002.

26. Arechavala-Gomeza V, Kinali M, Feng L, Guglieri M, Edge G, et al. (2010) Revertant fibres and dystrophin traces in Duchenne muscular dystrophy: implication for clinical trials. Neuromuscul Disord 20: 295–301.

27. Muller J, Vayssiere N, Royuela M, Leger ME, Muller A, et al. (2001) Comparative evolution of muscular dystrophy in diaphragm, gastrocnemius and masseter muscles from old male mdx mice. J Muscle Res Cell Motil 22: 133–139.

28. Lin S, Gaschen F, Burgunder JM (1998) Utrophin is a regeneration-associated protein transiently present at the sarcolemma of regenerating skeletal muscle fibers in dystrophin-deficient hypertrophic feline muscular dystrophy. Neuropathol Exp Neurol 57: 780–790.

29. Desguerre I, Christov C, Mayer M, Zeller R, Becane HM, et al. (2009) Clinical heterogeneity of duchenne muscular dystrophy (DMD): definition of sub-phenotypes and predictive criteria by long-term follow-up. PLoS One 4: e4347.

30. Hoffman EP, Brown RH Jr, Kunkel LM (1987) Dystrophin: the protein product of the Duchenne muscular dystrophy locus. Cell 51: 919–928.

31. Koenig M, Monaco AP, Kunkel LM (1988) The complete sequence of dystrophin predicts a rod-shaped cytoskeletal protein. Cell 53: 219–228.

32. Muntoni F, Torelli S, Ferlini A (2003) Dystrophin and mutations: one gene, several proteins, multiple phenotypes. Lancet Neurol 12: 731–40. Review.

33. Nowak KJ, Davies KE (2004) Duchenne muscular dystrophy and dystrophin: pathogenesis and opportunities for treatment. EMBO Rep 5: 872–876. Review.

34. Klingler W, Jurkat-Rott K, Lehmann-Horn F, Schleip R (2012) The role of fibrosis in Duchenne muscular dystrophy. Acta Myol 31: 184–95.

35. Whitehead NP, Yeung EW, Allen DG (2006) Muscle damage in mdx (dystrophic) mice: role of calcium and reactive oxygen species. Clin Exp Pharmacol Physiol 33: 657–662. Review.

36. Allen DG, Gervasio OL, Yeung EW, Whitehead NP (2010) Calcium and the damage pathways in muscular dystrophy. Can J Physiol Pharmacol 88: 83–91.

37. Kieny P, Chollet S, Delalande P, Le Fort M, Magot A, et al. (2013) Evolution of life expectancy of patients with Duchenne muscular dystrophy at AFM Yolaine de Kepper centre between 1981 and 2011. Ann Phys Rehabil Med. 56: 443–454.

38. Nakamura K, Fujii W, Tsuboi M, Tanihata J, Teramoto N, et al. (2014) Generation of muscular dystrophy model rats with a CRISPR/Cas system. Sci Rep 4: 5635–56.

39. Sacco A, Mourkioti F, Tran R, Choi J, Llewellyn M, et al. (2010) Short telomeres and stem cell exhaustion model Duchenne muscular dystrophy in mdx/mTR mice. Cell 143: 1059–1071.

40. Mourkioti F, Kustan J, Kraft P, Day JW, Zhao MM, et al. (2013) Role of telomere dysfunction in cardiac failure in Duchenne muscular dystrophy. Nat Cell Biol. 15: 895–904.

9

Effects of Changes in Food Supply at the Time of Sex Differentiation on the Gonadal Transcriptome of Juvenile Fish. Implications for Natural and Farmed Populations

Noelia Díaz¤, Laia Ribas, Francesc Piferrer*

Institut de Ciències del Mar, Consejo Superior de Investigaciones Científicas (CSIC), Barcelona, Spain

Abstract

Background: Food supply is a major factor influencing growth rates in animals. This has important implications for both natural and farmed fish populations, since food restriction may difficult reproduction. However, a study on the effects of food supply on the development of juvenile gonads has never been transcriptionally described in fish.

Methods and Findings: This study investigated the consequences of growth on gonadal transcriptome of European sea bass in: 1) 4-month-old sexually undifferentiated fish, comparing the gonads of fish with the highest vs. the lowest growth, to explore a possible link between transcriptome and future sex, and 2) testis from 11-month-old juveniles where growth had been manipulated through changes in food supply. The four groups used were: i) sustained fast growth, ii) sustained slow growth, iii) accelerated growth, iv) decelerated growth. The transcriptome of undifferentiated gonads was not drastically affected by initial natural differences in growth. Further, changes in the expression of genes associated with protein turnover were seen, favoring catabolism in slow-growing fish and anabolism in fast-growing fish. Moreover, while fast-growing fish took energy from glucose, as deduced from the pathways affected and the analysis of protein-protein interactions examined, in slow-growing fish lipid metabolism and gluconeogenesis was favored. Interestingly, the highest transcriptomic differences were found when forcing initially fast-growing fish to decelerate their growth, while accelerating growth of initially slow-growing fish resulted in full transcriptomic convergence with sustained fast-growing fish.

Conclusions: Food availability during sex differentiation shapes the juvenile testis transcriptome, as evidenced by adaptations to different energy balances. Remarkably, this occurs in absence of major histological changes in the testis. Thus, fish are able to recover transcriptionally their testes if they are provided with enough food supply during sex differentiation; however, an initial fast growth does not represent any advantage in terms of transcriptional fitness if later food becomes scarce.

Editor: Balasubramanian Senthilkumaran, University of Hyderabad, India

Funding: Research was supported by the Spanish Ministry of Science projects "Epigen-Aqua" (AGL2010-15939) and "Aquagenomics" (CDS2007-0002) to FP. ND was supported by a scholarship from the Government of Spain (BES-2007-14273) and then by a contract by the Epigen-Aqua project. LR was supported by an Aquagenomics postdoctoral contract and Epigen-Aqua project. The funders had no role in study design, data collection and analysis, decision to publish, or preparation of the manuscript.

Competing Interests: The authors have declared that no competing interests exist.

* Email: piferrer@icm.csic.es

¤ Current address: Max Planck Institute for Molecular Biomedicine, Regulatory Genomics Lab, Münster, Germany

Introduction

Food availability and energetic demands fluctuate in most habitats. Animals are capable of sensing their inner energy levels and the external energy availability and thus act accordingly by long-term investments in processes like growth, immune functions or reproduction when food availability is not a problem, or by ensuring survival when food is scarce [1]. In fish, there is a tight relationship between food availability and reproduction [1,2] since it can alter the timing and duration of spawning, fecundity and egg size [3,4], or the timing of the reproductive cycles [5]. Favorable

feeding conditions produce early maturation of individuals [6] while a decrease in food availability causes a decrease in energy transfer to the gonads [7], but this relationship may present important differences between species since fish constitute a vast phylogenetic group with different behaviors and reproduction types [8].

The European sea bass (*Dicentrarchus labrax*) is a gonochoristic species with a polygenic sex determining system [9] presenting a long sexually undifferentiated process with sexual dimorphism at the time of sex differentiation onset (~150 days post hatch, dph, for females and ~180 dph for males) [10–13]. However, this dimorphism is more related to the attained length than to age [14]. The relationship between growth and sex differentiation has been previously studied in sea bass [9,12–19]. There is a relationship between body weight and sex since not only sea bass females are larger than males, but also both males and females in batches with higher percent females were bigger than males and females of batches with a lower female percent [9]. Further, early size-gradings of the population (at 66 and 123–143 dph, [13]; at 70 dph, [15]; or at 82 dph, [17]) selecting for the largest fish resulted in ~90% of females, but the opposite, i.e., selecting for the smallest fish, produced only ~65% males at one year of age, meaning that while the largest fish are essentially all females, among the smallest fish there are both males and females [13].

Recently, two experiments on growth rate alteration by manipulating food supply during the sex differentiation period were conducted with the European sea bass in our laboratory [19]. The first experiment showed that transiently but severely reducing food supply starting towards the end, middle or even at the beginning of the sex differentiation period – and thus negatively affecting growth – did not affect resulting sex ratios, indicating that gender was already fixed before the above mentioned period started [19]. The second experiment involved four groups of fish, which, through controlling food supply, were made to experience different growth rates during the sex differentiation period. Two groups, one fast-growing and the other slow-growing, originated from the fast-growing fish at 127 dph. The other two groups, also one fast-growing and the other slow-growing, originated from the slow-growing fish at 127 dph. In this case, there were differences in the final sex ratio of the population as fast-growing-derived groups presented more number of females (~40%) than the slow-growing-derived groups (~10%). Thus, the differences in the final sex ratio were not related to the growth rate during the sex differentiation but to whether fish derived from the fast- or slow-growing fish at 127 dph. These results confirmed the results of the first experiment and indicated that before the first signs of sex differentiation appear the relationship between growth and sex is already established, confirming that in the European sea bass larger sizes are associated with female development [13,19].

Partition of consumed energy into growth, energy storage and gonads according to temporal food availability, metabolic demands and reproductive needs have been studied since a long time ago [20]. Recently, with the advent of new technologies, the underlying mechanisms including associated changes in global gene expression can be investigated. However, transcriptomic analyses in fish have traditionally addressed nutrition and reproduction topics separately. Hence, while growth studies have put efforts towards the effects of diet substitutions [21–24], stocking density and food ration [25–27] and comparing domesticated vs. transgenic fish [28], on the other hand, reproduction-related transcriptome analyses have focused on describing changes associated with gonad maturation and differences between sexes [29–32], environmental effects [33] or hormonal treatment effects [34,35].

However, a study directly analyzing the effects of food supply on reproduction and, particularly, on the development of juvenile gonads has never been described in fish. In mitten crab (*Eriocheir sinensis*) during early development, when crabs store significant amounts of energy in the hepatopancreas, Jiang and collaborators [36] found four genes in the hepatopancreas and 13 genes in testis related to nutritional control, and three genes in the hepatopancreas and eight in the testes related to regulation of reproduction. Among the former, arginine kinase, zinc-finger proteins or leptin were upregulated in the hepatopancreas transcriptome as a sign of energy storage for further energy-demand of the reproductive

processes. Genes involved in the regulation of reproduction, such as cyclins, kinetochore spindle formation or the heat shock protein 70, were upregulated in testis and promoted an increase in cell division during spermatogenesis. In rats, dietary energy intake changes (restrictions and excesses) but also food availability had profound effects in gonads from both sexes at different levels (biochemical, endocrine, behavioral and genetic) [37]. Moreover, a transcriptomic analysis of the gonads of these rats facing diet restriction or excess showed how females were more affected by ration changes than males. Males were also better adapted to an intermittent fasting by increasing the probability of an eventual fertilization, while females were able to sense the food restriction and behaved as sub-fertile females [38].

The present study is based on samples collected in experiment 2 of Díaz et al. [19] described above and had two objectives: 1) to analyze the transcriptional differences in sexually differentiating European sea bass gonads from the naturally fastest growing vs. the slowest growing fish at 127 dph, i.e., before the first histological signs of sex differentiation at 150 dph [13], but after the first signs of molecular sex differentiation at 120 dph [39], to explore a possible link of transcriptomic signatures with future sex; and 2) the consequences of food availability between 133–337 dph (juvenile growth) on the subsequent testes transcriptome by analyzing the effects of growth acceleration and deceleration.

Materials and Methods

Animals and rearing conditions

As stated above, the fish that were transcriptomically analyzed in this study are the same fish described in Experiment 2 of Díaz et al. [19]. Briefly, European sea bass eggs obtained from a commercial hatchery were collected at one day post fertilization (dpf) on March 2009, transported to our experimental aquarium facilities and hatched following established procedures for this species [40] with minor changes, as previously described [18,19]. Fish were reared under natural conditions of photoperiod, pH (~7.9), salinity (~37.8 ppt), oxygen saturation (85–100%) and with a water renewal rate of 30% vol·h^{-1}. In order to avoid temperature influences on the sex ratio, the thermal regime used and previously described [19] included egg spawning at 13–14°C and larval rearing at 16±1°C until 60 dpf. Then, temperature was increased to 21°C at a rate of 0.5–1°C·day^{-1} and maintained until the first fall, when it was let to follow the natural temperature. The rearing density was kept low to avoid any possible distorting effect on sex ratios (details in [19]). Fish were manually fed three times a day with artemia AF, then artemia EG enriched with amino acid (INVE Aquaculture, Belgium) and dry food (ProAqua, S.A., Spain) of the appropriate pellet size as fish grew. Unless otherwise stated, juveniles and adults were fed *ad libitum* two times a day.

Fish were treated in agreement with the European Convention for the Protection of Animals used for Experimental and Scientific Purposes (ETS Nu 123, 01/01/91). Our facilities are approved for animal experimentation by the Ministry of Agriculture and Fisheries (certificate number 08039-46-A) in accordance of Spanish law (Real Decreto 223 of March 1988) and the experimental protocol was approved by the Spanish National Research Council (CSIC) Ethics Committee within project AGL2010-15939. Animals were sacrificed by an overdose of 2-phenoxyethanol (2PE) followed by severing of the spinal cord.

Experimental design

Details of the experimental design can also be found in Díaz et al. [19]. Briefly, fish were individually size-graded at 127 dph, i.e., at ~4 cm standard length (SL), before the histological process of sex differentiation started (~8 cm SL), and separated into two groups according to the SL they had attained and comprising the two extremes of the normal curve distribution: a fast growth group (group F), with a mean size of 5.0 cm SL (range 4.2–6.4 cm), and a slow growth group (group S), with a mean size of 3.5 cm SL (range 2.6–3.7 cm). After checking that fish of each group was of the desired size, then at 133 dph (time 1, T1, i.e., when fish were 4 months old) each group was further subdivided into two tanks (n = 79 fish per tank). On one hand, the F group was subdivided into two groups with initial similar mean SL and BW: the fast-fast group (FF), in which growth rates from that moment onwards were as before, and the fast-slow group (FS), in which the growth rate was reduced to match what had been the growth rate of group S until then. On the other hand, the S group was also subdivided into two groups with initial similar mean SL and BW: the slow-fast group (SF), in which the growth rate was increased to match what had been the growth rate of group F until then, and the slow-slow group (SS), in which the growth rates from that moment onwards were as before (see Fig. 1A for a diagram of the experimental design). Food supply changed as follows: prior to T1, all fish were fed *ad libitum* three times a day. Then, groups FF and SF (accelerated growth) were fed *ad libitum* four times a day, with an amount of food per day equivalent of 3–6% of their mean BW. On the other hand, groups FS and SS (decelerated growth) were feed only two times a day with an amount of food per day equivalent of 1.5–3% of their mean BW. The growth rate of all groups was carefully monitored by periodic samplings and the amount of food adjusted if necessary. Animals were sacrificed when they were 337 dph juveniles (T2, i.e., when fish were 11 months old) (range of fish per tank at that moment: 52–70). There was no mortality associated to treatments.

Samplings

Details on the follow-up of growth, including sexual growth dimorphism, gonadosomatic, hepatosomatic and carcass indices, as well as sex ratio and the degree of gonadal development of these fish have been previously described [19]. Fish used for objective 1 were sexually undifferentiated. For objective 2, when possible, sex was visually determined and confirmed histologically if necessary [19]. Only males were considered for objective 2 since the goal was to study the relationship between growth during sex differentiation on the subsequent juvenile testis transcriptome. Histological results indicated that testis contained no spermatozoa. We did not do a similar study on females because some of the resulting populations were highly male-biased (~90% males) and thus we did not have enough individual females in all groups. The number of fish used for each group and the biometry is shown in Table S1. Here, we focus only on the RNA extraction for transcriptomic analysis of gene expression and for microarray validation by qRT-PCR.

RNA extraction and cDNA synthesis

Total RNA was extracted from sexually undifferentiated gonads (mean SL ~4 cm) at 133 dph (T1) and sexually differentiated juvenile testis at 337 dph (T2).

A classical chloroform-isopropanol-ethanol RNA extraction protocol after a Trizol (Live Technologies, Scotland, UK) homogenization was used. RNA quality and concentration were measured by a ND-1000 spectrophotometer (NanoDrop Technologies) and checked on a 1% agarose/formaldehyde gel. RNA integrity was measured by a Bioanalyzer 2100 (RNA 6000 Nano LabChip kit Agilent, Spain). Samples with a 100–200 ng/µl RNA concentration and RIN>7 were used for microarray hybridizations.

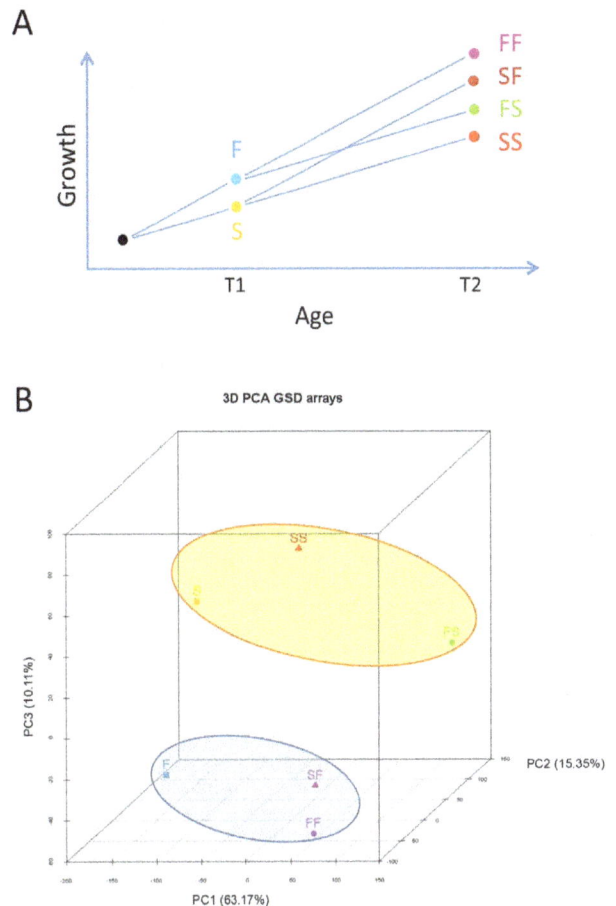

Figure 1. Experimental design and overall transcriptomic results. A) Experimental design involving European sea bass subjected to food restriction at different times during the sex differentiation period. B) Principal component analysis representation of transcriptomic results of the six groups. Time 1 fast- (F) and slow- (S) growing groups, and time 2 F- (FF and FS) and S- (SF and SS) derived groups.

In parallel, 200 ng of total RNA were treated by *E. coli* DNAse H and retrotranscribed (100 ng) to cDNA using SuperScript III RNase Transcriptase (Invitrogen, Spain) and Random hexamer (Invitrogen, Spain) following manufacturer's instructions.

Microarray. Experimental design

Hybridizations were performed at the Universitat Autònoma de Barcelona (UAB, Spain). The experiment included, on one hand, the comparison of 15 undifferentiated gonads from two groups (seven individual gonads from the F group against eight individual gonads from the S group) at 133 dph (T1), to explore transcriptomic differences between two groups with different growth rates since they were selected from the opposite extremes of the normal distribution curve. On the other hand, individual testes from each one of the four groups (groups FF, SF, FS and SS) differentially fed and sampled at 337 dph (T2), were analyzed to investigate the growth acceleration or growth deceleration effects on their transcriptome. Thus, 35 microarrays (one per fish) were used to analyze the gonadal transcriptome of the six groups considered in this study. To avoid batch effects samples were evenly distributed among the slides.

Microarray. RNA sample preparation and array hybridization

RNA labelling, hybridizations, and scanning were performed according to the manufacturer's instructions. Briefly, total RNA (100 ng) was amplified and Cy3-labeled with Agilent's One-Color Microarray-Based Gene Expression Analysis (Low Input Quick Amp Labelling kit) along with Agilent's One-Color RNA SpikeIn Kit. Then cRNA was purified with RNeasy mini spin columns (Qiagen), quantified with the Nanodrop ND–1000 and verified using the Bioanalyzer 2100. Each sample (1.65 μg) was hybridized to the *Dicentrarchus labrax* array (Agilent ID 023790) at 65°C for 17 h using Agilent's GE Hybridization Kit. Washes were conducted as recommended by the manufacturer using Agilent's Gene Expression Wash Pack with stabilization and drying solution. Arrays were scanned with Agilent Technologies Scanner, model G2505B. Spot intensities and other quality control features were extracted with Agilent's Feature Extraction software version 10.4.0.0. The experiment has been submitted to Gene Expression Omnibus (GEO)-NCBI database (GSE54362) and the platform that validates the microarray has the accession number (GPL13443).

Quantitative real time PCR (qRT-PCR)

Microarray validations were carried out by qRT-PCR analysis. Two genes from each one of the six possible microarray comparisons (see Table S2) were used for qRT-PCR validation, including one up- and one downregulated gene. Primers were designed using Primer 3 Plus (http://www.bioinformatics.nl/cgi-bin/primer3plus/primer3plus.cgi) against the annotated gene sequences directly from the European sea bass genome (Tine et al., 2014, unpublished), always trying to design the primers between exons to avoid DNA contamination (Table S2). Primer amplification efficiencies were tested by linear regression analysis from a cDNA dilution series and by running a melting curve (95°C for 15 s, 60°C for 15 s and 95°C for 15 s). Efficiency ($E = 10^{(-1/slope)}$, with values between 1.80 and 2.20), standard curves ranging from –2.9 to –3.9 and linear correlations (R^2) higher than 0.92 were recorded (Table S2). cDNA was diluted 1:10 for all the target genes and 1:500 for the endogenous control (the housekeeping gene *r18S*).

qRT-PCR was analyzed by an ABI 7900HT (Applied Biosystems) under standard cycling conditions. Briefly, an initial UDG decontamination cycle (50°C for 2 min), an activation step (95°C for 10 min) was followed by 40 cycles of denaturation (95°C for 15 s) and one annealing/extension step (60°C for 1 min). A final dissociation step was also added (95°C for 15 s and 60°C for another 15 s). Each sample was run in triplicate in 384-well plates in a final 10 μl volume (2 μl of 5x PyroTaq EvaGreen qPCR Mix Plus (ROX) from Cultek Molecular Bioline, 4 μl distilled water, 2 μl primer mix at a 10 μM concentration and 2 μl of cDNA). Negative controls were run per duplicate and *r18S* was used to calculate intra- and inter-assay variations. SDS 2.3 software (Applied Biosystems) was used to collect raw data and RQ Manager 1.2 (Applied Biosystems) was used to calculate gene expression. qRT-PCR data was analyzed by adjusting for E and normalizing to the *r18S* reference gene [41].

Statistical analysis of data. Microarray raw data normalization

Feature Extraction output data was corrected for background using normexp method [42] and was quantile normalized [43]. Reliable probes showed raw foreground intensity at least two times higher than the respective background intensity and were not saturated nor flagged by the Feature Extraction software. Our sea bass custom-made microarray presents most of the probes (64.7%) in duplicate but also with more than three identical probes for some genes. Median intensities per gene were calculated and a probe was considered reliable when at least half of its replicates were reliable as defined above. An empirical Bayes approach on linear models (Limma) [44] was used to perform a differential expression analysis. A False Discovery Rate (FDR) method was used to correct for multiple testing. Differentially expressed (DE) genes were filtered by fixing an absolute fold change (FC) of 1.5 and an adjusted *P*-value <0.01. MA data analyses were performed with the Bioconductor project (http://www.bioconductor.org/) in the R statistical environment (http://cran.rproject.org/) [45].

Statistical analysis of data. qRT-PCR statistics

Quantitative qRT-PCR statistical analysis was performed using 2DCt from the processed data [41]. 2DCt results were then checked for normality, homoscedasticity of variance and a one-way ANOVA test was used to assess differences between treatments using SPSS v.19 software.

Gene annotation and enrichment analysis

Gene data (names, abbreviations, synonyms and functions) were determined using Genecards (http://www.genecards.org/) and Uniprot (http://www.uniprot.org/). The web based tool AMIGO (http://amigo.geneontology.org/cgi-bin/amigo/go.cgi) [46] was used to look for the sequences of the DE genes found at the MA. After obtaining these sequences, Blast2GO software (http://www.blast2go.com) [47] was used to enrich GO term annotation and to analyze the subsequent altered KEGG pathways (http://www.genome.jp/kegg/), which were also further explored by DAVID (http://david.abcc.ncifcrf.gov/; [48,49]). Completing the analysis, Blast2GO with Fisher's Exact Test with Multiple Testing Correction of the False Discovery Rate [50] was used to analyze our DE genes using the custom-made microarray as background.

Protein names from the DE genes were then uploaded to the web-based tool STRING v9.1 (http://string-db.org/) [51] to analyze physical and functional protein interactions. Furthermore, an FDR test was applied to determine if the protein list was enriched (higher values meaning higher significances). A Mean Linkage Clustering (MLC or UPGMA), a simple agglomerative hierarchical clustering included in STRING was performed to group proteins. This method clusters proteins based on pairwise similarities in relevant descriptor variables.

Results

Overall assessment of transcriptomic results

Visualization of the spatial distribution of the microarray data of the six studied groups along the three major axis of the PCA is shown in Figure 1B. Component 1 contributed to 63.17% of the variation while the first three components together explained 88.63% of the variation. Two clusters could be observed, one containing group F and the F-derived groups with an accelerated growth (groups FF and SF), and the other formed by group S and the S-derived groups with growth deceleration (groups SS and FS).

The number of DE genes found in the only possible comparison at T1 as well as in the six possible comparisons between the four groups at T2 is shown in Table 1. The comparison with larger number of genes was FS vs. SS, while the FF vs. SF comparison gave no DE genes. From each one of the comparisons with DE genes, the most upregulated and the most downregulated genes (a total of twelve) were selected for a qRT-PCR validation (see details and quality control data of the designed primers in Table S2). All

Table 1. Differentially expressed genes in the different comparisons.

Group comparisons	Total # of genes	# Upregulated genes			# Downregulated genes		
		Total	Real	NA	Total	Real	NA
F vs. S	76	41	20	11	35	20	6
FF vs. SS	155	71	43	9	114	70	30
FS vs. FF	1092	662	316	47	431	153	111
SF vs. SS	94	42	26	1	53	37	9
SF vs. FS	938	507	184	162	604	303	40
FF vs. SF	0	0	0	0	0	0	0
FS vs. SS	2014	1452	717	108	562	261	202

NA, non annotated genes.

genes tested showed the same fold change tendency, thus validating the microarray results (Table 2). Among the tested genes four of them (*cct6a*, *rps15*, *fabp3* and *rpl9*) showed statistical differences ($P<0.05$) when analyzed by qRT-PCR. In the comparisons containing DE genes, analysis of the associated GO terms related to biological processes (BP), molecular function (MF) and cell component (CC) provided further information on the molecular signatures of each treatment (Table S3). Seven selected BP subcategories based on prior knowledge that they take place in the gonads are shown in Figure 2. Metabolic process, response to stimulus and signaling were, in that order, the most represented subcategories. Regarding the MF and CC subcategories, no clear differences were seen among the different comparisons. The most represented MF subcategories among the comparisons were binding and catalytic activities.

Transcriptome of sexually undifferentiated gonads of initial fast-growing vs. initial slow-growing fish (group F vs. group S comparison)

All fish from the F group clustered together and all but one fish from the S group did the same as shown in the heatmap (Figure 3A). Of the total 40 DE genes, among the 20 upregulated there were genes related to transcription, immune response or cytoskeleton structure, whereas among the 20 downregulated ones

there were genes mainly related to mitochondrial functions (Table S4).

Further analyzing the BP subcategories for the up- and downregulated GO terms (Figure 2A and 2B, respectively) showed how the number of GO terms for all the subcategories was always low when compared with T2 group comparisons. Tyrosine-protein kinase gene, a gene involved in male gonad development, was upregulated in the F group. In general, genes related to the response to stimulus and metabolic processes were downregulated. This applied also to genes related to growth such as growth hormone (*gh*) and adrenomedullin, which is related to male gonad development and response to stimulus.

DAVID analysis of DE genes showed two gene clusters within the upregulated genes related to cytoskeleton organization and lumen (enrichment scores 2.6 and 1.14, respectively), and five clusters within the downregulated genes mainly related to mitochondrion, binding, membrane structure and ion binding (enrichment scores 3.36, 1.56, 1.47, 0.31 and 0.22, respectively). KEGG pathway analysis of DE genes showed nine affected pathways: three were upregulated and included T-cell receptor signaling and linoleic acid metabolism, and six were downregulated and mainly related to the metabolism of xenobiotics and amino acid degradation (Table S5).

Table 2. Microarray validation by qRT-PCR.

Comparison	Gene	Microarray FC	Microarray adj. *P*-value	qRT-PCR FC	qRT-PCR SEM	qRT-PCR Student *t*-test
F vs. S	*flna*	2.85	0.004	1.801	0.509	0.332
	tspan1	−5.90	0.007	−2.47	0.185	0.356
FF vs. SS	*cct6a*	2.33	0.001	2.98	0.551	0.009
	rps15	−13.28	0.000	−1.46	0.407	0.001
FS vs. FF	*ggps1*	11.58	0.006	22.63	16.673	0.284
	fabp3	−15.34	0.007	−7.51	0.125	0.001
SF vs. SS	*rpl9*	2.61	0.001	18.79	7.023	0.047
	pcca	−14.03	0.000	−42.13	0.009	0.970
SF vs. FS	*lpl*	13.93	0.006	2.23	0.440	0.364
	tspan13	−10.27	0.006	−240.52	0.002	0.204
FS vs. SS	*ca1*	36.70	0.004	2.49	0.591	0.631
	agpat5	−13.38	0.000	−2.38	0.292	0.849

A

B

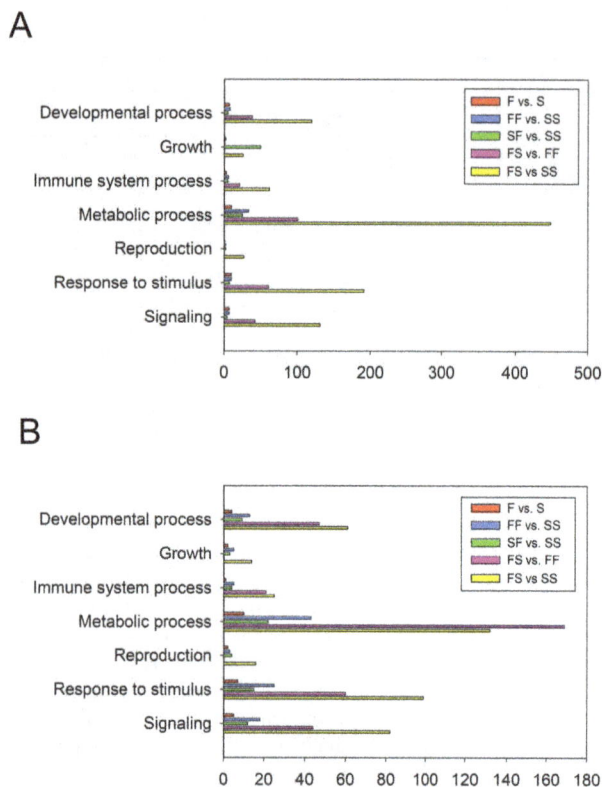

Figure 2. The seven selected GO subcategories for the Biological process (BP) category. A) Upregulated GO-terms and B) downregulated GO terms in the different studied comparisons.

Only seven interactions were found among the proteins coded by these DE genes; however, they were enriched in interactions ($P<0.001$). Proteins from the upregulated DE genes were required for the 60S ribosomal subunit biogenesis and mRNA stability and included proteins such as Ilf3, Nop56, Nop58 and Noc2l, with combined scores of protein-protein interactions ranging from 0.573 to 0.924 (a value of 1 represents the highest possible relationship). Nevertheless, when analyzing the proteins from the downregulated DE genes, three clusters of relationships were observed and related to: 1) respiratory electron transport (Uqcrc2 and Etfa; combined score: 0.969), 2) amino acid degradation (Bckdha and Ivd; combined score: 0.915), and 3) glutathione-mediated detoxification pathway (Gstk1 and Ggh; combined score: 0.899).

Transcriptome of juvenile testes of sustained fast-growing vs. sustained slow-growing fish (group FF vs. group SS comparison)

There were 113 DE genes when comparing the testis of FF and SS groups (43 up- and 70 downregulated genes; see Table S6 for a detailed list). A heatmap visualization of the data (Figure 3B) clearly separated individuals according to their group of origin.

The three most regulated GO terms in the BP category were related to metabolic processes and response to stimulus, followed by developmental process in the upregulated GO terms, and related to signaling for the downregulated GO terms (Figure 2A and 2B, respectively).

DAVID analysis of the DE genes yielded seven up- and 20 downregulated gene clusters mainly related to the Rps and Rpl

ribosomal protein families. KEGG analysis showed twelve altered pathways: three that were upregulated in the FF fish and showed an opposite behavior to what had been observed for the F vs. S comparison (for example, the drug metabolism and the xenobiotics and glutathione metabolisms). There were also nine downregulated pathways related to accelerated growth and metabolism (see Table S7 for a detailed list of the pathways). Although low representation of sequences was found for each pathway, after a FDR test, ribosome was the most enriched pathway among both the up- and downregulated pathways, while proteasome was also highly enriched among the downregulated DE genes.

A Fisher's Exact Test with Multiple Testing Correction of FDR analysis of the most specific terms showed that eight biological processes, three molecular functions and three cell components GO terms were over-represented when using the whole microarray as a background (see Table S8 for further details). Most of the GO terms were related to the ribosome structure and the translation process.

Protein-protein interaction analysis showed that upregulated proteins clustered in three different groups, where the largest one was related to the Rps (six different Rps) and Rpl (seven different Rpl) ribosomal protein families. These groups of proteins are found at the small and large ribosomal subunits, respectively (combined scores ranging from 0.401 to 0.999; Figure 4A). The other two clusters were conformed by the Iars2 and Cct6a proteins, which are related to translation and folding, as well as the 60S ribosomal subunit biogenesis-related proteins (Ube2a, Nop58, Sf3b1 and Cpsf1). On the other hand, downregulated protein interactions clustered in four groups, being the largest formed by the Rpl protein family (nine different Rpl proteins), but also forming part of the small ribosomal subunit and of the proteasome accessory complex (Figure 4B). The other three clusters were conformed by: 1) Agpat2 and Agpat5, which are involved in phospholipid metabolism; 2) Psmd13, Psmd8 and Psmc1, which are involved in ubiquitinated protein degradation; and 3) Prl and Ren, which are mainly involved in growth regulation and apoptosis.

The effects of accelerating growth: Transcriptome of juvenile testes of growth-accelerated fish vs. sustained slow-growing fish (group SF vs. group SS comparison)

Despite significant differences ($P<0.01$) in SL and BW in favor of fish from group SF when compared to fish of the SS group (Table S1), the two groups had a similar sex ratio with a clear male bias (90.6 and 92.2% males, respectively; reported in [19]). However, the transcriptional comparison of the SF vs. the SS group had a low or moderate number of DE genes in the testis transcriptome. A heatmap analysis (Figure 5A) visually representing the 63 DE genes, 26 up- and 37 downregulated genes (see Table S9 for further details), showed that these two groups clustered separately.

BP subcategories were analyzed for the up- and downregulated GO terms (Figure 2A and 2B, respectively). Metabolic process GO terms were the most upregulated and contained five genes that were mainly related to amino acid metabolism (ren, psme1, psmc1, trdmt1 and agpat5). However, renin and prolactin (prl), genes involved in positive regulation of growth, male gonad development, response to hormone stimulus and signaling (hormone-mediated or through G-protein coupled receptors) were downregulated.

DAVID analysis of the data with the highest stringency showed seven clusters for the upregulated genes (enrichment score of 23.68 to 0.42), being protein biosynthesis and translational elongation the most enriched ones. Among the downregulated genes, four

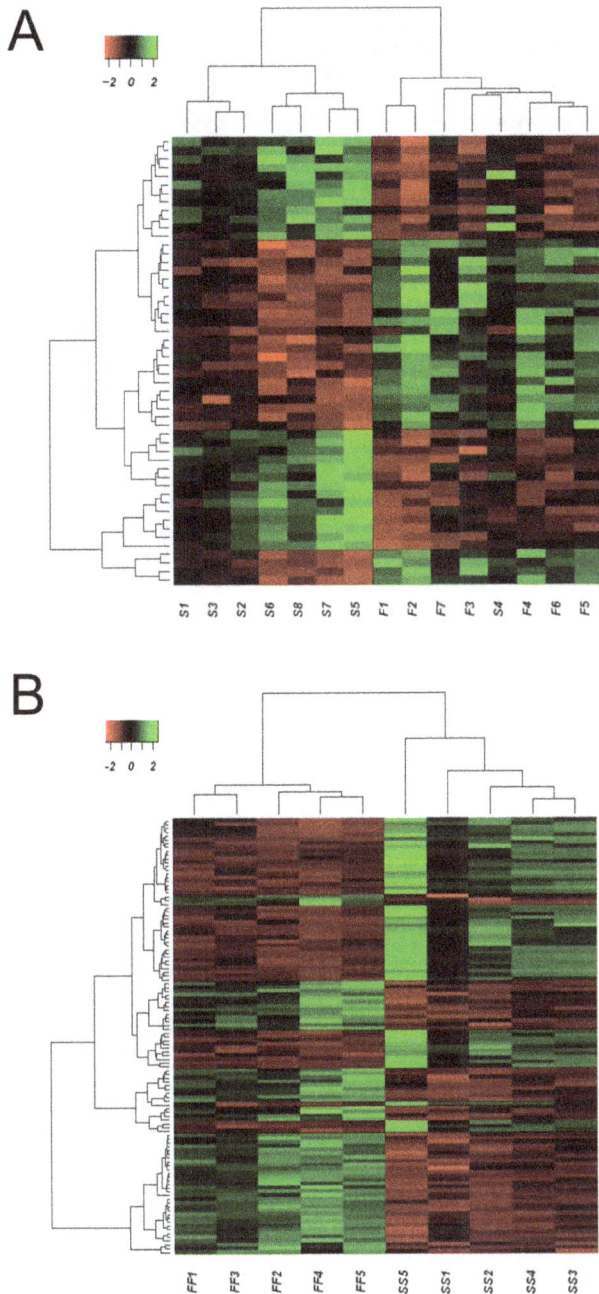

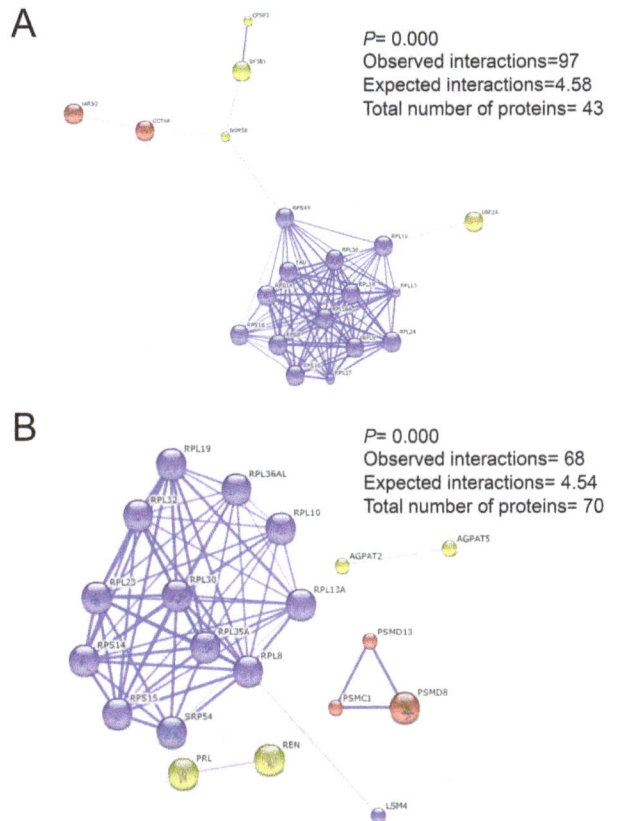

P= 0.000
Observed interactions=97
Expected interactions=4.58
Total number of proteins= 43

P= 0.000
Observed interactions= 68
Expected interactions= 4.54
Total number of proteins= 70

Figure 4. Protein-protein predicted confidence interactions for the FF vs. SS group comparison. A) The 43 proteins from the upregulated DE genes. B) The 70 proteins from the downregulated DE genes.

Figure 3. Individual heatmap representation of the transcriptome analysis. A) undifferentiated gonads of F versus S growing fish. B) differentiated testis of sustained fast (FF) or slow (SS) growing fish. Only DE genes are represented in the figure. High to low expression is shown by a degradation color from green to red, respectively. The scale bar shows Z-score values for the heatmap.

represented when compared against our microarray background and were mainly related to ribosome structure and translational elongation and termination (Table S11).

Protein-protein interaction analysis with an MLC clustering method showed proteins from the upregulated DE genes clustering together and enriched in interactions ($P<0.001$). These proteins were either ribosomal (Rpl and Rps ribosomal protein families) or ribosome associated proteins (e.g., Ef1a1, Ubc or Fau). On the contrary, proteins from the downregulated DE genes were not enriched in interactions ($P = 0.067$) since just one interaction was present between Prl and Cort proteins (data not shown), which are known to be involved in growth control and signaling.

The effects of accelerating growth: Transcriptome of juvenile testes of growth-accelerated fish vs. sustained fast-growing fish (group SF vs. group FF comparison)

Transcriptional analysis of the SF vs. FF group returned zero DE (Table 1) even when we looked for genes with a lower *P*-value (0.05) and lower FC (1.2). The two groups had different sex ratios (90.6% and 67.6% males, respectively) due to their initially different growth rates before size-grading, and FF fish were bigger in BW but not in SL at the time of sampling. However, from a transcriptional point of view, they had no differences, indicating a full recovery from the early naturally slow growth rates.

clusters (enrichment scores from 1.48 to 0.06) were present and mostly related to peptidase activity. KEGG analysis showed 16 pathways altered due to the growth acceleration (three up- and twelve downregulated KEGG pathways; Table S10) that were mostly related to amino acids, glutathione and lipid metabolism. Fisher's Exact Test with Multiple Testing Correction of FDR of the most specific terms showed eight biological processes, three molecular functions and two cellular components that were over-

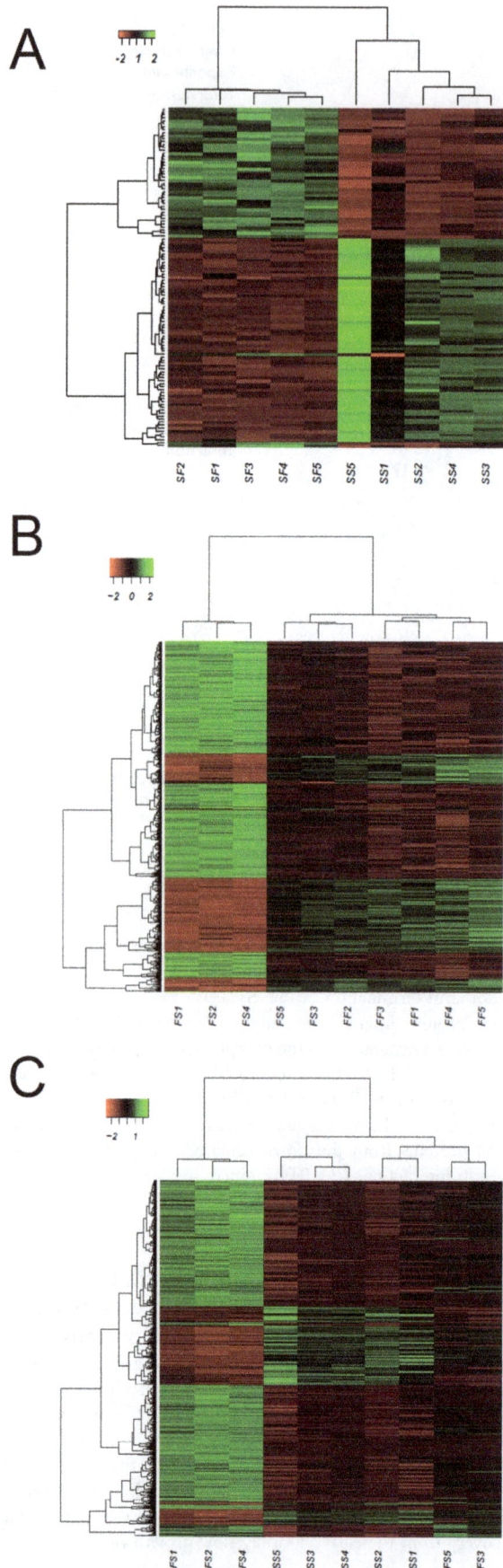

Figure 5. Individual heatmap representation of the transcriptome analysis of European sea bass one-year old testis. A) SF vs. SS comparison. B) FS vs. FF comparison. C) FS vs. SS comparison. Only DE genes are represented in the figure. High to low expression is shown by a degradation color from green to red, respectively. The scale bar shows Z-score values for the heatmap.

The effects of decelerating growth: Transcriptome of juvenile testes of growth-decelerated fish vs. sustained fast-growing fish (group FS vs. group FF comparison)

Fish that experienced the same initial fast-growing rate also had a similar sex ratio (67.6% and 61.4% males, respectively) when compared to the S-derived groups ($P<0.001$), which were highly male-biased (>90%). However, when comparing growth between decelerated fish (FS) vs. sustained fast-growing fish (FF) there were differences in the final growth due to the different feeding regimes (FF>FS in SL and BW) during the sex differentiation period.

Differences at the transcriptomic level were found (469 DE genes: 316 up- and 153 downregulated genes; Table S12). A heatmap visualization of the data (Figure 5B), showed that two FS individuals (FS3 and FS5) shared a transcriptomic pattern with those of the FF group.

The three most regulated GO terms in the BP category were related to metabolic processes, response to stimulus and developmental process in the upregulated GO terms while signaling was for the downregulated subcategory (Figure 2A and 2B, respectively).

DAVID analysis showed 37 clusters from the upregulated genes (enrichment scores from 3.66 to 0.07) and functions were mainly related to proteolysis, regulation of ubiquitin, proteasome and protein modifications processes. On the contrary, downregulated genes (37 clusters; enrichment score from 1.82 to 0.0) had functions mostly related to biosynthesis of phospho- and glycerolipids, anabolic processes and RNA processing and splicing. These DE genes were part of 56 affected pathways (41 upregulated and 15 downregulated; Table S13). Upregulated pathways were the most altered ones after filtering for high stringency and were related to pyrimidine metabolism ($P<0.001$), RNA polymerase ($P<0.05$), oxidative phosphorylation ($P<0.05$), terpenoid backbone biosynthesis ($P<0.05$), epithelial cell signaling ($P<0.05$), purine metabolism ($P<0.05$), glutathione metabolism ($P<0.05$), glycosylphophatidylinositol (GPI)-anchor biosynthesis ($P<0.05$). With this high stringency filtering criteria, only proteasome ($P<0.001$) and ubiquitin mediated proteolysis ($P<0.05$) appeared as being affected among the downregulated pathways.

The Fisher's Exact Test with Multiple Testing Correction of FDR of the most specific terms showed twelve biological processes, eight molecular functions and three cellular components that were over-represented when compared against our microarray as a background and were related to mitochondria and transport activity, while receptor activity was found under-represented (Table S14).

The protein-protein interaction analysis showed that proteins corresponding to both DE up- (four different clusters; Figure S1) and downregulated (ten different clusters; Figure S2) genes were enriched in interactions ($P<0.001$). Upregulated protein clusters were conformed by: 1) proteasome-related proteins (e.g., Psma, Cct6a, Skp1 or Ube2v2), 2) signaling and cholesterol storage-proteins (e.g., Dmd, Mtor or Lpl), 3) transcription regulator proteins (e.g., Max, Pdcd10 or Itgb4), and 4) mitochondrial membrane respiratory chain (e.g., Mt-co1, Mt-nd1 or Mt-nd4). Downregulated proteins clustered in ten different groups but containing a few proteins, with the most enriched ones playing a role in: 1) signaling and protein degradation (e.g., Mapk14, Htra2),

2) translation initiation and protein folding (e.g., Eif4g1, Pdfn1) or 3) transcription (e.g., Polr2h, Gtf3a).

The effects of decelerating growth: Transcriptome of juvenile testes of growth-decelerated fish vs. sustained slow-growing fish (group FS vs. group SS comparison)

The analysis of testes from fish that suffered from a growth deceleration during the sex differentiating period (FS) compared to fish with a sustained slow growth (SS group) showed differences ($P<0.001$) in the final sex ratio (61.4% and 92.2%, respectively) and in the final growth (FS>SS for both SL and BW). These results were further corroborated by the large transcriptomic differences found (978 DE genes: 717 up and 261 downregulated genes; Table S15). Heatmap visualization of data (Figure 5C) showed that two fish from the FS group (FS3 and FS5) clustered with the SS fish.

Analysis of the BP subcategories from the up- and downregulated GO terms (Figure 2A and 2B, respectively) showed how decelerating growth caused the highest changes in all subcategories. None of the genes from these GO terms of the FS vs. SS comparison coincided with those of the other decelerating comparison (FS vs. FF).

Further analysis of the data revealed that these 978 DE genes classified in 71 altered KEGG pathways (62 up- and nine downregulated), and were mostly related to RNA translation and elongation (Table S16). Moreover, clustering analysis of these genes with the highest stringency yielded 82 clusters for the up- and 48 clusters for the downregulated genes. The most enriched upregulated gene clusters were mainly related to the peroxisome, RNA splicing or nucleotide biosynthetic process, while the most enriched downregulated clusters were mostly related to protein catabolism processes, DNA modifications such as methylation, and response to nutrients.

A Fisher's Exact Test with multiple corrections for FDR of the most specific terms gave two BP, one MF and three CC GO terms that were over-represented when comparing to our custom microarray. These GO terms were mainly related to the ribosome structure and translation (Table S17). On the other hand, there was one GO term under-represented and related to the regulation of the immune system.

The highest representation of protein-protein interactions for this comparison (FS vs. SS) showed after a MLC clustering an enrichment in interactions ($P<0.001$), and presented six clusters for the upregulated proteins (Figure S3) related to: 1) ribosomal protein families (Rpl and Rps), 2) post-replication repair of damaged DNA and proteasome (e.g., Rad18 and Psm6, respectively), 3) response to stress (e.g., Tp53, Apex1, Ing1), 4) 60S ribosomal biogenesis and mRNA synthesis (e.g., Nop58, Nop16, Polr2f) and 5) respiratory chain (e.g., Ndufa1, Nduf53, Atp5g1). The twelve clusters for the downregulated proteins (Figure S4) were mainly related to: 1) regulation of metabolic pathways and chromosome stability (e.g., Csnk2b, Mapre1, Tubgcp2), 2) translation initiation (e.g., Eif4a2, Eif4e, Eif3d) and 3) regulation of cell responses (e.g., Prl, Irfl1, Wipi1).

The comparison between fast- and slow-growing groups vs. the FS group, a group that shows high transcriptomic activity but still some transcriptomic similarities with both groups (FF and SS), showed 253 shared DE genes (Figure 6). These were mainly grouped by five main functions: positive regulation of ubiquitination, RNA splicing and mRNA processing, regulation of apoptosis, glycerolipid and phospholipid metabolic process, and regulation of phosphorylation. Among these common 253 DE genes, there were just five genes that showed an opposite pattern of expression: atf4, prelid1, rps17, psma6 and dmd, which were more expressed in the

F-derived groups (FF>FS>SS) and mainly related to the proteasome complex and ribosomal structure. Renin (ren) and prolactin (prl), two genes involved in the positive regulation of growth, growth hormone and G-protein coupled receptor signaling pathways as well as in male gonad development were downregulated in both comparisons (FS vs. SS and FF; Figure 6) with a low expression of these genes in the initial fast-growing groups. This inhibition was also observed in fish with forced accelerated growth when compared to the slow-growing fish (SF vs. SS). These results indicated that these pathways are inhibited when the food availability is altered. Regarding the signaling function, apart from ren and prl, there were four coincident genes with the same pattern of expression in both comparisons (FS vs. SS and FF). Two of them, atf4 and ppkp2, were upregulated in the FS group and involved in unfolded protein response and cell-cell signaling, while the other two, errb and fkbp14, were downregulated in the FS group and involved in steroid hormone- mediated signaling pathway and also in unfolded protein response.

Discussion

The relationship between growth and sex has long been known for the European sea bass at the time of gonadal differentiation, where the largest fish are essentially all females whereas both sexes are found among the smallest fish, although males predominate. Early size-grading experiments (between 66–143 dph) have confirmed this [13,15,17] by obtaining ~90% females among the largest selected fish. Moreover, in a previous study we found that altering growth rates during the sex differentiation period in both size-graded and non-size-graded populations did not alter the sex ratios [19]. Here, it is presented for the first time a microarray analysis of undifferentiated gonads from 4-month-old sea bass with opposite growing rates just after size-grading (T1), and on differentiated testis (11-month-old juveniles, T2). To the best of our knowledge, the question of whether naturally occurring differences in somatic growth are somehow translated in observed transcriptomic differences in the gonads during sex differentiation has never been explored in fish. Nevertheless, what could be called a related type of work was performed in mitten crabs [36], where it was separately analyzed the relationship between nutrition and reproduction, by examining the hepatopancreas and testes transcriptomes, respectively. Interestingly, and regardless the differences among experimental designs and the model organisms used, some traits found in the study with crabs, as the differential expression of some heat shock proteins, cell death suppressors, RNA-dependent DNA polymerases or controllers of splicing, were also found in our study. Similarly to what has been previously reported in fish liver and muscle transcriptomes [22,24,26,52], it is then clear that the juvenile testis is also affected by changes in food supply. Interestingly, there are then common transcriptomic responses with the above mentioned tissues, but not with the brain transcriptomic responses [52].

Fast growing vs. slow growing fish before the onset of SD

Differences between naturally fast- vs. slow-growing (F vs. S) European sea bass of the same family early in development (T1) were not reflected in major transcriptomic changes in their undifferentiated gonads since only 40 DE genes were found. Translation was an active process in F fish gonads since genes coding for ribosomal structure, protein translation and folding were highly expressed. Immune response (positive regulation of apoptosis) and reproduction-related pathways were upregulated, although the only gene directly related to reproduction, the

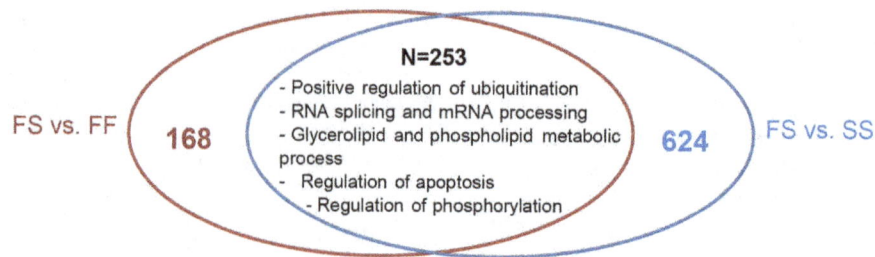

Figure 6. Venn diagram analysis of the DE genes by comparing (FS vs. FF) vs. (FS vs. SS). N represents the total number of common genes and the main categories in which genes are clustered.

tyrosine-protein kinase-like (*ptk*), which is associated with male gonad development function, is also present in other biological functions like cytoskeleton reorganization or cell proliferation. In contrast, gonads from fish that showed the slowest growth before size-grading (group S) were undergoing catabolism processes and protein recycling, since pathways related to negative regulation of growth, protein and amino acid catabolism, protein modification and fatty acid biosynthesis were highly expressed. The lack of abundance in reproduction-related genes among the DE genes present in the F vs. S comparison may be because the custom-made microarray used in this study was more enriched for immunology and growth-related terms rather than for reproduction related terms as it was based on the availability of the public sequences at the time. However, as can be seen in the results of this study, essentially most of the important genes related to sex differentiation found in this and other piscine species are present in this array, which ensures that it fulfills the requirements for such a type of study.

These results indicate that large intrafamily differences in somatic growth within a group of 4-month-old European sea bass due to natural variation are not necessarily translated in a large number of DE genes in their sexually undifferentiated gonads. This is relevant because occurs despite the fact that the groups made selecting the largest fish contain more future females than the groups made selecting the smallest fish, as shown before [13,15,17], and as evidenced by actual subsequent differences in the sex ratios of these groups, since the number of females in the F-derived groups was ~40% while in the S-derived groups it was only ~10% [19]. Thus, there is indeed a clear relationship between growth at the beginning of sex differentiation at 4 months and future sex ratio. Since prior to 4 months growth was not manipulated, growth and future sex can be related, but not food supply and future sex at 4 months. What is also not possible is to relate a given transcriptomic profile at the time of sex differentiation with future sex on an individual fish basis; that is impossible because in order to analyze the transcriptome the fish must be sacrificed. Changes at 4 months probably precede subsequent changes that would account for the differences in sex ratio between the F- and S-derived groups observed when juveniles.

Effects of unrestricted growth on the juvenile testis transcriptome

Food intake is one of the main factors influencing growth rates in aquatic production [25] and food restriction is directly associated with reduced growth rates in fish including the European sea bass [19]. The lack of transcriptomic differences between the SF and FF groups highlights the balance between protein synthesis and degradation, i.e., protein turnover, as one of the most important active processes in the gonads. Protein

turnover relies on proteins mainly obtained from the diet, since high protein contributions from diet or low protein turnover (catabolism) are translated into higher growth rates [53]. In fact, we found lower expression of genes related to catabolism in the accelerated growth group (SF) when compared to the group with sustained growth (FF). The genes related to protein turnover, together with genes involved in the immune system, were also downregulated in the juvenile testis, as it had been found before in the liver of Atlantic salmon (*Salmon salar*) fed with a supplemental diet [26], and it was suggested that this is so because they are involved in the regulation of the decrease in whole body metabolic demands, resulting in less energy wastage and an enhancement in growth performance [24].

Groups with unrestricted growth during the sex differentiation period (FF and SF) showed in common an increased expression of genes related to high protein translation and folding, mainly of proteins related to ribosome structure. This, together with the lack of histological differences between groups, shows how gonads from slow-growing fish can still recover after a period of slow growth if food supply is not a limiting factor thereafter. This is remarkable since it shows the plasticity of the gonads during the sex differentiation period to environmental effects, since fish present a capacity of exploiting a situation of food abundance and recover from initial slow growth.

Effects of restricted food supply on the juvenile testis transcriptome

As also found in the Atlantic salmon liver [26] and white muscle transcriptome [24,54] after food deprivation, protein synthesis and degradation decreased in European sea bass juvenile testes, since both processes are very demanding in terms of energy requirements [53]. It is known that defective or damaged proteins (proteolysis process) are constantly degraded by the proteasome following two main pathways: lysosome or ubiquitin-proteasome pathways [55,56]. Our results show that in European sea bass gonads proteolysis was mainly achieved by the ubiquitin-proteasome pathway rather than the lysosome pathway, as described before in rainbow trout and gilthead sea bream, *Sparus aurata* [22,54,57], since we observed a larger representation of genes involved in the ubiquitin-proteasome pathway. The ubiquitin-proteasome pathway, mainly responsible for protein degradation, was downregulated in the group FS when compared to the group FF and contrasting with the SF and SS groups. Also, food supply restriction was accompanied by a downregulation of genes related to protein synthesis and degradation, and with the immune system, in agreement with previous observations made in the Atlantic salmon liver [26] and in white muscle [24] transcriptomes after food deprivation, although some genes of the complement system [58] were still upregulated. Moreover, a decrease in the lipoprotein levels, in transcription, in tRNA synthesis, and in protein synthesis and elongation in group FS, as a consequence of food deprivation, was coincident with previous results in fasted cod (*Gadus morhua*) as an energy-conserving mechanism [59], and is a common and strong downregulated response of the energy-generating processes in the adipose tissue [60]. This may be because during food deprivation hormonal signals such as growth hormone or insulin levels are translating food restriction signals into a protein turnover change by decreasing the protein synthesis and increasing catabolism [61] to save energy [1,62].

Several studies in fish have analyzed the responses to starvation by measuring catch-up growth [63] or analyzing the effects on the muscle and liver transcriptomes [22,64] but, to our knowledge, this is the first time that a similar study is performed in fish gonads. Two individuals from group FS (FS3 and FS5) clustered with individuals from the FF and SS groups, showing transcriptome similarities. This suggest that these FS fish still conserved some traits of the pre-T1 period but also traits related to the T1–T2 period when food supply was reduced, reflecting the existence of an adaptation to sudden feeding changes [1]. Most of these changes were related to protein turnover. This may be due to the adaptive capacity of fish to sense environmental fluctuations that in turn drive changes in their metabolism, since protein turnover is a highly energy-demanding process [24,53].

The comparison of groups FS vs. FF or SS evidenced some common features of expression where processes such as apoptosis, ubiquitin catabolism, peroxisome, kinase activity or regulation of cellular growth were increased. On the other hand, processes such as proteolysis, regulation of protein modifications, RNA processing, regulation of transcription factors, chromatin assembly, response to nutrients or gamete generation and reproduction were decreased, opposite to what has been found for sea bream heart transcriptome, where transcription was enhanced and transcription inhibitors downregulated [22]. These observations indicate that FS fish had to cope with the dramatic reduction of food intake by saving energy at different levels [1,19,24,26]. This is also supported by the fact that pathways related to metabolism such as lipid mobilization, or purine and pyrimidine metabolism were upregulated, as well as the pathways related to stress response such as the mTOR signaling pathway, which is involved in DNA damage and nutrient deprivation. In contrast, amino acid metabolism, xenobiotic removal or glucose metabolism were downregulated when comparing the FS to the FF and SS groups, showing how the naturally fast-growing fish fed with a non-restrictive diet and later subjected to food restriction did not obtain enough dietary energy and therefore had to start mobilizing lipids and activating gluconeogenesis. Moreover, in agreement with what has been found in both white and red muscle of gilthead sea bream under food restriction [22], mitochondria and ATP transport GO terms were enriched in juvenile testes, a fact that has been proposed as a link between food restriction and stress response mediated by cortisol [22].

No matter which one of the extreme groups the FS group was compared to, processes related to translation and protein regulation such as unfolded protein response, proteasome and postregulation of damaged DNA were highly active. Moreover, processes such as translation initiation, protein folding, transcription and cell-cell signaling were also taking place. Together, these results indicated that although food was scarce and growth was decelerated the transcriptional and the translational machineries of the testis were still active in the FS group. Furthermore, the steroid biosynthesis pathway was downregulated in the FS group when compared to the SF and SS groups, suggesting that the adaptation to the growth decrease could be affecting the energy dedicated to future gonad maturation, although without apparent major consequences, since there were no histological differences when fish were sampled. Reproduction-related processes such as steroid biosynthesis, steroid hormone-mediated signaling and cholesterol storage were affected by growth deceleration, since they were downregulated in the FS group when compared to FF group. However, the FS group showed an increase in GO terms related to spermatogenesis/male gamete generation. This suggests that the FS fish, although being the most different group from a transcriptomic point of view due to food restriction during the sex differentiation period, were still allocating some of the energy in preparation for gonad maturation, which in farmed European sea bass takes place during the second year of life.

Conclusions

To the best of our knowledge, this is the first study evaluating the effects of different growth rates on gonadal development in fish with a transcriptomic approach.

The transcriptome of sexually-undifferentiated gonads was not drastically affected by initial natural differences in growth rates (fish from the opposite sides of the normal distribution curve). In addition, regardless the maturation status of the gonad (T1 and T2), as it has also been shown previously for liver and muscle, the slow-growing fish transcriptome showed an altered protein turnover with a higher catabolism, represented by a reduction in transcription and translation, a decreased immunological response, and a metabolism based on lipids and gluconeogenesis. On the other hand, the transcriptome of fast-growing fish reflects an enhancement of anabolic processes such as transcription, translation, protein synthesis and elongation and a metabolism based on glucose.

In differentiated juvenile gonads, the highest effects on the testis transcriptome were observed when forcing a naturally fast-growing fish to decelerate its growth through food restriction, since those fish showed high transcriptomic differences when compared to the sustained fast-growing fish and even more differences when compared to the fish with sustained slow-grow. These results suggest that food availability during the sex differentiation period is indeed able to modulate the testis transcriptome.

Interestingly, individuals with an initial slow grow but later with an accelerated growth due to increased food supply during sex differentiation showed a recovered transcriptome. These results suggest that fish are able to recover transcriptionally their testes if they are provided with enough food supply during the sex differentiation period. Nevertheless, the opposite is not true, since a natural initial fast growth does not ensure any advantage in terms of transcriptional fitness if later food becomes scarce. These results have implications for natural fish populations subjected to fluctuating food supply in a scenario of global change, as well as for populations or a part thereof of farmed fish under suboptimal feeding regimes, since they provide information on the possible consequences that these situations may have for the reproductive physiology of fish.

Supporting Information

Figure S1 Protein-protein predicted confidence interactions for the FS vs. FS group comparison. The interactions of 266 proteins from the upregulated DE genes are shown. The expected and observed interactions are shown with the significance level.

Figure S2 Protein-protein predicted confidence interactions for the FS vs. FF group comparison. The interactions of 129 proteins from the downregulated DE genes are shown. The expected and observed interactions are shown with the significance level.

Figure S3 Protein-protein predicted confidence interactions for the FS vs. SS group comparison. The interactions of 602 proteins from the upregulated DE genes are shown. The expected and observed interactions are shown with the significance level.

Figure S4 Protein-protein predicted confidence interactions for the FS vs. SS group comparison. The interactions of 206 proteins from the downregulated DE genes are shown. The expected and observed interactions are shown with the significance level.

Table S1 Biometric data of the individuals used for the transcriptomic analysis.

Table S2 Quantitative RT-PCR primer characteristics.

Table S3 List of the number of GO terms found for each category for all the comparisons studied.

Table S4 DE gene list for the F vs. S group comparison.

Table S5 Affected KEGG pathways in the F vs. S group comparison.

Table S6 DE gene list for the FF vs. SS group comparison.

Table S7 Affected KEGG pathways in the FF vs. SS group comparison.

Table S8 Two-tails Fisher's exact test with Multiple Testing Corrections of FDR results for the FF vs. SS group comparison.

Table S9 DE gene list for the SF vs. SS group comparison.

Table S10 Affected KEGG pathways in the SF vs. SS group comparison.

Table S11 Two-tails Fisher's exact test with Multiple Testing Corrections of FDR results for the SF vs. SS group comparison.

Table S12 DE gene list for the FS vs FF group comparison.

Table S13 Affected KEGG pathways in the FS vs. FF group comparison.

Table S14 Two-tails Fisher's exact test with Multiple Testing Corrections for FDR results for the FS vs. FF group comparison.

Table S15 DE gene list for the FS vs. SS group comparison.

Table S16 Affected KEGG pathways in the FS vs. SS group comparison.

Table S17 Two-tails Fisher's exact test with Multiple Testing Correction for FDR results for the FS vs. SS group comparison.

Acknowledgments

Thanks are due to S. Joly for technical assistance and to the staff of our experimental aquarium facilities (ZAE) for assistance with fish rearing.

Author Contributions

Conceived and designed the experiments: FP. Performed the experiments: ND LR. Analyzed the data: ND LR FP. Contributed reagents/materials/analysis tools: FP. Wrote the paper: ND LR FP.

References

1. Schneider JE (2004) Energy balance and reproduction. Physiology and Behavior 81: 289–317.
2. Castellano JM, Roa J, Luque RM, Diéguez C, Aguilar E, et al. (2009) KiSS-1/ kisspeptins and the metabolic control of reproduction: physiologic roles and putative physiopathological implications. Peptides 30: 139–145.
3. Volkoff H, Xu M, MacDonald E, Hoskins L (2009) Aspects of the hormonal regulation of appetite in fish with emphasis on goldfish, Atlantic cod and winter flounder: Notes on actions and responses to nutritional, environmental and reproductive changes. Comparative Biochemistry and Physiology, Part: A 153: 8–12.
4. Morgan MJ, Wright PJ, Rideout MN (2013) Effect of age and temperature of two gadoid species. Fisheries Research 138: 42–51.
5. Yoneda M, Wright PJ (2005) Effects to varying temperature and food availability on growth and reproduction in first-time spawning female Atlantic cod. Journal of Fish Biology 67: 1225–1241.
6. Kjesbu OS (1994) Time of start of spawning in Atlantic cod (*Gadus morhua*) females in relation to vitellogenic oocyte diameter, temperature, fish length and condition. Journal of Fish Biology 45: 719–735.
7. Marshall CT, Yaragina NA, Lambert Y, Kjesbu OS (1999) Total lipid energy as a proxy for total egg production by fish stocks. Nature 402: 288–290.
8. Devlin RH, Nagahama Y (2002) Sex determination and sex differentiation in fish: an overview of genetic, physiological, and environmental influences. Aquaculture 208: 191–364.
9. Vandeputte M, Dupont-Nivet M, Chavanne H, Chatain B (2007) A polygenic hypothesis for sex determination in the European sea bass – *Dicentrarchus labrax*. Genetics 176: 1049–1057.
10. Roblin C, Bruslé J (1983) Gonadal ontogenesis and sex differentiation in the sea bass, *Dicentrarchus labrax*, under fish-farming conditions. Reproduction, Nutrition and Development 23: 115–127.
11. Blázquez M, Piferrer F, Zanuy S, Carrillo M, Donaldson EM (1995) Development of sex control techniques for European sea bass (*Dicentrarchus labrax* L.) aquaculture: Effects of dietary 17 alpha-methyltestosterone prior to sex differentiation. Aquaculture 135: 329–342.
12. Mylonas CC, Anezaki L, Divanach P, Zanuy S, Piferrer F, et al. (2005) Influence of rearing temperature during the larval and nursery periods on growth and sex differentiation in two Mediterranean strains of *Dicentrarchus labrax*. Journal of Fish Biology 67: 652–668.
13. Papadaki M, Piferrer F, Zanuy S, Maingot E, Divanach P, et al. (2005) Growth, sex differentiation and gonad and plasma levels of sex steroids in male- and female-dominant populations of *Dicentrarchus labrax* obtained through repeated size grading. Journal of Fish Biology 66: 938–956.
14. Blazquez M, Carrillo M, Zanuy S, Piferrer F (1999) Sex ratios in offspring of sex-reversed sea bass and the relationship between growth and phenotypic sex differentiation. Journal of Fish Biology 55: 916–930.
15. Koumoundouros G, Pavlidis M, Anezaki L, Kokkari C, Sterioti K, et al. (2002) Temperature sex determination in the European sea bass, *Dicentrarchus labrax* (L., 1758) (Teleostei, Perciformes, Moronidae): Critical sensitive ontogenetic phase. Journal of Experimental Zoology 292: 573–579.
16. Saillant E, Chatain B, Menu B, Fauvel C, Vidal MO, et al. (2003) Sexual differentiation and juvenile intersexuality in the European sea bass (*Dicentrarchus labrax*). Journal of Zoology 260: 53–63.

17. Saillant E, Fostier A, Haffray P, Menu B, Laureau S, et al. (2003) Effects of rearing density, size grading and parental factors on sex ratios of the sea bass (*Dicentrarchus labrax* L.) in intensive aquaculture. Aquaculture 221: 183–206.

18. Navarro-Martín L, Blázquez M, Viñas J, Joly S, Piferrer F (2009) Balancing the effects of rearing at low temperature during early development on sex ratios, growth and maturation in the European sea bass (*Dicentrarchus labrax*). Limitations and opportunities for the production of highly female-biased stocks. Aquaculture 296: 347–358.

19. Díaz N, Ribas L, Piferrer F (2013) The relationship between growth and sex differentiation in the European sea bass (Dicentrarchus labrax). Aquaculture 408–409: 191–202.

20. Adams S, McLean R, Parrotta J (1982) Energy partioining in largemouth bass under conditions of seasonally fluctuating prey availability transactions. American Fisheries Society 111: 549–558.

21. Geay F, Ferraresso S, Zambonino-Infante JL, Bargelloni L, Quentel C, et al. (2011) Effects of the total replacement of fish-based diet with plant-based diet on the hepatic transcriptome of two European sea bass (*Dicentrarchus labrax*) half-sib families showing different growth rates with the plant-based diet. BMC Genomics 12: 522.

22. Calduch-Giner JA, Sitja-Bobadilla A, Davey GC, Cairns MT, Kaushik S, et al. (2012) Dietary vegetable oils do not alter the intestine transcriptome of gilthead sea bream (*Sparus aurata*), but modulate the transcriptomic response to infection with *Enteromyxum leei*. BMC Genomics 13: 470.

23. Campos C, Valente LMP, Borges P, Bizuayehu T, Fernandes JMO (2010) Dietary lipid levels have a remarkable impact on the expression of growth-related genes in Senegalese sole (*Solea senegalensis* Kaup). The Journal of Experimental Biology 213: 200–209.

24. Tacchi L, Bickerdike R, Douglas A, Secombes CJ, Martin SAM (2011) Transcriptomic responses to functional feeds in Atlantic salmon (*Salmo salar*). Fish and Shellfish Immunology 31: 704–715.

25. Salas-Leiton E, Anguis V, Martín-Antonio B, Crespo D, Planas JV, et al. (2010) Effects of stocking density and feed ration on growth and gene expression in the Senegalese sole (*Solea senegalensis*): Potential effects on the immune response. Fish and Shellfish Immunology 28: 296–302.

26. Martin SAM, Douglas A, Houlihan DF, Secombes CJ (2010) Starvation alters the liver transcriptome of the innate immune response in Atlantic salmon (*Salmo salar*). BMC Genomics 11: 418.

27. Yi SK, Gao ZX, Zhao HH, Zeng C, Luo W, et al. (2013) Identification and characterization of microRNAs involved in growth of blunt snout bream (*Megalobrama amblycephala*) by Solexa sequencing. BMC Genomics 14: 754.

28. Overtuf K, Skhrani D, Devlin RH (2010) Expression profile for metabolic and growth-related genes in domesticated and transgenic coho salmon (*Oncorhynchus kisutch*) modified for increased growth hormone production. Aquaculture 307: 111–122.

29. Sun F, Liu S, Gao X, Jiang Y, Perera D, et al. (2013) Male-biased genes in catfish as revealed by RNA-seq analysis of the testis transcriptome. PLoS ONE 8: e68452.

30. Ravi P, Jiang J, Liew WC, Orban L (2014) Small-scale transcriptomics reveals differences among gonadal stages in Asian seabass (*Lates calcarifer*). Reproductive Biology and Endocrinology 12: 5.

31. Tao W, Yuan J, Zhou L, Sun L, Sun Y, et al. (2013) Characterization of gonadal transcriptomes from Nile Tilapia (*Oreochromis niloticus*) reveals differentially expressed genes. Plos ONE 8: e63604.

32. Rolland AD, Lareyre JJ, Goupil AS, Montfort J, Ricordel MJ, et al. (2009) Expression profiling of rainbow trout testis development identifies evolutionary conserved genes involved in spermatogenesis. BMC Genomics 10: 546.

33. Bozinovic G, Oleksiak MF (2011) Omics and environmental science genomic approaches with natural fish populations from polluted environments. Environmental Toxicology and Chemistry 30: 283–289.

34. Schiller V, Wichmann A, Kriehuber R, Muth-Kohne E, Giesy JP, et al. (2013) Studying the effects of genistein on gene expression of fish embryos as an alternative testing approach for endocrine disruption. Comparative Biochemistry and Physiology C-Toxicology and Pharmacology 157: 41–53.

35. Schiller V, Wichmann A, Kriehuber R, Schafers C, Fischer R, et al. (2013) Transcriptome alterations in zebrafish embryos after exposure to environmental estrogens and anti-androgens can reveal endocrine disruption. Reproductive Toxicology 42: 210–223.

36. Jiang H, Yin Y, Zhang X, Hu S, Wang Q (2009) Chasing relationships between nutrition and reproduction: A comparative transcriptome analysis of hepato-pancreas and testis from Eriocheir sinensis. Comparative Biochemistry and Physiology, Part D: Genomics and Proteomics 4: 227–234.

37. Martin P, Kohlmann K, Scholtz G (2007) The parthenogenetic Marmorkrebs (*Marbled crayfish*) produces genetically uniform offspring. Naturwissenschaften 94: 843–846.

38. Martin B, Pearson M, Brenneman R, Golden E, Wood W III, et al. (2009) Gonadal Transcriptome alterations in response to dietary energy intake: sensing the reproductive environment. PloS One 4: e4146.

39. Blazquez M, Navarro-Martin L, Piferrer F (2009) Expression profiles of sex differentiation-related genes during ontogenesis in the European sea bass

40. Moretti A, Pedini M, Cittolin G, Guidastri R (1999) Manual on hatchery production of seabass and gilthead seabream. FAO, Roma.

41. Schmittgen TD, Livak KJ (2008) Analyzing real-time PCR data by the comparative CT method. Nature Protocols 3: 1101–1108.

42. Ritchie ME, Silver J, Oshlack A, Holmes M, Diyagama D, et al. (2007) A comparison of background correction methods for two-colour microarrays. Bioinformatics 23: 2700–2707.

43. Bolstad B (2001) Probe level quantile normalization of high density oligonucleotide array data. Unpublished manuscript from the Division of Biostatistics, University of California, Berkely.

44. Smyth G (2005) Limma: linear models for microarray data. In: Gentleman R, Carey VJ, Huber W, Irizarry RA, Dudoit S, editors. Bioinformatics and computational biology solutions using R and Bioconductor. Springer New York. pp. 397–420.

45. Gentleman R, Carey V, Bates D, Bolstad B, Dettling M, et al. (2004) Bioconductor: open software development for computational biology and bioinformatics. Genome Biology 5: R80.

46. Carbon S, Ireland A, Mungall CJ, Shu S, Marshall B, et al. (2009) AmiGO: online access to ontology and annotation data. Bioinformatics 25: 288–289.

47. Conesa A, Gotz S, Garcia-Gomez JM, Terol J, Talon M, et al. (2005) Blast2GO: a universal tool for annotation, visualization and analysis in functional genomics research. Bioinformatics 21: 3674–3676.

48. Huang DW, Sherman BT, Lempicki RA (2009) Systematic and integrative analysis of large gene lists using DAVID Bioinformatics Resources. Nature Protocols 4: 44–57.

49. Huang DW, Sherman BT, Lempicki RA (2009) Bioinformatics enrichment tools: paths toward the comprehensive functional analysis of large gene lists. Nucleic Acids Research 37: 1–13.

50. Benjamini Y, Hochberg Y (1995) Controlling the false discovery rate: a practical and powerful approach to multiple testing. Journal of the Royal Society Series B 57: 289–300.

51. Franceschini A, Szklarczyk D, Frankild S, Kuhn M, Simonovic M, et al. (2013) STRING v9.1: protein-protein interaction networks, with increased coverage and integration. Nucleic Acids Research 41: D808–815.

52. Drew RE, Rodnick KJ, Settles ML, Wacyk J, Churchill EJ, et al. (2008) Effect of starvation on the transcriptomes of the brain and liver in adult female zebrafish (*Danio rerio*). Physiological Genomics 35: 283–295.

53. Houlihan DF, Carter CG, McCarthy ID (1995) Protein synthesis in fish; Hochachka M, editor. Amsterdam: Elsevier.

54. Martin SAM, Blaney S, Bowman AS, Houlihan DF (2002) Ubiquitin-proteasome-dependent proteolysis in rainbow trout (*Oncorhynchus mykiss*), effect of food deprivation. European Journal of Applied Physiology 445: 257–266.

55. Tanaka K, Chiba T (1998) The proteasome. a protein/destroying machine. Genes to Cells 3: 499–510.

56. Craiu A, Akopian T, Goldberg A, Rock KL (1997) Two distinct proteolytic processes in the generation of a major histocompatibility complex class I-presented peptide. Proceedings of the National Academy of Sciences 94: 10850–10855.

57. Palstra A, Beltran S, Burgerhout E, Brittijn S, Magnoni L, et al. (2013) Deep RNA sequencing of the skeletal muscle transcriptome in swimming fish. PLoS ONE 8: e53171.

58. Boshra H, GElman AE, Sunyer JO (2004) Structural and functional characterization of complement C4 and C1s/like molecules in teleost fish. Insights into the evolution of classical and alternative pathways. Journal of Immunology 171: 349–359.

59. Kjaer MA, Vegusdal A, Berge GM, Galloway TF, Hillestad M, et al. (2009) Characterisation of lipid transport in Atlantic cod (*Gadus morhua*) when fasted and fed high or low fat diets. Aquaculture 288: 325–336.

60. Higami Y, Pugh T, Page G, Allison D, Prolla T, et al. (2004) Adipose tissue energy metabolism: altered gene expression profile of mice subjected to long-term caloric restriction. FASEB Journal 18: 415–417.

61. Gabillard JC, Kamangar BB, Monstserrat N (2006) Coordinated regulation of the GH/IGF system genes during refeeding in rainbow trout (*Oncorhynchus mykiss*). Journal of Endocrinology 191: 15–24.

62. Dobly A, Martin SAM, Blaney SC, Houlihan DF (2004) Protein growth rate in rainbow trout (*Oncorhynchus mykiss*) is negatively correlated to liver 20S proteasome activity. Comparative Biochemistry and Physiology, Part A: Molecular and Integrative Physiology 137: 75–85.

63. Rescan PY, Montfort J, Ralliere C, Le Cam A, Esquerre D, et al. (2007) Dynamic gene expression in fish muscle during recovery growth induced by a fasting-refeeding schedule. BMC Genomics 8: 438.

64. Salem M, Kenney PB, Rexroad CE, Yao JB (2006) Molecular characterization of muscle atrophy and proteolysis associated with spawning in rainbow trout. Comparative Biochemistry and Physiology, Part D: Genomics and Proteomics 1: 227–237.

acclimated to two different temperatures. Journal of Experimental Zoology, Part B: Molecular and Developmental Evolution 312B: 686–700.

10

A Transcriptome for the Study of Early Processes of Retinal Regeneration in the Adult Newt, *Cynops pyrrhogaster*

Kenta Nakamura[2], Md. Rafiqul Islam[1], Miyako Takayanagi[1], Hirofumi Yasumuro[1], Wataru Inami[1], Ailidana Kunahong[1], Roman M. Casco-Robles[1], Fubito Toyama[3]*, Chikafumi Chiba[2]*

1 Graduate School of Life and Environmental Sciences, University of Tsukuba, Tsukuba, Ibaraki, Japan, **2** Faculty of Life and Environmental Sciences, University of Tsukuba, Tsukuba, Ibaraki, Japan, **3** Graduate School of Engineering, Utsunomiya University, Utsunomiya, Tochigi, Japan

Abstract

Retinal regeneration in the adult newt is a useful system to uncover essential mechanisms underlying the regeneration of body parts of this animal as well as to find clues to treat retinal disorders such as *proliferative vitreoretinopathy*. Here, to facilitate the study of early processes of retinal regeneration, we provide a *de novo* assembly transcriptome and inferred proteome of the Japanese fire bellied newt (*Cynops pyrrhogaster*), which was obtained from eyeball samples of day 0–14 after surgical removal of the lens and neural retina. This transcriptome (237,120 *in silico* transcripts) contains most information of cDNAs/ESTs which has been reported in newts (*C. pyrrhogaster*, *Pleurodeles waltl* and *Notophthalmus viridescence*) thus far. On the other hand, *de novo* assembly transcriptomes reported lately for *N. viridescence* only covered 16–31% of this transcriptome, suggesting that most constituents of this transcriptome are specific to the regenerating eye tissues of *C. pyrrhogaster*. A total of 87,102 *in silico* transcripts of this transcriptome were functionally annotated. Coding sequence prediction in combination with functional annotation revealed that 76,968 *in silico* transcripts encode protein/peptides recorded in public databases so far, whereas 17,316 might be unique. qPCR and Sanger sequencing demonstrated that this transcriptome contains much information pertaining to genes that are regulated in association with cell reprogramming, cell-cycle re-entry/proliferation, and tissue patterning in an early phase of retinal regeneration. This data also provides important insight for further investigations addressing cellular mechanisms and molecular networks underlying retinal regeneration as well as differences between retinal regeneration and disorders. This transcriptome can be applied to ensuing comprehensive gene screening steps, providing candidate genes, regardless of whether annotated or unique, to uncover essential mechanisms underlying early processes of retinal regeneration.

Editor: Panagiotis A. Tsonis, University of Dayton, United States of America

Funding: This work was supported by grants from the Ministry of Education, Culture, Sports, Science and Technology (23124502; 221S0002) and the Japan Society for the Promotion of Science (24650229; 25870096; 21300150; 24240062). The funders had no role in study design, data collection and analysis, decision to publish, or preparation of the manuscript.

Competing Interests: The authors have declared that no competing interests exist.

* Email: fubito@is.utsunomiya-u.ac.jp (FT); chichiba@biol.tsukuba.ac.jp (CC)

Introduction

The newt has long been recognized as a master of regeneration from which the principles for the regeneration of body parts following traumatic injury can be learnt. This animal has an outstanding ability, when metamorphosed or sexually matured, or even as an aged adult, to regenerate missing body parts – including a part of a limb, the jaw, the tail (the spinal cord), the brain, the heart, the eye (the lens and the retina) – from remaining tissues at the site of injury. This takes place through dedifferentiation–redifferentiation/-transdifferentiation (or reprogramming) of terminally differentiated cells as well as recruitment of endogenous stem/progenitor cells, their proliferation and patterning, and physiological integration of regenerates into the body system [1,2]. Although ample endeavors have been made to understand these surprising phenomena for over a century, the underlying cellular mechanisms and molecular networks are still largely uncertain primarily because of our technical limitations. However, studies in this field are now moving forward with increasing speed by

incorporating highly efficient technologies to analyze gene functions [3,4] and transcriptomes [5,6]. For studies of the newt whose genomic size is too large to be sequenced (estimated ~20 Gbp [7]), construction of an *in silico* transcriptome by RNAseq/*de novo* assembly is certainly cost effective, and the resulting data sets are highly informative for bench screening candidate genes.

Regeneration of the adult newt retina is a good system to study the early processes of regeneration, particularly the mechanisms underlying reprogramming, cell-cycle re-entry/proliferation and patterning, because of its simplicity. The cell source for retinal regeneration in the posterior eye is the retinal pigment epithelium (RPE) cells only. The RPE, which is a highly specialized monolayer lining the back of the neural retina, expresses specific molecular markers such as RPE65, allowing the RPE cells and RPE-derived cells to be tracked in an early phase of retinal regeneration [8]. In addition, this system has an obvious medical target, namely *proliferative vitreoretinopathy* (*PVR*) in which RPE cells proliferate and transform in response to retinal injury, leading to the loss of vision [9].

Among all newt species, the Japanese fire bellied newt (*Cynops pyrrhogaster*) is the best choice for the study of retinal regeneration since: 1) a surgical procedure to induce retinal regeneration has been established; 2) morphological stages of retinal regeneration have been defined in detail; 3) both *in vivo* and *in vitro* functional gene assay systems are being developed [4,8,10,11]. However, to facilitate the study of retinal regeneration in this species, one considerable obstacle still remains: limited information on genes. Thus, in the current study, to overcome this problem we carried out mRNA-seq/*de novo* assembly in this species, providing an *in silico* reference transcriptome specialized for the study of early processes of retinal regeneration.

Materials and Methods

Ethics statement

This study using the Japanese fire bellied newt *C. pyrrhogaster* was permitted by the University of Tsukuba Animal Use and Care Committee (AUCC). Surgical removal of the neural retina (retinectomy) and sacrifice were carried out under anesthesia [anesthetic: FA100 (4-allyl-2-methoxyphenol); DS Pharma Animal Health, Osaka, Japan] to minimize suffering [8]. No other *in vivo* experiments were done.

The field study did not involve endangered or protected species. The newts were originally captured by a provider (Mr. Kazuo Ohuchi, Misato, Saitama, 341-0037 Japan; http://homepage3. nifty.com/monmo51-kaeru/index.html) using a net from canals along the rice paddies located within ~25 km in diameter around a Miyayama area (35.130013,140.013842) in Kamogawa city, Chiba prefecture, Japan [4]. No specific permissions were required for the location of capture.

Newt strain

The newts, which have been captured since 2008 (~300/year) (see Ethics statement), have been stocked/cultured in both the laboratory (Univ. of Tsukuba) and a field 'Imori-no-Sato' (Kaizuka/Kamitakai, Toride city, Ibaraki prefecture, Japan; http://imori-net.org/) [4]. This population belongs to Kanto group in Northern lineage [12] and is called 'Toride-Imori'. In this study, sexually mature adult Toride-Imori (total body length: male, ~9 cm; female, 11–12 cm) which had been reared in the laboratory were used.

Housing condition

In the laboratory, the animals had been reared in containers/ aquarium tanks (≤1 newt/base area of 50 cm^2 square; the water depth was 5–15 cm; a stone/a piece of kitchen sponge was placed therein, serving as land) at ~18°C under a natural light condition; they had been fed with frozen mosquito larvae (Akamushi; Kyorin Co., Ltd., Japan) every day and the containers/aquarium tanks had been kept clean [4].

Anesthesia

For retinectomy, the animals were anesthetized as follows: 0.3 ml of the anesthetic FA100 (4-allyl-2-methoxyphenol) was poured in 300 ml tap water (~22°C) in a bottle (φ of the bottom: ~7 cm; height: ~5 cm) with a lid; the bottle was sealed immediately and shaken several times so that the solution is mixed well; the newts (up to 5) were placed in this solution (i.e., 0.1% FA100) and the bottle was sealed again immediately; the bottle containing the animals was placed in the dark at room temperature (~22°C) for 2 h, allowing dark adaptation of the

retina which makes the adherence between the neural retina and the RPE weaker. After this treatment, they were rinsed in distilled water (DW) and wiped with a paper towel. Under this condition, they did not show the pupillary reflex during surgery, and not awake for at least 4 h.

For sacrifice, intact animals and those of day-14 or later after retinectomy whose wound has closed were anesthetized as done for retinectomy. However, in the case of animals between 4 h and day-14 after retinectomy, an alternative anesthetic method was applied to avoid damage of RPE cells due to invasion of the hypotonic anesthetic solution into the eye chamber: the animals were injected with 100 μl of 20% FA100 (dissolved in PBS) into their abdominal cavity through a fine needle (27Gx3/4″, NN-2719S, Terumo, Tokyo, Japan) connected to a syringe (1 ml, SS-01T, Terumo); they fell asleep in 30 min. For animals within 4 h after retinectomy, additional anesthetic treatment was not done.

Retinectomy

To induce retinal regeneration, the neural retina was removed, together with the lens, from the left eye (the eyeball size: 2 mm in diameter) of a living animal as follows ([8]; see Figure 1A): the mouth cavity of an anesthetized animal was stuffed with a roll of the absorbent cotton so that the eyeball is pushed out from the eyelid; the animal was held on the silicon bottom of an operating chamber so that the left eye faces up, using a U-shaped pin which was mounted on the neck of the animal and stuck onto the silicon bottom of the chamber; the animal on the chamber was placed under a binocular, and the dorsal half of the left eye was cut open along the position slightly below the boundary between the cornea and sclera using a blade and fine scissors; both the neural retina and the lens were carefully removed by a fine injection needle (27Gx3/4″, NN-2719S, Terumo) and forceps, while gently infusing a sterile saline solution (in mM: NaCl, 115; KCl, 3.7; CaCl$_2$, 3; MgCl$_2$, 1; D-glucose, 18; HEPES, 5; pH 7.5 adjusted with 0.3N NaOH) into the vitreous chamber through the same injection needle which was connected to a syringe (1 ml, SS-01T, Terumo) via a filter cassette (0.20 μm pore size, Cellulose acetate, DISMIC-25CS, ADVANTEC, Japan); at this time the retinal margin containing the *ora serrata* (the tissue harboring the retinal stem/progenitor cells) which remained along the base of the ciliary epithelium was also removed by forceps. After operation, the eye flap consisting of the iris and cornea was carefully placed back to its original position. The operated animals were placed on a paper towel (lightly wet with DW) in a plastic container [≤5 newts/ container (width: 14 cm×depth: 19 cm×height: 4 cm)] and allowed to recover, and then reared in an incubator (~22°C; the day-night cycle was 12 h:12 h) until use (up to 14 days in this study). In the mean time, the containers were kept clean and the animals were not fed to control the speed of regeneration. The stage of retinal regeneration and corresponding day post-operation (po) were determined according to previous criteria ([8]; see Figure 1B).

Collection of eyeballs

To collect eyeballs, the anesthetized animals (dried on a paper towel) were decapitated. The head was fixed on the silicon bottom of the operating chamber with a marking pin, and placed under the binocular. The eyeballs were carefully enucleated with fine scissors and forceps. In the case of retinectomized animals, especially before day-14 po, this operation was made with minute attention because their wounded eyes were very fragile.

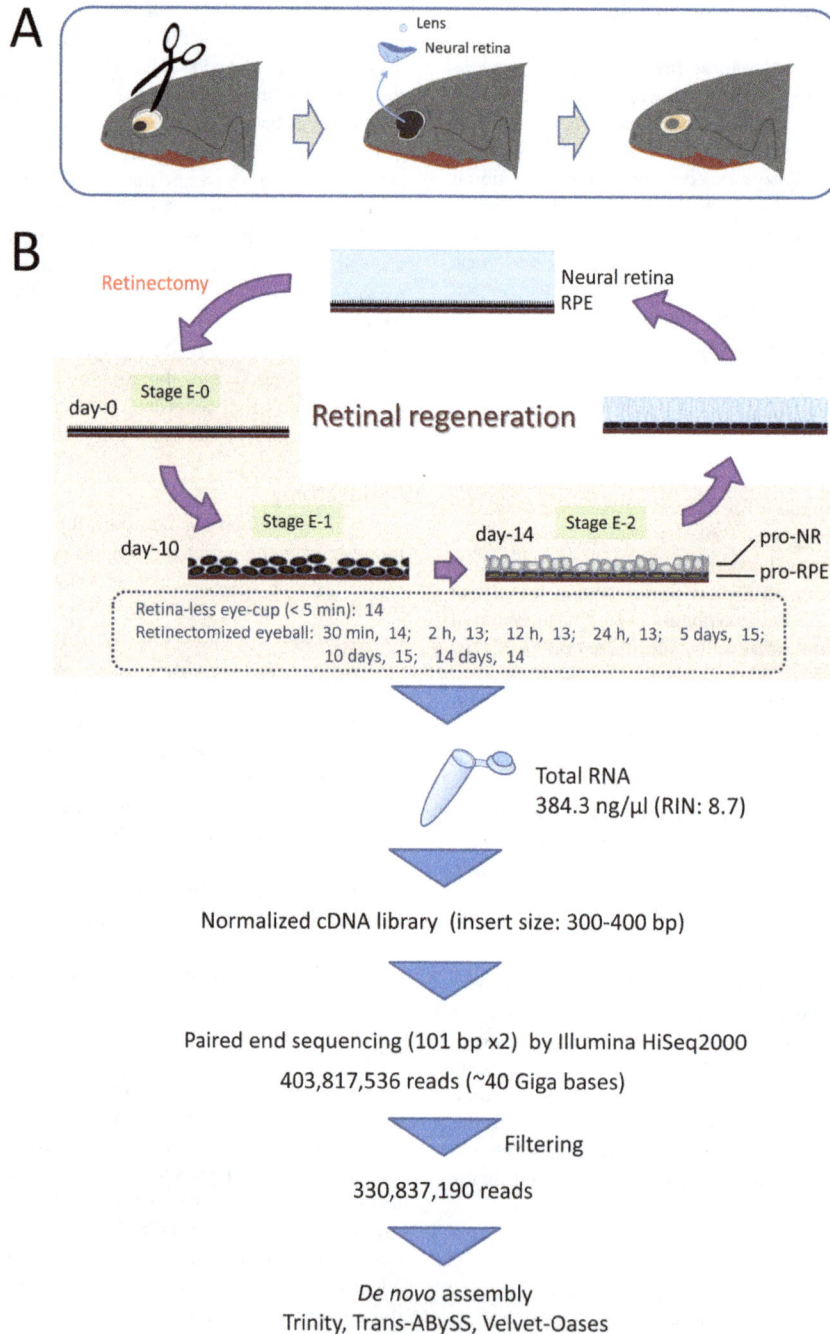

Figure 1. Workflow from retinectomy to *de novo* assembly. A. Retinectomy. **B.** Sample collection, mRNA-seq and *de novo* assembly. *Stage E-0*: The RPE immediately after retinectomy. *Stage E-1*: Almost all RPE cells that have lost their epithelial characteristics and formed aggregates have entered the S-phase of the cell-cycle. *Stage E-2*: Partially depigmented cells are segregated into two rudiment layers (pro-NR and pro-RPE), which give rise to a new neural retina and the RPE layer itself. Under the current experimental conditions, regenerating retinas at Stage E-1 and E-2 are obtained at day-10 and -14 po [8].

Tissue samples for analyses

For *de nove* assembly transcriptome, the workflow is illustrated in Figure 1B. To obtain the transcriptome involved in early processes of retinal regeneration, especially for the study of reprogramming, cell-cycle re-entry/proliferation and patterning in full, retinectomized eyeballs (time po: 30 min, 2 h, 12 h, 24 h, 5 days, 10 days, 14 days) and the retina-less eye-cups (RLECs) of normal eyeballs were used. These retinectomized eyeballs should

have contained RPE cells which just received somewhat onset-signals for regeneration [13,11] as well as those which have undergone mitosis (stage E-1) and segregation into two layers, i.e., rudiments for a new neural retina and RPE (stage E-2) ([8]; see Figure 1B for the morphological stages of retinal regeneration). RLECs were used as a source of intact RPE cells (stage E-0). This sample was prepared as described previously [8] with some modifications: a normal eyeball was placed, the cornea side up,

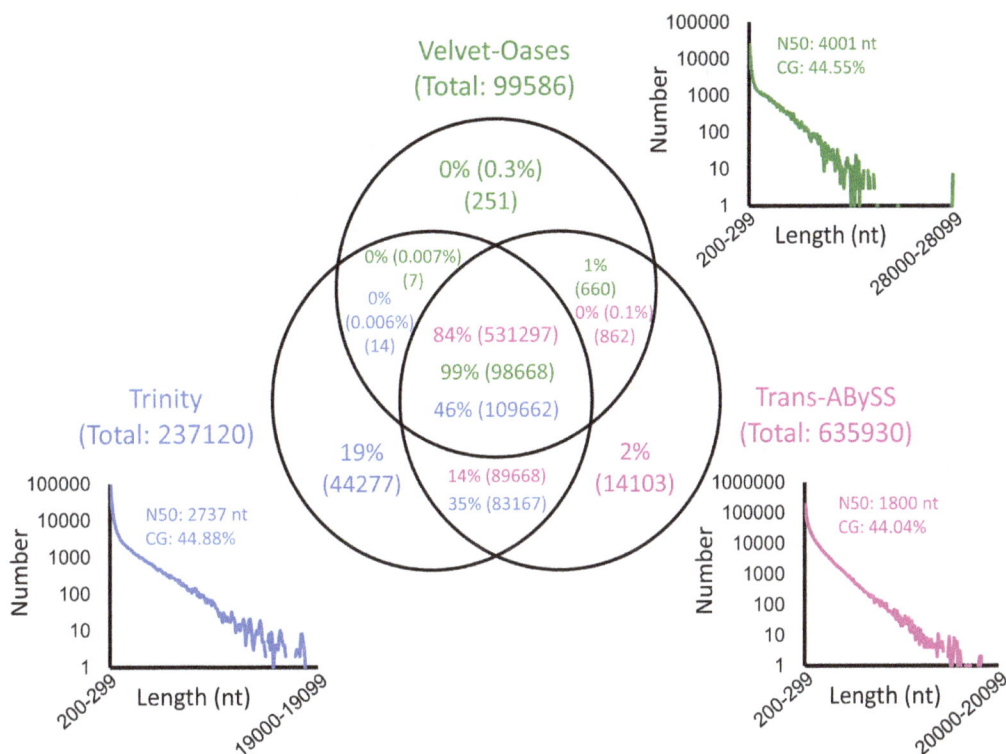

Figure 2. Comparisons between transcriptomes deduced by different assemblers. Scatter plot graphs show the length distribution of *IS*-transcripts deduced by Trinity (blue), Trans-ABySS (magenta) and Velvet-Oases (green). As suggested by the total number and N50 length of *IS*-transcripts, Velvet-Oases tended to give long *IS*-transcripts at the expense of their total number, while Trans-ABySS deduces a large number of short *IS*-transcripts. Trinity was intermediate with respect to these parameters. The Venn diagram shows how much one assembler covers the *IS*-transcripts deduced by the other assemblers. The number of *IS*-transcripts is given in parentheses. Trinity covered almost all the *IS*-transcripts deduced by either Velvet-Oases or Trans-ABySS.

onto a filter membrane (MF™ Membrane Filter, 0.45 μm HA, HAWP01300, Merck Millipore, Darmstadt, Germany) in a 35 mm plastic dish (Falcon 353001, Becton Dickinson, NJ 07417-1886), and then cut along the equator by manipulating a blade and scissors; RNase-free PBS chilled on ice was gently poured onto the eyeball; after the dish was filled up with this saline, the anterior half of the eyeball containing the iris and lens was removed to make the eye-cup; the neural retina was carefully separated from the RPE with a fine pin and forceps, and then removed from the rest of the eye-cup, i.e., the RLEC, by cutting the optic nerve with fine scissors. This operation was completed within 5 min.

For quantitative PCR (qPCR), both the right (intact) and left (retinectomized) eyeballs of animals at day-10 and -14 po were used; that is, the right eyeballs were used for day-0 po sample containing intact RPE (stage E-0), and the left ones were for samples of day-10 po (stage E-1) or day-14 po (stage E-2). Tissue samples for RNA isolation were prepared as follows. After the animal was decapitated under anesthesia, the right and left eyeballs were collected in different dishes (filled with RNase-free PBS) on ice. The right eyeball was immediately dissected into the RLEC, and the RPE sheet together with the choroid tissues was isolated by separating these from the sclera using a fine pin and forceps. On the other hand, the left eyeball of day-10 or -14 po, which was put on the filter membrane and soaked in chilled RNase-free PBS as done for the right eyeball, was carefully opened from the wound at the time of retinectomy using a fine pin and scissors; after the anterior part of the eyeball containing the iris and the ciliary marginal zone was carefully removed, RPE -derived cells in the posterior eye were collected

together with the choroid tissues as done for the right eyeball. After blood cells in the choroid were removed as much as possible by shaking them in the dish, each day samples were transferred into different tubes (containing RNase-free PBS) on ice with a pipette (3.5 ml Transferpipette, Sarsted, D-51588 Nümbrecht, Germany). For one round of RNA purification, this process was repeated for at least 5 animals, collecting 5 each-day samples (good samples only) at once.

Library construction and sequencing for *de novo* assembly transcriptome

A total of 14 RLECs and 97 retinectomized eyeballs (30 min, 14; 2 h, 13; 12 h, 13; 24 h, 13; 5 days, 15; 10 days, 15; 14 days, 14) were harvested, under a conventional nuclease free condition, one by one in a 50 ml tube containing liquid nitrogen, and stored in −80°C until use (see the workflow in Figure 1B). Total RNA was purified from them all by the Isogen protocol (Nippon Gene, Tokyo, Japan) according to the manufacturer's instructions and evaluated with an Agilent 2100 Bioanalyzer (Agilent Technology, Santa Clara, CA 95051). Using a qualified RNA sample (384.3 ng/μl, RIN: 8.7), a normalized cDNA library (insert size: 300–400 bp) was constructed with a TrueSeq RNA Sample Prep Kit (Illumina, San Diego, CA 92122) followed by Duplex Specific Nuclease normalization (Illumina) according to the manufacturer's instructions. Subsequently, paired end sequencing (101 bp read×2) was carried out by Illumina HiSeq2000 and a set of raw read data [403,817,536 reads; 40,786 Mbases; % of high quality base (≥Q30): 92.74; Mean quality score: 35.92] was

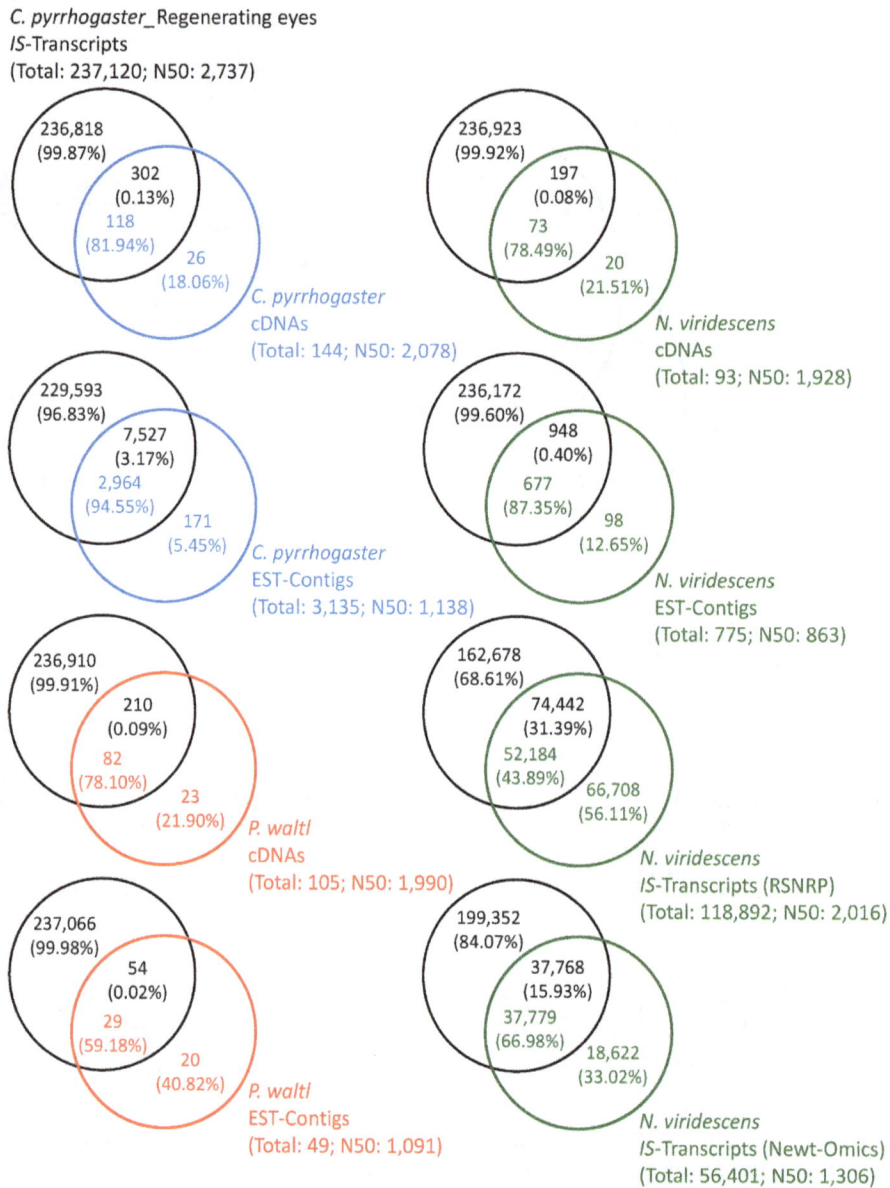

*C. pyrrhogaster*_Regenerating eyes
IS-Transcripts
(Total: 237,120; N50: 2,737)

Figure 3. Comparisons with other newt transcriptomes. Black circles: *IS*-transcripts of this study. Blue circles: cDNAs and EST-contigs reported in *C. pyrrhogaster*. Red circles: cDNAs and EST-contigs reported in *P. waltl*. Green circles: cDNAs, EST-contigs and *IS*-transcripts reported in *N. viridescens*. The values in each circle (written in corresponding color) mean the number and ratio of *IS*-transcripts, cDNAs or EST-contigs.

produced after base calling and Chastity filtering (CASAVA ver.1.8.1., Illumina). These raw reads were filtered to remove reads with adaptors and low-quality sequences (reads with unknown sequences 'N') as follows: first, the software cutadapt (version 1.0; http://code.google.com/p/cutadapt/) [14] was used to trim adapters, and then the trimmed reads and reads containing 'N' are discarded using in-house scripts. Finally, 330,837,190 clean reads (101 bp each) were yielded for the following *de novo* assembly.

De novo assembly

To obtain contigs and *in silico* transcripts (*IS*-transcripts), the clean reads, obtained after filtering the raw reads, were assembled using *de novo* assemblers. There is no consensus in terms of the best *de novo* assemblers. So, three widely used algorisms were

applied: Trinity [15] (version 2012-10-05; http://trinityrnaseq. sourceforge.net/), Trans-ABySS [16] [for Trans-Abyss, version 1.4.4 (http://www.bcgsc.ca/platform/bioinfo/software/trans-abyss); for ABySS [17], version 1.3.5 (http://www.bcgsc.ca/platform/bioinfo/software/abyss] and Velvet-Oases [18] [for Velvet [19], version 1.2.01 (http://www.ebi.ac.uk/~zerbino/velvet/); for Oases, version 0.2.02 (http://www.ebi.ac.uk/~zerbino/oases/)]. Trinity was run with '–min_kmer_cov 2'. By applying the '–min_kmer_cov 2' parameter (the default is 1), only k-mers that occur more than once are assembled. This parameter is used to reduce memory requirements and runtimes. Generally, this will eliminate super rare transcripts and sequencing errors, but will not usually affect assembly quality. ABySS was run with k-mer values form 51 to 95 in steps of 2, and then all 23 assemblies from AbySS were merged into one assembly using Trans-ABySS. Velvet was run with k-mer values from 45 to 95 in steps of 5, and

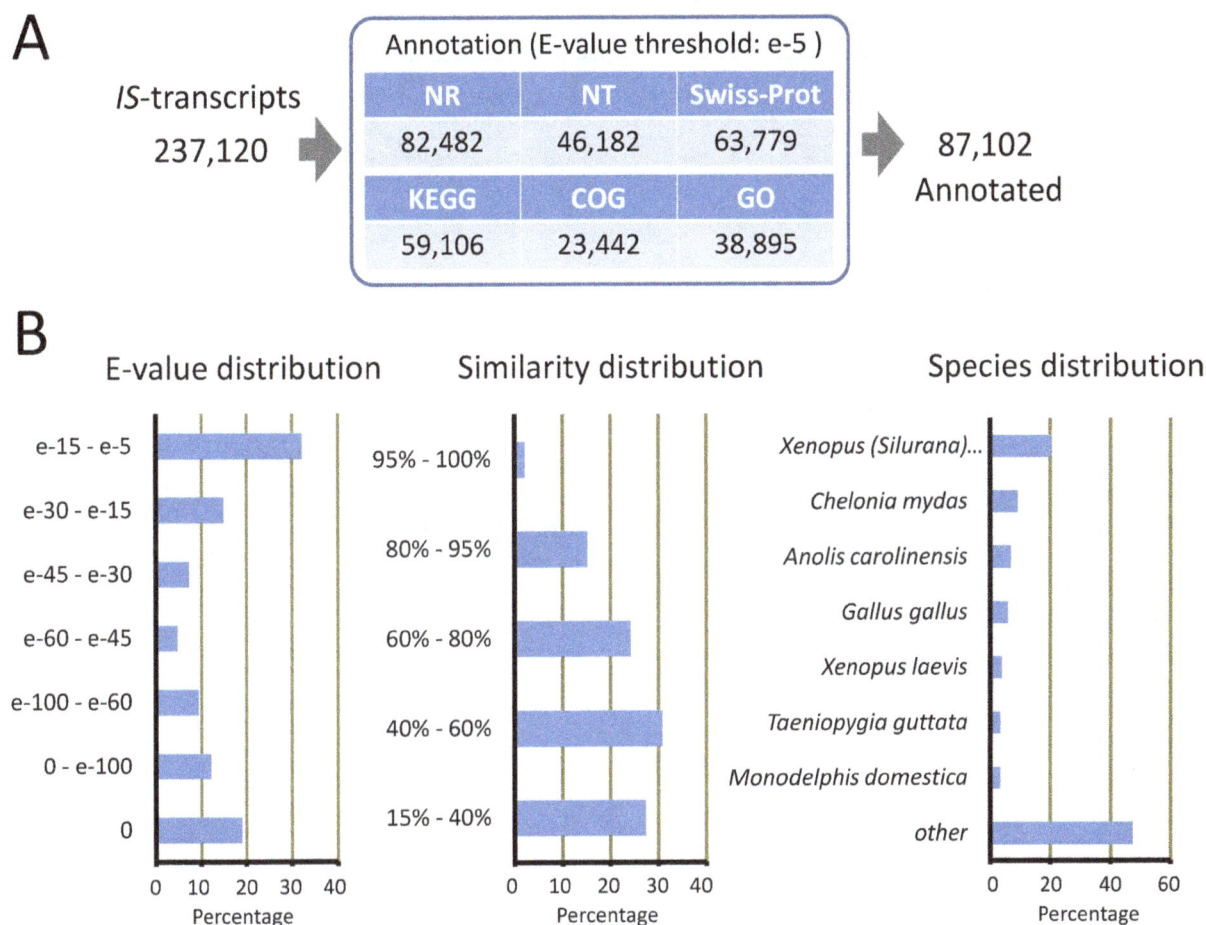

Figure 4. Functional annotation. A. Summary of annotation. **B.** E-value-, similarity (% identity)- and species-distribution in NR annotation. Species-distribution indicates that many of the *C. pyrrhogaster* IS-transcripts are close to genes of amniotes as well as those of amphibians such as *Xenopus (Silurana)* and *Xenopus laevis*, rather than fishes (e.g., *Danio rerio*, 2.9%; *Oryzias latipes*, 1.9%; these are included in 'other'). Interestingly, within amphibians, the newt seems to adhere to *X. (Silurana)* rather than to *X. laevis*.

then the contigs produced by Velvet at each k-mer value were further assembled into transcripts using Oases. Finally, the best result (k-mer = 75) was selected as output. Although the transcript data sets assembled at different k-mer values were generally merged using Oases-M, we just treated Velvet-Oases as a single k-mer assembler without Oases-M.

In this study, *IS*-transcripts deduced by Trinity were further analyzed by annotation as well as coding sequence (CDS) and protein/peptide prediction.

Annotation

Functional annotation of *IS*-transcripts was carried out by aligning them first to protein databases such as NCBI NR (release-20130408), Swiss-Prot (release-2013_03), KEGG (release 63.0) and COG (release-20090331) by blastx program (E-value threshold: e-5), and then to NCBI NT by blastn program (E-value threshold: e-5). Gene Ontology (GO) annotation was carried out by the Blast2GO program (v2.50) with NR annotation, and the data was classified by WEGO software [20]. Metabolic pathway analysis was carried out with the help of the KEGG database.

Coding sequence (CDS) and protein/peptide prediction

Both the nucleotide sequence (5′-3′) and protein/peptide CDS of *IS*-transcripts were predicted using proteins with highest ranks

in the functional annotation by blastx program with NR, Swiss-Prot, KEGG and COG (see above). *IS*-transcripts that could not be aligned to any databases were scanned by ESTScan (v3.0.2) to predict CDSs [21]. This program compensates for the frame shift errors.

Quantitative PCR (qPCR)

Each day samples (5 samples/tube), which were prepared and harvested under a conventional nuclease free condition (see above), were immediately used to purify total RNA (NucleoSpin RNA kit; Macherey-Nagel GmbH & Co. KG, Düren, Germany). First strand cDNAs were synthesized (SuperScript™ II Reverse Transcriptase; Life Technologies, Carlsbad, CA) with 30 ng total RNA, diluted 100x, and then used as a template for qPCR. qPCR was performed by a LightCycler Nano system (Roche Applied Science, Penzberg, Germany) according to the manufacturer's instructions of FastStart Essential DNA Green Master (Roche) or FastStart Essential DNA Probes Master (Roche), with 45 cycles. ID numbers of target *IS*-transcripts (*Ef1α, RPE65, CRALBP/ RLBP1, ZO1, Otx2, Musashi1a/c, Cyclin D1, CDK4, Histone H3, c-Myc, Klf4, Sox2, N-Cadherin, α-SMA, Vimentin, Pax6, Chx10/ Vsx2, FGFR1, FGFR3, Mitf, Wnt2b*), and their PCR primers and probes [selected from the Roche Universal Probe Library (https:// www.roche-applied-science.com)] are listed in Table S1. The

Table 1. GO classification.

Biological process

Class	Count
cellular process	29,570
single-organism process	25,093
metabolic process	21,542
biological regulation	19,602
regulation of biological process	18,451
response to stimulus	14,838
multicellular organismal process	13,061
developmental process	11,328
signaling	11,013
cellular component organization or biogenesis	10,934
localization	10,576
positive regulation of biological process	8,574
establishment of localization	8,345
negative regulation of biological process	7,416
immune system process	3,363
locomotion	3,320
reproduction	2,907
reproductive process	2,768
biological adhesion	2,223
multi-organism process	2,222
growth	1,992
rhythmic process	442
cell killing	97

Cellular component

Class	Count
cell	29,452
cell part	29,452
organelle	21,715
membrane	15,252
organelle part	13,910
membrane part	9,970
macromolecular complex	8,397
membrane-enclosed lumen	6,813
extracellular region	2,349
cell junction	1,989
synapse	1,713
extracellular region part	1,520
synapse part	1,205
extracellular matrix	894
extracellular matrix part	336
nucleoid	72

Molecular function

Class	Count
binding	26,423
catalytic activity	13,617
transporter activity	2,690
molecular transducer activity	2,323

Table 1. Cont.

Molecular function	
Class	**Count**
enzyme regulator activity	2,160
receptor activity	2,005
nucleic acid binding transcription factor activity	1,913
protein binding transcription factor activity	1,264
structural molecule activity	1,085
electron carrier activity	223
channel regulator activity	207
antioxidant activity	90
chemorepellent activity	66
receptor regulator activity	58
translation regulator activity	50
chemoattractant activity	30
metallochaperone activity	10
morphogen activity	3
nutrient reservoir activity	3
protein tag	1

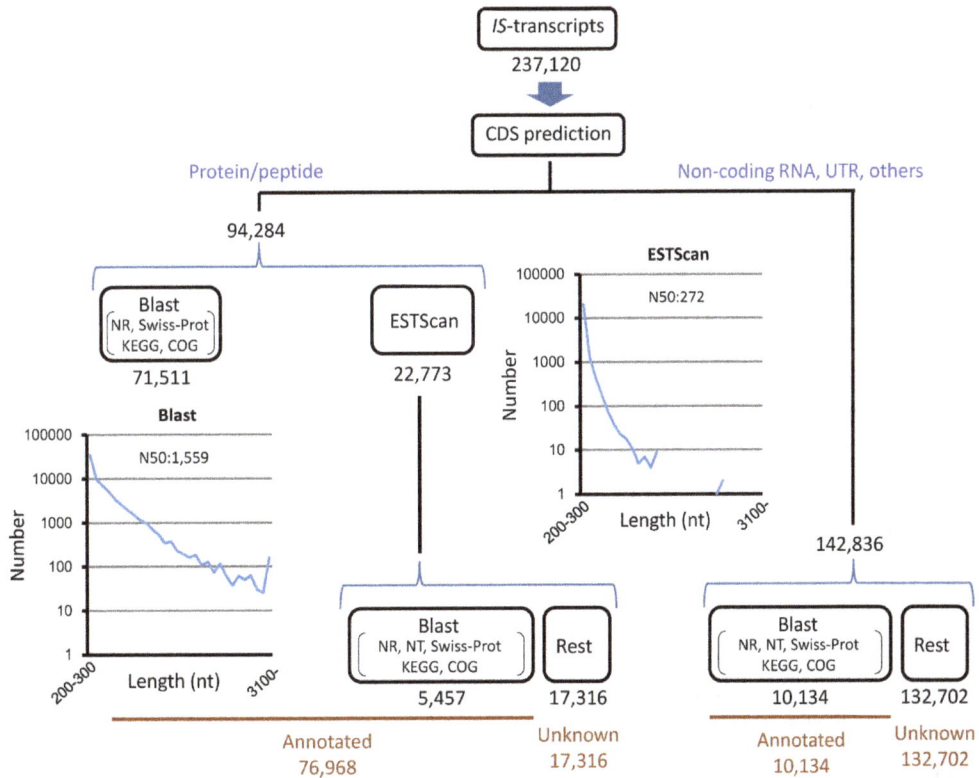

Figure 5. CDS prediction. Blastx predicted CDSs in 71,511 *IS*-transcripts (for the length distribution, see the graph 'Blast'; for protein/peptide sequences, see Table S5) while ESTscan predicted CDSs in 22,773 *IS*-transcripts (for the length distribution, see the graph 'ESTScan'; for protein/peptide sequences, see Table S6).

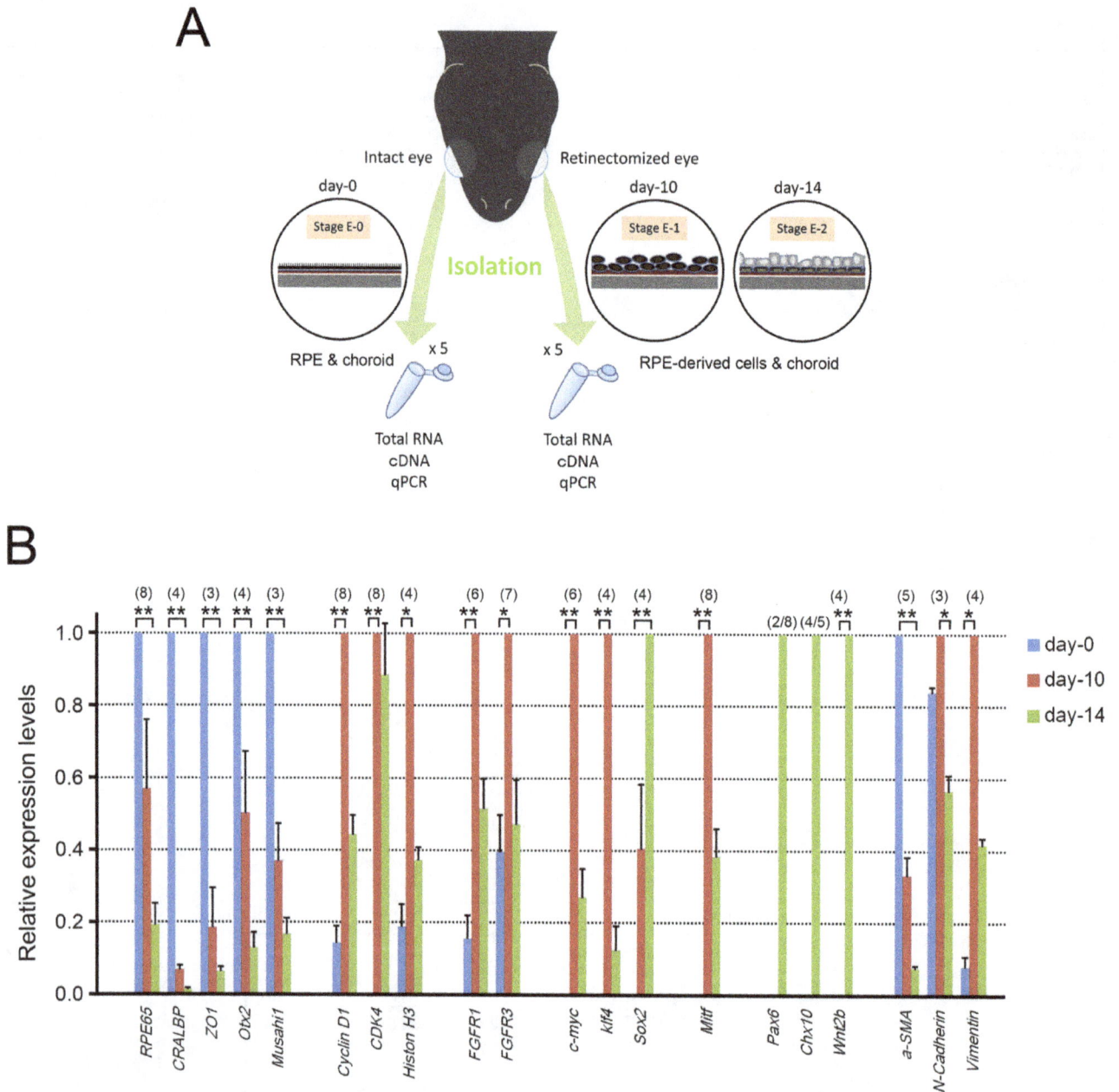

Figure 6. Validation by qPCR. A. Workflow of sample preparations for qPCR. RPE-choroid tissues harvested from the right eyes (intact eye) of 5 retinectomized animals at day-10 or day-14 were used for the normal RPE, i.e., the day-0 (Stage E-0) sample. RPE-derived cells harvested together with the choroid from the left eyes (retinectomized eye) of the same animals were used for the day-10 (Stage E-1) or day-14 (Stage E-2) sample. 3 different day samples (day-0, -10 and -14) were randomly grouped as one set of samples for qPCR. For more details, see Methods. **B.** qPCR analysis. 20 selected genes which have been suggested or inferred to be regulated in an early phase of retinal regeneration were found in the current *in silico* transcriptomes. Changes in their relative expression level until day-14 po were examined. Results represent means and SE. The number in parenthesis indicates the number of independent sample sets (see Methods) except for *Pax6* and *Chx10*, whose expression was detected in 2 of 8 sample sets and 4 of 5 sample sets, respectively at day-14 po only. Statistical significance based on Sheffe's test following the Friedman test (*: $p < 0.05$; **: $p < 0.01$), except for *Pax6* and *Chx10*.

DDBJ/GenBank accession number of the cDNA corresponding to each *IS*-transcript is also mentioned in Table S1. Most cDNAs were cloned and their sequences, determined by the Sanger method, were deposited in DDBJ.

For each gene, qPCR, which was always run simultaneously with day-0, -10 and -14 po samples, was repeated using more than three sets of independently collected samples. The relative expression level of each transcript was calculated as follows: the

amount of transcript was first compensated for *Ef1α* (the mean from several rounds of qPCR) in the same sample, and then normalized against that of the day at which the average value from all samples was highest. The data were presented as the mean ± SEM. Statistical differences were evaluated by Sheffe's test following the Friedman test.

Results and Discussion

Three transcriptomes deduced by different *de novo* assemblers

There is no consensus in terms of the best *de novo* assemblers. So, we tested three widely used algorisms, Trinity [15], Trans-ABySS [16] and Velvet-Oases [18]. Each algorism produced a large number of *IS*-transcripts: Trinity, 237,120 [N50 (the length of an *IS*-transcript whose order is 50% of all *IS*-transcripts): 2,737 nt]; Trans-ABySS, 635,930 (N50: 1,800 nt); Velvet-Oases, 99,586 (N50: 4,001 nt) [Figure 2; note that these sets of *IS*-transcripts (transcriptomes) can be downloaded from our repository site 'IMORI' (http://antler.is.utsunomiya-u.ac.jp/imori/)]. These three transcriptomes were compared with each other by blast (NCBI; v2.2.27+x64-linux; E-value threshold: e-5), revealing that Trinity covered almost all *IS*-transcripts in Trans-ABySS (~98%) and Velvet-Oases (~99%) (Figure 2). Thus, the Trinity data set seemed to be more comprehensive. So, here we applied the following analyses to this Trinity-deduced transcriptome.

Trinity-deduced transcriptome in this study considerably covers transcript information in newts

To validate the current Trinity-deduced transcriptome, we first investigated what extent it covers transcript information in newts. For this, we compared this transcriptome, by blastn (E-value threshold: e-5), with other transcriptomes (cDNAs, ESTs, and *IS*-transcripts) in three species (*C. pyrrhogaster*, *Pleurodeles waltl*, *Notophthalmus viridescence*), which were collected from databases in DDBJ/NCBI and from sequence repository sites for *N. viridescence*: Red Spotted Newt Resource Page (RSNRP) http://sandberg.cmb.ki.se/redspottednewt/ [5] and Newt-Omics http://newt-omics.mpi-bn.mpg.de/ [6] (Figure 3). ESTs were assembled into contigs by the CAP3 program [22], before comparison. This analysis revealed that the current transcriptome in early regenerating eyes (day 0–14 po) contains >78% information of cDNAs/ESTs (i.e., mRNA) reported in these three species so far, except for 59% of ESTs in *P. waltl*; notably, in *C. pyrrhogaster* ~82% of cDNAs and ~95% of EST-contigs were matched (Figure 3). In addition, the current transcriptome covered ~44% and ~67% of the *IS*-transcripts for *N. viridescence* (RSNRP and Newt-Omics, respectively). On the other hand, these *N. viridescence* data covered only ~31% and ~16% of the current transcriptome, respectively. Thus, the current transcriptome comprehensively covers transcript information in *C. pyrrhogaster*.

Characterization of the current transcriptome

Functional annotation. To further characterize the current transcriptome (237,120 *IS*-transcripts), physiological functions of each *IS*-transcript were predicted by annotation, which carried out by blast with NR, NT, Swiss-Prot, KEGG, COG and GO databases. Finally, a total of 87,102 *IS*-transcripts (range: 201–19,064 nt; N50: 1,202 nt) were annotated (Figure 4A; for details, see Table S2). For example, in NR annotation, 82,482 *IS*-transcripts (94.7% of all annotated transcripts) were enriched; 67.7% of them had an E-value of <e-5, and 41.8% had a similarity of >60%, indicating that many of the NR-annotated *IS*-transcripts contain the sequence information of genes highly homologues to those found in NR database (Figure 4B); species distribution seemed to reflect the position of this animal in phylogeny (Figure 4B). 38,895 of the NR-annotated *IS*-transcripts were also dealt GO terms (Figure 4A). As shown in GO

classification (Table 1) as well as in COG- and KEGG-classification (Table S3 and S4), these *IS*-transcripts were assigned to various functional categories and pathways. On the other hand, the remaining 150,018 *IS*-transcripts (range: 201–11,726 nt; N50: 290 nt) were not annotated.

Inferred proteome. We carried out CDS and protein/peptide prediction independently of functional annotation (Figure 5). Protein/peptide sequences were first predicted in 71,511 of 237,120 *IS*-transcripts from the top hit sequence in blastx with NR, Swiss-Prot, KEGG, and COG databases (range: 20–6,124 aa; N50: 211 aa; Table S5), and then in 22,773 of the remnant *IS*-transcripts by ESTScan (range: 29–2,448 aa; N50: 90 aa; Table S6). Finally, these data were combined to the above annotation results. By this procedure, most of the predicted protein/peptides (76,968/94,284) were annotated; 5,457 ESTScan-predicted protein/peptides (range: 31–2448 aa; N50: 107 aa), of which *IS*-transcripts had a range of 201–11,070 nt (N50: 327 nt), were mostly unknown, uncharacterized, unnamed or hypothetical proteins. On the other hand, 142,836 *IS*-transcripts were unlikely to encode protein/peptides; 10,134 (range: 201–15,043 nt; N50: 454 nt) of them were annotated, mostly showing similarity to noncoding RNA and genomic sequences including the untranslated region (UTR) in other species.

17,316 *IS*-transcripts (range: 201–7,076 nt; N50: 262 nt) whose CDSs were predicted by ESTScan (Table S7) and the remaining 132,702 transcripts (range: 201–11,726 nt; N50: 290 nt) were not annotated ('Rest' in Figure 5). These were mostly short (~80% of the former and ~70% of the latter were 200–400 nt), but interestingly identical sequences [149 (range: 68–1915 aa; N50: 123 aa) of the former and 996 (range: 202–9052 nt; N50: 790 nt) of the latter] were also found in the newt limb blastema (R. M. Casco-Robles, unpublished data). Therefore, all of them cannot be associated with errors in *de novo* assembly, as revealed in *N. viridescence* [6].

In future studies, we need confirm the presence of these molecules as well as their physiological functions in this species.

qPCR analysis of gene expression which has been inferred in early processes of retinal regeneration demonstrates the applicability of the current transcriptomes

The current transcriptomes contained much information of genes which have not been identified in *C. pyrrhogaster* but whose transcription has been hypothesized to be regulated in early processes of retinal regeneration. Here, to validate the applicability of the current transcriptomes, we carried out qPCR analysis of selected 20 genes (Table S1) with cDNA samples prepared from RPE-choroid tissues or RPE-derived cells-choroid tissues which were harvested from intact (day-0) or retinectomized (day-10 or day-14) eyes respectively (Figure 6A); for unidentified genes (*CRALBP/RLBP1, ZO1, Cyclin D1, CDK4, Histone H3, c-Myc, Klf4, N-Cadherin, α-SMA, Vimentin, Chx10/Vsx2, Mitf, FGFR1, FGFR3*), the sequence of the target transcript was confirmed by conventional molecular cloning and Sanger DNA sequencing using the same tissue samples as those used for qPCR.

The results (Figure 6B) revealed that the expression levels of many genes characterizing RPE cells (*RPE65, CRALBP/RLBP1, ZO1, Otx2, Musashi1a/c*) gradually decreased upon retinectomy, while the expression of genes for the cell-cycle (*Cyclin D1*, *CDK4*, *Histone H3*) and growth signaling (*FGFR1*, *FGFR3*) was up-regulated to reach a maximum level by day-10 po when almost all RPE cells had entered the cell cycle (Stage E-1). Intriguingly, pluripotency factors except for *Oct4* (i.e., *c-Myc*, *Klf4*, *Sox2*) were expressed between day-0 and day-10 po; *Oct4* was not found in the current transcriptomes. Up-regulation of these three factors and lack of *Oct4* expression were also reported in early lens and limb regeneration of *N. viridescence* [23]. A microphthalmia factor *Mitf*, a marker gene for immature or uncommitted RPE cells [24], was also expressed by day-10 po but then its expression level decreased. On the other hand, marker genes for retinal stem/progenitor cells (*Pax6*, *Chx10/Vsx2*) were expressed at day-14 po when two rudiment layers (pro-NR and pro-RPE) for the prospective neural retina and RPE just appear (Stage-E2). Interestingly, the expression of *Wnt2b* was coincidental. Wnt/β-catenin signaling has been suggested to promote differentiation of the RPE while protecting cell-fate switching of the uncommitted RPE into the neural retina in embryonic eye development [11,24].

In *PVR* leading to the loss of vision after retinal injury in humans, RPE cells transform into mesenchymal cells such as myofibroblasts, probably by passing through a stem-cell state [9]. In the newt, expression of mesenchymal markers (*N-Cadherin*, *Vimentin*) seemed to be up-regulated by day-10 po and then decreased while the relative expression level of a marker of myofibroblasts, α-*SMA* (smooth muscle actin), obviously decreased (Figure 6B).

Consequently, the current transcriptomes could be a good tool to identify or screen candidate genes which might be involved in early processes of retinal regeneration.

Implications for reprogramming of adult newt RPE cells into a multipotent state

In the adult newt, the RPE is a sole cell source for retinal regeneration in the posterior eye [8]. Upon retinectomy, RPE cells are detached from each other and leave the basement membrane, forming cell aggregates, while entering the S-phase of cell-cycle (Stage E-1). This event typically occurs between day-5 and day-10 po. The RPE-derived cells at Stage E-1 form the pro-NR and pro-RPE layers by day-14 po (Stage E-2).

In the previous studies [25,26], we demonstrated that an RNA-binding protein Musashi-1, whose expression is restricted in the nucleus of the intact RPE cells, changes its location into the cytoplasm upon retinectomy, although the amount of the transcripts tends to be decreased as suggested in the current qPCR. This pattern of Musashi-1 expression was observed in almost all of the RPE-derived cells at Stage E-1 uniformly. As the regeneration proceeds to Stage E-2, Musashi-1 expression was down-regulated along the pro-RPE layer, while sustained along the pro-NR layer [26]. Since the cytoplasmic expression of Musashi-1 is a character of neural stem/progenitor cells [27], these observations led us to an implication that the RPE cells are reprogrammed into a multipotent state of cells by Stage E-1, and specified into 2 cell populations forming the pro-NR and pro-RPE layers between Stage E-1 and E-2. However, the nature of the RPE-derived cells remains unclear.

The current qPCR revealed that gene expression suggesting multipotent properties of cells is up-regulated upon retinectomy, while gene expression for original RPE characters is down-regulated, giving us an insight for the cell reprogramming. Interestingly, certain pluripotency factors (*c-Myc*, *Klf4*, *Sox2*) as well as *Mitf* were first detected in day-10 samples (Stage E-1). On the other hand, *Pax6* was detected only in day-14 samples. In the previous study, we detected *Pax6* expression at day-10 po, being

earlier than *Chx10-1/Vsx1* (a retinal progenitor marker) at day-14 po [28]. This inconsistency may be due to the difference in the template for PCR: in the previous study, we used a PCR-amplified library of cDNAs as the template, and therefore the detection sensitivity should have been higher. Taken together, our results suggest that the intact adult newt RPE cells are not comparable to either immature/uncommitted state of RPE cells or retinal stem/progenitor cells, as well as a possibility of reprogramming of RPE cells into such multipotent cells upon retinectomy. On the next stage of this study, we must carry out immunohistochemistry and single-cell qPCR to address the expression of these factors in the RPE-derived cells at Stage E-1 (our study on this subject was published [29] while the current article was under review).

The current qPCR with day-14 samples detected a factor (*Chx10/Vsx2*) suggesting the presence of retinal progenitor cells and that (*Wnt2b*) inferred to be involved in the RPE differentiation. *Sox2*, which is also a marker of retinal stem/progenitor cells [30], increased in its expression level between day-10 and day-14, whereas *c-Myc*, *Klf4* and *Mitf* were declined. These expression patterns seem to support our previous implication that the RPE-derived cells are specified into two cell populations from which the pro-NR and pro-RPE layers are formed, leading a possibility that the cells in the pro-NR and pro-RPE layers might be correspond to the retinal progenitor cells and RPE cells which just started their differentiation, respectively. In addition, the expression of α-*SMA*, which is observed in the choroid tissues (data not shown), was obviously declined to a minimal level at day-14, revealing a contrast to human *PVR* in which the RPE-derived multipotent cells transform into myofibroblasts. Thus, in the adult newt system, the RPE-derived cells (possibly in a multipotent state) direct properly and seem to generate retinal progenitor cells for regeneration of a missing neural retina.

In the current study, we did not carry out further analyses because the purpose of the study was to establish a good transcriptome database. However, our study would move onto the next stage at histological and functional levels.

Conclusions

To facilitate investigations of adult newt retinal regeneration, we provided *in silico* transcriptomes covering information of genes which are expressed in eyeballs containing Stage E-0 to E-2 regenerating retinas. Gene expression patterns revealed by qPCR demonstrate the usefulness of the transcriptomes for the study of early processes of retinal regeneration. This tool can be applied to the next comprehensive gene screening steps to uncover essential mechanisms underlying reprogramming, cell-cycle re-entry/proliferation, and patterning in adult newt retinal regeneration.

Supporting Information

Table S1   Primers and probes for PCR targets.

Table S2   Summary of annotation.

Table S3   COG classification.

Table S4   KEGG classification.

Table S5   Proteins/peptides predicted by blast.

Table S6   Proteins/peptides predicted by ESTScan.

Table S7 Unknown proteins/peptides predicted by ESTScan.

Acknowledgments

The authors thank Hokkaido System Science (Sapporo, Hokkaido, Japan) and BGI Japan (Kobe KIMEC Center, Japan) for assistance with initial data analyses, and all members of the Japan Newt Research Community (JNRC) for their valuable comments and discussion.

Author Contributions

Conceived and designed the experiments: KN FT CC. Performed the experiments: KN MRI MT HY WI AK. Analyzed the data: KN RMCR FT CC. Contributed reagents/materials/analysis tools: FT CC. Wrote the paper: KN FT CC.

References

1. Alvarado AS, Tsonis PA (2006) Bridging the regeneration gap: genetic insights from diverse animal models. Nat Rev Genet 7: 873–884.
2. Eguchi G, Eguchi Y, Nakamura K, Yadav MC, Millán JL, et al. (2011) Regenerative capacity in newts is not altered by repeated regeneration and aging. Nat Commun DOI:10.1038/ncomms1389.
3. Tsonis PA, Haynes T, Maki N, Nakamura K, Casco-Robles MM, et al. (2011) Controlling gene loss of function in newts with emphasis on lens regeneration. Nat Protoc 6: 593–599.
4. Casco-Robles MM, Yamada S, Miura T, Nakamura K, Haynes T, et al. (2011) Expressing exogenous genes in newts by transgenesis. Nat Protoc 6: 600–608.
5. Abdullayeva I, Kirkham M, Björklund ÅK, Simon A, Sandberg R (2013) A reference transcriptome and inferred proteome for the salamander *Notophthalmus viridescens*. Exp Cell Res 319: 1187–1197.
6. Looso M, Preussner J, Sousounis K, Bruckskotten M, Michel CS, et al. (2013) A *de novo* assembly of the newt transcriptome combined with proteomic validation identifies new protein families expressed during tissue regeneration. Genome Biol 14: R16.
7. Litvinchuk SN, Rosanov JM, Borkin LJ (2007) Correlations of geographic distriibution and temperature of embryonic development with the nuclear DNA content in the Salamandridae (Urodela, Amphibia). Genome 50: 333–342.
8. Chiba C, Hoshino A, Nakamura K, Susaki K, Yamano Y, et al. (2006) Visual cycle protein RPE65 persists in new retinal cells during retinal regeneration of adult newt. J Comp Neurol 495: 391–407.
9. Chiba C (2014) The retinal pigment epithelium: An important player of retinal disorders and regeneration. Exp Eye Res 123: 107–114.
10. Susaki K, Chiba C (2007) MEK mediates *in vitro* neural transdifferentiation of the adult newt retinal pigment epithelium cells: Is FGF2 an induction factor? Pigment Cell Res 20: 364–379.
11. Yoshikawa T, Mizuno A, Yasumuro H, Inami W, Vergara MN, et al. (2012) MEK-ERK and heparin-susceptible signaling pathways are involved in cell-cycle entry of the wound edge retinal pigment epithelium cells in the adult newt. Pigment Cell Melanoma Res 25: 66–82.
12. Tominaga A, Matsui M, Yoshikawa N, Nishikawa K, Hayashi T, et al. (2013) Phylogeny and historical demography of *Cynops pyrrhogaster* (Amphibia: Urodela): Taxonomic relationships and distributional changes associated with climatic oscillations. Mol Phylogenet Evol 66: 654–667.
13. Mizuno A, Yasumuro H, Yoshikawa T, Inami W, Chiba C (2012) MEK-ERK signaling in adult newt retinal pigment epithelium cells is strengthened immediately after surgical induction of retinal regeneration. Neurosci Lett 523: 39–44.
14. Martin M (2011) Cutadapt removes adapter sequences from high-throughput sequencing reads. EmBnet Journal 17: 10–12.
15. Grabherr MG, Haas BJ, Yassour M, Levin JZ, Thompson DA, et al. (2011) Full-length transcriptome assembly from RNA-Seq data without a reference genome. Nat Biotechnol 29: 644–652.
16. Robertson G, Schein J, Chiu R, Corbett R, Field M, et al. (2010) *De novo* assembly and analysis of RNA-seq data. Nat Methods 7: 909–912.
17. Simpson JT, Wong K, Jackman SD, Schein JE, Jones SJM, et al. (2009) ABySS: A parallel assembler for short read sequence data. Genome Res 19: 1117–1123.
18. Schulz MH, Zerbino DR, Vingron M, Birney E (2012) *Oases*: robust *de novo* RNA-seq assembly across the dynamic range of expression levels. Bioinformatics 28: 1086–92.
19. Zerbino DR, Birney E (2008) Velvet: algorithms for de novo short read assembly using de Bruijn graphs. Genome Res 18: 821–829.
20. Ye J, Fang L, Zheng H, Zhang Y, Chen J, et al. (2006) WEGO: a web tool for plotting GO annotations. Nucl Acids Res 34: W293–W297.
21. Iseli C, Jongeneel CV, Bucher P (1999) ESTScan: a program for detecting, evaluating, and reconstructing potential coding regions in EST sequences. Proc Int Conf Intell Syst Mol Biol 138–48.
22. Huang XQ, Madan A (1999) CAP3: A DNA sequence assembly program. Genome Res 9: 868–877.
23. Maki N, Suetsugu-Maki R, Tarui H, Agata K, Del Rio-Tsonis K, et al. (2009) Expression of stem cell pluripotency factors during regeneration in newts. Dev Dyn 238: 1613–1616.
24. Bharti K, Gasper M, Ou J, Brucato M, Clore-Gronenborn K, et al. (2012) A regulatory loop involving PAX6, MITF, and WNT signaling controls retinal pigment epithelium development. PLoS Genet 8: e1002757.
25. Susaki K, Kaneko J, Yamano Y, Nakamura K, Inami W, et al. (2009) Musashi-1, an RNA-binding protein, is indispensable for survival of photoreceptors. Exp. Eye Res 88: 347–355.
26. Kaneko J, Chiba C (2009) Immunohistochemical analysis of Musashi-1 expression during retinal regeneration of adult newt. Neurosci Lett 450: 252–257.
27. Okano H, Kawahara H, Toriya M, Nakao K, Shibata T, et al. (2005) Function of RNA-binding protein Musashi-1 in stem cells. Exp Cell Res 306: 349–356.
28. Chiba C, Mitashov VI (2007) Cellular and molecular events in the adult newt retinal regeneration. In: Chiba C, editor. Strategies for Retinal Tissue Repair and Regeneration in Vertebrates: from Fish to Human. India: Research Signpost. 15–33.
29. Islam MR, Nakamura K, Casco-Robles MM, Kunahong A, Inami W, et al. (2014) The newt reprograms mature RPE cells into a unique multipotent state for retinal regeneration. Sci Rep 4: 6043.
30. Hever AM, Williamson KA, van Heyningen V (2006) Developmental malformations of the eye: the role of *PAX6*, *SOX2* and *OTX2*. Clin Genet 69: 459–470.

11

Angiogenic Potential of Vitreous from Proliferative Diabetic Retinopathy and Eales' Disease Patients

Ponnalagu Murugeswari[1¤a], Dhananjay Shukla[2¤b], Ramasamy Kim[2], Perumalsamy Namperumalsamy[2], Alan W. Stitt[3], Veerappan Muthukkaruppan[1]*

1 Department of Immunology and Cell Biology, Aravind Medical Research Foundation, Dr.G.Venkataswamy Eye Research Institute, Madurai, India, **2** Vitreous and Retina Service, Aravind Eye Care System, Madurai, India, **3** Centre for Experimental Medicine, Queens University, Belfast, United Kingdom

Abstract

Purpose: Proliferative Diabetic Retinopathy (PDR) and Eales' Disease (ED) have different aetiologies although they share certain common clinical symptoms including pre-retinal neovascularization. Since there is a need to understand if the shared end-stage angiogenic pathology of PDR and ED is driven by common stimulating factors, we have studied the cytokines contained in vitreous from both patient groups and analyzed the angiogenic potential of these samples *in vitro*.

Material and Methods: Vitreous samples from patients with PDR (n = 13) and ED (n = 5) were quantified for various cytokines using a cytokine biochip array and sandwich ELISA. An additional group of patients (n = 5) with macular hole (MH) was also studied for comparison. To determine the angiogenic potential of these vitreous samples, they were analyzed for their ability to induce tubulogenesis in human microvascular endothelial cells. Further, the effect of anti-VEGF (Ranibizumab) and anti-IL-6 antibodies were studied on vitreous-mediated vascular tube formation.

Results: Elevated levels of IL-6, IL-8, MCP-1 and VEGF were observed in vitreous of both PDR and ED when compared to MH. PDR and ED vitreous induced greater levels of endothelial cell tube formation compared to controls without vitreous (P< 0.05). When VEGF in vitreous was neutralized by clinically-relevant concentrations of Ranibizumab, tube length was reduced significantly in 5 of 6 PDR and 3 of 5 ED samples. Moreover, when treated with IL-6 neutralizing antibody, apparent reduction (71.4%) was observed in PDR vitreous samples.

Conclusions: We have demonstrated that vitreous specimens from PDR and ED patients share common elevations of pro-inflammatory and pro-angiogenic cytokines. This suggests that common cytokine profiles link these two conditions.

Editor: Rudolf Kirchmair, Medical University Innsbruck, Austria

Funding: Supported by grants from Department of Science and Technology (SR/SO/HS-04/2003), TIFAC-CORE in diabetic retinopathy, India, and Commonwealth Split-Site Scholarship programme. The funders had no role in study design, data collection and analysis, decision to publish, or preparation of the manuscript.

* Email: muthu@aravind.org

¤a Current address: Stem Cell Research Laboratory, Narayana Nethralaya Foundation, Bangalore, India
¤b Current address: Retina-Vitreous Service, Ratan Jyoti Nethralaya, Gwalior, India

Introduction

Proliferative Diabetic Retinopathy (PDR) and Eales' Disease (ED) are potentially blinding vitreoretinal diseases. ED is an idiopathic inflammatory venous occlusion that primarily affects the peripheral retina of healthy young men (20–30 years). Retinal changes in ED include periphlebitis, peripheral non-perfusion and neovascularization. In addition, visual loss is characteristically caused by bilateral recurrent vitreous hemorrhage [1], [2]. Although PDR and ED have different etiology, the symptoms and signs of these diseases run parallel to each other with some differences [3]. We have previously observed this commonality in the presence of inflammatory cytokines [interleukin-6 (IL-6), interleukin-8 (IL-8), monocyte chemoattractant protein-1 (MCP-1)] and vascular endothelial growth factor (VEGF) in vitreous [4].

Angiogenesis is a highly complex and coordinated process requiring multiple receptors and ligands in endothelial cells for which VEGF is a pivotal element in a variety of normal and pathological circumstances [5], [6]. VEGF is upregulated during retinal hypoxia and levels of this growth factor are raised in the vitreous of patients with PDR [7], [8] and VEGF-neutralizing antibodies are now being used for treatment of pathological ocular angiogenesis and macular oedema [9], [10], [11], [12].

Though the presence of pro-angiogenic and pro-inflammatory cytokines in the vitreous of PDR and ED patients has been demonstrated by several studies the functional potential of their vitreous has never been shown. Therefore, we have investigated the angiogenic potential of vitreous from these patients using the endothelial tube formation assay *in vitro*, and the effects of anti-VEGF and anti-IL6 neutralizing antibodies on this potential.

Materials and Methods

Ethics Statement

Patients were recruited in accordance with the Declaration of Helsinki and with the approval of the Institutional Review Board (IRB) of Aravind Eye Hospital. Written informed consent as per the procedure approved by IRB was obtained from the patients and maintained.

Study Subjects

Consecutive patients scheduled to undergo vitrectomy for proliferative ED, advanced PDR or Macular Hole (MH) were prospectively recruited in one of the three treatment groups. The clinical diagnosis of ED was made on the basis of peripheral venous sheathing in multiple quadrants in the fellow eye, or the operated eye, as observed intra-operatively. The patients with conditions which could secondarily cause venous sheathing, such as non-inflammatory retinal vein occlusions, diabetic or hypertensive retinopathy, sickle cell retinopathy, ocular inflammatory conditions like choroiditis, or pars planitis, or associated systemic infective/autoimmune/other inflammatory disease (as revealed by history, examination, or investigations), were excluded from the study. The indications for vitrectomy included best-corrected visual acuity less than 20/400 due to vitreous haemorrhage of at least 2 months duration with/without epiretinal membranes; rhegmatogenous/combined-mechanism retinal detachment; or a tractional retinal detachment involving the macula. Eventhough, haemorrhage is an important indication for vitrectomy; small samples of undiluted vitreous were obtained before isotonic infusion and aspiration. The samples which contained blood contamination were excluded from the study.

All the diabetic patients included in this study had type 2 diabetes mellitus. Like ED patients, they were also surgical inpatients, who presented with advanced PDR, as defined by the ETDRS study report number 12 [13]. The indications for vitrectomy were the same as mentioned above for ED patients.

Patients with idiopathic MH were chosen as control. MH is caused by vitreomacular traction, without any associated retinal ischaemia, vascular proliferation or inflammation, and is therefore least likely to be associated with local release of VEGF or other inflammatory cytokines. MH patients were evaluated to rule out a history or signs of ocular trauma or any other associated ocular pathology. The main preoperative investigation in MH was optical coherence tomography, to assess the morphology of MH for surgical feasibility and prognosis. MH patients had no systemic diabetes mellitus. Patients were recruited in accordance with the Declaration of Helsinki and with the approval of the Institutional Review Board of Aravind Eye Hospital. Informed consent was also obtained from the patients. All the cases had no history of a previous ocular surgery.

Thirteen PDR patients (11 males, 2 females), aged 54.2 ± 7.8 years (mean $\pm$ SD) having diabetes for 13.0 ± 9.7 years were included in the study. All 5 ED patients were males and 31 ± 10 years old. Five MH patients (1 female, 4 males) aged 60.4 ± 8.2 were included as controls. All the PDR and ED patients had retinal neovascularization and vitreous haemorrhage. As the etiology is different [14], the ED patients were males and relatively younger in age. Therefore, this difference in age and sex is admissible.

Sample Collection

At the beginning of vitrectomy, undiluted vitreous (200–700 μl) was aspirated via pars plana with a vitreous cutter, before opening the infusion port. These undiluted vitreous samples (13 PDR, 5 ED and 5 MH) were immediately frozen in aliquots in polypropylene tube at $-80°C$ until assay.

Cytokine Assay

Vitreous samples (PDR, n = 8; ED, n = 2) were quantified for a range of cytokines (IL-2, IL-4, IL-6, IL-8, IL-10, IL-1α, IL-1β, IFN-γ, VEGF, TNF-α, MCP-1, EGF) using the Evidence

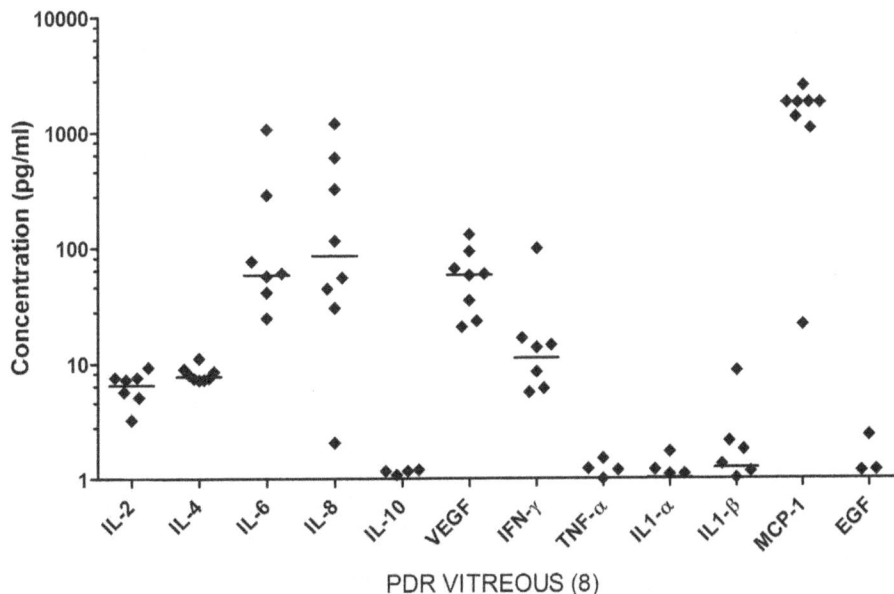

Figure 1. Scatter plot showing the distribution levels of 12 cytokines in vitreous from PDR (n = 8) patients, quantified using cytokine bio-chip array. Each sample represents the mean of duplicates. Solid line indicates the median. PDR-Proliferative diabetic retinopathy; ED- Eales' disease. IL - Interleukins, VEGF- Vascular endothelial growth factor, IFN-γ- Interferon gamma, TNF-α - Tumour necrosis factor alpha, MCP-1 - Monocyte chemoattractant protein-1, EGF - Epidermal growth factor.

Table 1. Levels of cytokines in PDR, ED and MH vitreous used for tubulogenesis assay.

Sample Code	IL6 (pg/ml)	IL8 (pg/ml)	VEGF (pg/ml)	IL-1β (pg/ml)	MCP1 (pg/ml)
*PDR-27	0	0	849.2	-	978.2
* PDR-29	0	0	0	0	0
* PDR-38	260.2	38.0	2401.4	0	2486.9
* PDR-40	225.1	89.4	682.7	0	2571.0
* PDR-45	0	228.3	5805.7	0	1863.6
† PDR-49	>530	30.1	130.2	0	1349.0
† PDR-53	0	2.1	20.6	2.1	21.6
† PDR-58	60.3	604.2	57.5	1.7	>900
† PDR-61	56.6	55.4	66.0	1.0	>900
† PDR-66	24.6	44.3	59.1	0.6	1071.0
* ED-09	0	290.2	105.7	23.8	2063.0
* ED-12	175.0	632.6	0	2.0	-
* ED-16	-	148.1	221.2	3.2	-
† ED-17	38.5	56.9	144.0	0.3	1017.9
† ED-19	158.9	961.0	204.9	2.6	>900
* MH-09	0	4.9	0	103.7	118.4
* MH-29	0	0	0	35.1	937.8
* MH-35	0	0	0	0	0
* MH-38	0	0	107.8	0	174.0
MH-44	-	-	-	-	-

The cytokine levels were analysed by cytokine bio-chip array for 5 PDR, 2 ED vitreous (indicated by †) and the remaining (PDR = 5, ED = 3, MH = 4) by sandwich ELISA (indicated by*). The values in the table represent the mean concentration duplicate of each sample. For the two ED (†) samples, cytokines not included in this table are: IL-2, 8.1 pg/ml; IL-4, 7.14 pg/ml; IFN- γ, 12.8 pg/ml and less than 1.6 pg/ml for IL-10, TNF-α, IL-1α. 0- Less than the detectable limit; - cytokine not analysed; > Higher than the maximum detectable limit. PDR - Proliferative diabetic retinopathy; ED - Eales' disease; MH - Macular Hole.

Investigator Cytokine Biochip Array (Randox, UK). For each sample, this assay was carried out in duplicate. It is a biochip with multiple cytokine antibodies coated on a solid substrate. After incubation with vitreous samples, cytokine specific enzyme-labelled secondary antibodies were added, and the cytokines were quantified by chemiluminescence using a charged coupled device (CCD) camera. The standard curves were constructed for each cytokine and the sample concentration was determined by the evidence investigator (Randox, UK). As a follow-up, IL-6, IL-8, MCP-1, and VEGF were also quantified by sandwich ELISA (BD Biosciences, R & D systems) according to the manufacturer's instructions for 5 PDR, 3 ED and 4 MH samples. The standard curve was prepared using recombinant human cytokines.

Tubulogenesis assay

Human dermal microvascular endothelial cells (HMECs) were purchased (PromoCell Gmbh, Heidelberg, Germany) and grown in T-25 Nunclon culture flasks (Nalge Nunc International,UK) with medium (PromoCell) supplemented with growth supplements (PromoCell) and Primocin (50 mg/ml) (Invitrogen,UK). The *in vitro* tubulogenesis assay was performed as previously described using confluent cells of passage four to six [15]. Briefly for each assay 2×10^4 HMECs in 5 µl were resuspended in 10 µl of vitreous and mixed with 15 µl of growth factor reduced matrigel (BD Biosciences,USA) (PDR, n = 6; ED, n = 5, MH n = 5). In place of vitreous 10 µl of medium was used in controls.

In another series of experiments, HMEC cultures were treated with ranibizumab (RZB) (0.125 mg/ml or 0.25 mg/ml; PDR,

n = 6; ED, n = 5) (Genentech,Inc., South San Francisco) or anti-IL-6 (0.1 µg/ml; PDR,n = 7;ED,n = 2) neutralizing antibody (R & D systems, UK). IL-6 was chosen in preference to other cytokines and chemokines, since it is known to be elevated in vitreous and plays a role in the pathogenesis of ocular diseases [4], [16], [17]. Moreover, previous studies have demonstrated that IL-6 levels in aqueous and vitreous fluids from PDR patients significantly correlate with disease severity [17], [18], [19]. The above mixture of matrigel, HMEC, and RZB was prepared collectively for 3 assays (n = 3) and anti-IL-6 mixture for 2 assays for each sample. Thirty microliter aliquots from the mixture were spotted for each assay onto Nunclon 48-well culture plates. After matrigel polymerization at 37°C for 30 minutes, blobs were covered with endothelial cell supplement medium. Tube formation was observed after 48 hours, phase images were captured in five regions per well and tube length was quantified using NIS-Elements software (Nikon, UK). The mean tube length of all the five regions of triplicate/duplicate was obtained.

Statistical Analysis

The Man-Whitney U test was used to analyse the differences between the experimental and control groups. Values are expressed as mean ± SD. All analysis was done using statistical software Stata 11.0. The results were considered significant at P<0.05.

Results

Cytokine Profile

We have earlier demonstrated that significantly higher concentrations of IL-6, IL-8, MCP-1 and VEGF were observed in

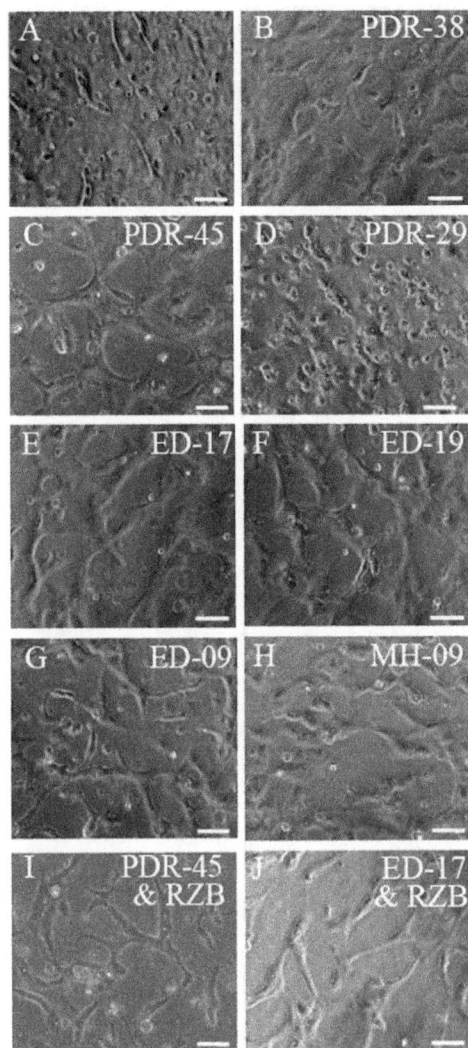

Figure 2. Representative phase images of tube formation induced by vitreous from PDR/ED/MH patients in human dermal microvascular endothelial cell (HMEC). 2×10^5 HMECs in triplicate were exposed to vitreous alone or with RZB (0.125 mg/ml). Tube formation was observed and images were captured after 48 hours incubation. Each panel shows a part of the representative well. The tube length was quantified by NIS-Elements software (Nikon). Scale bar = 100 μm. A. Control (without vitreous); B and C - PDR vitreous-induced tube formation, which had high levels of VEGF/IL-6/MCP-1; D-PDR vitreous with trace levels of cytokines showing a very few tube formation. E- G - ED vitreous-induced tube formation as in PDR. H - MH vitreous. I and J are images of vascular tubes in the presence of PDR/ED vitreous and anti-VEGF antibody, showing reduction in tube length compared to C and E respectively. Number in the images denotes the patient ID as in table 1. PDR - Proliferative diabetic retinopathy; ED- Eales' disease; MH- Macular Hole. RZB – Ranibizumab.

vitreous of PDR (n = 25) and ED (n = 10) than in that of MH patients (n = 25) [4]. The above study also showed strikingly similar cytokine profile in both PDR and ED vitreous and this finding is further confirmed by an extremely rapid cytokine biochip array (Figure 1, Table 1). Moreover, only trace levels of IL-10, TNF-α, IL1-α, IL1-β and EGF were observed in both groups.

Endothelial Tube Formation induced by patient vitreous

The tubulogenesis assay is an *in vivo* correlate of angiogenesis involving endothelial alignment, elongation and polygonal network [20]. Cytokine concentrations estimated by ELISA or Biochip for PDR, ED and MH vitreous, which have been used for tubulogenesis assay, are presented in Table 1. After 48 hours of culture the endothelial cells in the absence of vitreous formed a few short tubes and majority of them remained as individual cells (Fig. 2A). Mean tubular length was 122 μm in control (Fig. 3). In some of the vitreous samples, there were more prominent polygonal tubule network and these had correspondingly high concentrations of VEGF or IL-6 (Fig. 2B and C). On the other hand, vitreous with only trace amounts of cytokines produced minimal tubule formation similar to the control (endothelial cells without vitreous) (Fig. 2D and 3A). Among the 6 PDR vitreous samples tested, 5 showed a significant increase in tube formation (mean fold increase ranging from 0.3 to 0.5, P<0.05, Fig. 3A).

Interestingly a similar vascular response pattern was observed with ED vitreous containing either high (ED-17 and 19; Fig. 2E and F) or low levels of (ED-09, Fig. 2G) cytokines (Table 1). Among, the 5 vitreous samples from ED patients 4 showed significant increase in tube formation (mean fold increase ranging from 0.1 to 0.7) (P<0.05; Fig. 3B).

In MH patients though the retinal vascular reaction does not occur in vivo, we observed significant tube formation with vitreous from all these patients (Fig. 3C). The nature of the vascular assembly is shown in figure 2H for vitreous of MH-09 which contained substantial amount of MCP-1 and IL1-β and trace level of VEGF (Table-1). With limited number of samples studied, it was not possible to apply a test for correlation. However, the data indicate that tube formation in vitro is influenced by the levels of various pro-angiogenic factors present in the vitreous samples.

Effect of RZB on vitreous-induced Endothelial Tube Formation

HMECs were exposed to clinically-relevant concentrations of RZB in combination with the each of the vitreous samples from PDR and ED patients. Tube formation was observed and images were captured after 48 hours of incubation. Figures 2C and 2I show the nature of the vascular assembly in the presence or absence of RZB with PDR-45 vitreous. The tube length was reduced in all six cases with 0.125 mg/ml RZB and the mean fold decrease ranged from 0.2 to 0.7 though significantly in 5 cases (Fig. 4A). Interestingly, we have also observed the reduction in tube length when ED vitreous was treated with RZB (Fig. 2E and 2J). This anti-VEGF effect was observed in all five cases with a mean fold decrease ranging from 0.2 to 1.2, though showing a significant decrease in three cases (Fig. 4B).

Effect of IL-6 neutralizing antibody on vitreous-induced endothelial tube formation

Since IL-6 was observed at high levels in PDR as well as in ED vitreous, we carried out another set of experiments to test the effect of anti-IL6 neutralizing antibody (0.1 μg/ml) on tube formation. Among seven samples of PDR vitreous tested, tube length was reduced in 5 samples with a mean fold ranging from 0.06 to 0.3 (Fig. 5A, C and E). Even though a noticeable reduction was observed in tube length, there was no significant difference between the groups with anti-IL-6 treatment. Our study included only two ED samples for anti-IL-6 treatment due to its reduced availability of clinical samples and limited volume of vitreous; of the two an apparent reduction in tube length was observed in one sample (ED -17) (Fig. 5B, D and F).

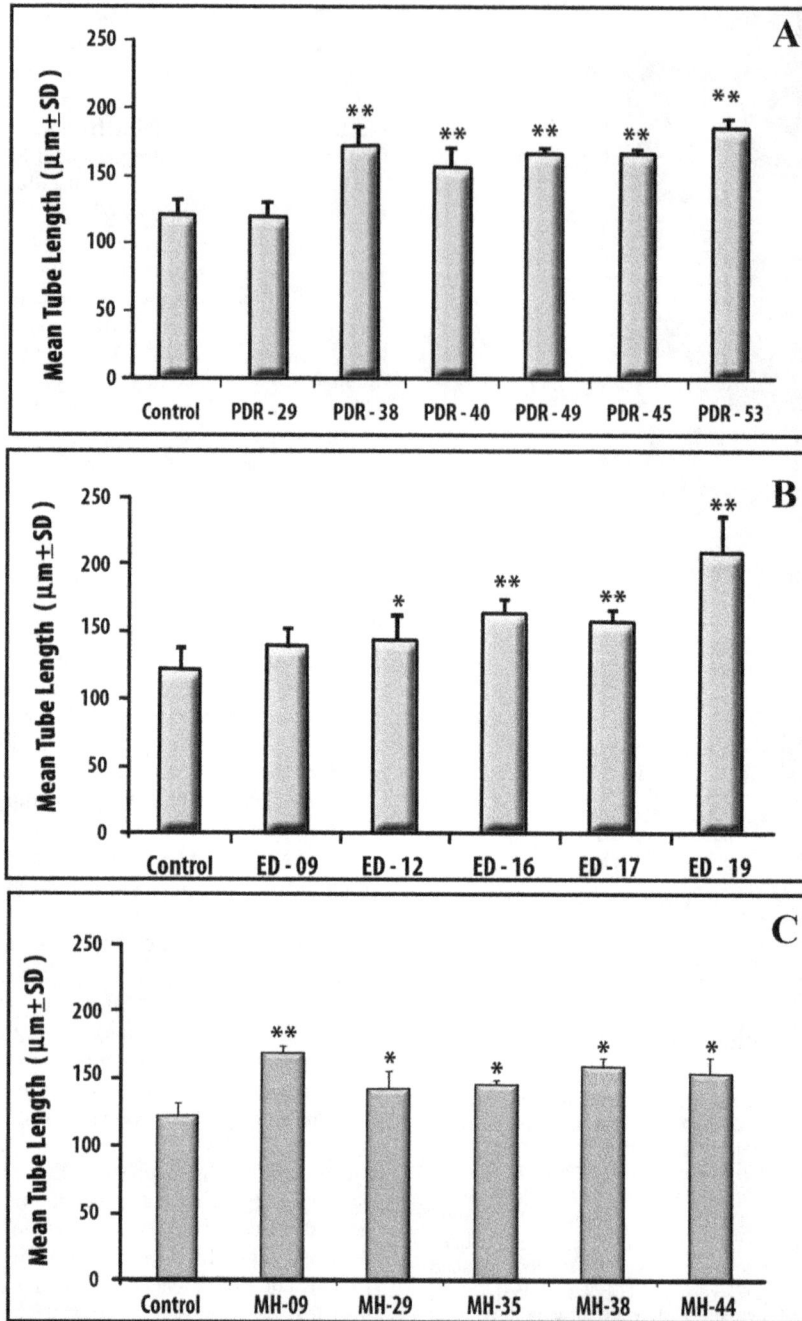

Figure 3. Angiogenic potential of vitreous in capillary tube formation. Experimental details are as in Fig. 2. (A) PDR vitreous, (B) ED vitreous, (C) MH vitreous. Bar graph shows the mean concentration of the triplicate of each sample. PDR-proliferative diabetic retinopathy; ED- Eales' disease; MH- Macular Hole. Number in X-axis denotes the patient number as in table 1. **P<0.001; *P<0.05.

Discussion

Angiogenic growth factors contribute to neovascularization that occurs in retinal diseases like DR and ED. In these vitreo-retinal diseases inflammatory processes are also considered to be critical, suggesting that a range of secreted factors in the vitreous cavity are associated with pathological processes [21], [22], [23]. The presence of inflammatory and angiogenic growth factors has been demonstrated in vitreous [4], [16], [24] from these patients. Though PDR & ED differ in their etiology the profile of proangiogenic factors in the vitreous is markedly similar. We have previously found no significant difference in the concentration of VEGF, IL-6, IL-8 and MCP-1 between these two diseases [4]. Moreover, the profile of twelve cytokines is also similar as shown by cytokine array (Fig. 1, Table 1) and this correlates with the other assays. This bio-chip approach is rapid and accurate taking <4 hours for analysis. It is known that some patients fail to respond to RZB therapy and in the future there may be a place for rapid vitreous analysis post-vitrectomy and establishing if certain growth factors or cytokines are absent or elevated. This would enable patient-specific tailoring the subsequent therapy.

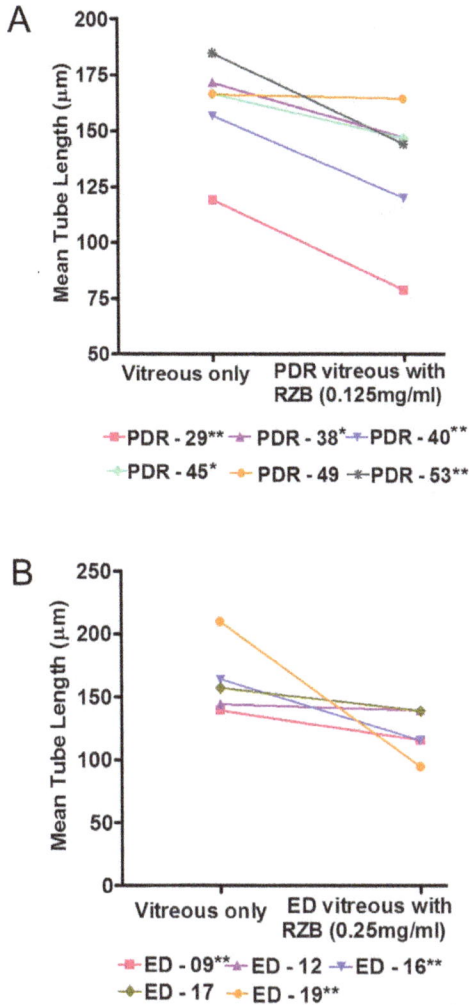

Figure 4. Effect of anti-VEGF antibody on vitreous-induced tube formation. HMECs were mixed with PDR vitreous in presence of RZB (in triplicate cultures). Tube formation was observed and images were captured after 48 hours incubation. (A) 0.125 mg/ml of RZB with PDR vitreous (B) 0.25 mg/ml of RZB with ED vitreous. The tube length was quantified by Nikon NIS-Elements software. Number in the legend denotes the patient ID as in table 1. PDR - Proliferative diabetic retinopathy; ED- Eales' disease; RZB-Ranibizumab. **P<0.001;*P<0.05.

Though there have been several reports on the presence of cytokines in vitreous of patients with vitreo-retinal diseases, to our knowledge, this is the first study to evaluate the angiogenic potential of vitreous from these patients. Using the *in vitro* tubulogenesis assay, we demonstrated the ability of PDR, ED, MH vitreous in inducing endothelial tube formation. In general, markedly increased number of vascular tubes and tubular network was observed with vitreous from the above patient groups than the control without vitreous. It is possible that the factors particularly VEGF, IL-6, IL-8 and MCP-1 might be responsible for vascular tube formation (Table 1, Fig. 3). Further, in addition to VEGF which is known to play a significant role in neovascularization in PDR [25], vitreous containing high levels of IL-6 and IL-8 or MCP-1 but with trace amount of VEGF also induced network of vascular tubes. (Table 1; Fig. 3; PDR-40, 49; ED-12, MH-09, 29). Therefore, in addition to VEGF, inflammatory cytokines are involved in inducing neovascularization.

There is substantial supporting evidence to indicate the interrelationship between the expression of inflammatory cytokines and vascular growth factors in different types of clinical conditions involving angiogenesis. The expression of IL-6, a multi-functional cytokine is elevated in tissues that undergo active angiogenesis, but it does not induce proliferation of endothelial cells. IL-6 has the ability to induce angiogenesis indirectly by the expression of VEGF [26] and possibly by increasing endothelial permeability [27].

Certain transcription factors like NF-Kappa B are known to activate synergistically transcription of cytokines such as IL-6 and chemokines IL-8 and MCP-1 [28], [29], [30]. Further, transcriptional activation of VEGF by IL-6 via STAT-3 pathway and transactivation of VEGF-R2 by IL-8 are associated with vascular permeability [31], [32]. Thus, IL-6, present in significant amount in vitreous of PDR and ED patients, may be able to function as indirect inducer of tube formation *in vitro*. Whether a similar IL-6 induced VEGF expression occurs in the endothelial cells in the tubulogenesis assay needs further studies.

RZB is a high affinity recombinant Fab which binds to the receptor–binding site of all biologically active forms of VEGF-A, thus preventing the activation of two related receptor tyrosine kinase, VEGFR-1 and VEGFR-2. Consequently, the endothelial cell proliferation, migration, permeability and vascular assembly are inhibited by RZB [33], [34], [35]. In this context our study was designed to evaluate the ability of RZB and anti-IL-6 antibodies to inhibit the angiogenic ability of PDR and ED vitreous. A marked reduction in vascular tube formation was observed when PDR vitreous was treated with RZB although this was not observed in all the samples. For example among the two PDR vitreous samples, though both contained low levels of VEGF, tube formation was significantly reduced by RZB in PDR-53 but not in PDR-49 and this difference may possibly be due to the presence of low levels of IL-6 and MCP-1 in the former but high levels of these cytokines in the latter. In general our results can be correlated with clinical studies wherein up to 38.6% of PDR patients did not respond to VEGF therapy [36].

In our context with IL-6 neutralizing antibody we observed tube length reduction in PDR; however, the reduction was not significant (Table-1, Fig. 5E). These samples contained high levels of MCP-1 and varying concentrations of other cytokines. In general, the present study suggests the involvement of inflammatory cytokines in addition to VEGF in inducing tube formation. Further studies are required to evaluate the inter-relationship between the expression of inflammatory cytokines and VEGF as well as their synergistic activity at the molecular level.

In conclusion, this is the first study to evaluate the angiogenic potential of vitreous from PDR and ED patients, demonstrating that VEGF present in ED vitreous is involved in inducing the vascular endothelial cell migration and assembly. Further, the importance of proinflammatory factors in addition to VEGF in retinal neovascularization is well-indicated. It is important to note that in patients, vitreous levels of growth factors and cytokines may not necessarily be due to the pathophysiology of the respective disease entity, but from intravitreal blood and associated cell sources such as thrombocytes. Nevertheless, in this study patient numbers are limited since the majority of vitreous samples we obtained needed to be excluded from the study due to blood contamination. Nevertheless, the study forms a basis for extending patient numbers and further investigating the cytokine profile in PDR and ED and how this influences key function endpoints such as pathological angiogenesis. Therefore, the current data is a useful platform for extending the investigation by using several functional endothelial cell migration, proliferation, and permeability assays [37], [38] to elucidate the importance of various factors (IL-8,MCP-1) and their

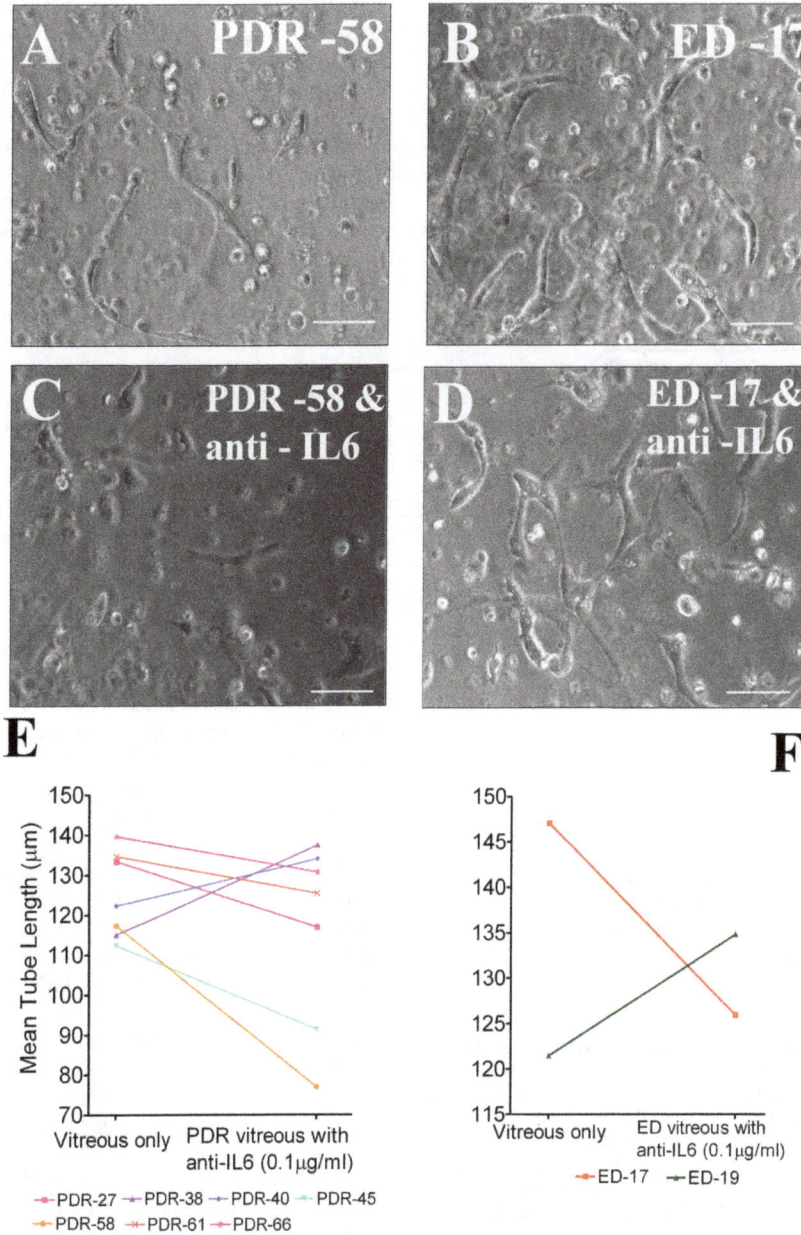

Figure 5. Effects of anti-IL-6 on vitreous-induced tube formation. HMECs were mixed with PDR or ED vitreous with or without anti-IL6 (0.1 µg/ml). Tube formation was observed and images were captured after 48 hours of incubation (in duplicate cultures). A and B are PDR or ED vitreous-induced tube formation. C and D are images of vascular tubes of corresponding vitreous with anti-IL6. E and F – shows the effect of anti-IL6 on vascular tube length for 6 PDR (E) and 2 ED (F) vitreous samples. The tube length was quantified by NIS-Elements software (Nikon). Number in the image denotes the patient ID as in table 1. PDR-Proliferative diabetic retinopathy; ED- Eales' disease. Scale bar =100 µm.

interaction in neovascularization, by using the patients' vitreous and a combination of neutralizing antibodies.

Acknowledgments

We thank Dr. Carmel McVicar and Dr. Medina Reinhold for their technical support in tubulogenesis assay and Mr. B. Vijaya Kumar in statistical analysis.

Author Contributions

Conceived and designed the experiments: VRM AS PM. Performed the experiments: PM. Analyzed the data: PM AS VRM. Contributed reagents/materials/analysis tools: AS DS RK PN. Wrote the paper: PM VRM AS DS. Helped in getting the fund from TIFAC-CORE: PN.

References

1. Das T, Biswas J, Kumar A, Nagpal PN, Namperumalsamy P, et al. (1994) Eales' disease. Indian J Ophthalmol 42: 3–18.

2. Murthy KR, Abraham C, Baig SM, Badrinath SS (1977) Eales' disease. Proc All Ind Ophthalmol. pp. 323.

3. Atmaca LS, Batioglu F, Atmaca Sonmez P (2002) A long-term follow-up of Eales' disease. Ocul Immunol Inflamm 10: 213–221.

4. Murugeswari P, Shukla D, Rajendran A, Kim R, Namperumalsamy P, et al. (2008) Proinflammatory cytokines and angiogenic and anti-angiogenic factors in vitreous of patients with proliferative diabetic retinopathy and eales' disease. Retina 28: 817–824.

5. Ferrara N, Hillan KJ, Gerber HP, Novotny W (2004) Discovery and development of bevacizumab, an anti-VEGF antibody for treating cancer. Nat Rev Drug Discov 3: 391–400.

6. Ferrara N, Damico L, Shams N, Lowman H, Kim R (2006) Development of ranibizumab, an anti-vascular endothelial growth factor antigen binding fragment, as therapy for neovascular age-related macular degeneration. Retina 26: 859–870.

7. Aiello LP, Avery RL, Arrigg PG, Keyt BA, Jampel HD, et al. (1994) Vascular endothelial growth factor in ocular fluid of patients with diabetic retinopathy and other retinal disorders. N Engl J Med 331: 1480–1487.

8. Abu El-Asrar AM, Nawaz MI, Kangave D, Mairaj Siddiquei M, Geboes K (2013) Angiogenic and vasculogenic factors in the vitreous from patients with proliferative diabetic retinopathy. J Diabetes Res 2013: 539658.

9. Rosenfeld PJ, Brown DM, Heier JS, Boyer DS, Kaiser PK, et al. (2006) Ranibizumab for neovascular age-related macular degeneration. N Engl J Med 355: 1419–1431.

10. Boyer DS, Hopkins JJ, Sorof J, Ehrlich JS (2013) Anti-vascular endothelial growth factor therapy for diabetic macular edema. Ther Adv Endocrinol Metab 4: 151–169.

11. Koytak A, Altinisik M, Sogutlu Sari E, Artunay O, Umurhan Akkan JC, et al. (2013) Effect of a single intravitreal bevacizumab injection on different optical coherence tomographic patterns of diabetic macular oedema. Eye (Lond) 27: 716–721.

12. Osaadon P, Fagan XJ, Lifshitz T, Levy J (2014) A review of anti-VEGF agents for proliferative diabetic retinopathy. Eye (Lond) 28: 510–520.

13. ETDRS (1991) Fundus Photographic risk factors for progression of diabetic retinopathy. ETDRS Report Number 12. Ophthalmology. pp. 823–833.

14. Namperumalsamy P, Shukla D (2013) Eales' disease. In: Ryan SJ, editor-in-chief. Textbook of RETINA. St.Louis Elsevier. pp. 1479–1485.

15. Stitt AW, McGoldrick C, Rice-McCaldin A, McCance DR, Glenn JV, et al. (2005) Impaired retinal angiogenesis in diabetes: role of advanced glycation end products and galectin-3. Diabetes 54: 785–794.

16. Yoshimura T, Sonoda KH, Sugahara M, Mochizuki Y, Enaida H, et al. (2009) Comprehensive analysis of inflammatory immune mediators in vitreoretinal diseases. PLoS One 4: e8158.

17. Koskela UE, Kuusisto SM, Nissinen AE, Savolainen MJ, Liinamaa MJ (2013) High vitreous concentration of IL-6 and IL-8, but not of adhesion molecules in relation to plasma concentrations in proliferative diabetic retinopathy. Ophthalmic Res 49: 108–114.

18. Funatsu H, Yamashita H, Shimizu E, Kojima R, Hori S (2001) Relationship between vascular endothelial growth factor and interleukin-6 in diabetic retinopathy. Retina 21: 469–477.

19. Funatsu H, Yamashita H, Noma H, Mimura T, Nakamura S, et al. (2005) Aqueous humor levels of cytokines are related to vitreous levels and progression of diabetic retinopathy in diabetic patients. Graefes Arch Clin Exp Ophthalmol 243: 3–8.

20. Arnaoutova I, George J, Kleinman HK, Benton G (2009) The endothelial cell tube formation assay on basement membrane turns 20: state of the science and the art. Angiogenesis 12: 267–274.

21. Joussen AM, Poulaki V, Le ML, Koizumi K, Esser C, et al. (2004) A central role for inflammation in the pathogenesis of diabetic retinopathy. FASEB J 18: 1450–1452.

22. Adamis AP, Berman AJ (2008) Immunological mechanisms in the pathogenesis of diabetic retinopathy. Semin Immunopathol 30: 65–84.

23. Abcouwer SF (2013) Angiogenic Factors and Cytokines in Diabetic Retinopathy. J Clin Cell Immunol Suppl 1.

24. Bromberg-White JL, Glazer L, Downer R, Furge K, Boguslawski E, et al. (2013) Identification of VEGF-independent cytokines in proliferative diabetic retinopathy vitreous. Invest Ophthalmol Vis Sci 54: 6472–6480.

25. Tolentino MJ, Miller JW, Gragoudas ES, Chatzistefanou K, Ferrara N, et al. (1996) Vascular endothelial growth factor is sufficient to produce iris neovascularization and neovascular glaucoma in a nonhuman primate. Arch Ophthalmol 114: 964–970.

26. Cohen T, Nahari D, Cerem LW, Neufeld G, Levi BZ (1996) Interleukin 6 induces the expression of vascular endothelial growth factor. J Biol Chem 271: 736–741.

27. Maruo N, Morita I, Shirao M, Murota S (1992) IL-6 increases endothelial permeability in vitro. Endocrinology 131: 710–714.

28. Marumo T, Schini-Kerth VB, Busse R (1999) Vascular endothelial growth factor activates nuclear factor-kappaB and induces monocyte chemoattractant protein-1 in bovine retinal endothelial cells. Diabetes 48: 1131–1137.

29. Goebeler M, Gillitzer R, Kilian K, Utzel K, Brocker EB, et al. (2001) Multiple signaling pathways regulate NF-kappaB-dependent transcription of the monocyte chemoattractant protein-1 gene in primary endothelial cells. Blood 97: 46–55.

30. Grosjean J, Kiriakidis S, Reilly K, Feldmann M, Paleolog E (2006) Vascular endothelial growth factor signalling in endothelial cell survival: a role for NFkappaB. Biochem Biophys Res Commun 340: 984–994.

31. Wei LH, Kuo ML, Chen CA, Chou CH, Lai KB, et al. (2003) Interleukin-6 promotes cervical tumor growth by VEGF-dependent angiogenesis via a STAT3 pathway. Oncogene 22: 1517–1527.

32. Petreaca ML, Yao M, Liu Y, Defea K, Martins-Green M (2007) Transactivation of vascular endothelial growth factor receptor-2 by interleukin-8 (IL-8/CXCL8) is required for IL-8/CXCL8-induced endothelial permeability. Mol Biol Cell 18: 5014–5023.

33. Chen Y, Wiesmann C, Fuh G, Li B, Christinger HW, et al. (1999) Selection and analysis of an optimized anti-VEGF antibody: crystal structure of an affinity-matured Fab in complex with antigen. J Mol Biol 293: 865–881.

34. Olsson AK, Dimberg A, Kreuger J, Claesson-Welsh L (2006) VEGF receptor signalling - in control of vascular function. Nat Rev Mol Cell Biol 7: 359–371.

35. Schnichels S, Hagemann U, Januschowski K, Hofmann J, Bartz-Schmidt KU, et al. (2013) Comparative toxicity and proliferation testing of aflibercept, bevacizumab and ranibizumab on different ocular cells. Br J Ophthalmol 97: 917–923.

36. Arevalo JF, Wu L, Sanchez JG, Maia M, Saravia MJ, et al. (2009) Intravitreal bevacizumab (Avastin) for proliferative diabetic retinopathy: 6-months follow-up. Eye (Lond) 23: 117–123.

37. Reinhold HS, Buisman GH (1975) Repair of radiation damage to capillary endothelium. Br J Radiol 48: 727–731.

38. Williams MC, Wissig SL (1975) The permeability of muscle capillaries to horseradish peroxidase. J Cell Biol 66: 531–555.

12

Physical Activity during Pregnancy and Offspring Neurodevelopment and IQ in the First 4 Years of Life

Marlos R. Domingues[1]*, Alicia Matijasevich[2,3], Aluísio J. D. Barros[2], Iná S. Santos[2], Bernardo L. Horta[2], Pedro C. Hallal[1,2]

1 Sports Department, Federal University of Pelotas, Pelotas, Rio Grande do Sul, Brazil, **2** Postgraduate Programme in Epidemiology, Federal University of Pelotas, Pelotas, Rio Grande do Sul, Brazil, **3** Department of Preventive Medicine, School of Medicine, University of Sao Paulo, Sao Paulo, Brazil

Abstract

Background: Maternal physical activity during pregnancy could alter offspring's IQ and neurodevelopment in childhood.

Methods: Children belonging to a birth cohort were followed at 3, 12, 24 and 48 months of age. Physical activity during pregnancy was assessed retrospectively at birth. Neurodevelopment was evaluated by Battelle's Development Inventory (12, 24 and 48 months) and IQ by the Weschler's Intelligence Scale (48 months). Neurodevelopment was based on Battelles' (90th percentile) and also analyzed as a continuous outcome. IQ was analyzed as a continuous outcome. Potential confounders were: family income, mother's age, schooling, skin color, number of previous births and smoking; and newborns': preterm birth, sex and low birth weight.

Results: From birth to 48 months, sample size decreased from 4231 to 3792. Crude analysis showed that IQ at 48 months was slightly higher (5 points) among children from active women. The Battelle's score at 12 and 24 months was higher among offspring from active mothers. After controlling for confounders, physical activity during pregnancy was positively associated to the Battelle's Inventory at 12 months IQ, however, at 48 months no association was observed.

Conclusion: Physical activity during pregnancy does not seem to impair children's neurodevelopment and children from active mothers presented better performance at 12 months.

Editor: Aimin Chen, University of Cincinnati, United States of America

Funding: The 2004 birth cohort study is currently supported by the Wellcome Trust Initiative entitled Major Awards for Latin America on Health Consequences of Population Change, grant entitled: "Implications of early life and contemporary exposures on body composition, human capital, mental health and precursors of complex chronic diseases in three Brazilian cohorts (1982, 1993 and 2004)", grant N° 086974/Z/08/Z. Previous phases of the study were supported by the World Health Organization, National Support Program for Centers of Excellence (PRONEX), the Brazilian National Research Council (CNPq), the Brazilian Ministry of Health, and the Children's Mission. The funders had no role in study design, data collection and analysis, decision to publish, or preparation of the manuscript.

Competing Interests: The authors have declared that no competing interests exist.

* Email: marlosufpel@gmail.com

Introduction

Physical activity during pregnancy is known to result in health benefits as much as among non-pregnant individuals. For example, women who exercise while pregnant experience less muscular discomforts and depressive symptoms, keep their weight gain within normal ranges and present lower blood pressure and blood sugar levels [1]. Evidence also shows that children may benefit from maternal exercise, as preterm birth is less frequent among exercisers [2,3]. Besides, active mothers are more likely to present healthier lifestyles that may influence future health outcomes such as future diabetes and hypertension [1,4,5].

However, as it is observed in the general population, physical activity (PA) level during gestation is below the recommended by health guidelines [6,7]. In Brazil, specifically in this population, leisure-time physical activity prevalence is very low and few women attain the current guidelines of 150 minutes per week [6]. Many reasons are reported by women to not exercise [8], and even among those who were previously active, unexplained fear of harming herself or fetus is observed [6,7]. Although international guidelines recommend PA during pregnancy [9,10], in many countries physical activity counseling is not routine during antenatal care.

In the 1980's, when physical activity in pregnancy became a research subject, one of the first findings was that children from exercising mothers were thinner, as they were born with less body fat [11]. Although lighter and thinner, these babies did not present low birth weight, but one of the concerns was that lower body fat could negatively affect neurodevelopment, as the nervous system depends largely on fat to develop. Future studies did not support such concern and evidence showed that children from exercising mothers perform equal or slightly better on neurodevelopment scales [11,12,13].

The aim of this study was to evaluate if leisure-time physical activity (LTPA) during pregnancy could alter offspring's IQ and neurodevelopment during childhood in a Brazilian birth cohort.

Materials and Methods

A birth cohort started in 2004 in the city of Pelotas (southern Brazil), when all births (from January 1st to December 31st) were identified and those live borns whose family lived in the urban area of the city recruited to the study. The study was hospital based (where more than 99% of births happen), but also included women delivering elsewhere, as they were referred to a hospital soon after birth. Mothers were interviewed and children were measured (perinatal interview) in the first 24 hours after delivery. Mothers were visited at home or went to the research center for future assessments when children were at the ages of 3, 12, 24 and 48 months. At the perinatal interview (hospital) we included all live births and stillbirths when birth weight was above 500 g from women living in the urban area of Pelotas. Along with maternal measurements and interviews, children were also measured and evaluated for several health outcomes, including mental health and neurodevelopment. The methods of the cohort are best described elsewhere [14,15].

Physical activity (PA) during pregnancy was assessed retrospectively during the perinatal interview using a questionnaire (developed and tested by the research team) to collect information on activities performed during leisure time in each trimester of gestation. The following variables were used to describe PA during pregnancy: PA during pregnancy (yes/no), considering any leisure activity in any trimester; any PA during first, second or third trimester; PA during each trimester (yes/no) considering the cutoff point of 150 minutes per week; and tertiles of minutes spent in PA, with a fourth category for women that reported no activity at all. A deeper description of PA patterns in this sample was previously published [6]. Household, commuting and occupational activities were not assessed.

Child development was assessed using two instruments, the screening version of Battelle's Development Inventory [16], that indicates suspected developmental delay, and IQ that was measured with the Wechsler Preschool and Primary Scale of Intelligence (WPPSI) [17]. Battelle's Inventory was administered at the ages of 12, 24 and 48 months, while the IQ test was used at the 48 months assessment. The Battelle's Inventory is a screening tool appropriate to be used from birth to 8 years and helps to identify disabilities among children, school readiness, interaction communication skills, attention, motor skills, and memory among other characteristics of children's neurodevelopment.

The screening version of Battelle's Development Inventory provides a continuous score that was categorized based on the 90th percentile (P90 - cutoff point) of the score obtained in our sample (best development results). An alternative analysis of the Battelle's score was also performed using the score as a continuous outcome. The IQ was analyzed as a continuous variable.

Statistical analysis (Stata 11.0) described the sample (crude analysis) according to demographics, behavior, development scores and physical activity variables. Multivariable models (adjusted analysis) were used to control for potential confounders, Poisson regression was used to measure the association between Battelle's outcomes and physical activity while a linear regression was used to study the influence of LTPA on IQ scores. The variables used for analysis were: family income (quintiles), maternal schooling (four groups), smoking during pregnancy (yes/no), maternal age and skin color (white/black/mixed), maternal occupation characteristics during pregnancy (standing for long periods and lifting heavy weights at work), number of previous births, maternal depressive symptoms at 48 months (based on the Edinburgh Postnatal Depression Scale), birth weight, child's sex and preterm birth (based on an algorithm that considered last menstrual period or ultrasound scans or Dubowitz score).

The study protocol was approved at each follow-up by the Medical Research Ethics Committee of the Federal University of Pelotas Medical School, affiliated with the Brazilian Federal Medical Council. Written informed consent was obtained from all mothers before all interviews. Whenever the mother was not in charge of taking care of the child, and the interviewee was another person, such as a close relative, legal guardian or caretaker, an informed consent was signed by this person before the interview.

Results

The first interview of the birth cohort (at hospital) included 4231 newborns. From birth to 48 months, due to losses to follow-up at 3, 12, 24 months of age, this number gradually decreased to 3792.

Table 1 describes the sample with respect to maternal and child characteristics (maternal age, schooling, skin color, physical activity during pregnancy, family income, smoking, depressive symptoms and child's sex, birth weight and preterm birth). Mothers were mostly white, in the 20–35 years age group, 27.5% smoked during pregnancy and only 13.3% of mothers reported any leisure-time physical activity (LTPA) during pregnancy. The proportion of mothers attaining the cutoff point of 150 minutes of physical activity per week decreased from 6.6% in the first trimester of gestation to 3.5% in the third. The value of PA tertiles in minutes for each trimester is shown in Table 2 along with weekly time range of physical activity in each trimester. Nearly 15% of children were born preterm and 9% presented low birth weight (<2500 g). The average child's IQ at 48 months was 99.6 (sd = 16.7) points.

In Table 3 the development variables (Battelle's Inventory and IQ scores) are described according to LTPA information. Performances in the Battelle test for 12 and 24 months favor offspring from active women. At 48 months there was no clear pattern in Battelle's test to draw any conclusions. The crude analysis of IQ at 48 months showed that children from active women scored, on average, five points higher. All 8 PA-related variables studied presented significant results in favor of active women in crude analysis for IQ.

We have also analyzed the Battelle's score as a continuous variable (data not shown) and observed significant results in favor of active women in crude analysis. For the score measured at 12 (mean score = 56.5 points) and 24 months (mean score = 78.2 points), on average children from active women presented a significant advantage of 1 (one) point in all PA-related variables. At 48 months (mean score = 37.0) most scores were still higher for active women, but none of the differences were significant.

After controlling for confounders (Table 4), only neurodevelopment at 12 months was associated with LTPA - children from women who were active during pregnancy [PR = 1.51 (95%CI: 1.17–1.94); p = 0.001] and active in first [PR = 1.33 (95%CI: 1.01–1.77); p = 0.04], second [PR = 1.50 (95%CI: 1.10–1.98); p = 0.009] and third [PR = 1.41 (95%CI: 1.01–1.97); p = 0.04] trimester presented higher scores at 12 months in the Battelle's test. And also, minutes of physical activity (in tertiles) were associated to being at or above the P90 at 12 months (p<0.001). At 24 months, all results were still higher for active women, but statistic significance was lost after adjustment. At 48 months, adjustment for confounders did not change the results for BDI, while differences in mean IQ were reduced to the point of not being significant any more.

In multivariable analysis of the continuous score (linear regression), we controlled for potential confounders (family

Table 1. Maternal and children's characteristics. Pelotas 2004 Birth Cohort (N = 4147).

	N(%)
Mother's age	
≤19	792 (19.1)
20–35	2800 (67.6)
36– +	553 (13.3)
Maternal Schooling (at birth)	
0–4	639 (15.5)
5–8	1691 (41.2)
9–11	1362 (33.2)
12– +	414 (10.1)
Skin color	
White	3030 (73.0)
Black	828 (20.0)
Mixed	289 (7.0)
Smoking during pregnancy	
Yes	1142 (27.5)
No	3005 (72.5)
Heavy lifting at work	
Yes	343 (8.3)
No	3804 (91.7)
Long standing at work	
Yes	937 (22.6)
No	3210 (77.4)
Preterm Birth	
Yes	602 (14.6)
No	3533 (85.4)
Low birth weight	
Yes	372 (9.0)
No	3772 (91.0)
Sex of the newborn	
Boys	2157 (52.0)
Girls	1990 (48.0)
Maternal Depression at 48 m (Edimburgh)	
Yes (10+)	1107 (29.5)
No (0–9)	2641 (70.5)
LTPA during pregnancy	
Yes	553 (13.3)
No	3594 (86.7)
LTPA first trimester	
Yes	440 (10.6)
No	3707 (89.4)
LTPA second trimester	
Yes	363 (8.8)
No	3784 (91.2)
LTPA third trimester	
Yes	278 (6.7)
No	3869 (93.3)
Minutes of LTPA 1st trimester	
0–149	3874 (93.4)
150– +	273 (6.6)
Minutes of LTPA 2nd trimester	

Table 1. Cont.

	N(%)
0–149	3945 (95.1)
150– +	202 (4.9)
Minutes of LTPA 3rd trimester	
0–149	4003 (96.5)
150– +	144 (3.5)
Physical activity tertiles (min.)	
First	146 (3.5)
Second	213 (5.1)
Third	194 (4.7)
Inactive women	3594 (86.7)

income, schooling, smoking, skin color and preterm birth) and at 12 and 24 months all results favored active women but none was significant. At 48 months the scores for children from active women were slightly worst but, as in crude analysis, no significant results were observed.

Discussion

We evaluated potential longitudinal effects of physical activity during pregnancy on children's development and IQ in the early years of infancy. This is the first study of its kind developed in a middle-income country and the birth cohort is being followed from 2004 to present time.

Our results agree with previous longitudinal studies [11,12,13], as we did not identify negative effects of physical activity during pregnancy on infant's development. Most crude results favor active women, but many were not statistically significant after controlling for potential confounders.

In a comparison between active and inactive women, Clapp et al. did not observe clinical significant between-group differences in performance on either the Bayley Scales of Infant Development or mental development scales, indicating that the offspring of exercising mothers have normal development in the first year of life [13]. Later, in a follow-up study, at 5 years of age, the motor, integrative, and academic readiness skills were similar between children from active and inactive women. However, children from exercising women performed significantly better on the Wechsler scales and tests of oral language skills [11]. Another study [12] indicated that neonatal behavior may be distinct as early as during the first week of life. Children from exercising women performed better in 2 of the 6 evaluations of the Brazelton Scale 5 days after

birth. Performances were better in the ability to orient to environmental stimuli and ability to regulate their state or quiet themselves after sound and light stimuli. Meanwhile, the scores reflecting habituation, motor organization, autonomic stability and behavioral state range were not significantly different.

Physical activity in Brazil is highly associated to the sociodemographic characteristics (income and education), in general population and especially among pregnant women [6]. Development outcomes, such as IQ quotient, are known to be affected by maternal characteristics, mainly maternal education and by the home environment and stimulation [18,19,20,21]. Children from active women presented higher IQ scores in all comparisons with inactive women, but maternal schooling and socioeconomic position may affect children's IQ and neurodevelopment [18,19,20,21] and also are highly associated to physical activity, especially in Brazil [6]. In an alternative analysis, we observed that from the lower schooling category (0–4 years) to the highest category (12 years or more), the IQ score increases linearly 25% while a fivefold increase in mean minutes of physical activity during pregnancy was detected. Therefore, schooling is a potential confounder on the association between performance in development tests and maternal physical activity during pregnancy. Indeed, if schooling is not considered during multivariable analysis, all beneficial results are significant favoring active women.

Although our observational study cannot discuss causation, biologically, the potential effects of physical activity during pregnancy on neurodevelopment could possibly be explained by different pathways. First, glucose metabolism changes during pregnancy - gestational diabetes and maternal insulin resistance may change negatively intellectual and psychomotor development

Table 2. Distribution of leisure-time physical activity in minutes (standard deviation) according to trimester of pregnancy and tertile of weekly activity, among women reporting any physical activity in the period.

	First trimester	Second trimester	Third trimester
First tertile	79.9 (38.4)	60.0 (31.8)	69.4 (39.2)
Second tertile	169.2 (78.2)	125.0 (46.8)	122.6 (59.8)
Third tertile	324.5 (140.7)	272.3 (110.9)	248.7 (114.7)
Time range of weekly physical activity among active women	20–840	10–630	15–630

Pelotas 2004 Birth Cohort (2004).

Table 3. Battelle's Development Inventory score (P90) and IQ scores (at 48 months) according to leisure-time physical activity (LTPA) information. Pelotas 2004 Birth Cohort (2004–2008).

	P90 of Battelle 12 m	p	P90 of Battelle 24 m	p	P90 of Battelle 48 m	p	IQ (mean)	p
Any LTPA during pregnancy		<0.001		0.06		0.91		<0.001
Yes	15.1		10.9		6.2		103.9	
No	8.4		8.4		6.3		99.1	
Any LTPA 1st trimester		0.001		0.06		0.39		<0.001
Yes	13.9		11.2		5.3		103.6	
No	8.7		8.4		6.4		99.3	
Any LTPA 2nd trimester		<0.001		0.04		0.70		0.001
Yes	15.6		11.8		5.8		103.9	
No	8.7		8.4		6.3		99.4	
Any LTPA 3rd trimester		0.001		0.18		0.79		0.0007
Yes	15.1		11.0		6.7		103.2	
No	8.9		8.6		6.3		99.5	
LTPA≥150 min/wk 1st trimester		0.03		0.09		0.51		<0.001
Yes	13.1		11.6		5.3		104.6	
No	9.0		8.5		6.4		99.4	
LTPA≥150 min/wk 2nd trimester		0.007		0.02		0.64		0.0001
Yes	14.9		13.5		5.5		103.8	
No	9.0		8.5		6.3		99.6	
LTPA≥150 min/wk 3rd trimester		0.02		0.68		0.55		0.02
Yes	15.4		9.7		7.5		103.1	
No	9.1		8.7		6.3		99.7	
Physical activity tertiles (min.)		<0.001		0.12		0.82		<0.001
First	23.9		11.7		6.7		102.0	
Second	11.2		9.0		7.3		105.7	
Third	12.5		12.3		4.6		103.3	
Inactive women	8.4		8.4		6.3		99.1	

Table 4. Multivariable analyses of the relationship between LTPA during pregnancy and neurodevelopmental variables.

	Battelle 12 m		Battelle 24 m		Battelle 48 m		IQ (β)	
	PR (95%CI)	p	PR (95%CI)	p	PR (95%CI)	p		p
Any LTPA during pregnancy		0.001		0.75		0.62		0.48
Yes	1.51 (1.17-1.94)		1.05 (0.79-1.39)		0.91 (0.63-1.32)		0.52	
No	1.00		1.00		1.00			
Any LTPA 1st trimester		0.04		0.44		0.24		0.59
Yes	1.33 (1.01-1.77)		1.12 (0.84-1.51)		0.77 (0.50-1.19)		0.44	
No	1.00		1.00		1.00			
Any LTPA 2nd trimester		0.009		0.550		0.44		0.73
Yes	1.48 (1.10-1.98)		1.10 (0.80-1.53)		0.83 (0.53-1.32)		−0.31	
No	1.00		1.00		1.00			
Any LTPA 3rd trimester		0.04		0.89		0.85		0.78
Yes	1.41 (1.01-1.97)		1.03 (0.70-1.50)		1.05 (0.65-1.69)		−0.29	
No	1.00		1.00		1.00			
LTPA≥150 min/wk 1st trimester		0.31		0.46		0.36		0.21
Yes	1.20 (0.84-1.71)		1.14 (0.80-1.63)		0.78 (0.45-1.34)		1.26	
No	1.00		1.00		1.00			
LTPA≥150 min/wk 2nd trimester		0.07		0.21		0.50		0.90
Yes	1.42 (0.98-2.07)		1.28 (0.87-1.88)		0.81 (0.44-1.50)		−0.14	
No	1.00		1.00		1.00			
LTPA≥150 min/wk 3rd trimester		0.14		0.86		0.60		0.72
Yes	1.40 (0.90-2.19)		0.95 (0.56-1.61)		1.18 (0.64-2.18)		−0.48	
No	1.00		1.00		1.00			
Physical activity tertiles (min.)		<0.001		0.86		0.88		0.76
First	2.53 (1.83-3.50)		1.19 (0.74-1.92)		0.99 (0.52-1.90)		−0.72	
Second	0.97 (0.62-1.53)		0.77 (0.47-1.27)		1.04 (0.61-1.76)		1.16	
Third	1.31 (0.86-1.98)		1.23 (0.82-1.84)		0.709 (0.35-1.39)		0.82	
Inactive women	1.00		1.00		1.00			

LTPA: leisure-time physical activity - PR: prevalence rates −95%CI: 95% confidence intervals.
Multivariable analyses included in the model: family income, schooling, smoking, skin color, maternal age, number of previous births, maternal occupational characteristics and preterm birth. Prevalence rates for the Battelle's outcomes derive from Poisson regression. Beta values for IQ scores derive from linear regression and indicate differences between inactive and active women. Pelotas 2004 Birth Cohort (2004-2008).

in children and result in attention disorders and lower motor coordination [22]. Previous studies have reported that intra uterine environment (maternal blood glucose) may affect children's cognitive abilities [23]. Physical activity's role in insulin control is recognized and could affect intrauterine glucose metabolism [22,24]. Second, placental development is distinct between active and inactive women [25,26] - physical activity results in better circulation and higher placental volumes, which could improve oxygen availability in uterus and influence neurologic development. Third, although aware of potential reverse causality, depression is frequently associated to physical inactivity [27] and inactive women are more likely to be depressed during pregnancy [28]. Gestational depression affects negatively psychological development in early infancy [29] and maternal well being and anxiety are associated to both physical activity and children's behavior [28,30]. A sex-stratified analysis was carried out (data not shown) to assess potential gender differences and, although few results were changed, we noticed that all associations were stronger in magnitude among boys. Our results may be another indication that pregnancy characteristics perhaps affect distinctively girls' and boys' neurodevelopment during infancy [29].

It seems that the beneficial effects of physical activity, if real, weaken as the child ages, because the stronger (and significant) effects observed in our study were restrict to the first year of life.

Among the limitations of our study we must highlight the following issues: 1) the retrospective evaluation of physical activity could result in recall bias, however, our goal was to identify habitual activities to understand how usual behavior could influence the outcomes; 2) lack of intensity information was an option of the researchers because current intensity is already problematic to be assessed, thus past intensity was not a reliable information; 3) we also did not collect data on occupational, commuting or household activities. We only had information about standing or heavy lifting at worksite; however these variables were not associated with any of the outcomes, but were included in the multivariable model. On the other hand, few studies are available presenting longitudinal effects of LTPA during pregnancy, and our population sample also collaborates to the quality of our data with respect to potential selection bias. Loss to follow-up was no differential by physical activity status. From 2004 to 2008

we lost 8.4% of women, however the physical activity prevalence did not change significantly (13.3% vs. 13.5%).

Unfortunately, the amount of women reporting physical activity throughout the whole pregnancy or reporting physical activity during the third trimester is very small. Also, the percentage of women attaining the physical activity guidelines (150 minutes per week) in any of the trimesters was very little, and even with our large population sample, some of the associations tested could have been affected by lack of statistical power. For example, less than 4% of women achieved the recommended amount of PA in the third trimester.

As in any health study that considers physical activity as an exposure, we cannot rule out positive effects of different lifestyle characteristics that usually are associated to physical activity and were not evaluated. Physical activity is a voluntary behavior and it is plausible that, women who chose to exercise during gestation, also made other healthier choices in several aspects of pregnancy (food choices, for example) and during their children's early infancy. The maternal profile of women who are more health concerned may affect child's development in different manners and many characteristics cannot be considered in population studies or included in statistical analysis, resulting in potential residual negative confounding.

Based on our results, we conclude that LTPA during pregnancy does not seem to affect negatively children's neurodevelopment as children from active mothers presented better results for most of the studied outcomes. After controlling for confounders, children from women who were active in the first, second and third trimester of pregnancy presented significant better results at 12 months of age. Thus, PA should be advised to pregnant women based on all benefits already known [1,10]. Child's improved neurodevelopment, especially in the first year of life, may be another positive effect of an active lifestyle during gestation.

Author Contributions

Conceived and designed the experiments: MRD AM AJDB IS. Performed the experiments: MRD. Analyzed the data: MRD BH PH. Contributed reagents/materials/analysis tools: MRD AM AJDB IS BH PH. Wrote the paper: MRD AM AJDB IS BH PH.

References

1. Pivarnik JM, Chambliss HO, Clapp JF, Dugan SA, Hatch MC, et al. (2006) Impact of physical activity during pregnancy and postpartum on chronic disease risk. Med Sci Sports Exerc 38: 989–1006.
2. Domingues MR, Barros AJ, Matijasevich A (2008) Leisure time physical activity during pregnancy and preterm birth in Brazil. Int J Gynaecol Obstet 103: 9–15.
3. Domingues MR, Matijasevich A, Barros AJ (2009) Physical activity and preterm birth: a literature review. Sports Med 39: 961–975.
4. Devlieger R, Guelinckx I, Vansant G (2008) The impact of obesity on female reproductive function. Obes Rev 9: 181–182 author reply 183.
5. Guelinckx I, Devlieger R, Beckers K, Vansant G (2008) Maternal obesity: pregnancy complications, gestational weight gain and nutrition. Obes Rev 9: 140–150.
6. Domingues MR, Barros AJ (2007) Leisure-time physical activity during pregnancy in the 2004 Pelotas Birth Cohort Study. Rev Saude Publica 41: 173–180.
7. Haakstad LA, Voldner N, Henriksen T, Bo K (2009) Why do pregnant women stop exercising in the third trimester? Acta Obstet Gynecol Scand 88: 1267–1275.
8. Duncombe D, Wertheim EH, Skouteris H, Paxton SJ, Kelly L (2009) Factors related to exercise over the course of pregnancy including women's beliefs about the safety of exercise during pregnancy. Midwifery 25: 430–438.
9. Committee on Obstetric Practice (2002) ACOG committee opinion. Exercise during pregnancy and the postpartum period. Number 267, January 2002. American College of Obstetricians and Gynecologists. Int J Gynaecol Obstet 77: 79–81.
10. Haskell WL, Lee IM, Pate RR, Powell KE, Blair SN, et al. (2007) Physical activity and public health: updated recommendation for adults from the

American College of Sports Medicine and the American Heart Association. Circulation 116: 1081–1093.
11. Clapp JF 3rd (1996) Morphometric and neurodevelopmental outcome at age five years of the offspring of women who continued to exercise regularly throughout pregnancy. J Pediatr 129: 856–863.
12. Clapp JF 3rd, Lopez B, Harcar-Sevcik R (1999) Neonatal behavioral profile of the offspring of women who continued to exercise regularly throughout pregnancy. Am J Obstet Gynecol 180: 91–94.
13. Clapp JF 3rd, Simonian S, Lopez B, Appleby-Wineberg S, Harcar-Sevcik R (1998) The one-year morphometric and neurodevelopmental outcome of the offspring of women who continued to exercise regularly throughout pregnancy. Am J Obstet Gynecol 178: 594–599.
14. Barros AJ, Santos Ina S, Victora CG, Albernaz EP, Domingues MR, et al. (2006) [The 2004 Pelotas birth cohort: methods and description.]. Rev Saude Publica 40: 402–413.
15. Santos IS, Barros AJ, Matijasevich A, Domingues MR, Barros FC, et al. (2010) Cohort Profile: The 2004 Pelotas (Brazil) Birth Cohort Study. Int J Epidemiol.
16. Newborg J, Stock J, Wnek L, Guidabaldi J, Svinicki J (1988) Battelle Developmental Inventory. Itasca, IL: Riverside Publishing.
17. Wechsler D (1967) Wechsler Preschool and Primary Scale of Intelligence. New York: Psychological Corporation; 1967.
18. Huisman M, Araya R, Lawlor DA, Ormel J, Verhulst FC, et al. Cognitive ability, parental socioeconomic position and internalising and externalising problems in adolescence: findings from two European cohort studies. Eur J Epidemiol 25: 569–580.

19. Santos DN, Assis AM, Bastos AC, Santos LM, Santos CA, et al. (2008) Determinants of cognitive function in childhood: a cohort study in a middle income context. BMC Public Health 8: 202.

20. To T, Guttmann A, Dick PT, Rosenfield JD, Parkin PC, et al. (2004) Risk markers for poor developmental attainment in young children: results from a longitudinal national survey. Arch Pediatr Adolesc Med 158: 643–649.

21. Barros AJ, Matijasevich A, Santos IS, Halpern R (2010) Child development in a birth cohort: effect of child stimulation is stronger in less educated mothers. Int J Epidemiol 39: 285–294.

22. Ornoy A, Ratzon N, Greenbaum C, Wolf A, Dulitzky M (2001) School-age children born to diabetic mothers and to mothers with gestational diabetes exhibit a high rate of inattention and fine and gross motor impairment. J Pediatr Endocrinol Metab 14 Suppl 1: 681–689.

23. Fraser A, Nelson SM, Macdonald-Wallis C, Lawlor DA (2012) Associations of Existing Diabetes, Gestational Diabetes, and Glycosuria with Offspring IQ and Educational Attainment: The Avon Longitudinal Study of Parents and Children. Exp Diabetes Res 2012: 963735.

24. Silverman BL, Rizzo TA, Cho NH, Metzger BE (1998) Long-term effects of the intrauterine environment. The Northwestern University Diabetes in Pregnancy Center. Diabetes Care 21 Suppl 2: B142–149.

25. Clapp JF (2006) Influence of endurance exercise and diet on human placental development and fetal growth. Placenta 27: 527–534.

26. Clapp JF 3rd, Kim H, Burciu B, Lopez B (2000) Beginning regular exercise in early pregnancy: effect on fetoplacental growth. Am J Obstet Gynecol 183: 1484–1488.

27. Strohle A (2009) Physical activity, exercise, depression and anxiety disorders. J Neural Transm 116: 777–784.

28. Poudevigne MS, O'Connor PJ (2006) A review of physical activity patterns in pregnant women and their relationship to psychological health. Sports Med 36: 19–38.

29. Gerardin P, Wendland J, Bodeau N, Galin A, Bialobos S, et al. Depression during pregnancy: is the developmental impact earlier in boys? A prospective case-control study. J Clin Psychiatry 72: 378–387.

30. Hernandez-Martinez C, Arija V, Balaguer A, Cavalle P, Canals J (2008) Do the emotional states of pregnant women affect neonatal behaviour? Early Hum Dev 84: 745–750.

13

Maternal Obesity and Tobacco Use Modify the Impact of Genetic Variants on the Occurrence of Conotruncal Heart Defects

Xinyu Tang[1], Todd G. Nick[1], Mario A. Cleves[2], Stephen W. Erickson[3], Ming Li[1], Jingyun Li[1], Stewart L. MacLeod[2], Charlotte A. Hobbs[2]*

1 Biostatistics Program, Department of Pediatrics, College of Medicine, University of Arkansas for Medical Sciences, Little Rock, Arkansas, United States of America, **2** Division of Birth Defects Research, Department of Pediatrics, College of Medicine, University of Arkansas for Medical Sciences, Little Rock, Arkansas, United States of America, **3** Department of Biostatistics, College of Public Health, University of Arkansas for Medical Sciences, Little Rock, Arkansas, United States of America

Abstract

Conotruncal heart defects (CTDs) are among the most severe birth defects worldwide. Studies of CTDs indicate both lifestyle behaviors and genetic variation contribute to the risk of CTDs. Based on a hybrid design using data from 616 case-parental and 1645 control-parental triads recruited for the National Birth Defects Prevention Study between 1997 and 2008, we investigated whether the occurrence of CTDs is associated with interactions between 921 maternal and/or fetal single nucleotide polymorphisms (SNPs) and maternal obesity and tobacco use. The maternal genotypes of the variants in the glutamate-cysteine ligase, catalytic subunit (*GCLC*) gene and the fetal genotypes of the variants in the glutathione S-transferase alpha 3 (*GSTA3*) gene were associated with an elevated risk of CTDs among obese mothers. The risk of delivering infants with CTDs among obese mothers carrying AC genotype for a variant in the *GCLC* gene (rs6458939) was 2.00 times the risk among those carrying CC genotype (95% confidence interval: 1.41, 2.38). The maternal genotypes of several variants in the glutathione-S-transferase (*GST*) family of genes and the fetal genotypes of the variants in the *GCLC* gene interacted with tobacco exposures to increase the risk of CTDs. Our study suggests that the genetic basis underlying susceptibility of the developing heart to the adverse effects of maternal obesity and tobacco use involve both maternal and embryonic genetic variants. These results may provide insights into the underlying pathophysiology of CTDs, and ultimately lead to novel prevention strategies.

Editor: Dana C. Crawford, Case Western Reserve University, United States of America

Funding: This work is supported by the National Institute of Child Health and Human Development (NICHD) under award number 5R01HD039054-12 and the National Center on Birth Defects and Developmental Disabilities (NCBDDD) under award number 5U01DD000491-05. The funders had no role in study design, data collection and analysis, decision to publish, or preparation of the manuscript.

Competing Interests: The authors have declared that no competing interests exist.

* Email: HobbsCharlotte@uams.edu

Introduction

Congenital heart defects (CHDs) are among the most common and severe birth defects worldwide, with reported estimated prevalence of 9.1 per 1,000 live births after 1995 [1]. Conotruncal heart defects (CTDs), a class of CHDs, affect the cardiac outflow tracts and great arteries which include truncus ateriosus, interrupted aortic arch type B, transposition of great arteries, double outlet right ventricle, conoventricular septal defect, tetralogy of Fallot, and pulmonary atresia with ventricular septal defect.

Nonsyndromic CTDs are due to a multifactorial etiology involving a complex interplay between genetic susceptibilities and environmental factors [2,3]. One such interplay is between maternal folic acid supplement use and genetic variants in folate-related pathways. Based on the finding that maternal periconceptional folic acid intake decreases the occurrence of CTDs [4,5], multiple studies have investigated associations between CTDs and polymorphisms in folate-related genes [6–8].

It has also been reported that genetic variants in folate-related pathways modify the association between birth defects and maternal intake of folic acid containing supplements [9].

Developmental toxicology studies using animal models have repeatedly demonstrated the unquestioned importance of genetic variation in determining risks to environmental factors [10]. Inbred strains of mice, representing different mouse genomes, vary in their susceptibility to teratogenic and xenobiotic agents. In reproductive age women, obesity and tobacco use have been associated with multiple adverse outcomes including intrauterine growth retardation [11,12], prematurity [13,14], and birth defects [15–17]. Obesity and cigarette smoking are also associated with alterations in folate and glutathione metabolism resulting in decreased folate [18,19], increased homocysteine [20–24], and decreased glutathione [25–30] that may compromise the *in-utero* environment. Some studies have demonstrated that maternal genetic variants modulate the association between pregnancy smoking exposure and fetal growth restriction [31,32]. It is possible that variants in genes that encode for critical enzymes in folate, homocysteine and glutathione pathways modify the adverse impact of obesity and tobacco on the developing heart.

In this study, we used a hybrid design which combines genetic and lifestyle data from case-parental and control-parental triads to investigate whether CTDs are associated with interactions between maternal and/or fetal single nucleotide polymorphisms (SNPs) and maternal obesity and tobacco use. In contrast to most published reports of Gene × Environment (G × E) interactions [4,33,34], we have evaluated maternal and fetal genetic effects simultaneously. A total of 1536 SNPs in 62 target genes were selected for this study from folate-related metabolic pathways.

Materials and Methods

Study Population

Families were recruited for the National Birth Defects Prevention Study (NBDPS) with estimated dates of delivery between October 1997 and August 2008 (www.nbdps.org). Detailed information about the NBDPS is outlined in Yoon et al [35]. Families were identified through population-based birth defects surveillance systems in 10 states: Arkansas, California, Iowa, Massachusetts, New Jersey (through 2002), New York, Texas, Georgia, North Carolina (beginning 2003), and Utah (beginning 2003). In this study, cases were singleton live-born infants with CTDs. NBDPS cardiac cases were reviewed by pediatric cardiologists using a classification strategy developed by investigators within the NBDPS. This strategy targeted etiologic investigations of CHDs that encourage explicit case definitions and aggregates of defects with a focus on simple, isolated phenotypes and associations [36]. Cases with recognized or strongly suspected monogenic or chromosomal conditions were excluded. Controls were singleton live-born infants without any major structural birth defects [37]. Both case and control mothers completed phone interviews. The study was approved by the University of Arkansas for Medical Sciences' Institutional Review Board and the NBDPS with protocol oversight by the Centers for Disease Control and Prevention Center for Birth Defects and Developmental Disabilities. All of the study subjects gave written informed consent. For minors, informed written consent was obtained from their legal guardian.

Maternal Interview

Participation in the NBDPS included a one-hour interview with mothers of cases and controls, conducted in English or Spanish, by interviewers using a computer-assisted telephone questionnaire [35]. In this study, we investigated how maternal obesity and tobacco use modify maternal and fetal genetic effects on the risk of CTDs. Obesity, using the Institute of Medicine definition [38], was defined as a body mass index (BMI) ≥ 30.0 and normal weight between 18.5 and 25.0. Smokers were defined as women who smoked cigarettes during the 3 months after conception and nonsmokers otherwise. Maternal use of folic acid supplements is a known risk factor for the occurrence of CTDs and warranted a separate manuscript [9].

DNA Collection

After completion of maternal interviews, a buccal cell collection kit was sent to participants to obtain cheek cell samples from case/control and parents. The collection kit included informed consent forms, instructions, $20 money order, materials for completing the specimen collection and prepaid US mail packets for specimen return [35].

Gene/SNP Selection

As previously described [39], a custom panel of 1536 SNPs in 62 genes involved in folate metabolism was developed jointly between our lab and Illumina. Candidate genes were required to encode an enzyme in one of the candidate metabolic pathways and be expressed in liver and/or heart tissue [40]. For each candidate gene, a maximally informative set of haplotype-tagging SNPs was selected using both linkage disequilibrium statistics and Illumina assay design scores. The custom genotyping panel was devised in 2005–2006. At that time there were two genes called RFC1 in the commonly used genetic databases. The genotype data presented here are for SNPs in the Replication Factor C (activator 1) 1 (RFC1) gene. This gene is an activator of DNA polymerase and is required for DNA synthesis and repair.

Genotyping and Quality Assessment

Genotyping was conducted on a total of 635 case and 1702 control families using 200 ng of WGA DNA on the Illumina Golden Gate platform [41]. Initial genotype calls were generated using Genome Studio's GenCall, Illumina's proprietary algorithm, with subsequent analysis performed using SNPMClust, a bivariate Gaussian model-based genotype clustering and calling algorithm developed in-house. A total of 297 individuals were removed due to study ineligibility (n = 33), high no-call rates (n = 63), or high rates of Mendelian inconsistency (n = 201). We found that the quality of genotype clustering varied substantially from SNP to SNP, which we attribute to the in silico design of the SNP panel based on data from phases I and II of the HapMap project, without the subsequent quality checks that would be applied to a standard commercial SNP panel. While the majority of SNPs exhibited well-segregated genotype clusters, a substantial percentage exhibited poor clustering and/or lower minor allele frequency (MAF) than expected. To ensure high-quality genotypes, we applied stringent quality control measures and excluded SNPs with obviously poor clustering behavior (60 SNPs), no-call rates >10% (328 SNPs), Mendelian error rates >5% (11 SNPs), MAF <5% (204 SNPs), or significant deviation from Hardy-Weinberg Equilibrium in at least one racial group ($p < 10^{-4}$, 12 SNPs). The final dataset included 4648 individuals, each with 921 SNPs.

Statistical Methods

Descriptive statistics were summarized for families. Comparisons of maternal characteristics were carried out between case and control families using the two-sample t-tests for continuous variables and chi-square tests for categorical variables. Because both case-parental and control-parental triads were enrolled and genotyped, a hybrid design provided optimal power by using a log-linear model [42]. This model has several advantages including incorporating case and control triads, providing population-based inferences related to maternal and fetal genetic effects, providing evidence related to fetal genetic effects that are robust to population admixture, and allowing inclusion of missing genotype data. To assess Gene × Environment (G × E) interactions, a log-linear model was fitted for each SNP as a function of mating types, an interaction between mating types and a maternal factor (E), case/control status (D), maternal genotype (D × M), fetal genotype (D × C), an interaction for maternal genotype (D × E × M), and an interaction for fetal genotype (D × E × C). The detailed information regarding the model is outlined in Hobbs et al [9]. Based on the log-linear model for counts and assuming a Poisson distribution, the relative risk (RR) (95% CI) of having one copy of the minor allele compared to no copy was estimated for each SNP. Because we assumed multiplicative risk of alleles, the estimated RR (95% CI) of having 2 copies of the minor allele is square of the estimated RR (95% CI) of having 1 copy of the minor allele. We also performed sensitivity analysis by restricting our models to the Caucasian families only. The Caucasian families were selected for the sensitivity analysis because among all the racial groups, we had the largest sample size for Caucasian families. For other racial groups, we did not have power to perform any subgroup analyses.

The Bayesian false-discovery probability (BFDP) [43–48] was used to evaluate the chance of false-positive associations using estimates of the RR and corresponding 95% CI obtained from log-linear models. In the results section, we reported associations where BFDP <0.80, which is a commonly used threshold suggested by Wakefield [43] for summary analyses. Patterns of linkage disequilibrium (LD) between significant SNPs in the same gene were constructed to assess the correlations among these SNPs using the parents of control families. Data were analyzed using statistical software LEM [49] for fitting log-linear models, R v3.0.1 (R Foundation for Statistical Computing, Vienna, Austria) for computing descriptive statistics, and BFDPs and HaploView 4.2 [50] for developing LD maps.

Results

Study Sample

A total of 616 case families and 1645 control families were included in our analysis. Due to different call rates among SNPs, case and control triads/dyads/monads numbers differ for any given SNP. Maternal demographics and lifestyle factors are summarized in **Table 1**. The case mothers were slightly older than the control mothers [28.3±6.1 vs. 27.5±6.0, p=0.002]. Case and control mothers did not differ on any of the other demographics and lifestyle characteristics (p>0.05).

Genetic Variants

As described above, 921 candidate SNPs were included in the final analyses. The information about the number of SNPs in each gene, chromosome and pathway is displayed in **Figure 1**. Hobbs et al. [9] identified the maternal genotypes of 17 SNPs and the fetal genotypes of 17 SNPs associated with CTD risks independently from interactions with maternal folic acid supplement use, obesity and tobacco use. Briefly, the maternal genotypes of 10 SNPs and the fetal genotypes of 2 SNPs in the glutamate-cysteine ligase, catalytic subunit (GCLC) gene were found to be significantly associated with the risk of CTDs through main genetic effects [9].

Gene × Environment Interactions

In Hobbs et al. [9], we identified SNPs that had a main effect on the occurrence of CTDs. To identify significant G × E interactions, we evaluated the combined effects of each maternal and infant SNP with maternal folic acid supplement use, obesity and tobacco use. Significant interactions are discussed below.

SNP × Folic Acid Supplement Use. In Hobbs et al. [9], we have studied the interactions between 921 SNPs and periconceptional folic acid supplement use, and found that the maternal genotypes of 19 SNPs and the fetal genotypes of 9 SNPs had BFDP <0.80 for testing the G × E interactions with maternal use of folic acid supplements. All these SNPs were not found to be significant based on the gene only models. Detailed information about the SNP × folic acid supplement use is outlined in Hobbs et al [9].

SNP × Obesity. A main effect of the association between CTDs and obesity compared to normal weight resulted in an estimated RR = 1.27 (95% CI: 0.99, 1.63; p=0.06). Because previous studies showed that obesity independent of pre-existing Type I or Type II diabetes was significantly associated with

elevated risk of CTDs [15,51], we computed a model for each SNP among normal-weight and obese women. The results are shown in **Figure 2** and **Table S1**.

Among 10 maternal SNPs with a significant gene × obesity interaction, three SNPs are in the GCLC gene, three SNPs are in the O-6-methylguanine-DNA methyltransferase (MGMT) gene, and the remaining four SNPs are in the glutathione S-transferase mu 2 (GSTM2), transcobalamin II (TCN2), cystathionine-beta-synthase (CBS), and DNA (cytosine-5-)-methyltransferase 3-like (DNMT3L) genes respectively. The most significant maternal gene × obesity interaction (i.e. most significant difference in the estimated RR for certain genotype among obese women compared to normal weight women) was embedded in the GCLC gene. The genotype of rs6458939 was not associated with the risk of CTDs among normal-weight women [estimated RR: 0.89, 95% CI: 0.70, 1.12]; however, the risk of delivering infants with CTDs among obese women carrying AC genotype for rs6458939 was estimated to be 2.00 (95% CI: 1.41, 2.83) times the risk among those carrying CC genotype. The RR (95% CI) for carrying two copies of the minor allele is square of the estimated RR (95% CI) for carrying one copy of the minor allele. Thus, only the RR (95% CI) is reported for carrying one copy of the minor allele compared to no copy of the minor allele. A similar pattern was observed for the maternal genotypes of two other SNPs in the GCLC gene. All three maternal SNPs in the GCLC gene were determined to be in high LD (D'≥0.98).

Among 10 fetal SNPs found to be associated with the risk of CTDs in combination with obesity, three SNPs reside in the glutathione S-transferase alpha 3 (GSTA3) gene, three SNPs are in the adenosylhomocysteinase-like 2 (AHCYL2) gene, three SNPs are in the DNA (cytosine-5-)-methyltransferase 3 alpha (DNMT3A) gene, and the other SNP resides in the GCLC gene. We observed the most significant difference in the fetal effect of the GSTA3 gene on the risk of CTDs between normal-weight and obese women. The risk of CTDs was elevated for infants with AG genotype for rs668163 compared to GG genotype among obese women (RR: 1.83, 95% CI: 1.33, 2.52), but was not significantly reduced among normal-weight women (RR: 0.94, 95% CI: 0.75, 1.18). Similar findings were observed for the fetal genotypes of two other SNPs in the GSTA3 gene. All three fetal SNPs in the GSTA3 gene were in high LD (D'≥0.96).

Similar results were observed when overweight women were compared to normal-weight women (data not shown) and based on Caucasian families only (data not shown). There was no main effect of any of the SNPs discussed above.

SNP × Tobacco Use. A main effect of the association between CTDs and maternal tobacco use compared to no tobacco use had an estimated RR = 1.08 (95% CI: 0.85, 1.37; p=0.54). We investigated the interactions between SNPs and maternal tobacco use on the occurrence of CTDs by fitting the G × E log-linear model with maternal tobacco use for each SNP. The maternal and fetal SNPs that had a significant interaction with maternal tobacco use are presented in **Figure 3** and **Table S2**.

Among 6 maternal SNPs found to have BFDP <0.80, two of them are located in the RFC1 gene, two SNPs are in the glutathione S-transferase alpha 4 (GSTA4) gene, and the other two SNPs are in the glutathione S-transferase alpha 1 (GSTA1) and glutathione S-transferase alpha 2 (GSTA2) genes respectively. The GSTA1, GSTA2 and GSTA4 genes are glutathione-S-transferase (GST) family of genes. The RRs of the maternal AC genotype for rs2397135 in the GSTA4 gene compared to CC genotype were estimated to be 0.92 (95% CI: 0.76, 1.11) among women without tobacco use, and 1.74 (95% CI: 1.21, 2.49) among women with

Table 1. Maternal characteristics for 616 case families and 1,645 control families enrolled in National Birth Defects Prevention Study between 1997 and 2008.

	Case	Control
	(N = 616)	(N = 1,645)
Age at delivery (years)	28.3±6.1	27.5±6.0
Race		
African American	49 (8%)	143 (9%)
Caucasian	401 (66%)	1,136 (69%)
Hispanic	123 (20%)	285 (17%)
Others	39 (6%)	78 (5%)
Missing information	4	3
Education		
<12 years	83 (14%)	217 (13%)
High school degree or equivalent	167 (27%)	413 (25%)
1–3 years of college	173 (28%)	454 (28%)
At least 4 years of college or Bachelor degree	190 (31%)	559 (34%)
Missing information	3	2
Household income		
Less than 10 Thousand	94 (16%)	236 (15%)
10 to 30 Thousand	150 (26%)	408 (27%)
30 to 50 Thousand Dollars	118 (20%)	348 (23%)
More than 50 Thousand	217 (37%)	538 (35%)
Missing information	37	115
Folic acid supplements		
No	299 (49%)	738 (45%)
Yes[1]	314 (51%)	907 (55%)
Missing information	3	0
BMI		
Underweight (BMI <18.5)	31 (5%)	74 (5%)
Normal weight (18.5<=BMI <25)	298 (50%)	880 (55%)
Overweight (25<=BMI <30)	141 (24%)	360 (23%)
Obese (>=30)	121 (20%)	281 (18%)
Missing information	25	50
Alcohol consumption		
No	460 (76%)	1,251 (76%)
Yes[2]	149 (24%)	390 (24%)
Missing information	7	4
Tobacco use		
No	498 (81%)	1,356 (82%)
Yes[3]	114 (19%)	288 (18%)
Missing information	4	1

Summary statistics were expressed as mean ± standard deviation for continuous variables, and frequency (percentage) for categorical variables.
[1] Folic acid supplements yes is defined as maternal use of folic acid supplements at least two months during the exposure window (i.e. one month prior to conception and two months after conception).
[2] Alcohol consumption yes is defined as drinking during the three months after pregnancy.
[3] Tobacco use yes is defined as cigarette smoking during the three months after pregnancy.

tobacco use, interaction BFDP = 0.68<0.80. The two maternal SNPs in the *RFC1* gene were in moderate LD (D' = 0.82).

Among 5 fetal SNPs identified to have BFDP <0.80, three SNPs are in the *GCLC* gene, one SNP (rs11727502) is in the *RFC1* gene, and the other SNP (rs10277237) is located in the nitric oxide synthase 3 (endothelial cell) (*NOS3*) gene. The RRs of

the fetal AG genotype for rs7742367 in the *GCLC* gene compared to AA genotype were estimated to be 1.14 (95% CI: 0.83, 1.40) among women without tobacco use, and 2.12 (95% CI: 1.48, 3.04) among women with tobacco use, interaction BFDP = 0.71<0.80. The two fetal SNPs (rs7742367 and rs10948751) in the *GCLC*

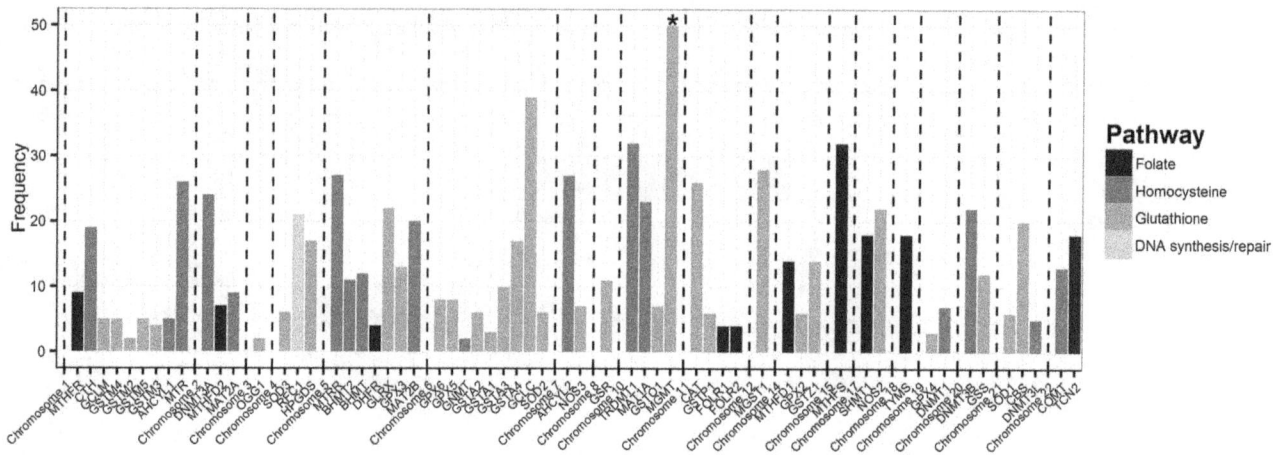

Figure 1. SNP, gene, chromosome and pathway information. Information about the number of SNPs from each gene, chromosome and pathway (* there are a total of 142 candidate SNPs on MGMT, truncated at 50 for display purpose). The average number of SNPs per gene is 15 ± 19. Among 60 genes, 10 (17%) genes are involved in folate pathway, 32 (53%) in glutathione pathway, and 17 (28%) in homocysteine pathway, and one gene RFC1 (2%) in DNA synthetic/repair pathway.

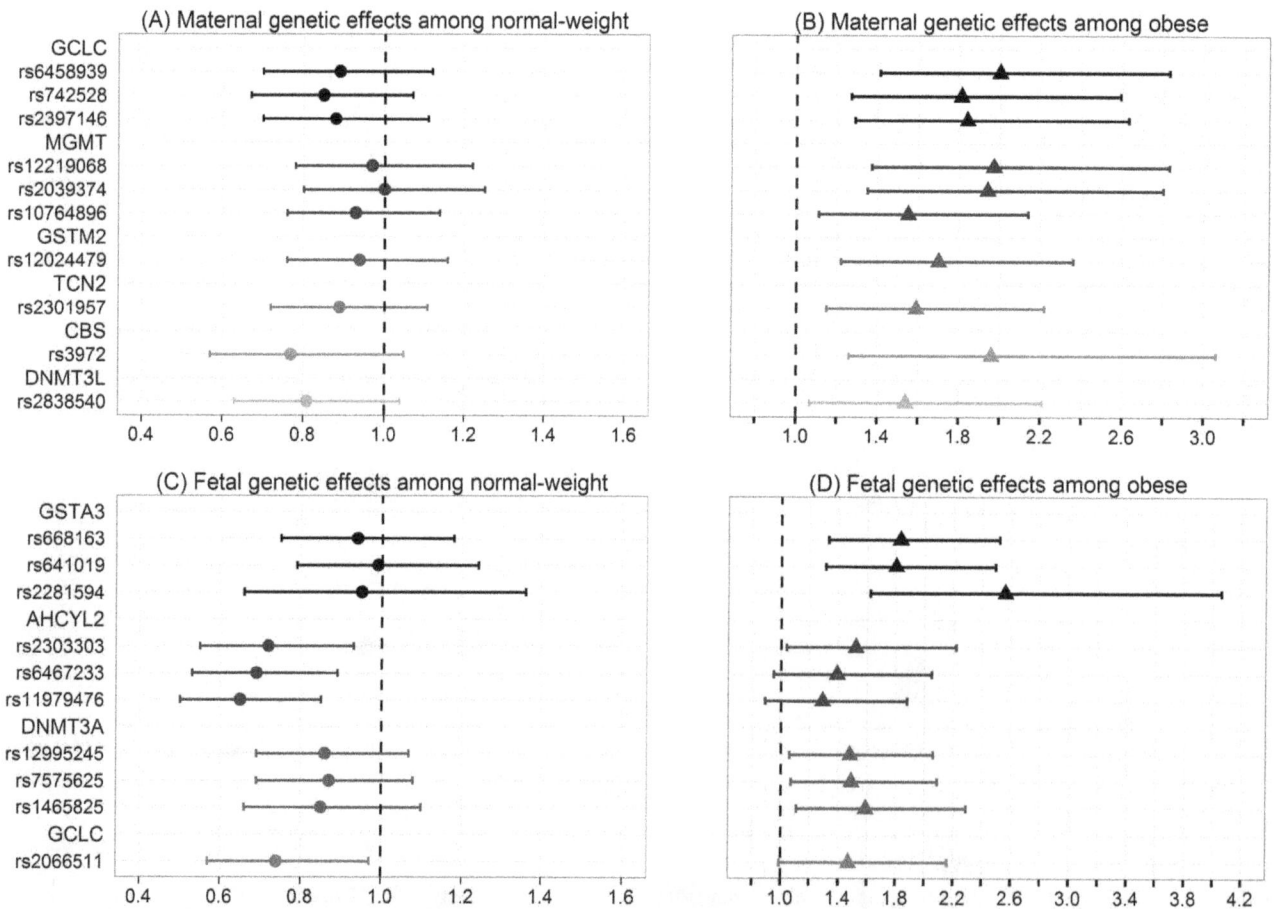

Figure 2. Interactions between gene and obesity. The estimated relative risks and their 95% confidence intervals for each maternal and fetal SNPs that were identified to have a significant interaction with maternal obesity on the risk of CTDs. (A) Maternal genetic effects among normal-weight women; (B) Maternal genetic effects among obese women; (C) Fetal genetic effects among normal-weight women; (D) Fetal genetic effects among obese women.

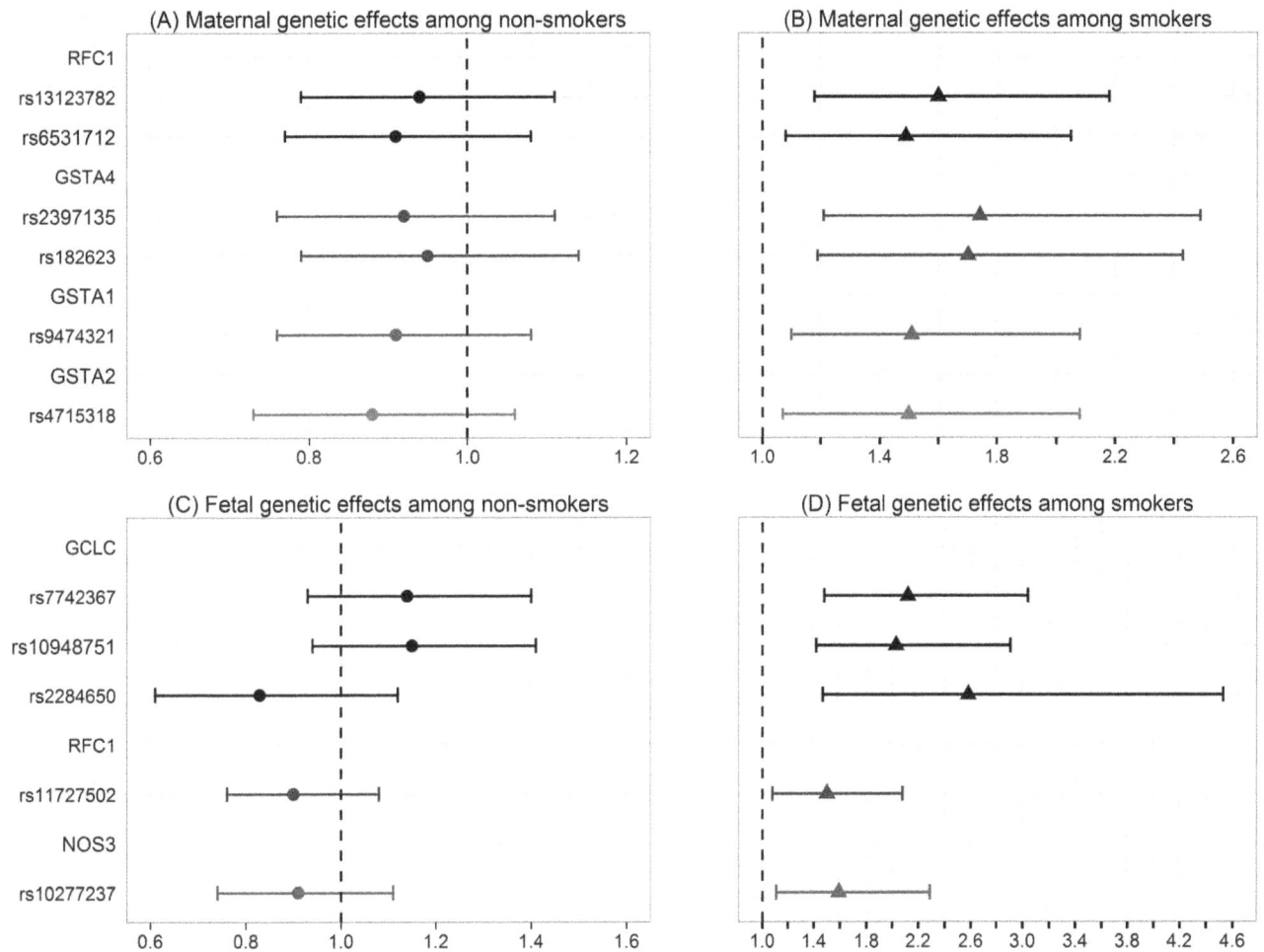

Figure 3. Interactions between gene and maternal tobacco use. The estimated relative risks and their 95% confidence intervals for each maternal and fetal SNPs that were identified to have a significant interaction with maternal tobacco use on the risk of CTDs. (A) Maternal genetic effects among non-smokers; (B) Maternal genetic effects among smokers; (C) Fetal genetic effects among non-smokers; (D) Fetal genetic effects among smokers.

gene were in high LD (D' = 0.99), and in moderate LD with the other SNP (rs2284650) respectively (D' = 0.89 and 0.93).

Similar results were found based on Caucasian families only (data not shown). There was no SNP main effect.

Discussion

This study provides evidence for a moderating effect of multiple SNPs related to one-carbon metabolism and glutathione metabolism on the relationship between CTDs and maternal obesity and tobacco use. Significant genes and their pathway information are displayed in **Figure 4**. Our findings support the hypothesis that CTDs are associated with G × E interactions between common genetic variants and lifestyle factors. Consistent with previous studies [52–54], our results provide supportive evidence that both maternal and fetal genotypes modify the susceptibility of the developing fetus to factors that may alter the *in-utero* environment.

Maternal obesity independent of diabetes is a well-accepted risk factor for multiple nonsyndromic birth defects including CTDs [15,51,55]. Among obese women, the maternal genotypes of 10 SNPs within 6 genes (*GCLC, MGMT, GSTM2, TCN2, CBS,* and *DNMT3L*) and the fetal genotypes of 7 SNPs within 3 genes

(*GSTA3, AHCYL2,* and *DNMT3A*) were associated with an increased risk of CTDs, whereas, among women of normal weight, none of the maternal genotypes had a significant impact on the RR, and the fetal genotypes of 4 SNPs in 2 genes (*AHCYL2* and *GCLC*) decreased the risk of CTDs.

Among women who used tobacco in the periconceptional period, the maternal genotypes of 6 SNPs in 4 genes (*RFC1, GSTA4, GSTA1,* and *GSTA2*) were associated with an increased risk of CTDs. The *GST* genes function in the detoxification of several environmental exposures, including carcinogens, medications, and factors increasing oxidative stress. The fetal genotypes of 5 SNPs in 3 genes (*GCLC, RFC1* and *NOS3*) increased the occurrence of CTDs among women who smoked. None of the maternal or fetal genotypes significantly impact the risk among women without tobacco use.

Our findings compel us to comment on some of the findings in the context of previous studies. Several maternal and fetal genotypes of SNPs in the glutathione transferase (*GSTA4, GSTA1, GSTA2, GSTM2,* and *GSTA3*) genes increased the impact of maternal obesity and tobacco use on the risk of CTDs. The Glutathione S-transferase complex of genes are involved in the detoxification of a large number of xenobiotics including chemical compounds in tobacco smoke [56]. Genetic variants in

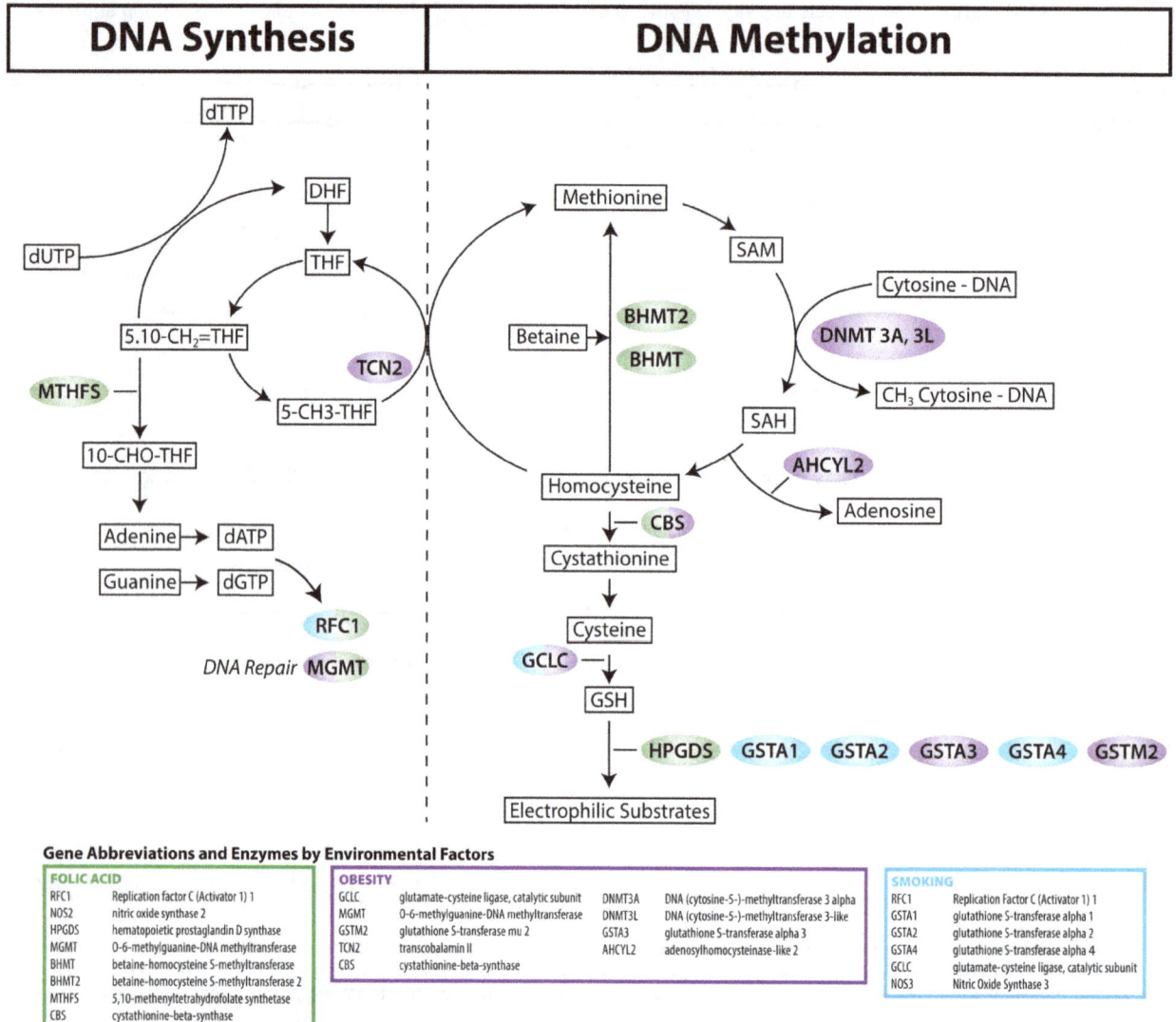

Figure 4. Significant genes and their pathways. Related pathways include the folate pathway, which is involved in DNA synthesis, and the methionine and glutathione pathways, which are necessary for DNA methylation and for conjugation of electrophilic substrates, respectively. Substrates and products of enzymatic reactions appear in boxes. Enzymes appear in color-coded bubbles according to their association with birth defects risk and maternal folic acid supplement use (Green), obesity (Purple) and tobacco use (Blue).

glutathione S-transferase genes may be a factor in determining individual susceptibility to embryotoxic effects of xenobiotics, including tobacco. Lending credence to the possibility that glutathione S-transferases in interaction with tobacco may impact the developing fetus, is a report that glutathione S-transferase genetic variants modified the impact of cigarette smoking on birth weight [32].

In earlier reports [57,58], we demonstrated that women who had infants with CHDs had lower GSH levels. GSH level is essential for cell progression and serves several functions that are crucial during embryogenesis. GSH is important in the defense against oxidative stress, detoxification of xenobiotics, regulation of cell cycle progression and apoptosis, and modulation of immune function [56]. GCLC is the gene that encodes for the catalytic subunit glutamate-cysteine ligase, which is the enzyme that catalyzes the conversion of cysteine with glutamate, generating glutathione. Several maternal and fetal genotypes of SNPs in the

GCLC gene increased the RR of CTDs among women who smoked and/or were obese.

Among women who did not use folic acid supplements, 8 SNPs in the hematopoietic prostaglandin D synthase gene (HPGDS) increased the risk of CTDs, but had no associated increased risk among folic acid supplement users [9]. HPGDS is a member of the sigma class of glutathione S-transferase gene family. The HPGDS gene catalyzes both isomerization of PGH (2) to PGD (2) and conjugation of glutathione to 1-chlore-2, 4-dinitrobenzene [59–61]. HPGDS is present in mast cells, macrophages and other cellular sources [62]. Little is known about the impact of prostaglandin synthase on the development of CHDs. Previous studies have shown that HPGDS plays a role in multiple physiological processes that are important during embryogenesis, including detoxification of xenobiotics [63], regulation of inflammatory pathways [59] and modification of reactive oxygen species [64]. Animal models and mechanistic studies are needed to validate and clarify the role of maternal *HPGDS* in the development of heart defects.

There are several strengths of this study including the use of log-linear models based on a hybrid design, its large sample size, population-based ascertainment of cases, and a rigorous analytic inquiry into potential G × E interactions on CTD risks. There are also limitations in our study. No adjustment was made for confounding effects from other exposures when fitting the model, which prohibited us from drawing further conclusions about the observed associations. Other limitations related to DNA and lifestyle data collection existed. The usage of buccal cell collection kits in collecting DNA samples created a disparity among the quality of the DNA samples, resulting in a reduction from 1534 to 921 candidate SNPs. Maternal obesity and tobacco use were self-reported, and not objectively measured.

It is now generally accepted that most CHDs are the result of a complex interplay between genetic and environmental factors. Our results suggest that, in the search for genetic variants related to CTDs, accounting for environmental and lifestyle factors may improve the ability to detect genetic effects. Because the typical effect size for maternal and fetal genetic variation acting on CHD phenotypes is small [65,66], large sample sizes and validation within independent populations are necessary to draw firm conclusions about how certain polymorphisms modulate the experience of maternal lifestyles. The present findings will need to be replicated and further studies should incorporate functional genetics and targeted deep sequencing while enhancing information collected on environmental and lifestyle factors. One limitation of our current study is that to properly assess the accuracy of our findings, they will need to be replicated on a fully distinct and independent sample [67]. Nevertheless, our study is the largest family-based case-control study of CTDs ever conducted in the U.S., and to our knowledge, there is no comparable, independent sample which includes the maternal and fetal genotype data, and maternal lifestyle and demographic data, necessary to validate our results.

Supporting Information

Table S1 Maternal and fetal SNPs with interactive effects with maternal obesity. For each significant SNP, the information about its pathway, chromosome, gene, allele, estimated relative risks and their 95% confidence intervals among normal weight and obese women, p-value and BFDP for the interaction term are presented.

Table S2 Maternal and fetal SNPs with interactive effects with maternal tobacco use. For each significant SNP, the information about its pathway, chromosome, gene, allele, estimated relative risks and their 95% confidence intervals among nonsmokers and smokers, p-value and BFDP for the interaction term are presented.

Acknowledgments

The authors wish to thank the generous participation of the numerous families that made this research study possible. We also thank the Centers for Birth Defects Research and Prevention in California, Georgia, Iowa, Massachusetts, New Jersey, New York, North Carolina, Texas, and Utah for their contribution of data and manuscript review. The authors thank Ashley S. Block and Zuzana Gubrij for assistance in preparation of the manuscript.

The contents of this manuscript are solely the responsibility of the authors and do not necessarily represent the official views of the Centers for Disease Control and Prevention.

Author Contributions

Conceived and designed the experiments: XT TGN MAC SWE ML JL SLM CAH. Performed the experiments: SWE CAH. Analyzed the data: XT TGN MAC ML JL SLM CAH. Contributed reagents/materials/analysis tools: XT TGN MAC SWE ML JL SLM CAH. Wrote the paper: XT TGN MAC SWE ML JL SLM CAH.

References

1. van der Linde D, Konings EE, Slager MA, Witsenburg M, Helbing WA, et al. (2011) Birth prevalence of congenital heart disease worldwide: a systematic review and meta-analysis. J Am Coll Cardiol 58: 2241–2247.

2. Pierpont ME, Basson CT, Benson DW Jr., Gelb BD, Giglia TM, et al. (2007) Genetic basis for congenital heart defects: current knowledge: a scientific statement from the American Heart Association Congenital Cardiac Defects Committee, Council on Cardiovascular Disease in the Young: endorsed by the American Academy of Pediatrics. Circulation 115: 3015–3038.

3. Jenkins KJ, Correa A, Feinstein JA, Botto L, Britt AE, et al. (2007) Noninherited risk factors and congenital cardiovascular defects: current knowledge: a scientific statement from the American Heart Association Council on Cardiovascular Disease in the Young: endorsed by the American Academy of Pediatrics. Circulation 115: 2995–3014.

4. Shaw GM, O'Malley CD, Wasserman CR, Tolarova MM, Lammer EJ (1995) Maternal periconceptional use of multivitamins and reduced risk for conotruncal heart defects and limb deficiencies among offspring. Am J Med Genet 59: 536–545.

5. Botto LD, Khoury MJ, Mulinare J, Erickson JD (1996) Periconceptional multivitamin use and the occurrence of conotruncal heart defects: results from a population-based, case-control study. Pediatrics 98: 911–917.

6. Pediatric Cardiac Genomics C, Gelb B, Brueckner M, Chung W, Goldmuntz E, et al. (2013) The Congenital Heart Disease Genetic Network Study: rationale, design, and early results. Circ Res 112: 698–706.

7. Shaw GM, Lu W, Zhu H, Yang W, Briggs FB, et al. (2009) 118 SNPs of folate-related genes and risks of spina bifida and conotruncal heart defects. BMC Med Genet 10: 49.

8. Lupo PJ, Goldmuntz E, Mitchell LE (2010) Gene-gene interactions in the folate metabolic pathway and the risk of conotruncal heart defects. J Biomed Biotechnol 2010: 630940.

9. Hobbs CA, Cleves MA, Macleod SL, Erickson SW, Tang X, et al. (2014) Conotruncal heart defects and common variants in maternal and fetal genes in folate, homocysteine, and transsulfuration pathways. Birth Defects Res A Clin Mol Teratol 100: 116–126.

10. Wlodarczyk BJ, Palacios AM, Chapa CJ, Zhu H, George TM, et al. (2011) Genetic basis of susceptibility to teratogen induced birth defects. Am J Med Genet C Semin Med Genet 157: 215–226.

11. Radulescu L, Munteanu O, Popa F, Cirstoiu M (2013) The implications and consequences of maternal obesity on fetal intrauterine growth restriction. J Med Life 6: 292–298.

12. Delpisheh A, Brabin L, Drummond S, Brabin BJ (2008) Prenatal smoking exposure and asymmetric fetal growth restriction. Ann Hum Biol 35: 573–583.

13. Cnattingius S, Villamor E, Johansson S, Edstedt Bonamy AK, Persson M, et al. (2013) Maternal obesity and risk of preterm delivery. JAMA 309: 2362–2370.

14. Wang T, Zhang J, Lu X, Xi W, Li Z (2011) Maternal early pregnancy body mass index and risk of preterm birth. Arch Gynecol Obstet 284: 813–819.

15. Gilboa SM, Correa A, Botto LD, Rasmussen SA, Waller DK, et al. (2010) Association between prepregnancy body mass index and congenital heart defects. Am J Obstet Gynecol 202: 51 e51–51 e10.

16. Hackshaw A, Rodeck C, Boniface S (2011) Maternal smoking in pregnancy and birth defects: a systematic review based on 173 687 malformed cases and 11.7 million controls. Hum Reprod Update Sep-Oct 17: 589–604.

17. Malik S, Cleves MA, Honein MA, Romitti PA, Botto LD, et al. (2008) Maternal smoking and congenital heart defects. Pediatrics 121: e810–816.

18. Sen S, Iyer C, Meydani SN (2014) Obesity during pregnancy alters maternal oxidant balance and micronutrient status. J Perinatol 34: 105–111.

19. Stark KD, Pawlosky RJ, Sokol RJ, Hannigan JH, Salem N Jr. (2007) Maternal smoking is associated with decreased 5-methyltetrahydrofolate in cord plasma. Am J Clin Nutr 85: 796–802.

20. Davis W, van Rensburg SJ, Cronje FJ, Whati L, Fisher LR, et al. (2014) The fat mass and obesity-associated FTO rs9939609 polymorphism is associated with elevated homocysteine levels in patients with multiple sclerosis screened for vascular risk factors. Metab Brain Dis 29: 409–419.

21. Linnebank M, Moskau S, Semmler A, Hoefgen B, Bopp G, et al. (2012) A possible genetic link between MTHFR genotype and smoking behavior. PLoS One 7: e53322.

22. Vaya A, Rivera L, Hernandez-Mijares A, de la Fuente M, Sola E, et al. (2012) Homocysteine levels in morbidly obese patients: its association with waist circumference and insulin resistance. Clin Hemorheol Microcirc 52: 49–56.

23. Sanchez-Margalet V, Valle M, Ruz FJ, Gascon F, Mateo J, et al. (2002) Elevated plasma total homocysteine levels in hyperinsulinemic obese subjects. J Nutr Biochem 13: 75–79.

24. Pfeiffer CM, Sternberg MR, Schleicher RL, Rybak ME (2013) Dietary supplement use and smoking are important correlates of biomarkers of water-soluble vitamin status after adjusting for sociodemographic and lifestyle variables in a representative sample of U.S. adults. J Nutr 143: 957S–965S.

25. Amirkhizi F, Siassi F, Djalali M, Shahraki SH (2014) Impaired enzymatic antioxidant defense in erythrocytes of women with general and abdominal obesity. Obes Res Clin Pract 8: e26–34.

26. Chitty KM, Lagopoulos J, Hickie IB, Hermens DF (2014) The impact of alcohol and tobacco use on in vivo glutathione in youth with bipolar disorder: An exploratory study. J Psychiatr Res 55: 59–67.

27. Fisher-Wellman KH, Weber TM, Cathey BL, Brophy PM, Gilliam LA, et al. (2014) Mitochondrial respiratory capacity and content are normal in young insulin-resistant obese humans. Diabetes 63: 132–141.

28. Igosheva N, Abramov AY, Poston L, Eckert JJ, Fleming TP, et al. (2010) Maternal diet-induced obesity alters mitochondrial activity and redox status in mouse oocytes and zygotes. PLoS One 5: e10074.

29. Fratta Pasini A, Albiero A, Stranieri C, Cominacini M, Pasini A, et al. (2012) Serum oxidative stress-induced repression of Nrf2 and GSH depletion: a mechanism potentially involved in endothelial dysfunction of young smokers. PLoS One 7: e30291.

30. Ermis B, Ors R, Yildirim A, Tastekin A, Kardas F, et al. (2004) Influence of smoking on maternal and neonatal serum malondialdehyde, superoxide dismutase, and glutathione peroxidase levels. Ann Clin Lab Sci 34: 405–409.

31. Delpisheh A, Brabin L, Topping J, Reyad M, Tang AW, et al. (2009) A case-control study of CYP1A1, GSTT1 and GSTM1 gene polymorphisms, pregnancy smoking and fetal growth restriction. Eur J Obstet Gynecol Reprod Biol 143: 38–42.

32. Danileviciute A, Grazuleviciene R, Paulauskas A, Nadisauskiene R, Nieuwenhuijsen MJ (2012) Low level maternal smoking and infant birthweight reduction: genetic contributions of GSTT1 and GSTM1 polymorphisms. BMC Pregnancy Childbirth 12: 161.

33. Shaw GM, Iovannisci DM, Yang W, Finnell RH, Carmichael SL, et al. (2005) Risks of human conotruncal heart defects associated with 32 single nucleotide polymorphisms of selected cardiovascular disease-related genes. Am J Med Genet A 138: 21–26.

34. Botto LD, Mulinare J, Erickson JD (2000) Occurrence of congenital heart defects in relation to maternal mulitivitamin use. Am J Epidemiol 151: 878–884.

35. Yoon PW, Rasmussen SA, Lynberg MC, Moore CA, Anderka M, et al. (2001) The National Birth Defects Prevention Study. Public Health Rep 116 Suppl 1: 32–40.

36. Botto LD, Lin AE, Riehle-Colarusso T, Malik S, Correa A, et al. (2007) Seeking causes: classifying and evaluating congenital heart defects in etiologic studies. Birth Defects Res A Clin Mol Teratol 79: 714–727.

37. Cogswell ME, Bitsko RH, Anderka M, Caton AR, Feldkamp ML, et al. (2009) Control selection and participation in an ongoing, population-based, case-control study of birth defects: the National Birth Defects Prevention Study. Am J Epidemiol 170: 975–985.

38. Institute of M (2009) Weight Gain During Pregnancy: Reexamining the Guidelines.

39. Chowdhury S, Hobbs CA, MacLeod SL, Cleves MA, Melnyk S, et al. (2012) Associations between maternal genotypes and metabolites implicated in congenital heart defects. Mol Genet Metab 107: 596–604.

40. Romero P, Wagg J, Green ML, Kaiser D, Krummenacker M, et al. (2005) Computational prediction of human metabolic pathways from the complete human genome. Genome Biol 6: R2.

41. Fan JB, Chee MS, Gunderson KL (2006) Highly parallel genomic assays. Nat Rev Genet 7: 632–644.

42. Weinberg CR, Umbach DM (2005) A hybrid design for studying genetic influences on risk of diseases with onset early in life. Am J Hum Genet 77: 627–636.

43. Wakefield J (2007) A Bayesian measure of the probability of false discovery in genetic epidemiology studies. Am J Hum Genet 81: 208–227.

44. Liu Y, Shete S, Etzel CJ, Scheurer M, Alexiou G, et al. (2010) Polymorphisms of LIG4, BTBD2, HMGA2, and RTEL1 genes involved in the double-strand break repair pathway predict glioblastoma survival. J Clin Oncol 28: 2467–2474.

45. Park SL, Bastani D, Goldstein BY, Chang SC, Cozen W, et al. (2010) Associations between NBS1 polymorphisms, haplotypes and smoking-related cancers. Carcinogenesis 31: 1264–1271.

46. Oh SS, Chang SC, Cai L, Cordon-Cardo C, Ding BG, et al. (2010) Single nucleotide polymorphisms of 8 inflammation-related genes and their associations with smoking-related cancers. Int J Cancer 127: 2169–2182.

47. Spitz MR, Gorlov IP, Dong Q, Wu X, Chen W, et al. (2012) Multistage analysis of variants in the inflammation pathway and lung cancer risk in smokers. Cancer Epidemiol Biomarkers Prev 21: 1213–1221.

48. Zienolddiny S, Haugen A, Lie JA, Kjuus H, Anmarkrud KH, et al. (2013) Analysis of polymorphisms in the circadian-related genes and breast cancer risk in Norwegian nurses working night shifts. Breast Cancer Res 15: R53.

49. Vermunt JK (1997) LEM: A General Program for the Analysis of Categorical Data. Department of Methodology and Statistics, Tilburg University.

50. Barrett JC (2009) Haploview: Visualization and analysis of SNP genotype data. Cold Spring Harb Protoc 2009(10): pdb.ip71. doi:10.1101/pdb.ip71.

51. Oddy WH, De Klerk NH, Miller M, Payne J, Bower C (2009) Association of maternal pre-pregnancy weight with birth defects: evidence from a case-control study in Western Australia. Aust N Z J Obstet Gynaecol 49: 11–15.

52. Wang C, Xie L, Zhou K, Zhan Y, Li Y, et al. (2013) Increased risk for congenital heart defects in children carrying the ABCB1 Gene C3435T polymorphism and maternal periconceptional toxicants exposure. PLoS One 8: e68807.

53. Wang W, Wang Y, Gong F, Zhu W, Fu S (2013) MTHFR C677T polymorphism and risk of congenital heart defects: evidence from 29 case-control and TDT studies. PLoS One 8: e58041.

54. Finnell RH, Shaw GM, Lammer EJ, Brandl KL, Carmichael SL, et al. (2004) Gene-nutrient interactions: importance of folates and retinoids during early embryogenesis. Toxicol Appl Pharmacol 198: 75–85.

55. Madsen NL, Schwartz SM, Lewin MB, Mueller BA (2013) Prepregnancy Body Mass Index and Congenital Heart Defects among Offspring: A Population-based Study. Congenit Heart Dis 8: 131–141.

56. Lu SC (2013) Glutathione synthesis. Biochim Biophys Acta 1830: 3143–3153.

57. Hobbs CA, MacLeod SL, Jill James S, Cleves MA (2011) Congenital heart defects and maternal genetic, metabolic, and lifestyle factors. Birth Defects Res A Clin Mol Teratol 91: 195–203.

58. Hobbs CA, Cleves MA, Zhao W, Melnyk S, James SJ (2005) Congenital heart defects and maternal biomarkers of oxidative stress. Am J Clin Nutr 82: 598–604.

59. Pinzar E, Miyano M, Kanaoka Y, Urade Y, Hayaishi O (2000) Structural basis of hematopoietic prostaglandin D synthase activity elucidated by site-directed mutagenesis. J Biol Chem 275: 31239–31244.

60. Inoue T, Irikura D, Okazaki N, Kinugasa S, Matsumura H, et al. (2003) Mechanism of metal activation of human hematopoietic prostaglandin D synthase. Nat Struct Biol 10: 291–296.

61. Urade Y, Mohri I, Aritake K, Inoue T, Miyano M (2006) Biochemical and structural characteristics of hematopoietic prostaglandin D synthase: From evolutionary analysis to drug designing. In: Morikawa K, Tate S, editors. Functional and Structural Biology on the Lipo-network. Kerala, India: Trivandrum: Transworld Research Network. pp. 135–164.

62. Oguma T, Asano K, Ishizaka A (2008) Role of prostaglandin D(2) and its receptors in the pathophysiology of asthma. Allergol Int 57: 307–312.

63. Vogel C (2000) Prostaglandin H synthases and their importance in chemical toxicity. Curr Drug Metab 1: 391–404.

64. Lee CJ, Goncalves LL, Wells PG (2011) Embryopathic effects of thalidomide and its hydrolysis products in rabbit embryo culture: evidence for a prostaglandin H synthase (PHS)-dependent, reactive oxygen species (ROS)-mediated mechanism. FASEB J 25: 2468–2483.

65. van Beynum IM, den HM, Blom HJ, Kapusta L (2007) The MTHFR 677C->T polymorphism and the risk of congenital heart defects: a literature review and meta-analysis. QJM 100: 743–753.

66. Goldmuntz E, Woyciechowski S, Renstrom D, Lupo PJ, Mitchell LE (2008) Variants of folate metabolism genes and the risk of conotruncal cardiac defects. Circ Cardiovasc Genet 1: 126–132.

67. Yong E, editor (2013) Genetic Test for Autism Refuted.

14

Gene Expression Differences among Three *Neurospora* Species Reveal Genes Required for Sexual Reproduction in *Neurospora crassa*

Nina A. Lehr[1], Zheng Wang[1,2], Ning Li[1], David A. Hewitt[3,4], Francesc López-Giráldez[1], Frances Trail[5,6], Jeffrey P. Townsend[1,2,7,8]*

1 Department of Ecology and Evolutionary Biology, Yale University, New Haven, Connecticut, United States of America, 2 Department of Biostatistics, Yale University, New Haven, Connecticut, United States of America, 3 Department of Botany, Academy of Natural Sciences, Philadelphia, Pennsylvania, United States of America, 4 Wagner Free Institute of Science, Philadelphia, Pennsylvania, United States of America, 5 Department of Plant Biology, Michigan State University, East Lansing, Michigan, United States of America, 6 Department of Plant, Soil and Microbial Sciences, Michigan State University, East Lansing, Michigan, United States of America, 7 Program in Computational Biology and Bioinformatics, Yale University, New Haven, Connecticut, United States of America, 8 Program in Microbiology, Yale University, New Haven, Connecticut, United States of America

Abstract

Many fungi form complex three-dimensional fruiting bodies, within which the meiotic machinery for sexual spore production has been considered to be largely conserved over evolutionary time. Indeed, much of what we know about meiosis in plant and animal taxa has been deeply informed by studies of meiosis in *Saccharomyces* and *Neurospora*. Nevertheless, the genetic basis of fruiting body development and its regulation in relation to meiosis in fungi is barely known, even within the best studied multicellular fungal model *Neurospora crassa*. We characterized morphological development and genome-wide transcriptomics in the closely related species *Neurospora crassa*, *Neurospora tetrasperma*, and *Neurospora discreta*, across eight stages of sexual development. Despite diverse life histories within the genus, all three species produce vase-shaped perithecia. Transcriptome sequencing provided gene expression levels of orthologous genes among all three species. Expression of key meiosis genes and sporulation genes corresponded to known phenotypic and developmental differences among these *Neurospora* species during sexual development. We assembled a list of genes putatively relevant to the recent evolution of fruiting body development by sorting genes whose relative expression across developmental stages increased more in *N. crassa* relative to the other species. Then, in *N. crassa*, we characterized the phenotypes of fruiting bodies arising from crosses of homozygous knockout strains of the top genes. Eight *N. crassa* genes were found to be critical for the successful formation of perithecia. The absence of these genes in these crosses resulted in either no perithecium formation or in arrested development at an early stage. Our results provide insight into the genetic basis of *Neurospora* sexual reproduction, which is also of great importance with regard to other multicellular ascomycetes, including perithecium-forming pathogens, such as *Claviceps purpurea*, *Ophiostoma ulmi*, and *Glomerella graminicola*.

Editor: Stefanie Pöggeler, Georg-August-University of Göttingen Institute of Microbiology & Genetics, Germany

Funding: This study was supported by a Gaylord Donnelley Environmental Fellowship of the Yale Institute of Biospheric studies to NL, and National Science Foundation (NSF) Grant MCB 0923797 to FT and JPT. The funders had no role in study design, data collection and analysis, decision to publish, or preparation of the manuscript.

Competing Interests: The authors have declared that no competing interests exist.

* Email: Jeffrey.Townsend@Yale.edu

Introduction

Many ascomycete fungi sexually reproduce by forming three-dimensional fruiting bodies that produce their sexual spores (ascospores) in sacs called asci. The genetics of fruit-body development in the filamentous fungi has been studied extensively in several fungi including *Aspergillus nidulans*, *Neurospora crassa*, *Sordaria macrospora*, *Fusarium graminearum*, and *Podospora anserina* [1–9]. Comparative genomics analyses suggest that meiosis machinery and ascospore production are generally conserved within ascomycetes [10–17]. Indeed, much of what we know about meiosis in plant and animal taxa has been deeply informed by studies of meiosis in *Saccharomyces* and, historically, *Neurospora* [18,19]. Although some genes involved in fungal fruiting body development have been identified by mutagenesis screens and characterized, the comparative genetic basis for the sexual cycle of these organisms in terms of interactions among gene networks has not been little explored [7,20]. Expression and comparative transcriptomics studies between *N. crassa* and *N. tetrasperma* have revealed key genes involved in asexual development and mating behaviors in these fungi [21,22]. However, inference and comparison of regulatory pathways for sexual development across these species has been difficult to achieve, partly due to the complex environmental stimuli relevant to sexual development and partly due to a lack of molecular detail regarding relevant gene interactions across sexual development. *Neurospora* species represent attractive models for elucidating the regulation of fruiting body development by transcriptional profiling and functional analysis due to their simple nutritional requirements, fast vegetative growth, and clearly recognizable stages during sexual development [23,24]. The most studied species is *N. crassa* [25], particularly well-known for its use in the original "one gene-one enzyme" experiments by Beadle and Tatum, for its eight-spored ordered asci which enabled centromere mapping of mutants, and for its experimentally tractable and intensively-studied genetic network for circadian rhythm [26,27].

The life histories of *Neurospora* spp. span the most common sexual strategies in the fungal kingdom, i.e. heterothallism (self-incompatibility with distinct mating types), pseudohomothallism (self-compatibility in which paired mating types coexist in one mycelium), and homothallism (self-compatibility regardless of mating type). Initiation of sexual reproduction is regulated by mating type genes and leads to cell fusion, nuclear pairing, nuclear fusion, meiosis, and the production of haploid ascospores. The determinant sequences for mating type, *mat A* and *mat a*, are at the same genetic locus in a given species but exhibit little to no similarity in nucleotide sequence; that is, they represent idiomorphs rather than alleles [28]. Heterothallic *Neurospora* species such as *N. crassa* and *N. discreta* [23] possess a bipolar mating system, and two strains with opposite mating types, *mat A* and *mat a*, must cross to initiate sexual development. Either mating type can produce female structures (protoperithecia with a trichogyne) as well as male reproductive structures (conidia) [29,30]. Pseudohomothallic species such as *N. tetrasperma* are generally self-compatible, even though they also require both mating types (*A* and *a*) to reproduce. In *N. tetrasperma*, both mating types are generally found within a single individual, although strains containing a single mating type exist in nature and can be isolated in lab. For *N. tetrasperma*, *mat A* and *mat a* nuclei associated in pairs in four heterokaryotic spores, packaged within an ascus for discharge [31]. Despite mating behavior differences that distinguish *N. tetrasperma* from *N. crassa* and *N. discreta*, phylogenetically *N. tetrasperma* and *N. crassa* are closely related and share the most recent common ancestor with *N. discreta* [32].

The genetic basis of sexual development has been the subject of many investigations, with unresolved controversy arising about the ancestral state of the life style of fungal sexual reproduction, heterothallism or homothallism, and its underlying genetic mechanism(s) [33–40]. For example, the heterothallic life style has been suggested to be ancestral within *Neurospora* [31], because the pseudohomothallic *N. tetrasperma* does not require a mating partner, but does require both *mat* idiomorphs to complete the sexual cycle. However, function of *mat* genes in *Neurospora* sexual development is not well understood, except their roles in heterothallic species such as *N. discreta* and *N. crassa* as regulators of pheromone expression to direct hyphal growth and fusion [4,41]. For heterothallic species as in *N. crassa*, the trichogynes, which originate from protoperithecia, change their direction of growth to approach conidia of the opposite mating type and fuse with them. Plasmogamy is followed by development of the perithecium, and later ascal development, karyogamy, and then formation of ascospores [42–44]. Two haploid nuclei of opposite mating type fuse in a young ascus resulting in a diploid zygote nucleus that immediately undergoes meioses, followed by a postmeiotic mitosis. The mature ascus delimits eight linearly arranged ascospores, each containing a single nucleus. After mitosis the ascospores become binucleate, gradually grow to their full size, and become pigmented.

Genomes of multiple species of *Neurospora* have been sequenced, and comparative genomic analyses have been focused on *N. crassa* and its closely related species [3,6,45]. Here we reveal candidate genes involved in fungal development and in the evolution of perithecia by comparisons of the gene expression levels within and across three *Neurospora* species. We assayed for large-scale differences in morphology and the transcriptomic landscape over the time course of sexual development, identifying putative genes involved in sexual development by comparative gene expression profiling. We also tested for knockout phenotypes of selected candidate genes by assaying knockout strains across sexual development for their ability to produce wild type perithecia. Our results provide insights into the links between gene expression and sexual development of *N. crassa* and related species, as well as contributing to our understanding of how fungi reproduce sexually.

Materials and Methods

2.1. Strains and culture conditions

Strains of complementary mating types *mat a* and *mat A* for *N. crassa* (FGSC4200, FGSC2489), *N. tetrasperma* (FGSC2509, FGSC2508) and *N. discreta* (FGSC8578, FGSC8579) were obtained from the Fungal Genetics Stock Center (FGSC) [46]. The strains were grown on Carrot Agar (CA), made as previously described [47]. The CA petri dish was covered with a cellophane membrane (Fisher Scientific Company) and plugs of agar with strains were deposited on the membrane and incubated at 26°C under constant artificial light from several Ecolux bulbs (F17T8.SP41-ECO, General Electric Company), which provided a net intensity of 14 µMol/m^2 S at the media surface. Conidia from the *mat a* strain on CA were collected and suspended in 2.5% Tween 60 (10^5–10^6 conidia/ml). Cultures of the *mat A* strain on CA were examined using a stereomicroscope for the formation of protoperithecia in 5–7 days, and areas with evenly distributed protoperithecia of a common size were delineated with a marker on the bottom of the plate to be harvested for stage-specific transcriptomics.

Crosses were performed by applying 2 ml of the suspension of *mat a* conidia in 2.5% Tween 60 (10^5–10^6 conidia/ml) to the surface of the *mat A* protoperithecia plates, at which point considerable disturbance to surface hyphae and other fungal tissues was unavoidable. Sexual development was monitored with a stereomicroscope until fully developed perithecia appeared [48]. Fungal material was harvested by scraping the surface with a razor blade in the areas, where protoperithecia or young perithecia similar in size were densely aggregated, right before the crossing and at 2, 24, 48 h after crossing. Sets of individual perithecia of similar morphological development were picked at 72, 96, and 144 h after crossing. For transcriptomic analysis, all tissues and perithecia were immediately and rapidly frozen in liquid nitrogen as they were sampled, then stored at −80°C.

2.2. Fixation and microscopy

Perithecium development was monitored for all three *Neurospora* species with a stereomicroscope over the time course of the sexual development. Cultures and crossings were performed as described in 2.1. Pieces of cellophane membrane (about 4 mm×2 mm) carrying 5–20 perithecia of similar size were cut from cultures and fixed in 1.5% formaldehyde and 0.025 M phosphate buffer for at least 48 h. The samples were embedded in resin and prepared for light microscopy as previously described [48]. Briefly, resin blocks were sectioned to a thickness of 1 to 2 μm using a glass knife and stained with 1% toluidine blue. A Leica DM LB microscope (Leica Microsystem Gmbh, Wetzlar, Germany) was used to capture images using a Zeiss AxioCam MRc color camera and AxioVision 4.8.2 (Göttingen, Germany). Image processing and annotation were performed using Adobe Photoshop CS3 (San Jose, CA).

To compare mature perithecia, scanning electron microscopy (SEM) was performed. Perithecia were collected, including the cellophane membrane they were growing on, and fixed immediately in 0.1 M sodium cacodylate buffer (pH 7.2) containing 2% glutaraldehyde over night. The samples were then washed with 0.1 M sodium cacodylate buffer, postfixed with 1% osmium tetroxide for 1 h, washed with distilled water, and dehydrated in an increasing series of ethanol. Then, all samples were critical point dried, mounted on an SEM stub, and sputter coated with gold. The samples were examined with a Scanning Electron Microscope (ISI SS-40).

2.3. RNA extraction, cDNA preparation, and transcriptomic sequencing

RNA was isolated from homogenized mycelia with TRI REAGENT (Invitrogen) and RNAeasy Kit (Qiagen) following protocol described by Clark et al. [49] and mRNA was purified using Dynabeads oligo(dT) magnetic separation (Invitrogen). The cDNA libraries for RNA sequencing were prepared according to the Illumina mRNA Sequencing Sample Preparation Guide. In brief, triplicates were prepared for each time point and pooled. Then, mRNA was purified from total RNA and 100 ng (9 μl) were fragmented with 10X fragmentation buffer (Ambion AM8740) and incubated at 70°C for 5 min prior to adding 1 ul stop buffer (Ambion). Fragmented mRNA was precipitated using 100% ethanol with glycogen (Ambion) at −80°C. Random hexamers (N_6, Invitrogen) were added to prime reverse transcription of the first strand cDNA separately for each sample, and to recover the second strand cDNA for all samples. After the ligation of standard adapters for Illumina sequencing, all samples were separated on a 2% low melting point agarose gel and processed cDNA fragments of lengths between 200 and 400 bp were selected by gel extraction and purified with Qiaquick gel extraction kit (Qiagen). The quantity of the samples was increased by a PCR using Pfx DNA polymerase (Invitrogen), and 15 cycles of PCR, each cycle comprising 98°C for 10 s, 65°C for 30 s, and 68°C for 30 s. The quantity and quality of the purified PCR products were checked at the Yale Center for Genome Analysis prior to sequencing. Single-end 35 bp reads of N_6-primed preparations were separately sequenced, each on eight lanes of an Illumina Genome Analyzer (Yale Center for Genome Analysis).

2.4. Data acquisition and analysis

The libraries were run on eight lanes of an Illumina Genome Analyzer, generating an average of 28 million single-end reads of 36 nucleotides each. Since transcriptomic tags also contain sequences that span exon junctions, the program Tophat v1.1.14 [50] was used to perform spliced alignments of the tags against the N. crassa OR74A genome (NC10; [51]), and those of N. tetrasperma FGSC 2508 (v2) and N. discreta FGSC 8579 (v1) obtained from JGI Genome Portal [52]. We scored results only for tags that mapped to a single unique location in the genome (–max-multihits option was set to 1) with less than three mismatches (–splice-mismatches option was set to 2). We used the default settings for all other Tophat options. We tallied tags aligning to exons of genes with the program HTSeq v0.4.5p6 (Unpublished; http://

www-huber.embl.de/users/anders/HTSeq/doc/) and the gene structure annotation file for the reference genome. LOX v1.4 [53] was applied to the tallies for each sample for each gene to estimate gene expression levels and credible intervals across developmental stages. LOX provided relative gene expression levels standardized by the lowest sample, with credible intervals, for all three species (Table S1).

Instead of using simple sequence similarity for homolog identification, we applied a phylogenetic approach that is reliable in calling homologs among closely related genomes, with a cost of power to identify potential homologs for recently evolved gene families and genes experiencing multiple duplication events in their evolutionary history. A total of 2352 orthologous genes were selected by the BranchClust method [54] for N. crassa, N. tetrasperma and N. discreta. BranchClust uses the Reciprocal Best Blast hit method and phylogenetic trees to select putatively orthologous genes using a default threshold e-value of 10^{-4}; we only choose the complete families in those three species. Transcriptomic sequencing revealed expression levels for all orthologous gene triads across sexual development (Table S2). Data is also deposited as accession 239 at the Filamentous Fungal Gene Expression Database (FFGED; [55]) and as accession GSE41484 for N. crassa [56], GSE60256 for N. tetrasperma, and GSE60255 for N. discreta at the National Center for Biotechnology Information Gene Expression Omnibus (GEO; http://www.ncbi.nlm.nih.gov/geo). More orthologs can be called among these fungi using BLAST approaches, and we also identified more single copy ortholog for these species using a BLAST-based method. However, from our other evolutionary study targeting several gene families with these species, we found that phylogeny-based ortholog calls by BranchClust were more robust and reliable.

A comparative heat map was constructed based on the level of expression of each gene for all three species over the entire time course of sexual development. Gene expression levels for each ortholog were normalized row-by-row by subtraction of the mean and division by the standard deviation. To cluster observed gene expression hierarchically, rows only were clustered by iteratively agglomerating the most similar two gene expression profiles and by averaging the agglomeration for the next iteration (Unweighted Pair Group Method with Arithmetic mean, UPGMA) within which similarity was assessed by the Pearson correlation coefficient of expression between genes. The Functional Catalogue (FunCat: [57]; http://mips.helmholtz-muenchen.de/proj/funcatDB/) annotation scheme was used to group genes according to their functions. The statistically significant overrepresentation of gene groups in functional categories relative to the whole genome was determined by a hypergeometric distribution P value calculation, facilitated by the MIPS FunCat online web application.

2.5. Comparative gene expression analysis of N. crassa, N. tetrasperma, and N. discreta

The LOX estimates across developmental stages for each orthologous gene for each species were assembled to compare gene expression levels across development of N. crassa, N. tetrasperma, and N. discreta. For each developmental stage, two values were calculated using the upper bound of the 95% confidence interval (CI) or the lower bound of the 95% CI from LOX. If the expression of a gene in species 1 was higher than species 2, we calculated the difference between the lower bound of expression in species 1 and the upper bound of expression in species 2. Conversely, if the expression of a gene in species 2 was higher than species 1, we calculated the difference between the lower bound of species 2 and the upper bound of species 1. These differences constitute those of which we can be highly confident. N. crassa was compared with *N. tetrasperma*, *N. crassa* with *N. discreta*, and lastly *N. tetrasperma* with *N. discreta*. The calculation described above was performed for all time points (2, 24, 48, 72, 96 and 144 h) for each gene. Finally, for each comparison, all genes were prioritized according to magnitude of difference in a descending order and the first 130 genes with the biggest changes across all time points and common to all comparisons were subsequently selected for knockout and phenotyping. The data for the comparative gene expression analysis can be found in the Tables S3–S5.

2.6. Assessing phenotypes of knock out mutants

Knockout strains for the top 130 candidate genes were compared to wild-type (WT) strains and screened for defects in fruiting body formation. Knockouts of candidate genes were obtained for both mating types in *Neurospora crassa* from the Fungal Genetics Stock Center (FGSC). These knockouts had been preliminarily assayed for phenotypes by several high-throughput screens as part of a *Neurospora* knockout project [58,59], and in that project had not exhibited a mutant phenotype in asexual growth or development. We performed a detailed, controlled screen crossing the two mating types of each deletion strain, cultured on CA in triplicate. The mat A cultures on CA were fertilized with conidia from the mat a strain. Perithecium formation was monitored with stereo- and light microscopy (Nikon Diaphot 300) for the presence and for the orientation of a single beak, black coloration and the typical vase-shape. Mature perithecia were examined in squash mounts for the presence of asci with normal pores, ascospores with normal shapes and numbers, and normal paraphyses. Normal spore development and ascus firing were examined by checking the lid of the Petri dish for black ascospores.

2.7 Cosegregation experiments of knockout mutants showing a phenotype different from the wild type

Knockouts of candidate genes obtained from FGSC were produced using a high-throughput gene deletion strategy in *N. crassa* strains with deletion mutations of *mus-51* and/or *mus-52*, mutations required for nonhomologous end-joining DNA repair. [58,59]. A high rate of spontaneous mutations has been observed in in *Δmus-51* and/or *Δmus-52* strains, and cosegregation experiments were used to demonstrate that the intended knockout deletion was responsible for the mutant phenotype [60]. A hygromycin resistance cassette at the location of the deletion mutation provides a selectable marker. The KO strains showing phenotypes in sexual development were crossed with wild-type strains (FGSC2489 *mat A* or FGSC4200 *mat a*). Individual ascospore progeny were isolated for resistance to hygromycin. Their phenotypes were then examined on SCM medium. Cosegregation of hygromycin resistance and the observed phenotype is necessary evidence that the observed phenotype was result of the deletion of the specified gene.

Results

3.1 Perithecium development in *N. crassa*, *N. tetrasperma*, and *N. discreta* on CA follows a common time course

To monitor the sexual development of all three *Neurospora* species, we crossed wild-type strains of both mating partners of each *Neurospora* species. The observed perithecial development of *N. crassa*, *N. tetrasperma*, and *N. discreta* aligned with the common time course known for these species [56]. We observed that during perithecial development, grayish to yellowish gray protoperithecia darkened, then blackened once mature perithecia

had formed, indicating the biosynthesis of melanin in the time course of perithecial development (Fig. 1). As expected, we also identified the fading of the orange pigmentation in the colony across the time course of sexual development, indicating a reduction in the asexual phase of the life cycle. Additionally, we found no obvious tissue differentiation between protoperithecia and perithecia within 24 h after crossing, except for a slight increase in size and a slight darkening in colour, indicating the biosynthesis of melanin as part of the successful fruiting body formation. Furthermore, the centrum parenchyma of thin-walled cells expanded with increasing perithecial size and differentiated filamentous structures and croziers formed after 48 h to 72 h. Asci containing developing ascospores were visible after 96 h, along with some narrow paraphyses. From 120 h to 144 h post crossing, a beak formed at the apex of the perithecium, and scanning electron microscopy of all three species revealed fully developed fruiting bodies with a beak on top of the perithecium (Fig. 2, panels A–C). Furthermore, squash mounts of perithecia 144 h after crossing revealed that the inside of the perithecium contained mature asci with ascospores (Fig. 2, panels D–F). A major phenotypic difference among species was the number of spores produced, which, as expected in *N. tetrasperma* was four, in contrast to eight in *N. crassa* and *N. discreta*. Additionally, *N. discreta* produced abundant conidia, which were visible associated with the perithecia after the sexual cycle has concluded (Fig. 2, panels C and F).

3.2. Transcriptional profiling of *N. crassa*, *N. tetrasperma*, and *N. discreta* reveals eight functional categories

To determine the gene expression of all three *Neurospora* species during the time course of sexual development, we performed Illumina next generation transcriptomic sequencing. We prepared samples by isolation of mRNA and performance of reverse transcription of genomic RNA with random hexamers (N_6), which were typically consistent in DNA concentration. Single-end 35-base reads of N_6-primed preparations were sequenced separately, each developmental stage of each species on one of eight lanes of an Illumina Genome Analyzer, at the Yale Center for Genomic Analysis. A total of 2352 single copy orthologous genes were selected by BranchClust [54] for *N. crassa*, *N. tetrasperma* and *N. discreta*. BranchClust uses the Reciprocal Best Blast hit method and phylogenetic trees to select putatively orthologous genes, with default threshold e-value of 10^{-4}. Our deep transcriptomic sequencing revealed expression in all three species across sexual development for all of these identified orthologous genes. To compare gene expression levels for these genes, we constructed a comparative heatmap with the Hierarchical Clustering Explorer Software [61] based on the level of expression of each gene for all three species over the entire time course of sexual development (Fig. 3) as described in section 2.4. Transcriptional profiling of *N. crassa*, *N. tetrasperma*, and *N. discreta* revealed eight clusters, which we then analyzed separately for functional category representation.

3.2.1 Among genes highly expressed in early stages of perithecium development, the ortholog set was enriched for genes involved in metabolic processes. Where previously identified, genes were assigned cellular or molecular functions by their Functional Catalogue (FunCat) [57] annotation. The statistical significance of overrepresentation of gene groups in functional categories relative to the whole genome was determined using the hypergeometric distribution, facilitated by the Munich Information Center for Protein Sequences (MIPS) FunCat online web application. Eight major clusters were identified based on expression patterns across sexual development for the identified

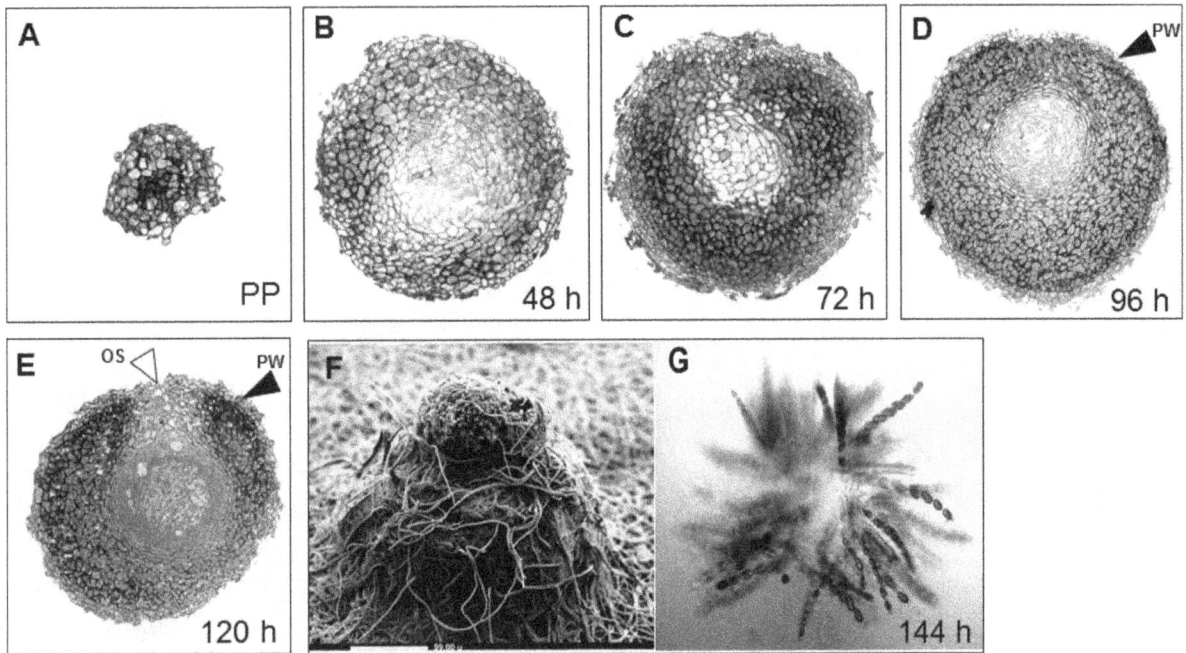

Figure 1. Sexual development of *N. crassa*. Cross sections of developing perithecia, from A) the protoperithecium (PP) through a time course of (B) 48 h, (C) 72 h, (D) 96 h, and (E) 120 h after fertilization. These images illustrate the development of several cell layers within the fruiting body such as the perithecium wall (PW, black arrowheads), composed of thick-walled cells, as well as the initial stages of formation of asci and ascospores and the ostiole (OS, white arrowheads) through which spores are released in a later stage. After (F) 144 h, the perithecium including its beak is fully developed, as shown by scanning electron microscopy (SEM). (G) A squash mount of a mature fruiting body, showing asci and ascospores.

orthologs. In cluster 1 (Fig. 3), 61 of the 671 proteins were unclassified. Of the classified proteins we identified, most were involved in metabolism, cellular transport, cell cycle and DNA processing, protein with binding function, cellular communication and transcription.

Genes in cluster 2 were mainly involved in metabolism, cellular transport, transport facilities and transport routes as well as

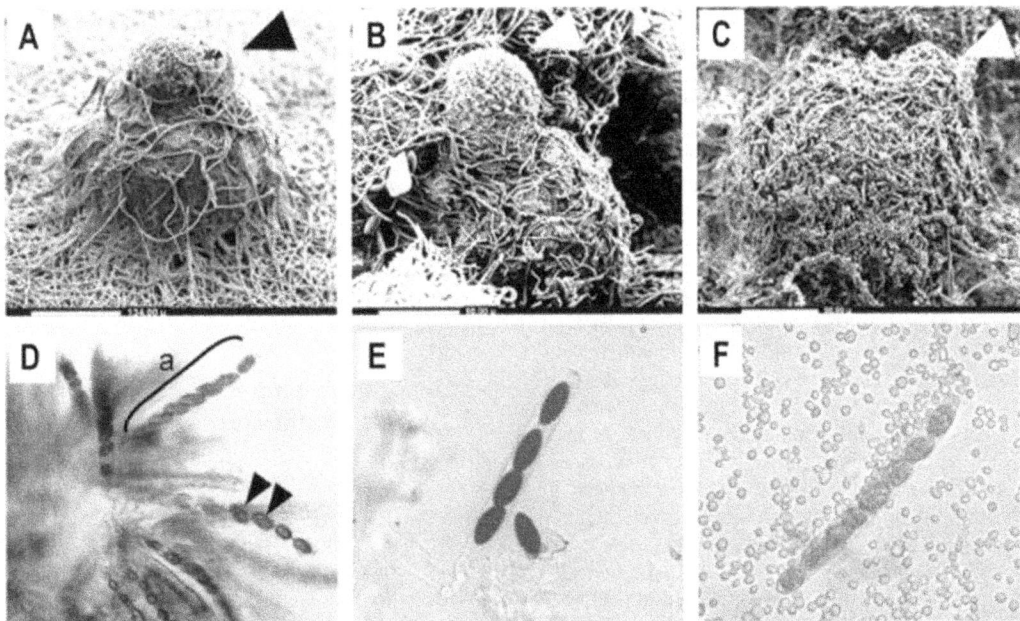

Figure 2. Key morphological characters of *N. crassa*, *N. tetrasperma* and *N. discreta* 144 h after crossing. Scanning electron micrographs, in which arrowheads indicate the perithecial beaks, of A) *N. crassa* (bar: 134 μm), B) *N. tetrasperma* (bar: 98 μm) and C) *N. discreta* (bar: 98 μm), and light micrographs of squash mounts of (D) *N. crassa* (a: ascus), (E) *N. tetrasperma*, and (F) *N. discreta* (with conidia). At 144 h, some spores are not fully mature.

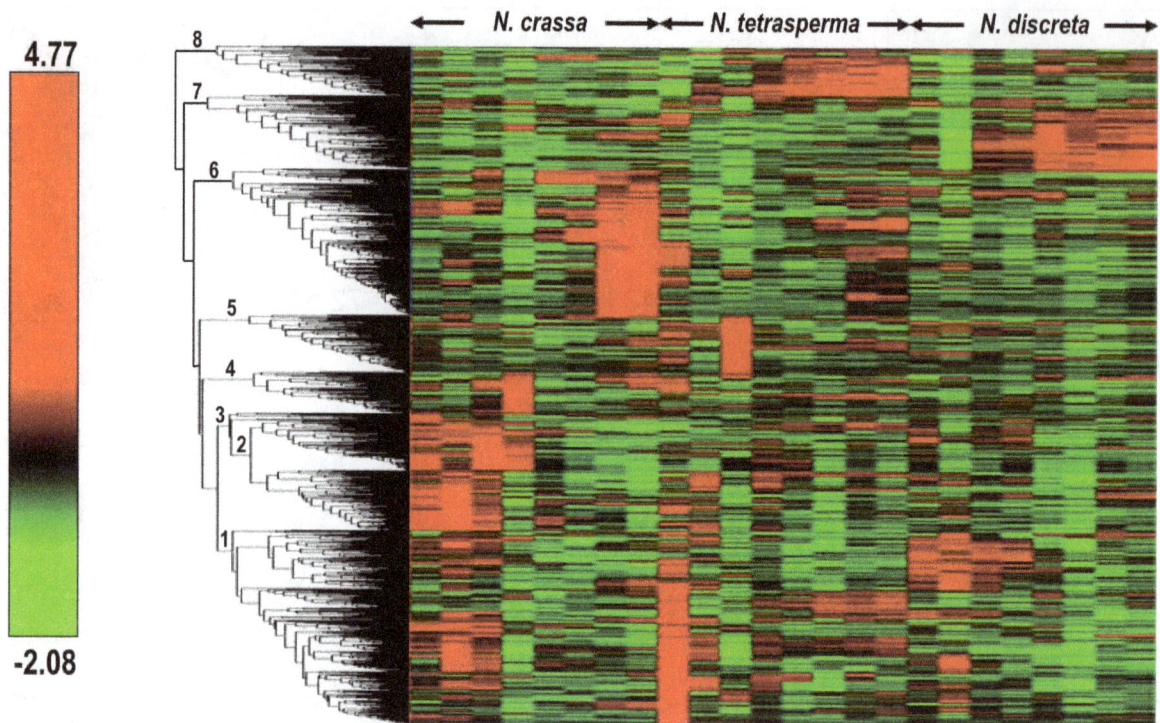

Figure 3. Comparative heat map of *N. crassa* **(N.c),** *N. tetrasperma* **(N.t) and** *N. discreta* **(N.d) gene expression.** Relative gene expression levels of the orthologous genes in all three species inferred by BranchClust, from Before Crossing (BC) to 144 h after crossing, were normalized row-by-row by subtraction of the mean and division by the standard deviation. Rows were hierarchically clustered by average linkage (UPGMA) applied to the Pearson correlation coefficient. Clusters 1–8 were analyzed for their function with FunCat. The scale of mean-centered relative gene expression ranges from −2.08–4.79, as indicated with the scale bar.

proteins with binding function, while only 16 of the proteins in cluster 2 were not classified. The metabolic genes were primarily involved in nucleotide/nucleoside/nucleobase metabolism, C-compound and carbohydrate metabolism, lipid, fatty acid and isoprenoid metabolism and metabolism of vitamins, cofactors, and prosthetic groups. The cellular transport-related genes were involved in transported compounds (substrates), protein transport, electron transport, vesicular transport such as the Golgi network and vacuolar/lysosomal transport. Genes encoding proteins with binding functions such as nucleic acid binding, metal binding and complex cofactor/cosubstrate binding. In cluster 3, only 19 out of 199 proteins were not classified, while the remaining proteins were mainly involved in metabolism, especially C-compound and carbohydrate metabolism. Genes in cluster 3 also tended to be involved in cellular transport, including cation and heavy metal ion transport as well as proteins with binding functions, e.g. nucleic acid binding, metal binding and RNA binding. Cluster 4 contained 139 proteins, which were involved mainly in metabolism, especially lipid, fatty acid and isoprenoid metabolism, cellular transport, and cell cycle including DNA restriction or modification. Overall, we observed an enrichment of genes involved in metabolic processes in early stages of perithecium development.

3.2.2 In later stages of the perithecium development, the ortholog set was enriched for genes involved in transcription, cell cycle, protein synthesis and cellular transport. Based on Functional Category (FunCat, [57]) annotations, cluster 5 comprises 198 proteins, of which the majority was involved in nucleotide/nucleoside/nucleobase metabolism, nucleotide/nucleoside/nucleobase metabolism and lipid, fatty acid, and isoprenoid metabolism, but also transcription such as RNA synthesis and RNA processing.

Interestingly, in clusters 6–8 there appears to be a species-specific expression in later stages of the development. Proteins of cluster 6 were mainly involved in transcriptional processes such as RNA synthesis, processing and modification, but also, cell cycle and DNA processing, as well as and metabolism. Cluster 7 contained 259 genes involved in transcription, protein binding, cell type differentiation and cellular transport. In cluster 8 only 13 genes were not identified. The cluster mainly comprised proteins involved in cell differentiation, cellular transport and transcription.

While in the early stages of the perithecium development, the fungi were observed to exhibit an enrichment of expressed genes involved in metabolic processes; in later stages of perithecium development, expressed genes from the ortholog set were enriched for genes involved in transcription, cell cycle, protein synthesis and cellular transport.

3.3 Comparative gene expression analyses revealed genes crucial for the successful development of fruiting bodies

Crossing of the *N. crassa* WT strains resulted in the successful formation of perithecia with perithecial beak (Fig. 4A). In order to identify genes that are required for the successful development of fruiting bodies, knockouts in *N. crassa* from the whole genome knockout project [58,59] of the 130 top candidate genes identified through pairwise comparison of orthologous gene expression in *N. crassa*, *N. tetrasperma*, and *N. discreta* were screened for mutant phenotypes in sexual development. Eight mutant phenotypes were observed that affected perithecium formation. Interestingly, development of NCU06874, encoding a HMG box (high mobility group)-containing protein was arrested after formation of protoperithecia on

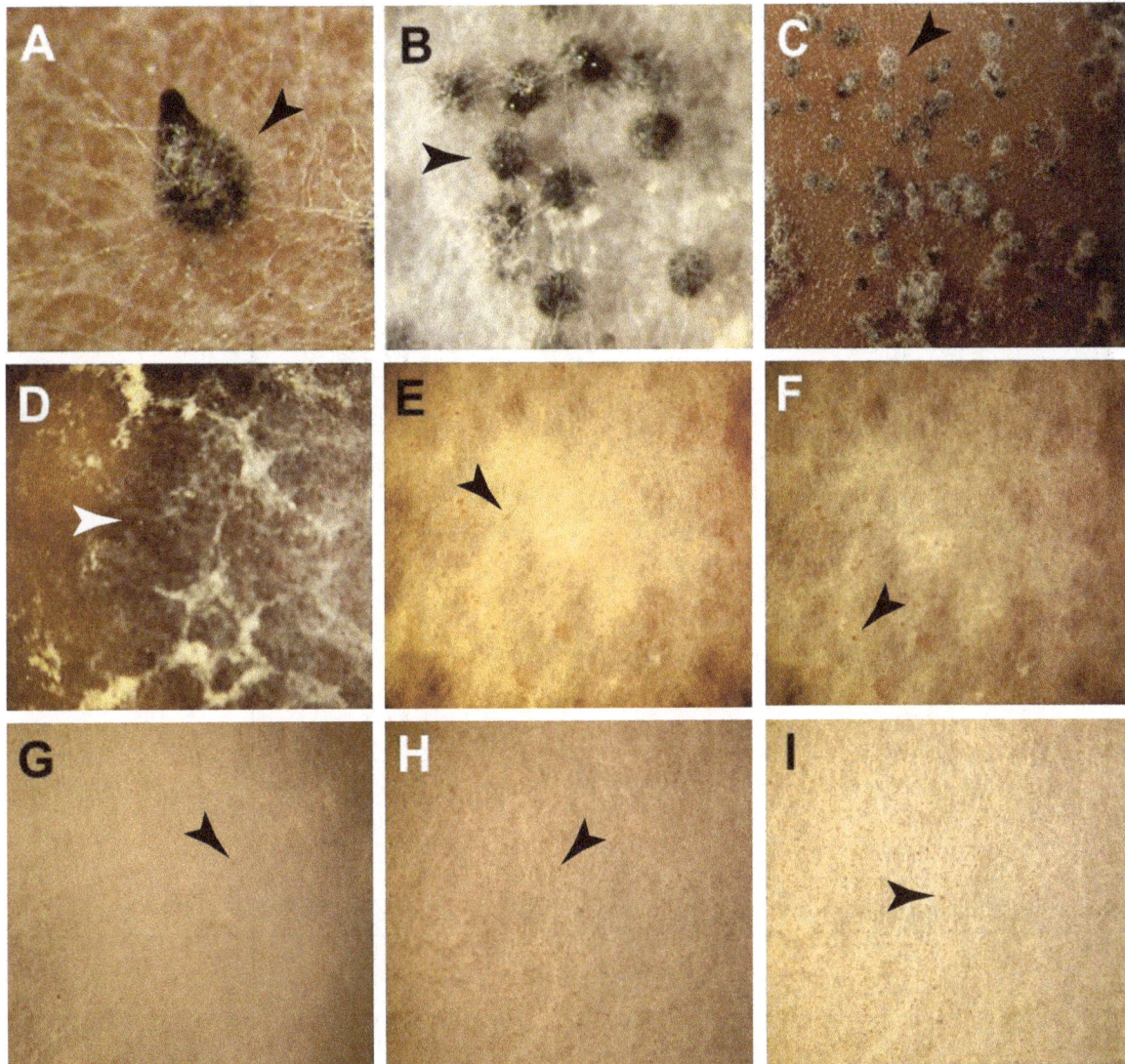

Figure 4. Phenotypes of perithecia from crosses in *N. crassa*. (A) WT control, (B) ΔNCU06316: perithecium development was arrested at early stage equally to 48–72 h, (C) ΔNCU07508: perithecium development was arrested at 48–72 h, (D) ΔNCU06874: no perithecia formed, only protoperithecia, though melanin was released into the medium (white arrow heads), (E) ΔNCU05609: no perithecia formed, only protoperithecia, (F) ΔNCU00175: no perithecia formed, only protoperithecia, (G) ΔNCU00427: no perithecia formed, only protoperithecia. (H) ΔNCU02089: protoperithecia failed to develop into perithecia, and (I) ΔNCU09525: protoperithecia only. Perithecia (large, black) and protoperithecia (small, yellowish-gray to gray) are indicated with black arrow heads.

a fluffy white mycelium. Putative functions of the remaining genes that affect fruiting body formation in *N. crassa* were determined by a Blast search [62] with the NCBI Blast server of the translated sequence against all known GenBank non-redundant protein sequences (Table 1). However, the deletion of either gene NCU06316, a putative argonaute siRNA chaperone complex subunit which in fission yeast is required for histone H3 Lys9 (H3-K9) methylation, heterochromatin, assembly and siRNA generation [1], or NCU07508, a putative type-2 protein geranylgeranyltransferase subunit, caused the perithecium development to arrest at an early stage between 48 and 72 h (Fig. 4, panels B and C). In the perithecia arising from these KO strains, we observed no perithecial beak and found that the developing perithecia remained spherical, indicating an arrest in development. Furthermore, the CA of these cultures was stained black along the mating zone, suggesting an increase in melanin biosynthesis (Fig. 4D).

The expression patterns of the genes whose knockouts showed an impact on fruiting body formation were monitored over the time course of sexual development starting from mature protoperithecia before crossing until 144 h, when wild-type cultures showed mature perithecia. The genes NCU06316 or 07508 exhibited a continuous increase in expression starting at 72 h (Fig. 5A). The increased expression correlated with the arrest of the fruiting body formation in an early stage between 48 and 72 h in deletion mutants.

Knockouts of NCU05609 (a proline-rich protein 8), 00175 (a repetitive proline-rich cell wall protein), 00427 (a YjeF_N domain-containing protein), 02089 (a hypothetical protein SMAC) and 09525 (a hypothetical protein SMAC) were arrested at the protoperithecium stage, indicating their importance in the formation of fruiting bodies. NCU09525 encodes a putative secreted protein and has been described in *Nectria haematococca*

Table 1. Putative functions of genes whose deletion impacts fruiting body formation.

ID	Annotation by homology	Max score	Total score	Query cover	E-value	Max iden
NCU06316	argonaute siRNA chaperone complex subunit Arb1 [*Colletotrichum fioriniae*], XP_007589964.1	246	246	76%	8e–72	37%
NCU07508	putative type-2 protein geranylgeranyltransferase subunit beta protein [*Botryotinia fuckeliana*], EMR87581.1	145	145	68%	1e–34	31%
NCU06874	HMG box-containing protein [*Magnaporthe oryzae*], XP_003718106.1	269	269	42%	7e–75	40%
NCU05609	proline-rich protein [*Coccidioides Posadasii*], ACU44647.1	46.6	46.6	64%	0.020	29%
NCU00175	repetitive proline-rich cell wall protein [*Colletotrichum higginsianum*], CCF45612.1	153	205	88%	1e–39	44%
NCU00427	YjeF_N domain-containing protein [*Magnaporthe oryzae*], ELQ39075.1	734	734	97%	0.0	53%
NCU02089	hypothetical protein SMAC_04986 [*Sordaria macrospora*], XP_003352871.1	1276	1276	99%	0.0	88%
NCU09525	Bacterial-type extracellular deoxyribonuclease [N. haematococca], XP_003051650.1	284	284	85%	2e–93	69%

as bacterial-type extracellular deoxyribonuclease [63]. NCU09525 and 00175 were highly expressed between 48 and 120 h, while expression of NCU05609 peaked at 48 and 96 h after crossing. NCU06874 and 00427 showed an increased expression 96 h after crossing. NCU02089 peaked at 48 h after crossing and remained fairly constant across the remainder of sexual development (Fig. 5C–D). Although only protoperithecia were observed in mutant crosses, the culture medium for the knockout of NCU06874 blackened, indicating the secretion of melanin, and implying that melanin biosynthesis remained functional despite the fact that no perithecia were formed. Genes encoding enzymes in the melanin synthesis pathway showed a highly similar pattern across sexual development within each *Neurospora* species. While melanin synthesis genes were highly expressed in *N. crassa* protoperithecia samples, they were expressed at a lower level in *N. tetrasperma* and *N. discreta* protoperithecial samples than in late perithecial samples (Table S1). Further research is needed to understand function of melanin synthesis early sexual development in different *Neurospora* species. The protein encoded by NCU06874 contains a High Mobility Group (HMG) box, a structure often found in proteins involved in the regulation of DNA-dependent processes and DNA repair. Recent studies have demonstrated that deletion of the ortholog of NCU06874 in *Podospora anserina* and *Fusarium graminearum* had no effect on perithecium development [64,65]. However, an effect on the distribution of perithecia in *P. anserina* was observed [64,65].

The gene NCU02089 belongs to cluster 5. In *N. crassa*, this gene is expressed fairly uniformly, showing only a slight peak in expression at 48 h. The corresponding ortholog in *N. tetrasperma* shows a similar expression pattern. In *N. discreta*, in contrast, the expression of NCU02089 remains uniform until 96 h, then increases until the end of the sexual development.

The genes NCU06316, 07508, 05609, 00427, 09525, and 06874 belong to cluster 6. In *N. crassa*, the genes NCU06316 and 07508 are highly expressed after 72 h. In *N. tetrasperma*, as in *N. crassa*, the ortholog of NCU07508 also increases in expression after 72 h, while in *N. discreta* the expression peaks at 48 h, and increases again after 96 h. The expression of the ortholog NCU06316 in *N. crassa* differs from the one in *N. tetrasperma* and *N. discreta*. Over the time course of sexual development, expression of this gene is very low. The gene NCU05609 shows an increasing expression over the time course of the sexual development, peaking at 48 and 96 h. In contrast, the corresponding ortholog in *N. tetrasperma* does not show any changes in gene expression, while in *N. discreta* the expression decreases until 24 h, then remains steady until the end of the sexual development. In *N. crassa*, the expression of the gene NCU00427 remains constant until 96 h after crossing and then increases towards the end of the development. The expression of the corresponding ortholog in *N. tetrasperma* peaks at 24 and 120 h, while in *N. discreta*, this gene remains constant over the time course of fruiting body formation. The expression of the gene NCU09525 peaks between 48 and 120 h in *N. crassa*, while in *N. tetrasperma* and *N.*

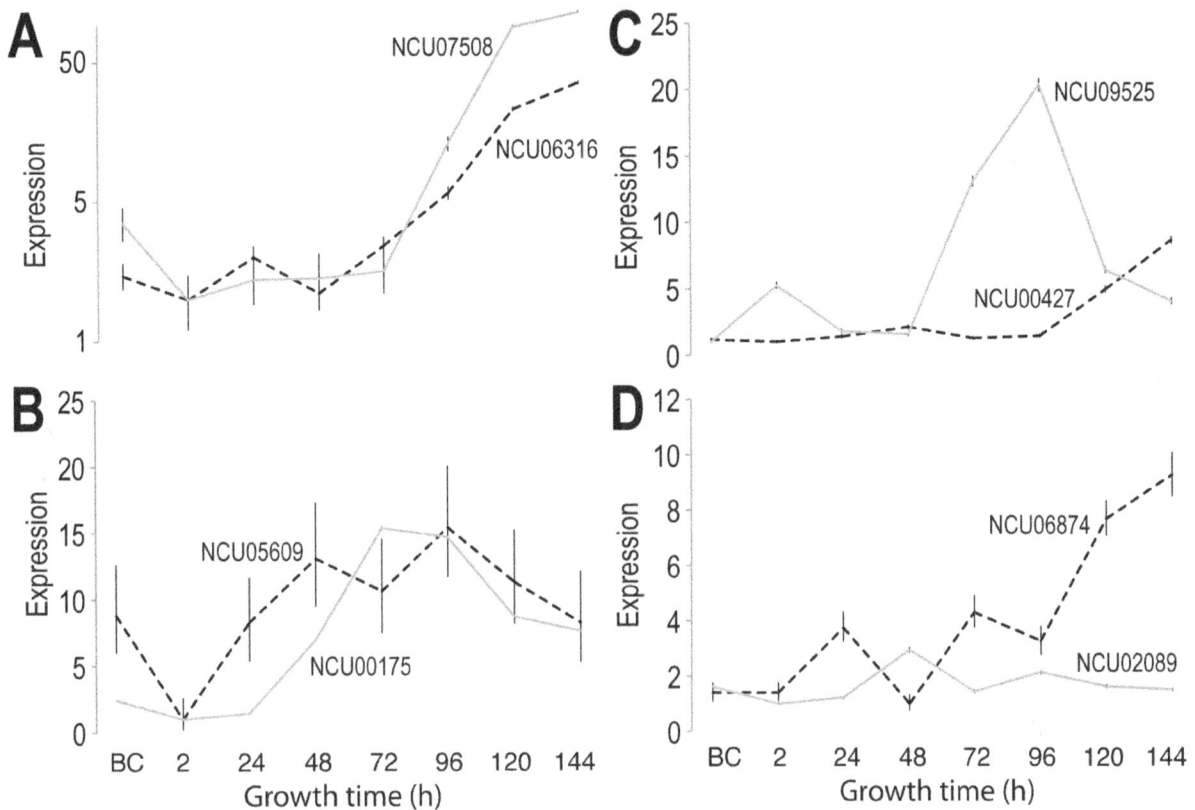

Figure 5. Expression patterns of eight genes with impact on successful perithecium formation in *N. crassa*. A. Expression of NCU06316 (solid line) and NCU07508 (dashed line), knockouts of which lead to early arrest of perithecial development at 48–72 h. Both genes exhibit a continuous increase of expression after 72 h. B. Expression of NCU00175 (solid line) and NCU05609 (dashed line), up-regulated during intermediate stages of perithecial development. C. Expression of NCU09525 (solid line) and NCU00427 (dashed line), both changing significantly in late perithecial development. D. Dynamic changes in expression of NCU02089 (solid line) and NCU06874 (dashed line) across perithecial development. Error bars indicate the inferred 95% credible interval.

discreta, its expression peaks initially at 2 h, potentially as a consequence of mycelial disruption caused by the spreading of conidia and mating of the strains, and then increases until 96 h after crossing followed by a slight decrease towards the end of the development. The gene NCU06874 shows peaks in gene expression at 24 and 72 h, and an increase of expression from 96 h until the end of perithecium development. In contrast, in both *N. tetrasperma* and *N. discreta*, the expression of this gene remains fairly constant over the time course of the sexual development.

The gene NCU00175 belongs to cluster 8 and shows an increased gene expression between 24 and 120 h in *N. crassa*. A similar pattern was observed for *N. tetrasperma*, where the expression increased at 48 and 120 h. In *N. discreta* however, the gene expression increased steadily until 96 h after crossing, remained constant.

Key differences in gene expression patterns of meiosis-related genes between *N. tetrasperma* and *N. crassa* as well as *N. discreta* were observed (Fig. 6). Many RNA processing genes and meiosis- and mitosis- specific genes have experienced multiple duplications. Only a few single copy meiosis-specific gene homologs have been confirmed with our methods. Among these genes, we observed differences in gene expression between homothallic *N. tetrasperma* and heterothallic *N. crassa* and *N. discreta*. The meiotic chromosome segregation protein 3 homologs, *N. tetrasperma* Neute1draft_73667 (JGI identifier for the *mat A* genome), *N.*

crassa NCU01858 (Broad Identifier), and *N. discreta* Ndisc8579_91937 (JGI identifier) showed a consistent gene expression pattern in *N. crassa* and *N. discreta*: two up-regulations at 2 h and 72 h (Fig. 6A). In contrast, the expression of the ortholog of chromosome segregation protein 3 in *N. tetrasperma* exhibited a steep down-regulation after strains were crossed and only a slight up-regulation 24 h after crossing. Furthermore, the translation factor pelota (Neute1draft_80736, NCU09161, and Ndisc8579_129376), which functions in meiotic cell division, exhibited a generally consistent expression across perithecial development in *N. crassa* and *N. discreta*, but a dynamic change in expression in *N. tetrasperma* during early perithecial development (Fig. 6B). A dramatic up-regulated expression of the meiosis-specific gene *spo11* (Neute1draft_98113 m, NCU01120, Ndisc8579_51404) started early in *N. tetrasperma* 24 h after crossing, but began in *N. crassa* and *N. discreta* 96 h after crossing (Fig. 6C). For sporulation-related genes, including *asm-1* (ascospore maturation-1), Neute1draft_115825, NCU01414, Ndisc8579_123166), *Rsp* (round spore, Neute1draft_126958, NCU02764, Ndisc8579_94666), *asd-3* (ascus development-3, Neute1draft_103712, NCU05597, Ndisc8579_126512), we also observed up-regulation of expression at the same stages between *N. crassa* and *N. discreta* (Fig. 6D–F). Up-regulation of homologs of *N. crassa asm-1* and *rsp* in *N. tetrasperma* occurred later than in *N. crassa* and *N. discreta*, but interestingly the homolog of *N. crassa asd-3* exhibited an early up-regulation for *N. tetrasperma* at

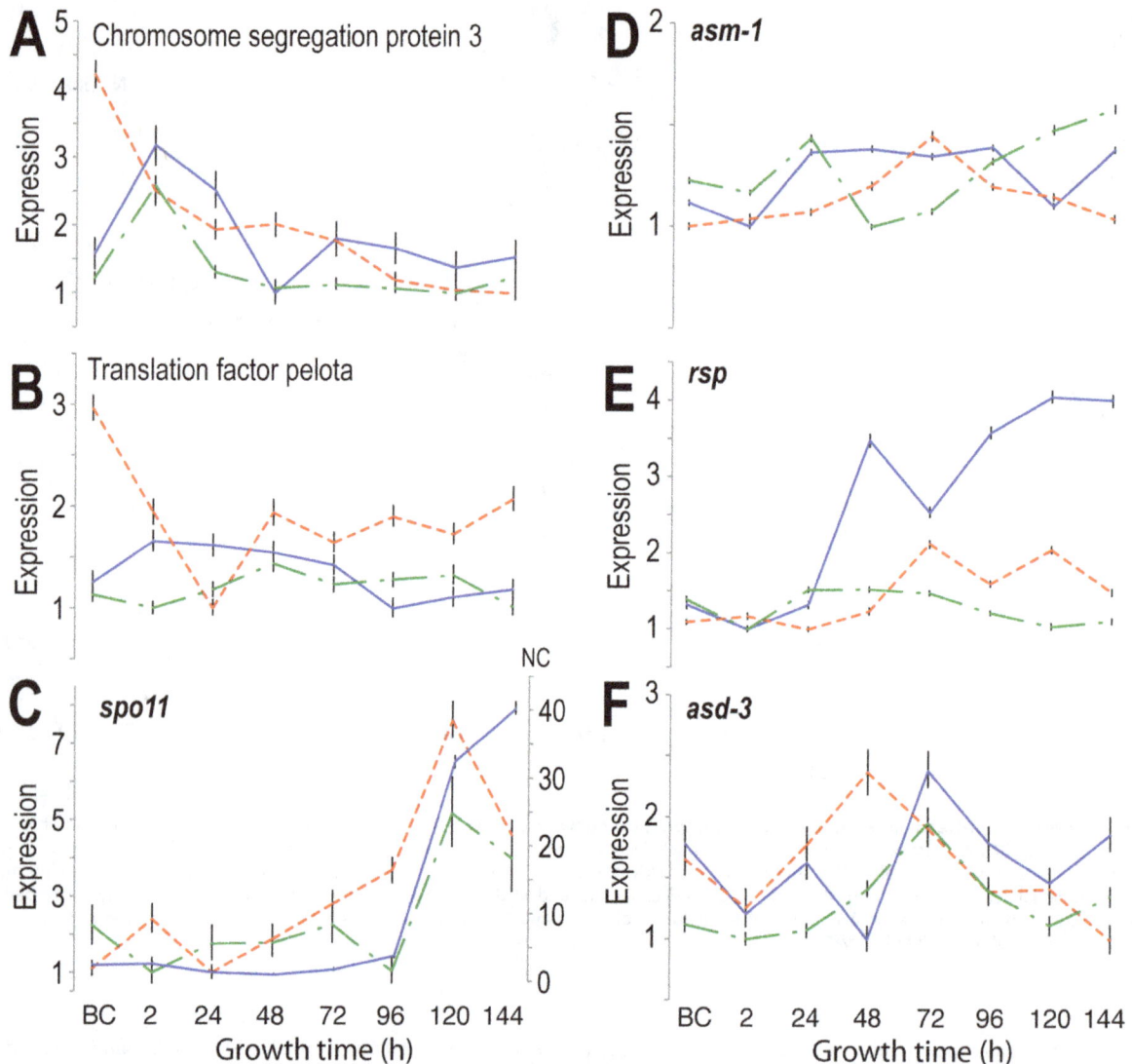

Figure 6. Comparative analysis of meiosis-related genes, exhibiting differences in gene expression between *N. tetrasperma* (red dashed line), *N. crassa* (blue solid line) and *N. discreta* (green dash-dotted line). A) Expression of the *N. crassa* gene encoding chromosome segregation protein 3 was up-regulated during early sexual development in both *N. crassa* and in the orthologous gene in *N. discreta*, but the *N. tetrasperma* ortholog was down-regulated across sexual development, B) Expression of the gene encoding translation factor pelota in *N. crassa* and expression of its ortholog in *N. discreta* was consistent, but the ortholog in *N. tetrasperma* was dramatically and dynamically differentially expressed. C) Expression of meiosis specific gene *spo-11* and its orthologs was up-regulated for all three species from 96 h after crossing, but the up-regulation started early for the *N. tetrasperma* ortholog. D) Expression of *asm-1* (ascus maturation) in *N. crassa* and its ortholog in *N. discreta* exhibited the same two-peaked pattern, featuring a peak of expression during early development of the perithecium and a second peak at a later stage of perithecial development, but the ortholog in *N. tetrasperma* exhibited only one peak 72 h after crossing. E) Expression of *N. crassa rsp* (round spore) and its ortholog in *N. discreta* exhibited up-regulation preceding 48 h, but upregulation in *N. tetrasperma* preceded 72 h. F). Expression of *N. crassa asd-3* (ascospore development) peaked at 72 h in *N. crassa*, as did its ortholog in *N. discreta*, but expression of the ortholog in *N. tetrasperma* peaked at 48 h. Error bars indicate the inferred 95% credible interval.

48 h. Similar up-regulation of *asd-3* for *N. crassa* and its homolog in *N. discreta* occurred at 72 h after crossing. The genes NCU01858 and 09161 belong to cluster 1, the genes NCU01120, 02764 and 05597 to cluster 6 and NCU01414 to cluster 7. Except NCU01858 and 09161, all genes shown in Fig. 6 belong to the candidate list.

3.4 Phenotypes of seven knockout mutants cosegregate with hygromycin resistance

To validate the linkage between the insertional mutation and the phenotype, we followed a strategy developed previously for *N.*

crassa KO strains [56], and backcrossed progeny (ascospores) were selected and germinated. Twelve to twenty single ascospore progeny displayed hygromycin resistance, and complete cosegregation of hygromycin resistance and identified phenotypes was observed for seven out of eight investigated genes (Table S6). The exception was the knockout of NCU09525, for which only a KO strain of *mat A* was available. No perithecia were produced in this mutant-wild type cross, preventing us from assessing segregation of the cross.

Discussion

Here, we compared the sexual development of three closely related *Neurospora* species, *N. crassa*, *N. tetrasperma*, and *N. discreta,* focusing on characterizing expression patterns for genes involved in sexual development and identifying new genes or new functions of annotated genes in regulating sexual reproduction, taking advantages of the distinct differences among these otherwise highly similar and closely related species. The comparison of gene expression levels across these species during fruiting body formation revealed eight genes that were shown to be crucial for the successful development of perithecia in *N. crassa*: their knockouts were unable to produce mature perithecia, and seven of the eight exhibited cosegregation of the phenotype and a hygromycin marker. We identified different expression patterns for meiosis-related genes between *N. tetrasperma* and *N. crassa* and *N. discreta* that correspond with observed differences in meiosis and sexual spore development between pseudohomothallic *N. tetrasperma* and heterothallic *N. crassa* and *N. discreta*. Since meiosis gene sets and sporulation machinery are largely conserved in presence as well as sequence within these genomes, consistent differences among functionally related genes in these species would be strong evidence of dependent associations for the reconstruction of gene networks and thus for understanding the genetic basis of meiosis and sexual sporulation in *Neurospora* and ascomycetes. Additionally, species-specific expression in later stages of development was observed when clustering the ortholog genes. Future efforts should expand analysis to identify a more complete set of genes involved in meiosis and sporulation based on genetics and reverse genetics on the model *N. crassa*.

We distinguished eight gene clusters based on similarity of overall expression patterns across sexual development in three *Neurospora* species. During early sexual development, genes involved in metabolism are enriched, and clusters that appear later during sexual development are enriched with genes involved in metabolism, energy, transcription, cell cycle, and DNA processing. The expression profiles of these clusters correlated with the morphological changes observed during sexual development: at first metabolic genes are upregulated, then transcription and cell cycle related genes are activated, and eventually protein synthesis and transport genes are transcribed to deliver proteins needed to their destination in order to form the complex three dimensional fruiting body.

Our study revealed eight genes that are required for the successful formation of fruiting bodies in *N. crassa*. Knockouts of two genes, NCU06316 and 07508 resulted in the arrest of perithecium formation between 48 and 72 h after crossing (Fig. 4, panels B and C). In correlation with the gene expression data (Fig. 6A), these genes are of great importance in the later stage of development when structures such as the perithecial beak or ascospores are developed. The developing perithecia for NCU06316 (a putative argonaute siRNA chaperone complex subunit) and 07508 (a putative type-2 protein geranylgeranyl-transferase subunit) remain round-shaped without formation of a beak. Six crosses of *N. crassa* mutant strains showed only protoperithecia and no formation of fruiting bodies. These results correlate with the gene expression observed over the time course of the sexual development in *N. crassa*. Here, an up-regulation of gene expression during the later stages of the development was observed. Our observations indicate that there are several genes that are required for the successful formation of perithecia.

Conclusions

We have used comparative transcriptomics as a tool to identify genes that are required for the successful formation of fruiting bodies. The impact of the deletion of the candidate genes was demonstrated in phenotypes observed in crosses exhibiting impaired perithecium formation at different stages of the development. Our findings shed light on the developmental process of fruiting body formation, evolution, and spore development in *Neurospora* species, thus establishing the foundation for future research particularly related to closely related pathogenic fungi. We suggest the potential utility of future research on the eight genes that we have discovered as essential contributors to fruit body development and as candidate genes for targets in the development of fungicides for control of plant pathogens.

Supporting Information

Table S1 Expression results of all genes of all three Neurospora species.

Table S2 Level of Expression (LOX) data of *N. crassa*, *N. tetrasperma* and *N. discreta* orthologues. LOX estimates the Level Of gene eXpression from high-throughput-expressed sequence datasets with multiple treatments or samples. The tables includes the corresponding upper and lower confidence intervals (CI) across developmental stages.

Table S3 Comparative gene expression analysis of *N. crassa*, *N. tetrasperma*, and *N. discreta*. The table includes the results from the calculations such as the differences of confidence intervals.

Table S4 Comparison of species using the IF statement function. In each comparison the highest value across all time points for each gene was selected (MAX) and the maximum values were then sorted in descending order.

Table S5 Prioritization of the maximum value across the time course of the sexual development for each gene in speecies comparisons.

Table S6 Cosegregation of hygromycin resistance and identified phenotypes in KO strains of interest.

Acknowledgments

We thank Dr. Joseph Wolenski and Barry Piekos for providing assistance in the Yale MCDB microscopy facility.

Author Contributions

Conceived and designed the experiments: JPT FT. Performed the experiments: NL ZW. Analyzed the data: NAL ZW NL FL-G. Contributed to the writing of the manuscript: NL ZW DAH FT JPT.

References

1. Busch S, Braus GH (2007) How to build a fungal fruit body: from uniform cells to specialized tissue. Mol Microbiol 64: 873–876.

2. Busch S, Schwier EU, Nahlik K, Bayram O, Helmstaedt K, et al. (2007) An eight-subunit COP9 signalosome with an intact JAMM motif is required for fungal fruit body formation. Proc Natl Acad Sci U S A 104: 8089–8094.

3. Coppin E, Berteaux-Lecellier V, Bidard F, Brun S, Ruprich-Robert G, et al. (2012) Systematic deletion of homeobox genes in *Podospora anserina* uncovers their roles in shaping the fruiting body. PLoS One 7: e37488.

4. Debuchy R, Berteaux-Lecellier V, Silar P (2010) Mating systems and sexual morphogenesis in Ascomycetes. In: Borkovich KA, Ebbole DJ, editors. Cellular and Molecular Biology of Filamentous Fungi. Washington, DC: ASM Press. pp. 501–535.

5. Harting R, Bayram O, Laubinger K, Valerius O, Braus GH (2013) Interplay of the fungal sumoylation network for control of multicellular development. Mol Microbiol 90: 1125–1145.

6. Jamet-Vierny C, Debuchy R, Prigent M, Silar P (2007) IDC1, a pezizomycotina-specific gene that belongs to the PaMpk1 MAP kinase transduction cascade of the filamentous fungus Podospora anserina. Fungal Genet Biol 44: 1219–1230.

7. Pöggeler S, Nowrousian M, Kück U (2006) Fruiting-body development in Ascomycetes, in The Mycota I. Growth, Differentiation and Sexuality. In: Kües U, Fischer R, editors. The Mycota I. (2nd edition). Growth, Differentiation and Sexuality. Berlin, Heideberg: Springer-Verlag. pp. 325–355.

8. Trail F, Gardiner DM (2014) Applicatioin of genomics to the study of pathogenecity and development in Fusarium. In: Nowrousian M, Esser K. editors. The Mycota: A comprehensive treatise on fungi as experimental systems for basic and applied research Fungal Genomics XIII. Second edition. Berlin, Heidelberg: Springer-Verlag. pp. 267–300.

9. Trail F (2013) Sex and fruiting in *Fusarium*. In: Brown D, Proctor R, editors. *Fusarium*: genomics, molecular and cellular biology. Norwich, UK: Horizon Scientific Press and Caister Academic Press.

10. Coppin E, Debuchy R, Arnaise S, Picard M (1997) Mating types and sexual development in filamentous ascomycetes. Microbiol Mol Biol Rev 61: 411–428.

11. DeLange AM, Griffiths AJ (1980) Meiosis in *Neurospora crassa*. I. The isolation of recessive mutants defective in the production of viable ascospores. Genetics 96: 367–378.

12. Eckert SE, Kubler E, Hoffmann B, Braus GH (2000) The tryptophan synthase-encoding trpB gene of *Aspergillus nidulans* is regulated by the cross-pathway control system. Mol Gen Genet 263: 867–876.

13. Eckert SE HB, Wanke C, Braus GH. (1999) Sexual development of *Aspergillus nidulans* in tryptophan auxotrophic strains. Arch Microbiol 172: 157–166.

14. Fischer R, Kües U (2003) Developmental processes in filamentous fungi. Genomics of Plants and Fungi. In: Prade RA, Bohnert HJ, editors. New York: Marcel Dekker. pp. 41–118.

15. Johnson TE (1979) A *Neurospora* mutation that arrests perithecial development as either male or female parent. Genetics 92: 1107–1120.

16. Leslie JF, Raju NB (1985) Recessive mutations from natural populations of *Neurospora crassa* that are expressed in the sexual diplophase. Genetics 111: 759–777.

17. Nowrousian M, Ringelberg C, Dunlap JC, Loros JJ, Kuck U (2005) Cross-species microarray hybridization to identify developmentally regulated genes in the filamentous fungus *Sordaria macrospora*. Mol Genet Genomics 273: 137–149.

18. Dodge BO (1927) Nuclear phenomena associated with heterothallism and homothallism in the Ascomycete *Neurospora*. Journal of Agricultural Research 35: 289–305.

19. Mao-Draayer Y, Galbraith AM, Pittman DL, Cool M, Malone RE (1996) Analysis of meiotic recombination pathways in the yeast *Saccharomyces cerevisiae*. Genetics 144: 71–86.

20. Lumbsch HT (2000) Phylogeny of filamentous ascomycetes. Naturwissenschaften 87: 335–342.

21. Samils N, Gioti A, Karlsson M, Sun Y, Kasuga T, et al. (2013) Sex-linked transcriptional divergence in the hermaphrodite fungus *Neurospora tetrasperma*. Proc Royal Soc B: Biol Sci 280: 20130862.

22. Wang Z, Kin K, Lopez-Giraldez F, Johannesson H, Townsend JP (2012) Sex-specific gene expression during asexual development of *Neurospora crassa*. Fungal Genet Biol 49: 533–543.

23. Perkins D, Raju NB (1986) *Neurospora discreta*, a new heterothallic species defined by its crossing behavior. Experimental Mycology 10: 323–338.

24. Turner E, Jacobson DJ, Taylor JW (2010) Reinforced postmating reproductive isolation barriers in *Neurospora*, an Ascomycete microfungus. J Evol Biol 23: 1642–1656.

25. Borkovich KA, Alex LA, Yarden O, Freitag M, Turner GE, et al. (2004) Lessons from the genome sequence of *Neurospora crassa*: tracing the path from genomic blueprint to multicellular organism. Microbiol Mol Biol Rev 68: 1–108.

26. Beadle GW, Tatum EL (1941) Genetic Control of Biochemical Reactions in *Neurospora*. Proc Natl Acad Sci U S A 27: 499–506.

27. Dunlap JC, Loros JJ (2004) The *Neurospora* circadian system. J Biol Rhythms 19: 414–424.

28. Metzenberg RL, Glass NL (1990) Mating type and mating strategies in *Neurospora*. Bioessays 12: 53–59.

29. Kronstad JW, Staben C (1997) Mating type in filamentous fungi. Annu Rev Genet 31: 245–276.

30. Raju N (1992) Genetic control of the sexual cycle in *Neurospora*. Mycol Res 96: 241–262.

31. Raju NB, Perkins DD (1994) Diverse programs of ascus development in pseudohomothallic species of *Neurospora*, *Gelasinospora*, and *Podospora*. Dev Genet 15: 104–118.

32. Dettman JR, Jacobson DJ, Taylor JW (2003) A multilocus genealogical approach to phylogenetic species recognition in the model eukaryote *Neurospora*. Evolution 57: 2703–2720.

33. Beatty NP, Smith ML, Glass NL (1994) Molecular characterization of mating-type loci in selected homothallic species of *Neurospora*, *Gelasinospora* and *Anixiella*. Mycol Res 98: 1309–1316.

34. Casselton LA (2008) Fungal sex genes-searching for the ancestors. Bioessays 30: 711–714.

35. Klix V, Nowrousian M, Ringelberg C, Loros JJ, Dunlap JC, et al. (2010) Functional characterization of MAT1-1-specific mating-type genes in the homothallic ascomycete *Sordaria macrospora* provides new insights into essential and nonessential sexual regulators. Eukaryot Cell 9: 894–905.

36. Nygren K, Strandberg R, Wallberg A, Nabholz B, Gustafsson T, et al. (2011) A comprehensive phylogeny of *Neurospora* reveals a link between reproductive mode and molecular evolution in fungi. Mol Phylogenet Evol 59: 649–663.

37. Paoletti M, Seymour FA, Alcocer MJ, Kaur N, Calvo AM, et al. (2007) Mating type and the genetic basis of self-fertility in the model fungus *Aspergillus nidulans*. Curr Biol 17: 1384–1389.

38. Strandberg R, Nygren K, Menkis A, James TY, Wik L, et al. (2010) Conflict between reproductive gene trees and species phylogeny among heterothallic and pseudohomothallic members of the filamentous ascomycete genus *Neurospora*. Fungal Genet Biol 47: 869–878.

39. Whittle CA, Nygren K, Johannesson H (2011) Consequences of reproductive mode on genome evolution in fungi. Fungal Genet Biol 48: 661–667.

40. Yun SH, Berbee ML, Yoder OC, Turgeon BG (1999) Evolution of the fungal self-fertile reproductive life style from self-sterile ancestors. Proc Natl Acad Sci U S A 96: 5592–5597.

41. Kim H, Wright SJ, Park G, Ouyang S, Krystofova S, et al. (2012) Roles for receptors, pheromones, G proteins, and mating type genes during sexual reproduction in *Neurospora crassa*. Genetics 190: 1389–1404.

42. Bistis GN (1981) Chemotropic Interactions between trichogynes and conidia of opposite mating-Type in *Neurospora crassa*. Mycologia 73: 959–975.

43. Harris JL, Howe HB Jr, Roth IL (1975) Scanning electron microscopy of surface and internal features of developing perithecia of *Neurospora crassa*. J Bacteriol 122: 1239–1246.

44. Raju N (1980) Meiosis and ascospore genesis in *Neurospora*. Eur J Cell Biol 23: 208–223.

45. Palma-Guerrero J, Hall CR, Kowbel D, Welch J, Taylor JW, et al. (2013) Genome wide association identifies novel loci involved in fungal communication. PLoS Genet 9: e1003669.

46. McCluskey K, Wiest A, Plamann M (2010) The Fungal Genetics Stock Center: a repository for 50 years of fungal genetics research. J Biosci 35: 119–126.

47. Hallen HE, Huebner M, Shiu SH, Guldener U, Trail F (2007) Gene expression shifts during perithecium development in *Gibberella zeae* (anamorph *Fusarium graminearum*), with particular emphasis on ion transport proteins. Fungal Genet Biol 44: 1146–1156.

48. Wang Z, Lehr N, Trail F, Townsend JP (2012) Differential impact of nutrition on developmental and metabolic gene expression during fruiting body development in *Neurospora crassa*. Fungal Genet Biol 49: 405–413.

49. Clark TA, Guilmette JM, Renstrom D, Townsend JP (2008) RNA extraction, probe preparation, and competitive hybridization for transcriptional profiling using *Neurospora crassa* long-oligomer DNA microarrays. Fungal Genet Rep 55: 18–28.

50. Trapnell C, Pachter L, Salzberg SL (2009) TopHat: discovering splice junctions with RNA-Seq. Bioinformatics 25: 1105–1111.

51. Galagan JE, Calvo SE, Borkovich KA, Selker EU, Read ND, et al. (2003) The genome sequence of the filamentous fungus *Neurospora crassa*. Nature 422: 859–868.

52. Grigoriev IV, Nordberg H, Shabalov I, Aerts A, Cantor M, et al. (2012) The genome portal of the Department of Energy Joint Genome Institute. Nucleic Acids Res 40: D26–32.

53. Zhang Z, Lopez-Giraldez F, Townsend JP (2010) LOX: inferring Level Of eXpression from diverse methods of census sequencing. Bioinformatics 26: 1918–1919.

54. Poptsova MS, Gogarten JP (2007) BranchClust: a phylogenetic algorithm for selecting gene families. BMC Bioinformatics 8: 120.

55. Zhang Z, Townsend JP (2010) The filamentous fungal gene expression database (FFGED). Fungal Genet Biol 47: 199–204.

56. Wang Z, Lopez-Giraldez F, Lehr N, Farre M, Common R, et al. (2014) Global gene expression and focused knockout analysis reveals genes associated with fungal fruiting body development in *Neurospora crassa*. Eukaryot Cell 13: 154–169.

57. Ruepp A, Zollner A, Maier D, Albermann K, Hani J, et al. (2004) The FunCat, a functional annotation scheme for systematic classification of proteins from whole genomes. Nucleic Acids Res 32: 5539–5545.

58. Colot HV, Park G, Turner GE, Ringelberg C, Crew CM, et al. (2006) A high-throughput gene knockout procedure for *Neurospora* reveals functions for multiple transcription factors. Proc Natl Acad Sci U S A 103: 10352–10357.

59. Dunlap JC, Borkovich KA, Henn MR, Turner GE, Sachs MS, et al. (2007) Enabling a community to dissect an organism: overview of the *Neurospora* functional genomics project. Adv Genet 57: 49–96.

60. Fu C, Iyer P, Herkal A, Abdullah J, Stout A, et al. (2011) Identification and characterization of genes required for cell-to-cell fusion in *Neurospora crassa*. Eukaryot Cell 10: 1100–1109.

61. Seo J, Shneiderman B (2002) Interactively exploring hierarchical clustering results. Computer 35: 80–86.

62. Altschul SF, Gish W, Miller W, Myers EW, Lipman DJ (1990) Basic local alignment search tool. J Mol Biol 215: 403–410.

63. Coleman JJ, Rounsley SD, Rodriguez-Carres M, Kuo A, Wasmann CC, et al. (2009) The genome of *Nectria haematococca*: contribution of supernumerary chromosomes to gene expansion. PLoS Genet 5: e1000618.

64. Ait Benkhali J, Coppin E, Brun S, Peraza-Reyes L, Martin T, et al. (2013) A network of HMG-box transcription factors regulates sexual cycle in the fungus *Podospora anserina*. PLoS Genet 9: e1003642.

65. Son H, Seo YS, Min K, Park AR, Lee J, et al. (2011) A phenome-based functional analysis of transcription factors in the cereal head blight fungus, *Fusarium graminearum*. PLoS Pathog 7: e1002310.

15

Maternal Parity and the Risk of Congenital Heart Defects in Offspring: A Dose-Response Meta-Analysis of Epidemiological Observational Studies

Yu Feng[1◑], Di Yu[1◑], Tao Chen[2], Jin Liu[2], Xing Tong[3], Lei Yang[1], Min Da[1], Shutong Shen[4], Changfeng Fan[1], Song Wang[1], Xuming Mo[1]*

1 Department of Cardiothoracic Surgery, The Affiliated Children's Hospital of Nanjing Medical University, Nanjing, Jiangsu, China, 2 Department of Epidemiology and Biostatistics, School of Public Health, Nanjing Medical University, Nanjing, Jiangsu, China, 3 Atherosclerosis Research Center, Key Laboratory of Cardiovascular Disease and Molecular Intervention, Nanjing Medical University, Nanjing, Jiangsu, China, 4 Department of Cardiology, The First Affiliated Hospital of Nanjing Medical University, Nanjing, Jiangsu, China

Abstract

Background: Epidemiological studies have reported conflicting results regarding maternal parity and the risk of congenital heart defects (CHDs). However, a meta-analysis of the association between maternal parity and CHDs in offspring has not been conducted.

Methods: We searched MEDLINE and EMBASE for articles catalogued between their inception and March 8, 2014; we identified relevant published studies that assessed the association between maternal parity and CHD risk. Two authors independently assessed the eligibility of the retrieved articles and extracted data from them. Study-specific relative risk estimates were pooled by random-effects or fixed-effects models. From the 11272 references, a total of 16 case-control studies and 3 cohort studies were enrolled in this meta-analysis.

Results: The overall relative risk of CHD in parous versus nulliparous women was 1.01 (95% CI, 0.97–1.06; $Q = 32.34$; $P = 0.006$; $I^2 = 53.6\%$). Furthermore, we observed a significant association between the highest versus lowest parity number, with an overall RR = 1.20 (95% CI, 1.10–1.31; ($Q = 74.61$, $P < 0.001$, $I^2 = 82.6\%$). A dose–response analysis also indicated a positive effect of maternal parity on CHD risk, and the overall increase in relative risk per one live birth was 1.06 (95% CI, 1.02–1.09); $Q = 68.09$; $P < 0.001$; $I^2 = 80.9\%$). We conducted stratified and meta-regression analyses to identify the origin of the heterogeneity among studies. A Galbraith plot was created to graphically assess the sources of heterogeneity.

Conclusion: In summary, this meta-analysis provided a robust estimate of the positive association between maternal parity and risk of CHD.

Editor: Zaccaria Ricci, Bambino Gesù Children's Hospital, Italy

Funding: The authors have no support or funding to report.

Competing Interests: The authors have declared that no competing interests exist.

* Email: mohsuming15@sina.com

◑ These authors contributed equally to this work.

Introduction

Congenital heart defects (CHD) are the most common human birth defects and the leading cause of perinatal mortality, with an incidence of approximately 4 to 50 per 1000 live birth or even higher [1]. The etiology of CHD is complex and may involve the interaction of environmental exposure and inherited factors [2]. A multitude of studies have identified both chromosomal and gene mutations as the cause of the syndromic version of the heart malfunction [1]. In contrast, the origin of non-syndromic CHD, which accounts for most congenital cardiac abnormalities, remains unknown.

Maternal phenylketonuria, diabetes mellitus, maternal teratogen exposure, and maternal therapeutic drug exposure during pregnancy may increase the risk of congenital malformations in offspring [3]. Apart from these influences, previous studies have indicated that inherent maternal characteristics, such as parity, may be responsible for certain categories of congenital defects. Some studies have observed a positive association between nulliparity and the risk of various birth defects [4–10]. In contrast, other studies have observed that multiparity is associated with an increased risk of specific birth defects [11–13]. The results for CHD are similar; no consensus has been reached, and some studies show positive associations while others find null results. The association between maternal parity and CHDs might be explained by unmeasured environmental risk factors which are more common among multiparous women than nulliparous women. Both biological and psychosocial interpretations can be proposed, including maternal stress, maternal uterus condition and serum levels of estradiol [14–17].

To date, an increasing number of studies has focused on the association between maternal parity and CHDs; however, the results have been ambiguous, possibly because of inadequate sample sizes. Therefore, we conducted a dose-response meta-analysis to quantitatively assess the effects of maternal parity on CHDs.

Methods

Literature Search

To identify relevant epidemiological studies, two independent researchers (Feng and Yu) conducted a computerized literature search in MEDLINE and EMBASE to retrieve articles that were catalogued between the databases' inception and March 8, 2014.The search terms for the exposure were: 'Parity', 'Pregnancy', 'Live Birth', 'Reproduction', 'Reproductive' and 'Reproductive Factors' and the search terms for the outcome were: 'Congenital Heart Defect', 'Heart Abnormality', 'Malformation Of Heart' and 'CHD'. In addition, we conducted a search for a broad range of environmental teratogens and CHDs and examined the relevant references and review articles; in this way, we could identify information from other related studies. We followed standards of quality for conducting and reporting meta-analyses [18].

Eligibility Criteria

We selected articles that (1) were original epidemiologic studies (i.e., case–control and cohort), (2) examined the association between maternal parity and CHDs overall or any one of the CHD subtypes in infants, (3) were published in the English language, (4) reported RRs (i.e., risk ratios or odds ratios) and associated 95% confidence intervals (CIs) or standard errors or provided the data necessary to recalculate these factors, and (5) defined CHDs or one of the CHD subtypes as an outcome. Articles that reported results from more than one population were considered to be separate studies. When multiple articles from the same study were provided, we used the article with the most applicable information and the largest number of cases. We excluded non-peer-reviewed articles, experimental animal studies, ecological assessments, correlation studies and mechanistic studies.

Data Extraction

Data extraction was carried out separately by two reviewers (Feng and Yu) working independently. When differences of opinion arose, they were resolved by a discussion between the two reviewers or by the involvement of a third reviewer (Chen) for adjudication. Parity was defined as the number of live births before the index delivery [19]. Nulliparous women were defined as those with no previous live births before the index delivery. Primiparous women were those with one live birth, and multiparous women were those with two or more prior live births. The studies that met the inclusion criteria were reviewed to retrieve the information of interest. The characteristics of interest included authors, year of

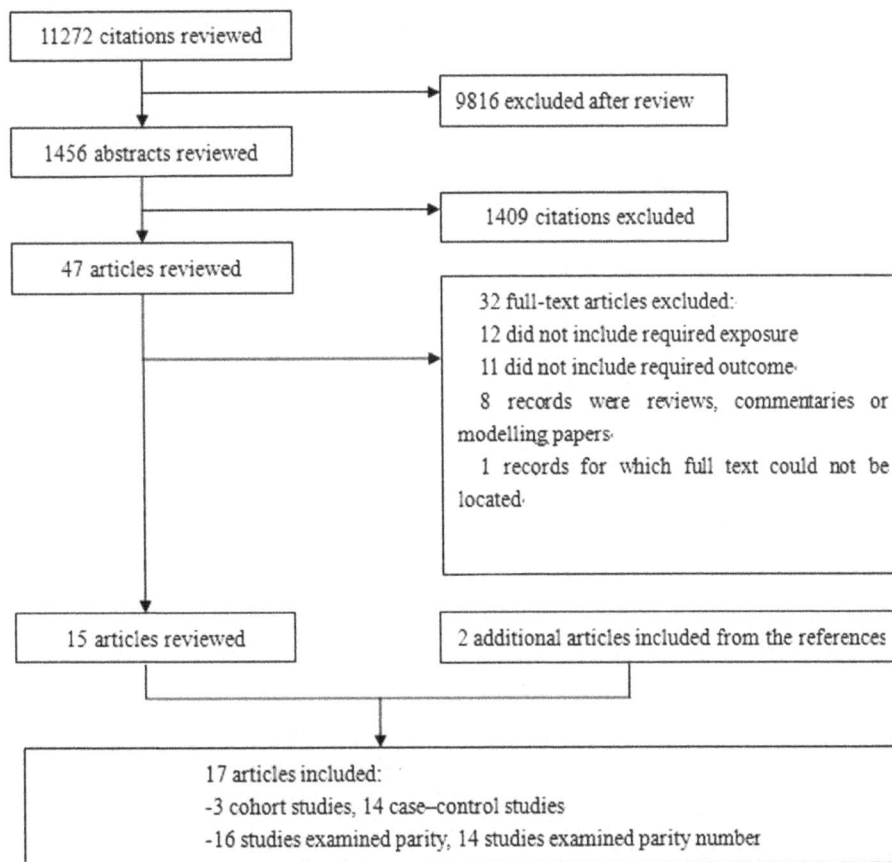

Figure 1. Study selection procedures for a meta-analysis of maternal parity and congenital heart defects (CHDs) in offspring.

publication, geographic region, periods of data collection, study design, sample size, case classification, exposure and outcome assessment (including parity as both a binary and categorical variable), adjusted estimates and their corresponding 95% CIs for parous versus nulliparous women, highest versus lowest number of previous births, and confounding factors that were controlled for by matching cases or adjustments in the data analysis. We back-calculated the point estimate and 95% CI if the original study did not report the risk estimates in this order. When no adjusted estimates were available, we extracted the crude estimate. If no estimate was provided in a given study, we recalculated odds ratios or risk ratios and 95% CIs from the presented raw data using standard equations.

To assess the study quality, we used a 9-star system on the basis of the Newcastle-Ottawa Scale [20]. This system judges a study based on three broad characteristics: the selection of study groups, comparability of study groups and ascertainment of the exposure or outcome of interest for case-control and cohort studies, respectively. The highest score was 9, and we defined a high quality study as one with a quality score greater than or equal to 7.

Statistical Analysis

We used study-specific relative risks as a summary statistic of the association between maternal parity and CHD risk. To simplify the procedure, a RR was used to represent all reported study-specific results from cohort studies and an OR to represent results from case-control studies. If a study did not use the lowest parity number as the reference category, the effective count method proposed by Hamling and colleagues [21] was used to recalculate the RRs.

For the dose–response analysis, which considers parity as a continuous variable, the method proposed by Greenland and colleagues [22] and Orsini and colleagues [23] was used to calculate study-specific slopes (i.e., linear trends) and 95% CIs. For studies which reported duration as a range, the midpoint, determined by calculating the average of the lower and upper bounds, was used. When the highest category was open-ended, the width of the open-ended interval was taken to be the same as that of the category immediately previous to it. When the lowest category did not have a lower bound, we considered the lower bound to be zero. We presented the dose–response results in forest plots on the basis of increments of 1 live birth with regard to parity.

Cochran's Q and I^2 statistics were used to test for heterogeneity among studies [24]. If there was evidence of heterogeneity ($P < 0.05$ or $I^2 \geq 56\%$), a random-effects model was used, which provided a more appropriate summary estimate for heterogeneous study-specific estimates. If the study revealed no evidence of heterogeneity, the fixed-effects analysis was used, an inverse variance weighting was applied to calculate summary RR estimates [25].

We conducted subgroup analyses based on study design (i.e., cohort versus case–control studies), geographical region (i.e., North America, Europe, and Asia), number of cases (i.e., ≤ 1000 versus > 1000), publication period (i.e., before 2010 versus 2010 or after), maternal age (i.e., ≤ 27 versus > 27), primary interest (i.e., whether the title or abstract refers to the reproductive factors as their research interest, yes versus no), and study quality (i.e., low versus high quality). We evaluated heterogeneity between subgroups by meta-regression. A P value less than 0.05 from the meta-regression was considered representative of a significant difference between subgroups. Finally, we conducted sensitivity analyses to explore whether a specific study strongly influenced the results by excluding one study at a time.

Publication bias was assessed via visual inspection of a funnel plot with asymmetry using both Egger's linear regression [26] and Begg's rank correlation [27] methods. Significant statistical publication bias was defined as a P value of <0.05 for the two above-mentioned tests. All statistical analyses were performed with STATA (version 11.0; StataCorp, College Station, Texas, USA).

Results

Study Characteristics

The search strategy generated 11272 citations; from these, 17 were used in the final analysis, representing 43880 incident cases (Figure 1). All of the studies were published between 1989 and 2013. There were 14 case–control studies [28–41] and 3 cohort studies [42–44]. The main characteristics of the included studies are presented in Table S1. As shown, 10 studies [28–30,32,33,35,36,39,43,44] were conducted in the United States or Canada, 6 in Europe [31,37,38,40–42], and 1 in Asia [34]. Among these studies, 16 investigated the association between maternal parity as a binary variable and CHD risk [28–43], and 14 examined the association of maternal parity number with CHD risk [28,30–34,36,38–44]. In the 3 cohort studies, cohort sizes varied from 22,365 [44] to 1,625,945 [43], and the number of CHD cases ranged from 4,123 [44] to 12,101 [43]. In the 16 case–control studies, the number of cases varied from 81 [28] to 7,575 [32], and the number of control subjects ranged from 302 [39] to 38,151 [41]. The highest parity number ranged from 2 [28] to more than 4 [39].

Parous versus Nulliparous

A total of 14 case–control studies and 2 cohort studies examined the association between parity as a binary variable and CHD risk. The overall relative risk of CHD for parous versus nulliparous women was 1.01 (95% CI, 0.97–1.06), with moderate heterogeneity ($Q = 32.34$; $P = 0.006$; $I^2 = 53.6\%$; **Table 1** and **Fig. 2**). There was no indication of publication bias based on the Egger test ($P = 0.295$) or visual inspection of the funnel plot (data not shown). In a sensitivity analysis, we sequentially excluded one study at a time and reanalyzed the data. The 16 study-specific relative risks for the parous versus nulliparous women ranged from a low of 1.01 (95% CI, 0.97–1.05; $Q = 34.59$; $P = 0.007$; $I^2 = 50.9\%$) after omission of the study by Padula and colleagues [36] to a high of 1.02 (95% CI, 0.99–1.06; $Q = 31.44$; $P = 0.018$; $I^2 = 45.9\%$) after omission of the study by Luo and colleagues [34]. As shown in **Table 1**, similar risks were observed between subgroup stratified by maternal age for association between maternal ever parity and CHD in offspring (P for heterogeneity = 0.12).

Highest versus Lowest Parity Number

A total of 11 case–control studies and 3 cohort studies examined the association between high and low parity and CHD risk. The estimate of the relative risk of CHD for the highest versus lowest parity categories was 1.20 (95% CI, 1.10–1.31). Statistically significant heterogeneity was detected ($Q = 74.61$, $P < 0.001$, $I^2 = 82.6\%$; Table 2 and Fig. 3) with no publication bias (Begg's test: $P = 0.443$, Egger's test: $P = 0.883$). The 13 study-specific relative risks when considering the parity number ranged from a low of 1.17 (95% CI, 1.07–1.27; $Q = 61.84$; $P = 0.000$; $I^2 = 80.6\%$) after omission of the study by Vereczkey and colleagues [40] to a high of 1.22 (95% CI, 1.12–1.34; $Q = 59.74$; $P = 0.000$; $I^2 = 79.9\%$) after omission of the study by Batra and colleagues [30].

Table 1. Summary risk estimates of the association between maternal ever parity and CHD risk in offspring.

Subgroup analysis	No. of studies	No. of cases	Summary RR (95% CIs)	P^1	I^2 (%)	P^2
Summary pooled estimate	16	39757	1.01(0.97–1.06)	0.006	53.6	
Geographical region						0.202
North America	9	31090	1.01(0.98–1.05)	0.313	14.5	
Europe	6	7974	1.14(0.98–1.33)	0.014	64.9	
Asia	1	693	0.82(0.71–0.97)	-	-	
Number of cases						0.438
≤1000	9	3691	1.14(0.98–1.32)	0.007	62.3	
>1000	7	36066	1.00(0.98–1.03)	0.165	34.5	
Publication period						0.719
Before 2010	8	26457	1.00(0.96–1.04)	0.119	39.1	
2010 or after	8	13300	1.05(0.95–1.17)	0.004	66.4	
Design						0.744
Case-control	14	20747	1.02(0.96–1.09)	0.027	46.9	
Cohort	2	19010	1.00(0.93–1.08)	0.006	86.6	
Maternal age(year)						0.12
≤27	8	18296	1.05(0.99–1.12)	0.169	32.4	
>27	7	21461	0.98(0.92–1.04)	0.063	49.8	
Primary interest						0.69
Yes	9	23805	1.01(0.94–1.09)	0.020	54.2	
No	7	15952	1.03(1.00–1.06)	0.144	39.2	
Quality assessment						0.362
High quality studies (scores≥7)	11	30300	1.03(0.99–1.07)	0.060	43.6	
Low quality studies (scores<7)	7	9457	0.96(0.92–1.00)	0.161	36.8	

[1]p-value for heterogeneity within each subgroup.
[2]p-value for heterogeneity between subgroups with meta-regression analysis.
Abbreviations: RR: relative risk; CI: confidence interval.

Dose–Response Analysis

A total of 11 case–control studies and 3 cohort studies were included in the dose-response analysis. The estimate of relative risk per live birth was 1.06 (95% CI, 1.02–1.09), and there was statistically significant heterogeneity ($Q = 68.09$; $P<0.001$; $I^2 = 80.9\%$; Table 2 and Fig. 4). Publication bias was not evident based on the Egger test ($P = 0.973$) or Begg test ($P = 0.101$), and no asymmetry was observed in the funnel plots. The 13 study-specific relative risks of parity ranged from a low of 1.05 (95% CI, 1.02–1.08; $Q = 56.94$; $P = 0.000$; $I^2 = 78.9\%$) after omission of the study by Vereczkey and colleagues [40] to a high of 1.06 (95% CI, 1.03–1.10; $Q = 51.12$; $P = 0.000$; $I^2 = 76.5\%$) after omission of the study by Cedergren and colleagues [42].

Heterogeneity Analysis

We conducted stratified and meta-regression analyses to identify the origin of the heterogeneity among studies. In subgroup analyses of parity as a binary variable and CHD risk, there was no indication of significant heterogeneity between subgroups according to meta-regression analyses (**Table 1**). However, significant heterogeneity existed in the dose-response analyses of the association between parity number and CHD risk. To clarify the sources of heterogeneity, we conducted a sensitivity analysis; however, I^2 did not decrease much by removing each study in turn. Subsequently, a meta-regression was performed with a Knapp-Hartung modification, and we found that differing numbers of cases may contribute to the heterogeneity

($p = 0.060$). We further created a Galbraith plot to graphically assess the sources of heterogeneity (Figures S1, S2). A total of 7 studies [30,32,34,40–42,44] were identified as the primary sources of heterogeneity (i.e., 6 studies [30,32,40–42,44] from the high versus low parity number analysis and 6 studies [30,34,40–42,44] from the dose-response analysis). Once the outlying studies were excluded, the heterogeneity was effectively removed (i.e., for the high versus low parity number analysis, $I^2 = 0.0\%$; for the dose-response analysis, $I^2 = 0.0\%$; however, the corresponding pooled RRs were not materially altered in any comparisons (i.e., for the high versus low parity analysis: RR = 1.23, 95% CI = 1.17–1.29; for the dose-response analysis: RR = 1.05, 95% CI = 1.03–1.07).

Discussion

To the best of our knowledge, this is the first quantitative meta-analysis evaluating the association between maternal parity and the risk of congenital heart defects. Overall, the findings of our meta-analysis suggested that maternal parity (i.e., the highest category compared to the lowest category, RR = 1.20, 95% CI = 1.10–1.31) was significantly associated with CHD risk. Meanwhile, in the dose-response meta-analysis, we found that the risk of CHD increased by 6% per live birth. However, there was no evidence that verified the association between parous versus nulliparous women (RR = 1.01, 95%CI = 0.97–1.06) and the risk of CHDs. Additionally, the results were consistent across most of the subgroup analyses (Table 1 and 2).

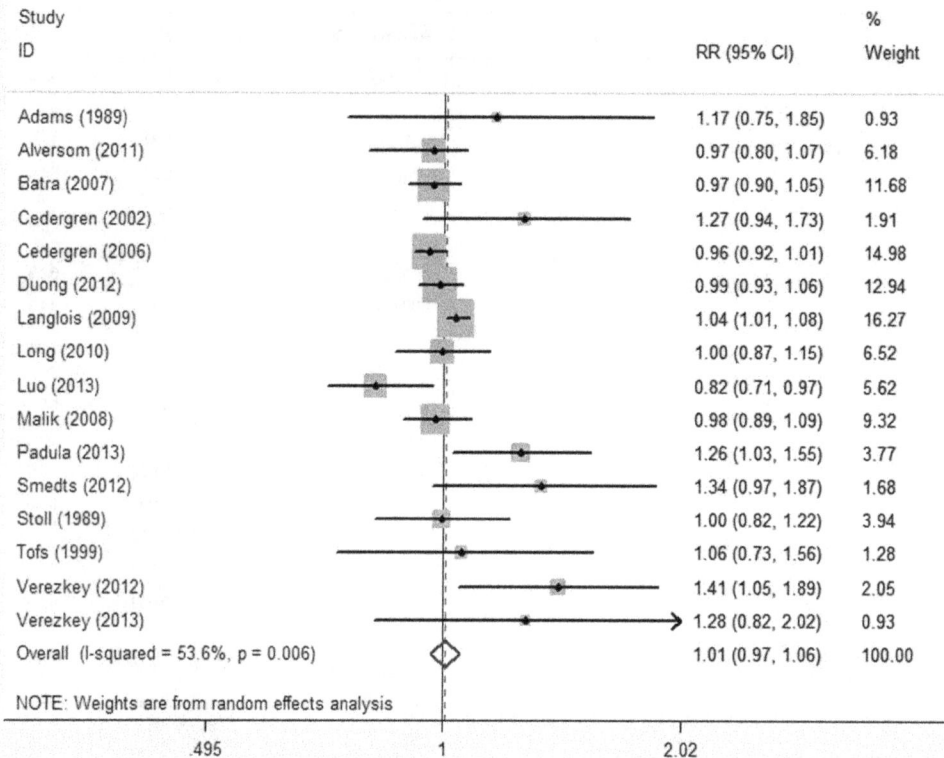

Study ID	RR (95% CI)	% Weight
Adams (1989)	1.17 (0.75, 1.85)	0.93
Alversom (2011)	0.97 (0.80, 1.07)	6.18
Batra (2007)	0.97 (0.90, 1.05)	11.68
Cedergren (2002)	1.27 (0.94, 1.73)	1.91
Cedergren (2006)	0.96 (0.92, 1.01)	14.98
Duong (2012)	0.99 (0.93, 1.06)	12.94
Langlois (2009)	1.04 (1.01, 1.08)	16.27
Long (2010)	1.00 (0.87, 1.15)	6.52
Luo (2013)	0.82 (0.71, 0.97)	5.62
Malik (2008)	0.98 (0.89, 1.09)	9.32
Padula (2013)	1.26 (1.03, 1.55)	3.77
Smedts (2012)	1.34 (0.97, 1.87)	1.68
Stoll (1989)	1.00 (0.82, 1.22)	3.94
Tofs (1999)	1.06 (0.73, 1.56)	1.28
Verezkey (2012)	1.41 (1.05, 1.89)	2.05
Verezkey (2013)	1.28 (0.82, 2.02)	0.93
Overall (I-squared = 53.6%, p = 0.006)	1.01 (0.97, 1.06)	100.00

NOTE: Weights are from random effects analysis

.495   1   2.02

Figure 2. Relative risk (RR) estimates for the association between ever parity and CHD risk. Meta-analysis random-effects estimates were used. The sizes of the squares reflect the weighting of the included studies. Bars represent 95% confidence intervals (CIs). The center of the diamond represents the summary effect; left and right points of the diamond represent the 95% confidence interval.

Although the specific biological mechanism underlying maternal parity and the risk of CHDs remains unclear, some relevant evidence has been published. Nutrient depletion was more likely to occur among mothers who had given birth to live fetuses than those who had never delivered. Folic acid is one of the most important vitamins, and the association between folic acid and birth defects has been widely studied. It has been confirmed that lack of it would cause severe congenital malformation [45], especially CHDs [46] and neural tube defects [47]. Additionally, mothers who gave birth to more fetuses were more likely to have shorter inter-pregnancy intervals, which have been verified to increase the risk of major congenital malformations, including CHDs [48]. Moreover, having young children who carry respiratory viruses in the household would increase the risk of an embryo's in utero exposure to viruses, such as rubella, which was confirmed to contribute to CHD more than half a century ago [49,50]. Moreover, changes in the intrauterine environment that affect embryonic development and eventually lead to birth defects may be explained by multiparity. In addition to biological interpretations, psychosocial explanations should also be explored. Multiparity would cause an increased burden on families and increased mental stress in parents. Moreover, Zhu et al [14] found that mothers who were exposed to stress during pregnancy were at an increased risk of having offspring with CHD.

When stratified by geographic region, a significant increase in CHD risk in North America and Europe was found to be associated with increases in parity number, and similar results

were found in a dose-response meta-analysis. However, the pooled RRs for North America and Europe differed when considering parity as a binary variable. Considering the fact that only one study from Asia was included, the influence of parity in this region needs further research. In the subgroup analysis to assess study quality, we observed statistically significant results in high quality studies that included analyses of both parity number and dose-response, while no significant association was found among low quality studies. For the subgroup analysis of study design, the pooled RR from case-control studies was different from cohort studies in the analysis of dose-response. Selection and information biases might account for the observed difference. Furthermore, compared to the cohort studies, the case-control studies had a lower median quality score (7 versus 8), which may have an influence on the results.

Some limitations of our study must be taken into account. First, a total of 14 case-control studies and 3 cohort studies were recruited into our meta-analysis, and we extracted our raw data primarily from case-control studies, which are susceptible to selection and information biases. Additionally, our meta-analysis was limited to studies published in English; the results may therefore have been affected by the lack of data from studies performed in other languages. Thus, general conclusions must be considered carefully and cannot be regarded as the final word on the matter. Second because we lacked a large data set, we did not conduct a subgroup analysis of CHD subtypes; however, different CHD subtypes have different etiologies. Maternal parity may be

Table 2. Summary risk estimates of the association between maternal parity number and CHD risk in offspring.

Subgroup analysis	Highest versus lowest						Dose-response analysis (per 1 live birth)					
	No. of studies	No. of cases	Summary RR (95% CIs)	P^1	I^2(%)	P^2	No. of studies	No. of cases	Summary RR (95% CIs)	P^1	I^2(%)	P^2
Summary pooled estimate	14	38027	1.21(1.11–1.31)	<0.001	83.8		14	38027	1.06(1.02–1.09)	<0.001	80.9	
Geographical region						0.924						0.379
North America	8	29621	1.18(1.07–1.30)	<0.001	84.3		8	29621	1.05(1.02–1.08)	<0.001	75	
Europe	5	7713	1.49(1.06–2.09)	<0.001	83.6		5	7713	1.17(1.02–1.34)	<0.001	83.1	
Asia	1	693	0.95(0.71–1.28)	–	–		1	693	0.92(0.82–1.04)	–	–	
Number of cases						0.140						0.060
≤1000	8	3430	1.41(1.12–1.79)	<0.001	66.1		8	3430	1.14(1.03–1.25)	<0.001	71.6	
>1000	6	34597	1.13(1.03–1.25)	<0.001	90.4		6	34597	1.04(1.00–1.07)	<0.001	88	
Publication period						0.340						0.665
Before 2010	7	23390	1.13(1.02–1.25)	<0.001	65.2		7	23390	1.03(1.02–1.05)	0.110	42	
2010 or after	7	14637	1.30(1.12–1.53)	<0.001	87.4		7	14637	1.09(1.02–1.16)	<0.001	82.7	
Design						0.829						0.626
Case-control	11	15539	1.22(1.07–1.39)	<0.001	70.1		11	15539	1.05(0.97–1.13)	<0.001	96.4	
Cohort	3	22488	1.21(1.09–1.34)	<0.001	88.8		3	22488	1.05(1.01–1.10)	<0.001	93.5	
Maternal age(year)						0.106						0.157
≤27	7	15229	1.37(1.16–1.63)	0.004	68.6		7	15229	1.12(1.04–1.20)	0.001	72.5	
>27	7	22798	1.11(0.97–1.27)	<0.001	88.8		7	22798	1.03(0.99–1.08)	<0.001	86.9	
Primary interest						0.774						0.737
Yes	10	21534	1.23(1.08–1.40)	<0.001	84.7		10	21534	1.06(1.02–1.11)	<0.001	84.1	
No	4	16493	1.18(1.01–1.37)	0.003	78.6		4	16493	1.05(1.00–1.10)	0.049	61.9	
Quality assessment						0.252						0.673
High quality studies (scores≥7)	8	28570	1.27(1.13–1.43)	0.000	88.6		8	28570	1.08(1.03–1.12)	0.000	83.6	
Low quality studies (scores<7)	6	9457	1.08(0.99–1.16)	0.735	0		6	9457	1.01(1.00–1.04)	0.251	24.4	

[1] p-value for heterogeneity within each subgroup.
[2] p-value for heterogeneity between subgroups with meta-regression analysis.
Abbreviations: RR: relative risk; CI: confidence interval.

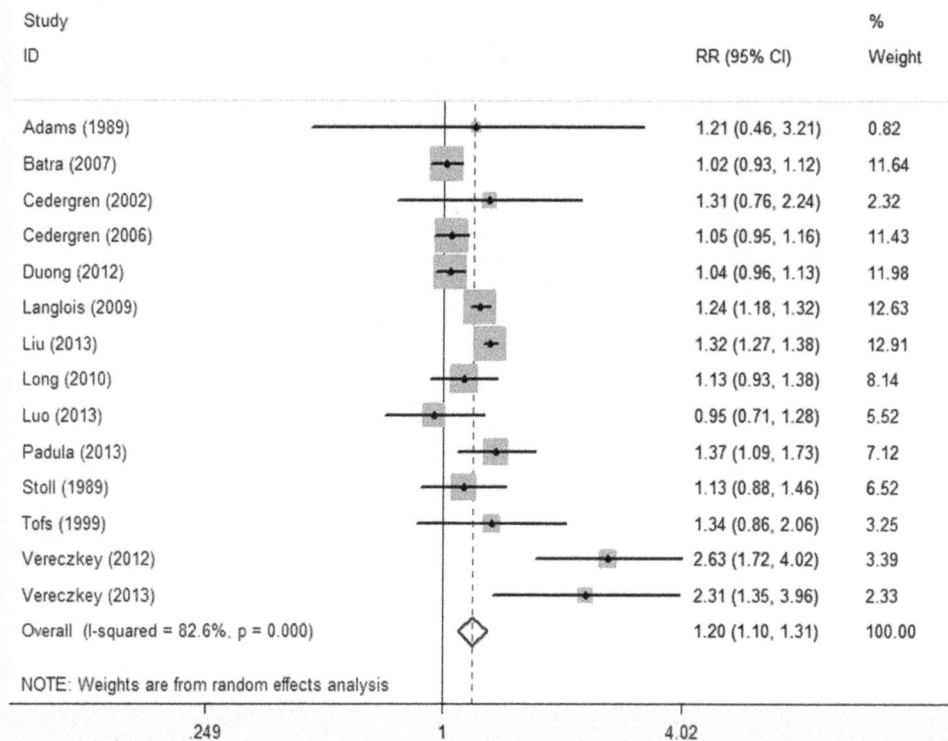

Figure 3. Relative risk (RR) estimates for the association between parity number (highest versus lowest) and CHD risk. Meta-analysis random-effects estimates were used. The sizes of the squares reflect the weighting of the included studies. Bars represent 95% confidence intervals (CIs). The center of the diamond represents the summary effect; left and right points of the diamond represent the 95% confidence interval.

not associated with all subtypes of CHD. Therefore, further research, including more high quality studies, is needed. Thirdly, although no evidence of publication bias was found, heterogeneity exists among the studies included in these analyses of both parity number and dose response; this heterogeneity may affect the interpretation of the overall results. In this study, we conducted sensitivity analyses to explore the sources of heterogeneity by deleting one study at a time from the pooled analysis. However, heterogeneity still could not be fully removed. Moreover, geographical region, sample size, CHD subtypes and other risk factors may result in heterogeneity. Therefore, we performed meta-regression and subgroup analyses to further investigate the sources of heterogeneity. In the dose-response analysis, we found that the heterogeneity stemmed partly from the number of cases. In contrast, no cause was found for the heterogeneity in the parity number meta-analysis. Furthermore, we created a Galbraith plot to assess the heterogeneity and to identify potentially outlying studies. A total of 6 were identified as the primary contributors to heterogeneity in both the analysis of parity number [30,32,40–42,44] and dose-response [30,34,40–42,44]. After excluding the outlying studies, the above-mentioned heterogeneity was effectively removed while the corresponding pooled RRs were not materially altered, indicating that the overall results regarding parity number and dose-response were statistically stable. Meanwhile, in the subgroup analysis to assess quality, heterogeneity was present in the high quality studies but not in the low quality ones. Of the 7 studies that were the main sources of heterogeneity, 5

[31,33,41,42,45] were high quality studies, which could explain the discrepancy. Finally, maternal age may be a major confounder, but in our study, similar risks were observed between subgroup stratified by maternal age for association between maternal parity and CHD in offspring (P for heterogeneity = 0.12 in maternal ever parity; P for heterogeneity = 0.106 and P for heterogeneity = 0.157 in maternal parity number). So we consider that maternal age may have no significant confounding effect on association between maternal parity and CHD in offspring. However, because of the limiting number of included studies, more studies are needed to validate our results.

Additionally, there are several important strengths of our study. First, to our knowledge, this is the first meta-analysis to report an association between maternal parity and CHDs. Moreover, our literature search was conducted on multiple databases, and the references from the retrieved articles were fully scrutinized to obtain any missing data. Therefore, our study included 43880 cases, enough to have sufficient statistical power to investigate the potential association between maternal parity and the risk of CHDs. Another strength of our study is that, although heterogeneity exists in our meta-analysis, we conducted a number of sensitivity, subgroup and Galbraith plot analyses and found that the results were stable.

In summary, this study provides evidence that maternal parity number was positively associated with the risk of CHDs. However, more prospective studies, particularly in developing countries, are needed to further investigate the association between maternal

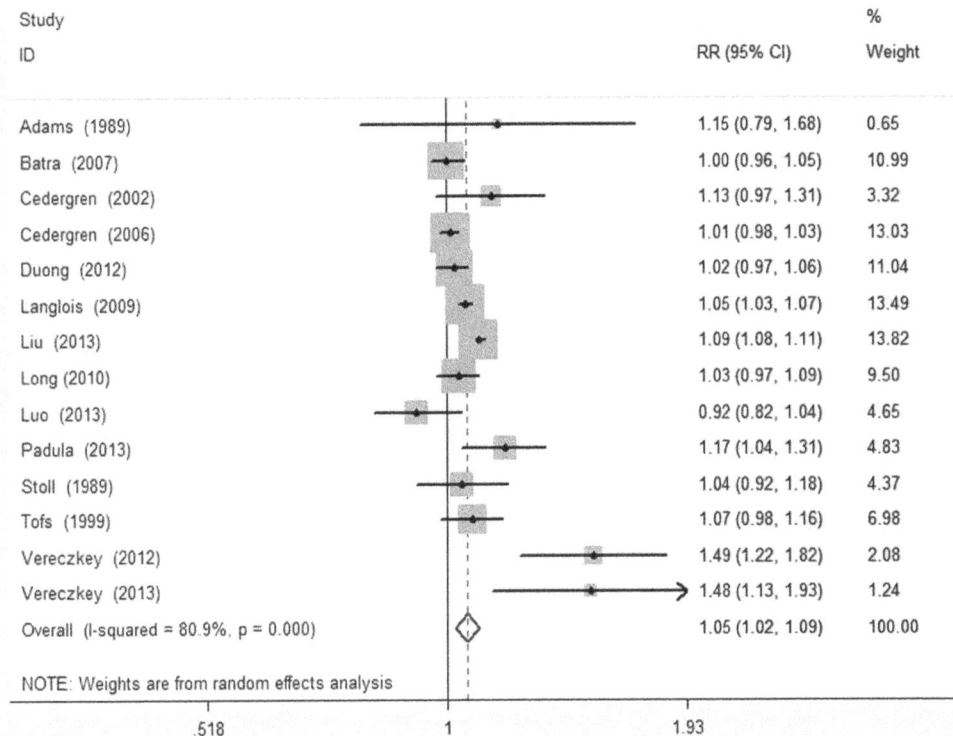

Figure 4. Relative risk (RR) estimates for the association between parity number (per 1 live birth) and CHD risk. Meta-analysis random-effects estimates were used. The sizes of the squares reflect the weighting of the included studies. Bars represent 95% confidence intervals (CIs). The center of the diamond represents the summary effect; left and right points of the diamond represent the 95% confidence interval.

parity and CHDs, especially with regard to the different subtypes of CHDs.

Supporting Information

Figure S1 Galbraith plots for parity number (highest versus lowest) and CHD risk.

Figure S2 Galbraith plots for parity number (per 1 live birth) and CHD risk.

Table S1 Characteristics of studies of maternal parity and CHD risk.

Checklist S1 PRISMA checklist.

Acknowledgments

We sincerely thank Dr. Hongcheng Zhu and Dr. Xi Yang for their help with this manuscript.

Author Contributions

Conceived and designed the experiments: XM. Performed the experiments: YF DY TC JL. Analyzed the data: YF DY XT. Contributed reagents/materials/analysis tools: LY MD SS CF SW. Wrote the paper: YF DY XM.

References

1. Pierpont ME, Basson CT, Benson DW Jr., Gelb BD, Giglia TM, et al. (2007) Genetic basis for congenital heart defects: current knowledge: a scientific statement from the American Heart Association Congenital Cardiac Defects Committee, Council on Cardiovascular Disease in the Young: endorsed by the American Academy of Pediatrics. Circulation 115: 3015–3038.
2. van der Bom T, Zomer AC, Zwinderman AH, Meijboom FJ, Bouma BJ, et al. (2011) The changing epidemiology of congenital heart disease. Nat Rev Cardiol 8: 50–60.
3. Jenkins KJ, Correa A, Feinstein JA, Botto L, Britt AE, et al. (2007) Noninherited risk factors and congenital cardiovascular defects: current knowledge: a scientific statement from the American Heart Association Council on Cardiovascular Disease in the Young: endorsed by the American Academy of Pediatrics. Circulation 115: 2995–3014.
4. Hay S, Barbano H (1972) Independent effects of maternal age and birth order on the incidence of selected congenital malformations. Teratology 6: 271–279.
5. Agopian A, Marengo L, Mitchell LE (2009) Descriptive epidemiology of nonsyndromic omphalocele in Texas, 1999–2004. Am J Med Genet A 149A: 2129–2133.
6. Bianca S, Ettore G (2003) Isolated esophageal atresia and perinatal risk factors. Dis Esophagus 16: 39–40.
7. Benjamin BG, Ethen MK, Van Hook CL, Myers CA, Canfield MA (2010) Gastroschisis prevalence in Texas 1999–2003. Birth Defects Res A Clin Mol Teratol 88: 178–185.
8. Pradat P, Francannet C, Harris JA, Robert E (2003) The epidemiology of cardiovascular defects, part I: a study based on data from three large registries of congenital malformations. Pediatr Cardiol 24: 195–221.

9. Werler MM, Bosco JL, Shapira SK, National Birth Defects Prevention S (2009) Maternal vasoactive exposures, amniotic bands, and terminal transverse limb defects. Birth Defects Res A Clin Mol Teratol 85: 52–57.

10. Carmichael SL, Shaw GM, Laurent C, Olney RS, Lammer EJ, et al. (2007) Maternal reproductive and demographic characteristics as risk factors for hypospadias. Paediatr Perinat Epidemiol 21: 210–218.

11. Vieira AR (2004) Birth order and neural tube defects: a reappraisal. J Neurol Sci 217: 65–72.

12. Hashmi SS, Waller DK, Langlois P, Canfield M, Hecht JT (2005) Prevalence of nonsyndromic oral clefts in Texas: 1995–1999. Am J Med Genet A 134: 368–372.

13. Canfield MA, Marengo L, Ramadhani TA, Suarez L, Brender JD, et al. (2009) The prevalence and predictors of anencephaly and spina bifida in Texas. Paediatr Perinat Epidemiol 23: 41–50.

14. Zhu JL, Olsen J, Sorensen HT, Li J, Nohr EA, et al. (2013) Prenatal maternal bereavement and congenital heart defects in offspring: a registry-based study. Pediatrics 131: e1225–1230.

15. Chubak J, Tworoger SS, Yasui Y, Ulrich CM, Stanczyk FZ, et al. (2004) Associations between reproductive and menstrual factors and postmenopausal sex hormone concentrations. Cancer Epidemiol Biomarkers Prev 13: 1296–1301.

16. Rovas L, Sladkevicius P, Strobel E, Valentin L (2006) Reference data representative of normal findings at three-dimensional power Doppler ultrasound examination of the cervix from 17 to 41 gestational weeks. Ultrasound Obstet Gynecol 28: 761–767.

17. Bernstein L, Depue RH, Ross RK, Judd HL, Pike MC, et al. (1986) Higher maternal levels of free estradiol in first compared to second pregnancy: early gestational differences. J Natl Cancer Inst 76: 1035–1039.

18. Stroup DF, Berlin JA, Morton SC, Olkin I, Williamson GD, et al. (2000) Meta-analysis of observational studies in epidemiology: a proposal for reporting. Meta-analysis Of Observational Studies in Epidemiology (MOOSE) group. JAMA 283: 2008–2012.

19. Baird JT Jr., Quinlivan LG (1972) Parity and hypertension. Vital Health Stat 11: 1–28.

20. Wells GA, Shea B, O'Connell D, Peterson J, Welch V, et al. (2013) The Newcastle-Ottawa Scale (NOS) for assessing the quality of nonrandomized studies in meta-analyses.comparison. Available: http://wwwohrica/programs/clinical_epidemiology/oxfordasp Accessed May 3, 2013.

21. Hamling J, Lee P, Weitkunat R, Ambuhl M (2008) Facilitating meta-analyses by deriving relative effect and precision estimates for alternative comparisons from a set of estimates presented by exposure level or disease category. Stat Med 27: 954–970.

22. Greenland S, Longnecker MP (1992) Methods for trend estimation from summarized dose-response data, with applications to meta-analysis. Am J Epidemiol 135: 1301–1309.

23. Orsini N, Li R, Wolk A, Khudyakov P, Spiegelman D (2012) Meta-analysis for linear and nonlinear dose-response relations: examples, an evaluation of approximations, and software. Am J Epidemiol 175: 66–73.

24. Higgins JP, Thompson SG, Deeks JJ, Altman DG (2003) Measuring inconsistency in meta-analyses. BMJ 327: 557–560.

25. Woolf B (1955) On estimating the relation between blood group and disease. Ann Hum Genet 19: 251–253.

26. Egger M, Davey Smith G, Schneider M, Minder C (1997) Bias in meta-analysis detected by a simple, graphical test. BMJ 315: 629–634.

27. Begg CB, Mazumdar M (1994) Operating characteristics of a rank correlation test for publication bias. Biometrics 50: 1088–1101.

28. Adams MM, Mulinare J, Dooley K (1989) Risk factors for conotruncal cardiac defects in Atlanta. J Am Coll Cardiol 14: 432–442.

29. Alverson CJ, Strickland MJ, Gilboa SM, Correa A (2011) Maternal smoking and congenital heart defects in the Baltimore-Washington Infant Study. Pediatrics 127: e647–653.

30. Batra M, Heike CL, Phillips RC, Weiss NS (2007) Geographic and occupational risk factors for ventricular septal defects: Washington State, 1987–2003. Arch Pediatr Adolesc Med 161: 89–95.

31. Cedergren MI, Selbing AJ, Kallen BA (2002) Risk factors for cardiovascular malformation–a study based on prospectively collected data. Scand J Work Environ Health 28: 12–17.

32. Duong HT, Hoyt AT, Carmichael SL, Gilboa SM, Canfield MA, et al. (2012) Is maternal parity an independent risk factor for birth defects? Birth Defects Res A Clin Mol Teratol 94: 230–236.

33. Long J, Ramadhani T, Mitchell LE (2010) Epidemiology of nonsyndromic conotruncal heart defects in Texas, 1999–2004. Birth Defects Res A Clin Mol Teratol 88: 971–979.

34. Luo YL, Cheng YL, Gao XH, Tan SQ, Li JM, et al. (2013) Maternal age, parity and isolated birth defects: a population-based case-control study in Shenzhen, China. PLoS One 8: e81369.

35. Malik S, Cleves MA, Honein MA, Romitti PA, Botto LD, et al. (2008) Maternal smoking and congenital heart defects. Pediatrics 121: e810–816.

36. Padula AM, Tager IB, Carmichael SL, Hammond SK, Yang W, et al. (2013) Ambient air pollution and traffic exposures and congenital heart defects in the San Joaquin Valley of California. Paediatr Perinat Epidemiol 27: 329–339.

37. Smedts HP, van Uitert EM, Valkenburg O, Laven JS, Eijkemans MJ, et al. (2012) A derangement of the maternal lipid profile is associated with an elevated risk of congenital heart disease in the offspring. Nutr Metab Cardiovasc Dis 22: 477–485.

38. Stoll C, Alembik Y, Roth MP, Dott B, De Geeter B (1989) Risk factors in congenital heart disease. Eur J Epidemiol 5: 382–391.

39. Torfs CP, Christianson RE (1999) Maternal risk factors and major associated defects in infants with Down syndrome. Epidemiology 10: 264–270.

40. Vereczkey A, Kosa Z, Csaky-Szunyogh M, Urban R, Czeizel AE (2012) Birth outcomes of cases with left-sided obstructive defects of the heart in the function of maternal socio-demographic factors: a population-based case-control study. J Matern Fetal Neonatal Med 25: 2536–2541.

41. Vereczkey A, Kosa Z, Csaky-Szunyogh M, Czeizel AE (2013) Isolated atrioventricular canal defects: birth outcomes and risk factors: a population-based Hungarian case-control study, 1980–1996. Birth Defects Res A Clin Mol Teratol 97: 217–224.

42. Cedergren MI, Kallen BA (2006) Obstetric outcome of 6346 pregnancies with infants affected by congenital heart defects. Eur J Obstet Gynecol Reprod Biol 125: 211–216.

43. Langlois PH, Scheuerle A, Horel SA, Carozza SE (2009) Urban versus rural residence and occurrence of septal heart defects in Texas. Birth Defects Res A Clin Mol Teratol 85: 764–772.

44. Liu S, Joseph KS, Lisonkova S, Rouleau J, Van den Hof M, et al. (2013) Association between maternal chronic conditions and congenital heart defects: a population-based cohort study. Circulation 128: 583–589.

45. Brentlinger PE (2001) Folic acid antagonists during pregnancy and risk of birth defects. N Engl J Med 344: 933–934; author reply 934-935.

46. Rosenquist TH, Ratashak SA, Selhub J (1996) Homocysteine induces congenital defects of the heart and neural tube: effect of folic acid. Proc Natl Acad Sci U S A 93: 15227–15232.

47. Czeizel AE, Dudas I, Vereczkey A, Banhidy F (2013) Folate deficiency and folic acid supplementation: the prevention of neural-tube defects and congenital heart defects. Nutrients 5: 4760–4775.

48. Grisaru-Granovsky S, Gordon ES, Haklai Z, Samueloff A, Schimmel MM (2009) Effect of interpregnancy interval on adverse perinatal outcomes–a national study. Contraception 80: 512–518.

49. Gibson S, Lewis KC (1952) Congenital heart disease following maternal rubella during pregnancy. AMA Am J Dis Child 83: 317–319.

50. Stuckey D (1956) Congenital heart defects following maternal rubella during pregnancy. Br Heart J 18: 519–522.

16

Lack of Involvement of CEP Adducts in TLR Activation and in Angiogenesis

John Gounarides[1], Jennifer S. Cobb[1], Jing Zhou[1], Frank Cook[1], Xuemei Yang[1], Hong Yin[1], Erik Meredith[2], Chang Rao[1], Qian Huang[3], YongYao Xu[3], Karen Anderson[4], Andrea De Erkenez[4], Sha-Mei Liao[4], Maura Crowley[4], Natasha Buchanan[4], Stephen Poor[4], Yubin Qiu[4], Elizabeth Fassbender[4], Siyuan Shen[4], Amber Woolfenden[4], Amy Jensen[4], Rosemarie Cepeda[4], Bijan Etemad-Gilbertson[4], Shelby Giza[4], Muneto Mogi[2], Bruce Jaffee[4], Sassan Azarian[4]*

1 Analytical Sciences, Novartis Institutes for Biomedical Research, Cambridge, MA, United States of America, 2 Global Discovery Chemistry, Novartis Institutes for Biomedical Research, Cambridge, MA, United States of America, 3 Developmental and Metabolic Pathways, Novartis Institutes for Biomedical Research, Cambridge, MA, United States of America, 4 Ophthalmology, Novartis Institutes for Biomedical Research, Cambridge, MA, United States of America

Abstract

Proteins that are post-translationally adducted with 2-(ω-carboxyethyl)pyrrole (CEP) have been proposed to play a pathogenic role in age-related macular degeneration, by inducing angiogenesis in a Toll Like Receptor 2 (TLR2)-dependent manner. We have investigated the involvement of CEP adducts in angiogenesis and TLR activation, to assess the therapeutic potential of inhibiting CEP adducts and TLR2 for ocular angiogenesis. As tool reagents, several CEP-adducted proteins and peptides were synthetically generated by published methodology and adduction was confirmed by NMR and LC-MS/MS analyses. Structural studies showed significant changes in secondary structure in CEP-adducted proteins but not the untreated proteins. Similar structural changes were also observed in the treated unadducted proteins, which were treated by the same adduction method except for one critical step required to form the CEP group. Thus some structural changes were unrelated to CEP groups and were artificially induced by the synthesis method. In biological studies, the CEP-adducted proteins and peptides failed to activate TLR2 in cell-based assays and in an in vivo TLR2-mediated retinal leukocyte infiltration model. Neither CEP adducts nor TLR agonists were able to induce angiogenesis in a tube formation assay. In vivo, treatment of animals with CEP-adducted protein had no effect on laser-induced choroidal neovascularization. Furthermore, in vivo inactivation of TLR2 by deficiency in Myeloid Differentiation factor 88 (Myd88) had no effect on abrasion-induced corneal neovascularization. Thus the CEP-TLR2 axis, which is implicated in other wound angiogenesis models, does not appear to play a pathological role in a corneal wound angiogenesis model. Collectively, our data do not support the mechanism of action of CEP adducts in TLR2-mediated angiogenesis proposed by others.

Editor: Michael E. Boulton, Indiana University College of Medicine, United States of America

Funding: This study was funded by the Novartis Institutes for Biomedical Research. The funder provided support in the form of salaries for all authors, but did not have any additional role in the study design, data collection and analysis, decision to publish, or preparation of the manuscript. The specific roles of these authors are articulated in the 'author contributions' section.

Competing Interests: The authors were all full-time employees of the Novartis Institutes of Biomedical Research, which funded this study, at the time this work was completed. There are no patents, products in development or marketed products to declare.

* Email: Sassan.Azarian@novartis.com

Introduction

Age-related macular degeneration (AMD) is a major cause of legal blindness in the elderly. The macula is a specialized area of the central retina that is enriched in photoreceptor cells and is responsible for high acuity vision. In AMD, progressive macular degeneration can impair critical daily functions such as reading, driving, and face recognition. Thus AMD can have a profound impact on quality of life. There are two forms of advanced AMD: dry and wet (neovascular) AMD [1]. AMD is thought to be a disease of the retinal pigment epithelium (RPE) cells, which provide critical support functions to adjacent photoreceptors [1]. In the early stage of disease, AMD retinas show progressive accumulation of extracellular deposits, drusen, as well as intracellular deposits, lipofuscin, at the level of the RPE. These deposits initially tend to accumulate in the macular area. Over time, RPE cells show pigmentary changes and begin to degenerate. In advanced stages, dry AMD patients exhibit substantial delineated areas of RPE atrophy, or geographic atrophy. Advanced wet AMD patients exhibit leaky blood vessels in the macula, in many cases emanating from the choriocapillaris [1].

Currently there are no treatments for dry AMD. In the Age-Related Eye Disease Study 1 (AREDS 1), dietary supplements comprised of anti-oxidants and select minerals reduced the risk of progression to advanced AMD by 25% [1]. Several therapeutic approaches are being tested in clinical trials [2] but there are no FDA-approved treatments in practice at this point. For wet AMD, anti-angiogenic treatments have been clinically proven to be efficacious [3]. However, not all patients respond to treatment and the burden of treatment is still relatively high. Thus there is a great medical need for novel treatments for AMD. The molecular details of pathogenesis in AMD are not fully established but several pathogenic mechanisms have been implicated [1]. For example, human molecular genetic data indicate the involvement of the alternative complement pathway. Another potential cause is proposed to be cumulative oxidative stress, based on preclinical studies and on the AREDS1 trial. One manifestation of oxidative stress is proposed to be the formation of CEP adducts, which are a type of advanced glycation end products [4].

Photoreceptor cells are highly enriched in docosahexaenoic acid (DHA), a labile fatty acid that is susceptible to breakdown by photo-oxidation and other forms of oxidative stress. The breakdown products include a reactive aldehyde, 4-hydroxy-7-oxohept-5-enoic acid, which can condense with primary amines to form a Schiff base. In the case of proteins, 4-hydroxy-7-oxohept-5-enoic acid condenses with lysineε-amines. Subsequent reactions result in a covalently attached CEP moity, yielding a stable CEP adduct [4]. In previous reports, antibodies raised against synthetic CEP reagents were used to identify, localize, and quantify CEP adducts by various immunological assays [5]. Elevated levels of CEP adducts were initially reported in proteomic studies of AMD donor eyes [5] and subsequently in AMD plasma [6,7]. Thus CEP adducts were implicated in AMD [5–7]. In later studies CEP adducts were reported to be pro-angiogenic, both *in vitro* and *in vivo*. These *in vivo* studies utilized the micropocket corneal neovascularization (CoNV) and the laser-induced choroidal neovascularization (CNV) models [8]. More recently, Toll-like receptor 2 (TLR2) was reported to mediate the CEP adduct-induced angiogenesis [9]. The angiogenic activity was reported to be independent of the vascular endothelial growth factor (VEGF) pathway [8,9].

There is a medical need for novel treatments for both wet and dry AMD. CEP adducts are implicated in both forms of AMD and represent an attractive potential target for drug discovery. Thus we initiated validation studies to assess the therapeutic potential of inhibiting CEP adducts.

Results

Synthesis of Tool Reagents

Several synthetic CEP adducts were generated according to reported procedures [10]. These tool reagents included protein (e.g. human serum albumin-CEP, or HSA-CEP), peptide (e.g. Ac-Gly-Lys-OMe-CEP, or dipeptide-CEP), and phospholipid (e.g. phosphatidyl ethanolamine-CEP, or PE-CEP) adducts, as listed in **Table S1**. The presence of CEP adducts and the stoichiometry of adduction was confirmed by ^{1}H-NMR and LC-MS/MS (**Figure S1 and Figure S2**). Invariably the presence of CEP moiety was established in the adducted samples and was never detected in the controls. CEP adduction was deemed successful by several measures. For example, ^{1}H-NMR analysis indicated the expected molecular signature of the CEP group in dipeptide-CEP but not the untreated dipeptide (**Figure S1B**). Likewise, LC-MS/MS analysis of enzymatic hydrolyzed adducted proteins detected lysine-CEP in HSA-CEP and MSA-CEP (mouse serum albumin-CEP) but not the respective controls (**Figure S1C**). For protein adducts, two controls were used: CTL1, which represents untreated protein; and CTL2, which represents treated unadducted protein. The latter control was treated exactly the same way as the corresponding CEP adduct except for one step, to prevent adduction (see Materials and Methods). No CEP moiety was detected in HSA-CTL2 or MSA-CTL2. There are a total of 59 lysines in HSA (UnitPro P02768) and 50 lysines in MSA

(UnitPro P07724). We identified 14 lysine-CEP sites in HSA-CEP and 40 lysine-CEP sites in MSA-CEP (**Figure S2**). These values are in fair agreement with the reported stoichiometries: 6 or 17 CEP-modified lysines for HSA-CEP [10,11], and 15 for MSA-CEP [10].

Does CEP-Adduction Affect Protein Structure?

When we analyzed HSA-CEP, HSA-CTL2, and HSA-CTL1 in structural studies we observed significant structural alterations in HSA-CEP in comparison with HSA-CTL1. However, HSA-CTL2 also incurred significant alterations, similar to but less extensive than HSA-CEP. On SDS-PAGE gels HSA-CEP and HSA-CTL2 but not HSA-CTL1 appeared to form oligomers in a ladder-like fashion (**Figure 1A**). Comparison of size exclusion chromatography (SEC) profiles indicated progressive loss of the HSA monomer in the order of HSA-CTL1 > HSA-CTL2 > HSA-CEP (**Figure 1B**). Circular dichroism (CD) also indicated a significant loss in secondary structure in HSA-CTL2 and HSA-CEP compared to HSA-CTL1 (**Figure 1C**).

Collectively these data show that our synthetic CEP adducts can incur two kinds of structural changes: a) CEP-independent changes and b) CEP-dependent changes. Comparison of our HSA-CTL1 and HSA-CTL2 illustrates the CEP-independent changes in protein structure, which indicate artificial changes introduced by the procedure for generating synthetic adducts. Comparison of our HSA-CTL2 and HSA-CEP illustrates the CEP-dependent changes that occur as a result covalent CEP groups in HSA-CEP, beyond the CEP-independent changes in HSA-CTL2.

Is TLR2 Activated By CEP Adducts?

We tested several CEP adducts in a cell-based TLR2 assay, essentially as described [9]. Pam3CSK4, a known TLR2 agonist, showed dose-dependent activation of TLR2 as monitored by production of NF-κB or IL-8 in cellular assays (**Figure 2**, top and middle panels). However, no TLR2 activation was detected by HSA-CEP (**Figure 2**), or several other CEP adducts tested: MSA-CEP, dipeptide-CEP, or PE-CEP (not shown). None of these adducts showed any cytotoxicity, as measured with CellTiter-Glo (CTG) kit (**Figure 2**). Next we used THP-1 cells, which naturally express several TLRs, including TLR2. While positive controls showed specific activation of the corresponding TLR, CEP adducts failed to activate TLR2 or any other TLR that was monitored (**Figure 2,** bottom panel). *In vivo*, CEP adducts did not induce biological effects that are mediated by TLR2. As **Figure 3A** shows, treatment of mice with Pam3CSK4 induced infiltration of neutrophils and macrophages in the retina. However, neither dipeptide-CEP nor MSA-CEP (not shown) induced retinal infiltration in the same assay. Representative images of this experiment are shown in **Figure 3B**. These results suggest that CEP adducts do not activate TLRs, including TLR2.

Are CEP Adducts or TLR2 Involved in Angiogenesis?

We used the tube formation *in vitro* assay to determine if CEP adducts are angiogenic, similar to a previously reported assay [9]. VEGF induced significant tube formation; however, neither CEP adducts nor Pam3CSK4 affected tube formation (**Table 1**). In addition, poly (I:C) (a TLR3 agonist) and LPS (a TLR4 agonist) failed to show any effect. Representative images of the tube formation assay are shown in **Figure 4**.

The CEP adducts were further evaluated in the laser-induced choroidal neovascularization (CNV) model, as was reported earlier [8]. Initial laser CNV studies were performed with C57BL/6N mice, which showed no effect of MSA-CEP (**File S1**). In light of

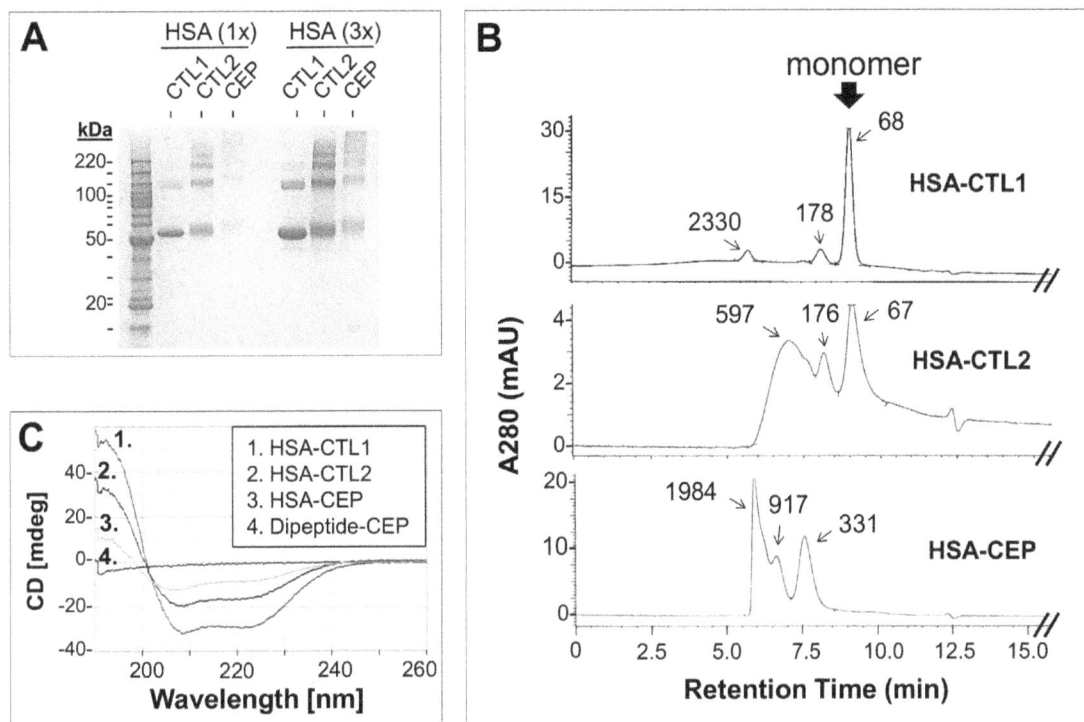

Figure 1. Structural Analyses of CEP Adducts. A) *SDS-PAGE analysis.* Aliquots of HSA-CTL1 (untreated), HSA-CTL2 (treated but unadducted) and HSA-CEP were subjected to reducing SDS-PAGE on 4–10% gels. Compared to HSA-CTL1, both HSA-CTL2 and HSA-CEP showed an increase in high-MW bands. B) *Size exclusion chromatography.* SEC under non-denaturing conditions indicated an increase in faster-eluting peaks in HSA-CTL2 and HSA-CEP compared to HSA-CTL1. C) *Circular dichroism.* CD analysis revealed a loss of secondary structure in HSA-CTL2 and HSA-CEP compared to HSA-CTL1.

the *rd8* mutation in the *Crb1* gene reported in this strain [12], we repeated the study with C57BL/6J mice which are wildtype for *Crb1* [12] and were used in the previously reported study [8]. We observed the same results with 2 experiments in each strain. Subretinal injection of VEGF significantly exacerbated CNV, while a VEGF-neutralizing antibody inhibited CNV (**Figure 5A and File S1**). This is consistent with VEGF being a major pro-angiogenic factor in CNV. However, subretinally administered MSA-CEP, at a dose nearly identical to that used in the previous report [8], had no effect in this model (**Figure 5A and File S1**). Representative images of the experiments in Figure 5A are shown in **Figure 5B**.

In studies with a corneal neovascularization (CoNV) mouse model, we observed that *Myd88*-deficiency had no significant effect on CoNV (**Figure 6**) when compared with similarly treated wild-type mice. Since *Myd88* deficiency abolishes TLR2 activity, this indicates that TLR2 is not required for angiogenesis in the abrasion-induced CoNV model. In contrast, CoNV is greatly dependent on VEGF-A: a) qPCR analysis showed a 30-fold increase in *VEGF-A* mRNA expression and b) treatment of abraded mice with VEGF-A neutralizing antibody showed significant reduction in CoNV area (**Figure S3**).

Discussion

Summary

Here we performed validation studies for the proposed CEP-TLR2 axis to assess the therapeutic potential for wet AMD treatment. Following a published procedure we generated synthetic CEP adducts and confirmed the presence of covalent

CEP groups. Structural analyses of a CEP adduct indicated changes in tertiary structure that were not observed in the naïve protein; however, similar structural changes were observed in the treated, unadducted control. Thus the physiological relevance of the observed structural changes is uncertain. Next we attempted to reproduce some of the reported biological effects of synthetic CEP adducts. When we tested our synthetic CEP adducts in *in vitro* and *in vivo* assays, we observed neither TLR2 activation nor pro-angiogenic activity. We conclude that our data do not support the CEP-TLR2 hypothesis.

Structural Changes in CEP Adducts

In our hands the published protocol for generating CEP adducts worked successfully, by the criterion of the presence of covalently-linked CEP groups in the adduct. This protocol worked robustly with all classes of reagents tested, including proteins, dipeptide, and lipid. Furthermore, the stoichiometry of adduction of each reagent was in fair agreement with the corresponding published stoichiometry.

In our attempt to understand the biological consequences of CEP adduction we initiated protein structural studies with HSA-CEP. Since the conversion of a significant number of positively-charged lysine side chains to negatively charged CEP groups (from the carboxylate group) would greatly alter its surface electrostatic potential, we anticipated and indeed observed structural changes in HSA-CEP in comparison to untreated HSA control (HSA-CTL1). We were surprised, however, to detect similar changes in HSA-CTL2, the treated unadducted control (**Figure 1**). Our interpretation is that two kinds of structural alterations can occur in synthetic CEP adducts: a) alterations that do not involve the

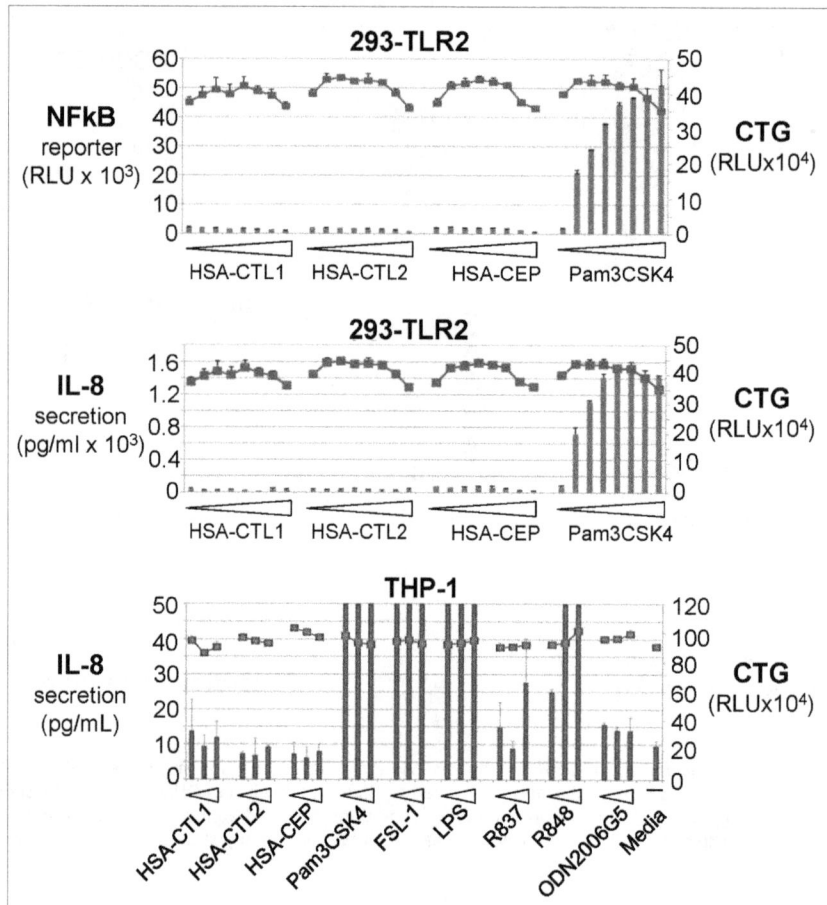

Figure 2. Cell-Based TLR Activation Assays. Various CEP adducts and TLR agonists were tested in HEK293 or THP-1 cells. Readouts were NFkB reporter signal or IL-8 secretion (columns), as indicated. In addition, the same wells were analyzed for viability with the CellTiter-Glo kit (axis on right; square symbols) to ensure that any lack of activation was not due to cell toxicity. HEK293 cells were treated with the following reagents: HSA-CTL1, HSA-CTL2, or HSA-CEP: 0, 3.9, 7.8, 15.6, 32.6, 62.5, and 125 and 250 μg/ml; Pam3CSK4: 0, 1.5, 3.2, 6.3, 12.5, 25, 50, 100 ng/mL. THP-1 cells were treated with the following reagents: HSA-CTL1, HSA-CTL2, or HSA-CEP: 62.5, 125, and 250 μg/mL; Pam3CSK4: 4, 20, and 100 ng/mL; FSL-1:0.4, 2, and 10 ng/mL; LPS: 4, 20, and 100 ng/mL; R837 or R848:0.4, 2, and 10 μM; ODN2006G5:0.2, 1, and 5 μM.

CEP group and were observed when comparing HSA-CTL2 to HSA-CTL1; and b) alterations that occur as a result of covalently-linked CEP groups and were observed when comparing HSA-CEP to HSA-CTL2. It is not clear which step(s) or reagents in the published adduction procedure led to the CEP-independent changes in HSA-CTL2. A candidate culprit is the organic solvent, dimethylformamide; organic solvents are known to affect the structure of some proteins. The adduction procedure entails exposure of protein to 30% dimethylformamide/PBS solution for 4 days at 37°C [10]. In searching the literature we found similar significant structural alteration in a CEP adduct published by another laboratory. A synthetic MSA-CEP adduct appeared to migrate as a continuous smear on denaturing SDS-PAGE and immunoblot, whereas the untreated MSA (the equivalent of MSA-CTL1) migrated as one predominant electrophoretic band (Figure 1 in [13]). A treated unadducted control, the equivalent of our MSA-CTL2, was not included in the report [13]. However since we used similar procedures for generating CEP adducts, in all likelihood the published MSA-CEP incorporated both CEP-independent and -dependent changes.

Thus far, no endogenous CEP adducts have been isolated directly from any biological sources and none have been characterized in the literature. For example, the stoichiometry of

CEP adduction (moles CEP per mole protein) and structural properties have not been reported for any endogenous CEP adducts. Hence at this point it would not be possible to verify that any synthetic CEP adduct is representative of endogenous ones with respect to protein structure. This caveat notwithstanding, we proceeded with biological studies to see if we could reproduce the biological effects of CEP adducts with regards to TLR2 activation and angiogenesis. In our approach we included HSA-CTL2 and MSA-CTL2 in the biological assays of the corresponding CEP adducts, so we might discern biological effects that are specific to the CEP group.

TLR2 Activation by CEP Adducts

We tested CEP-adducted protein or dipeptide in two cell-based assays: a) HEK293-TLR2 cells that specifically expressed TLR2 and, b) THP-1 cells that express multiple TLRs, including TLR2. In both assays, TLR2 activation was observed with a synthetic TLR2 agonist, Pam3CSK4, but not with synthetic CEP adducts. Specifically, we did not detect any effect of HSA-CEP in TLR2-expressing HEK293 cells, as was reported (Figure S14 in [9]). Not surprisingly the controls for our CEP adducts did not have any effects, either. Our cellular assays also did not register any effect of

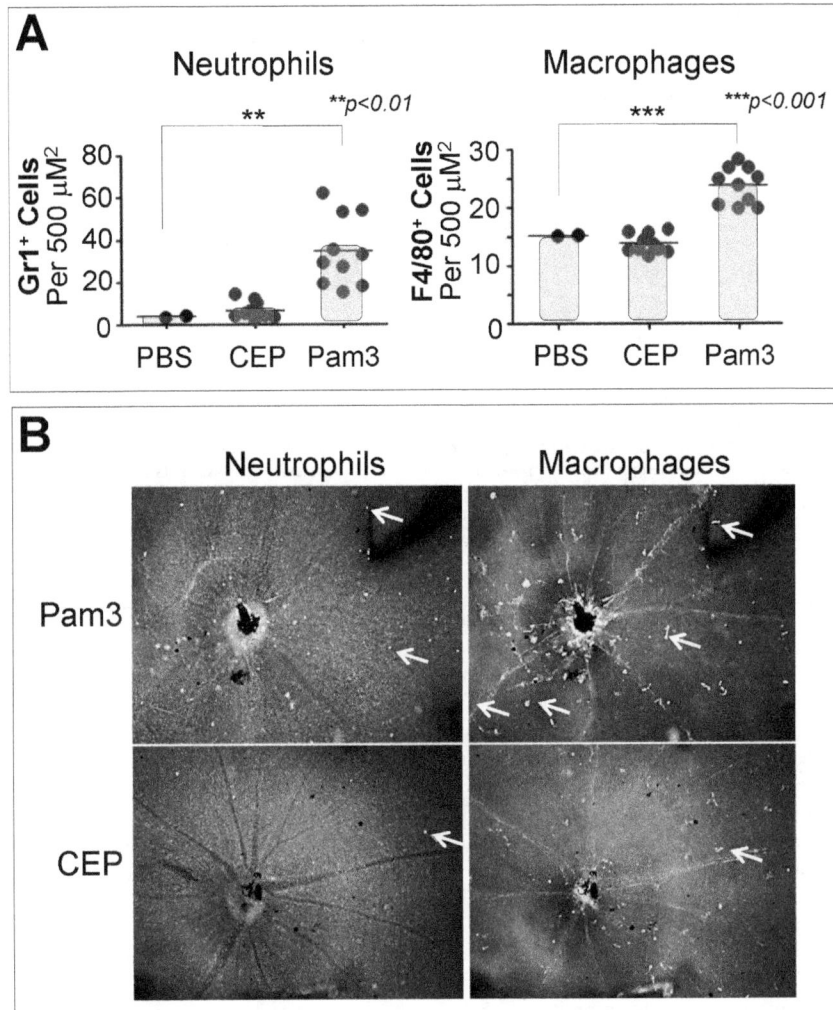

Figure 3. Retinal Leukocyte Infiltration Assay. A) Mice were injected intraperitoneally with either PBS, Pam3CSK4 (25 μg per animal, in PBS), or dipeptide-CEP (400 μg per animal, in PBS) and the retinas were analyzed 8 hours later. Retinal infiltration by neutrophils (Gr1+ cells) or macrophages (F4/80+ cells) was assessed by immunostaining with the respective markers and quantitated with Axiovision, as described in Materials and Methods. Statistical analysis was performed using the Student t-test. Only statistically significant differences are indicated in the graph. B) Shown are representative images of the experiment in **Figure 3A**. Arrows indicate examples of macrophages or neutrophils in the corresponding images. *CEP*, Dipeptide-CEP; *Pam3*, Pam3CSK4.

the dipeptide-CEP, which was reported to be pro-angiogenic in several cellular and *in vivo* assays in a TLR2-dependent manner [9]. However, this dipeptide-CEP was not tested in the same cell-based TLR assay used for HSA-CEP [9], so a direct comparison with our cell-based data is not possible.

We also evaluated CEP adducts in an *in vivo* model for TLR2 activation. In this model, treatment with Pam3CSK4 induced retinal leukocyte infiltration in wild-type mice, but not in *Myd88−/−* nor *TLR2−/−* mice (not shown). However, treatment of wild-type mice with CEP adducts did not result in measurable retinal leukocyte infiltration, indicating that TLR2 was not activated by CEP adducts *in vivo*.

CEP-TLR2 in Angiogenesis Assays

In vitro, neither HSA-CEP nor Pam3CSK4 (TLR2 agonist) showed any pro-angiogenic effect in the tube formation assay with human umbilical vein endothelial cells (HUVECs). Likewise, agonists to other TLRs (LPS, poly (I:C)) were not pro-angiogenic, whereas VEGF was.

In vivo, synthetic MSA-CEP did not exacerbate laser-induced CNV in a mouse model as reported [8]. This was the case with two substrains of C57BL/6 mice. In the initial two laser CNV studies we used C57BL/6N mice. Subsequently, it was reported that this substrain carries the *rd8* mutation in the *Crb1* gene [12]. We then performed two additional laser CNV studies with the C57BL/6J substrain, which has the wild-type *Crb1* gene [12]. The C57BL/6J substrain is the same one used in the previously published laser CNV study [8]. We observed similar results in all four laser CNV studies: CNV was exacerbated with exogenous VEGF, ameliorated with a VEGF-neutralizing antibody, and unaffected with MSA-CEP. Collectively, these data are consistent with a major angiogenic role for the VEGF pathway, but not for CEP adducts, in the laser-induced CNV model.

As an alternative *in vivo* model of ocular angiogenesis for exploring the CEP-TLR2 hypothesis, we used an abrasion-induced corneal neovascularization model (CoNV). This CoNV model is different from that used earlier with CEP adducts: in the corneal pocket CoNV model, a pellet containing synthetic HSA-

Table 1. Evaluation of CEP Adducts and TLR Agonists in the Tube Formation Assay.

	Reagent	Concentration	Average Tube Length (mm/mm²)	Std Dev	P value[a]
Experiment 1	untreated	–	1.97	+/−0.35	–
	LPS	10 ng/mL	3.38	+/−0.59	0.0072
	Poly (I:C)	10 µg/mL	0.16	+/−0.09	0.0005
	Pam3CSK4	500 nM	2.13	+/−0.24	0.9997
	VEGF[b]	3.1 ng/mL	10.61	–	<0.0001
Experiment 2	untreated	–	1.85	+/−0.27	–
	HSA-CEP	2 µg/mL	1.18	+/−0.32	0.1079
	HSA-CTL2	2 µg/mL	2.32	+/−0.45	0.3594
	VEGF	4 ng/mL	9.00	+/−1.64	<0.0001

GFP-transfected HUVEC, co-cultured with human fibroblasts, were treated on days 1, 2, 5, 7 and 9 with VEGF, TLR agonists, HSA-CEP, or HSA-CTL2. Control wells ("untreated") received media alone. Average tube length (mm/mm²) was determined by fluorescence measurements as described in Materials and Methods. Representative images from these two experiments are shown in **Figure 3**.
[a]One way ANOVA with Dunnett's multiple comparison test, compared to untreated sample.
[b]This VEGF control was measured in duplicate and not in triplicate, therefore no Std Dev is presented.

CEP was implanted in the cornea [8]. Since our synthetic CEP adducts were biologically inactive in all assays so far, we aimed to use a model where endogenous -not synthetic- CEP adducts might play a role. In the abrasion-induced CoNV model, angiogenesis occurs as part of the wound healing process, induced by mechanical abrasion of the cornea. Furthermore, it has been shown that macrophages are recruited to the cornea during early stages of neovascularization [14]. This resembles the back punch model, in which wound angiogenesis entails recruitment of macrophages [9]. CEP adducts were reported to be transiently present during this time, detected by immunocytochemistry with an antibody against synthetic CEP adducts. By immunolabeling, a substantial portion of CEP adducts was present in the recruited F4/80+ macrophages [9]. Treatment with dipeptide-CEP in this model accelerated wound closure and vascularization in a TLR2-dependent manner, as shown by the comparison of *TLR2−/−* and *TLR2+/+* mice. In the same vein, we used the abrasion-induced CoNV model and compared littermate *Myd88−/−* and *Myd88+/+* mice. (Myd88 is required for TLR2 function.) We found no difference in CoNV area between the two groups. This indicates that, in the abrasion-induced CoNV model, TLR2 and other Myd88-dependent TLRs are not involved in angiogenesis. It is not known whether endogenous CEP adducts are present in the abrasion-induced CoNV model. We also tested topical treatment with synthetic HSA-CEP, but observed no effect on CoNV (not shown). It seems therefore the CEP-TLR2 axis proposed for other wound angiogenesis models [9] does not apply to corneal abrasion-induced wound angiogenesis model tested here.

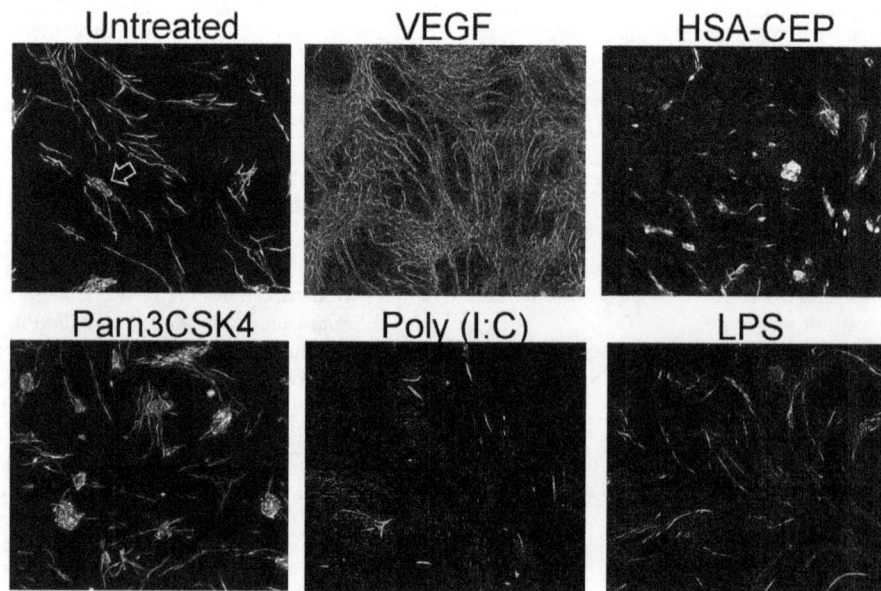

Figure 4. Tube Formation Assay. Shown are representative images of the experiments in presented numerically in **Table 1.** The figure shows images for untreated negative control (untreated), positive control (VEGF 165, 4 ng/mL), HSA-CEP (2 µg/mL), Pam3CSK4 (500 nM), LPS (10 ng/mL), poly (I:C) (10 µg/mL). The arrow in the untreated image shows an example of an island of unmigrated HUVEC cells, which is also seen in other images.

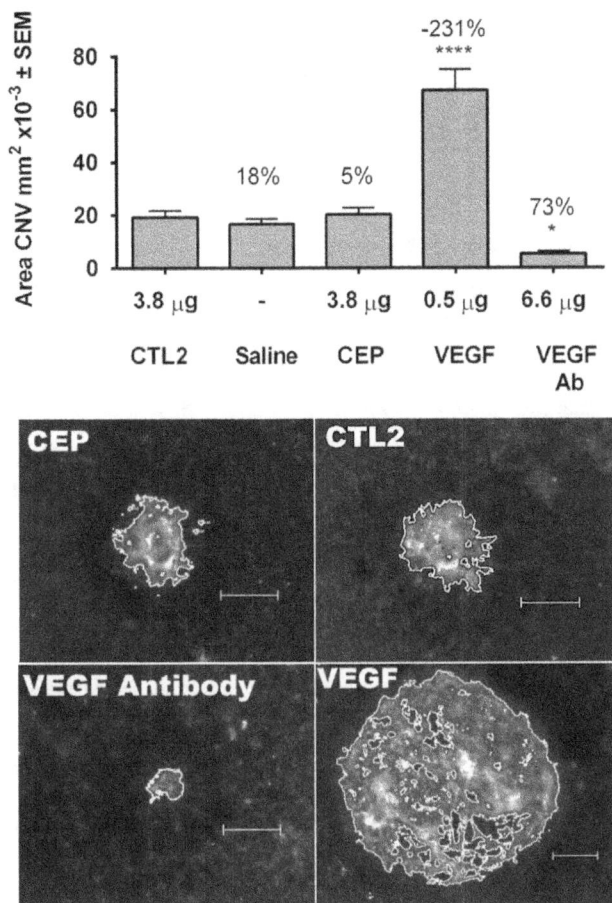

Figure 5. Mouse Laser-Induced CNV Assay. (A) Subretinal injection of MSA-CEP does not increase CNV area compared to mice injected with saline or MSA-CTL2. Bar graph shows mean area of CNV +/− SEM from first experiment evaluating the effect of subretinal injection of saline, 0.5 µg of rhVEGF165, 3.8 µg of MSA-CTL2, 3.8 µg of MSA-CEP, or 6.6 µg of 4G3 (an anti-mVEGF antibody) on laser-induced CNV in C57BL/6J mice. The number above each bar is the percentage inhibition relative to average CNV area in mice injected with MSA-CTL2. Subretinal injection of VEGF increases CNV area and subretinal injection of an anti-mVEGF antibody inhibits CNV area. * p<0.05, **** p<0.0001 by ANOVA with a Dunnett's post hoc analysis. (B) Representative fluorescent images of CNV lesions 7 days after laser from mice injected in the subretinal space with MSA-CEP, MSA-CTL2, VEGF or a VEGF Antibody as described above. Scale bar = 100 microns. *CTL2*, MSA-CTL2; *CEP*, MSA-CEP.

Figure 6. Mouse CoNV Assay. A) Adult *Myd88−/−* (KO) and littermate *Myd88+/+* (WT) mice (N = 7 to 8 animals/group) were subjected to corneal abrasion on Day 0. On Day 21 post-abrasion animals were euthanized and CoNV area (+/− SEM) was measured by fluorescence microscopy, as described in Materials and Methods. Statistical analysis was performed using two-way ANOVA. Within each genotype, the abraded group was significantly different from the naïve group (p<0.0001). However, comparison between the two genotypes showed no significant effect of the *Myd88* deficiency on CoNV area in response to abrasion. B) Representative images of the 4 groups shown in **Figure 6A**.

Synthetic vs. Endogenous CEP Adducts

It has been proposed that the biological effects of CEP adducts depend solely on the presence of CEP groups and not the host carrier [9]. This was not the case in our study: neither protein CEP adducts nor dipeptide CEP adducts produced any of the published biological effects [8,9] that were tested here. Our synthetic reagents were verified for the presence of covalently-attached CEP groups. There is no obvious explanation for these discrepancies. As innate immune receptors, TLRs recognize structures and patterns so it is conceivable that changes in structure, even if artificial, could elicit TLR activation. Thus one possibility is that there are structural differences between our synthetic CEP adducts and those used in previous studies, as the reagents were prepared at different laboratories.

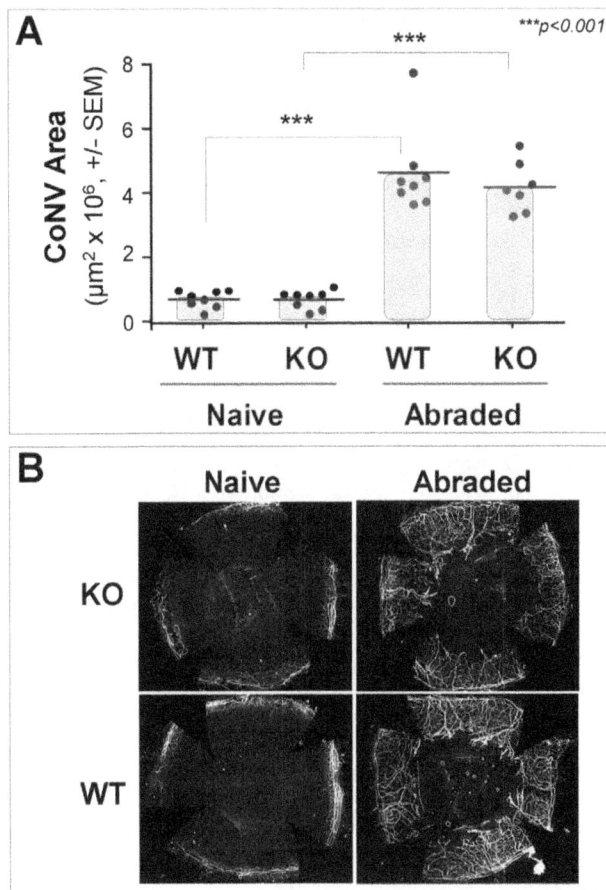

So far endogenous CEP adducts from any biological systems have not been isolated and therefore none have been characterized in structural studies. Furthermore, the abnormal electrophoretic patterns seen with synthetic CEP adducts reported here or by others [13] do not seem to resemble those of *in vivo* CEP adducts detected on immunoblots, reported either in human donor material (e.g. Figure 3 in [5]) or in the light-induced rat retinal degeneration model (e.g. Figure 1 in [15]). It is therefore uncertain which synthetic CEP adducts are representative of endogenous ones; perhaps none. The final answer awaits the isolation of endogenous CEP adducts and their characterization.

By extension, the biological effects reported with synthetic CEP adducts also need to be confirmed with endogenous CEP adducts. Our HSA-CTL2 did not show any biological activity, but neither did our HSA-CEP; hence in our study there was no concern about non-physiological biological effects, which HSA-CTL2 was intended for as a control. However, a treated unadducted control would be critically important when biological activity with a synthetic CEP adduct is observed. This is exemplified in a recent publication where the unadducted control, "sham-MSA", showed biological activity in some assays. In BALB/c macrophages, sham-MSA induced the upregulation of M1 markers and of an inflammatory gene, *KC*, 2 to 5 fold above that of control levels (Figures 1A and S2, respectively, in [16]). In some cases the sham-MSA effect represented 30%–40% in magnitude of the MSA-CEP effect: (*IL-1β*, *TNFα*, and *KC*), despite the absence of CEP groups in sham-MSA [16].

Treated unadducted controls were not used in earlier studies that reported a role for synthetic CEP adducts in TLR2 activation and in angiogenesis [8,9]. If it is possible that our CEP adducts are different from those used by others, it is also possible that our treated unadducted controls are different. The fact that our treated, unadducted controls (HSA-CTL2, MSA-CTL2) were biologically inactive does not necessarily apply to other studies in literature, as the example above illustrates. Thus the physiological relevance of synthetic CEP adducts is unclear, especially when untreated proteins were used as the only controls in the biological assays.

Polyclonal and monoclonal antibodies raised to synthetic CEP adducts have been reported to immunolabel biological samples from AMD patients in Western blots, immunohistochemical sections, and ELISAs [5–7,11]. The same antibodies reportedly immunolabelled biological samples in animal studies, e.g. [15,17]. However, the immunolabelled proteins were not confirmed to have any CEP moieties, by other independent assays that do not use antibodies (e.g. LC-MS/MS); i.e. the immunolabelled proteins were not confirmed to be bona fide CEP adducts. For example, several candidate CEP adducts were immunolabelled on Western blots of patient donor material and subsequently identified by LC-MS/MS analysis, however the presence of covalently-linked CEP groups was not confirmed by LC-MS/MS or other assays [5]. The fact that a synthetic CEP adduct (used a control) was immuno-labelled on the same western blot is not a surprise, as CEP antibodies were raised against a synthetic CEP adduct. At this point, the antibodies against synthetic CEP adducts [5] have not yet been validated for detection of endogenous CEP adducts. This is underscored by the electrophoretic changes in synthetic CEP adducts that do not seem to resemble those of endogenous CEP adducts, as explained above. Data generated by other (non-immunological) assays is needed to validate CEP antibodies.

Arguably the *in vivo* existence of CEP adducts requires confirmation, as well, since all evidence so far has been generated with these antibodies. For example, a proteomics study of AMD patient samples identified and quantified hundreds of proteins by LC-MS/MS [18]. Yet the same study reported that CEP adducts were below detection limits and "none were reliably identified" (supplemental information in [18]). On a promising note, an improved LC-MS/MS assay has been reported, with a sensitivity of 1 pmol or less of CEP-lysine in enzymatically-digested patient plasma samples [19]. According to ELISA data [7] the average levels of CEP adducts in AMD plasma is 37 pmol/mL, thus hopefully the improved LC-MS/MS assay can confirm the presence of CEP adducts *in vivo*.

In conclusion, our studies of synthetic CEP adducts did not validate the CEP-TLR2 axis in angiogenesis as proposed [8,9]. While the cause of the discrepancies is not clear, it does seem clear that the mere presence of CEP groups is not sufficient to elicit the reported biological responses. More data, ideally with endogenous CEP adducts, is needed to understand what properties of CEP adducts, if any, can lead to TLR2 activation and to angiogenesis. In light of the prevalence of AMD and the unmet medical needs, more research into the pathophysiology of CEP adducts is warranted.

As a postscript, after submission of our manuscript an independent report [20] also showed that CEP adducts alone do not induce TLR2 signalling nor related biological effects (e.g. Figures 1A and 1B), as we have reported here. Rather, the report claims that CEP adducts potentiate the effect of a synthetic TLR2 agonist, Pam3CSK4, in cultured murine bone-marrow derived macrophages [20].

Materials and Methods

Synthesis and Verification of CEP Adducts

Synthesis. All CEP adducts were synthesized as described [10]. The dipeptide, Ac-Gly-Lys-OMe, was obtained from BACHEM; HSA from AlbuminBio; MSA from AlbuminBio or Sigma; phosphatidyl ethanolamine was from Sigma. Controls for protein CEP adducts included untreated protein (CTL1) and treated unadducted protein (CTL2). The latter control was processed in the same synthesis procedure as that for CEP adducts, except 4,7-dioxoheptanoic acid 9-fluorenylmethyl ester was left out to avoid the covalent addition of CEP moiety. Protein CEP adducts, after final dialysis in PBS, were quantified by the Bradford assay and tested for endotoxins with the Endosafe-PTS kit (Charles River). For storage, samples were filtered through 0.2 μm, divided aseptically in 1-mL aliquots, and stored at −80 C.

Amino acids complete digestion. Enzymatic hydrolysis was adapted from [21]. An aliquot of 100 μg of protein is dissolved in 25 μl of PBS buffer (pH 7.4). Pronase E (Sigma, Cat. # P5147) (2 mg/ml in 10 mM potassium phosphate buffer, pH 7.4, 5 μl) was added. The sample was incubated at 37°C for 24 hours. Prolidase (Sigma, Cat. # P6675) and aminopeptidase (Sigma, Cat. # A8200) (both 2 mg/ml in 10 mM potassium phosphate buffer, pH 7.4, 5 μl) were added. The sample was incubated at 37°C for 48 hours. Amino acids from enzymatic hydrolysate (10 μl) was derivatized by Waters AccQ•Fluor (Waters, Cat.# WAT052880). Then 2 μl of derivatized samples was analyzed by Xevo-G2QTOF with Waters BEH C18 2.1 × 50 mm 1.7 μm column at 50°C at 1.0 mL/min, 0.1% formic acid in water, 0.04% formic acid in acetonitrile, 3–98% B in 9 min. An aliquot of 100 μg of protein is dissolved in 25 μl of PBS buffer (pH 7.4). Pronase E (Sigma, Cat. # P5147) (2 mg/ml in 10 mM potassium phosphate buffer, pH 7.4, 5 μl) was added. The sample was incubated at 37°C for 24 hours. Aminopeptidase (Sigma, Cat. # P6675) and prolidase (Sigma, Cat. # A9934) (both 2 mg/ml in 10 mM potassium phosphate buffer, pH 7.4, 5 μl) were added. The sample was incubated at 37°C for 48 hours. Amino acids from enzymatic hydrolysate (10 μl) was derivatized by Wasters AccQ Fluor (Waters, Cat.# Wat052880). Then 2 μl of derivatized samples was analyzed by Xevo-G2QTOF with Waters BEH C18 2.1 × 50 mm 1.7 μm column at 50°C at 1.0 mL/min, 0.1% formic acid in water, 0.04% formic acid in acetonitrile, 3–98% B in 9 min.

NMR of protein hydrolysate samples. Hydrolyzed protein samples were prepared for NMR analysis by the addition of 5 μL of D2O (CIL) to 15 μL of Hydrolysate solutions. Sodium 3-trimethylsilyl [2,2,3,3-d4]propionate (TMSP) as added as an internal chemical shift and quantitation reference. High-resolution ^{1}H-NMR spectra were acquired at 300±1 K, using a standard (D-90°-acquire) pulse sequences on a Bruker-600 Avance spectrometer (^{1}H frequency of 600.26 MHz). ^{1}H-NMR spectra were acquired with 256 free induction decays, 65,536 complex data points, a spectral width of 7.2 kHz, and a relaxation delay of 5 s. All spectra were processed by multiplying the FID by an exponential weighting function corresponding to a line broadening of 0.3 Hz. The CEP pyrrole resonances at ^{1}H$_δ$ 6.8 ppm, ^{1}H$_δ$

6.1 ppm and $^1H_\delta$ 5.9 ppm were integrated relative to the aromatic resonance of phenylalanine and tyrosine using the ACD 10.0 package (Advanced Chemistry Development, Toronto, Canada).

LC-MS/MS confirmation of CEP-lysine adduct. Carboxyethylpyrrole (CEP) adduct presence in CEP-conjugated murine serum albumin (MSA) and human serum albumin (HSA) has been confirmed by LC-MS/MS methodologies. CEP-MSA and CEP-HSA were hydrolyzed using protease cocktails (descriptions in above, AA complete digestion), followed by Accq-Tag Ultra derivatization (Catalog number 186003836, Waters Corporation, Milford, MA). Accq-Tagged CEP-lysine ionizes in electrospray positive mode and gives a protonated molecular ion of 439.1981, which can fragment and gives a characteristic 171.1 daughter ion from the Accq-Tag and a daughter ion of 206.1 from the carboxyethylpyrrole moiety. We employed multiple reaction monitoring, 206.1 precursor scan on a Triple Quadrupole mass spectrometer. Strong signal of 439.2 → 171.1 and 439.2 → 206.1 were observed using AB Sciex API4000 Triple Quad. Precursor ion of 439.2 was observed for 206.1 daughter ion in a precursor ion scan on the same API4000. Waters Xevo G2 Q-TOF mass spectrometer was employed for its high resolution power to further confirm the presence of CEP-lysine adducts. The molecular species of 439.1981 was observed in MS scan with 5 ppm mass accuracy across the chromatographic peak; the daughter ion 206.1181 was observed in the MS/MS scan with 10 ppm mass accuracy in the MSE approach.

Peptide Mapping for CEP Modification Location

Sample preparation. All solvents (HPLC grade) and chemicals were purchased from Sigma-Aldrich (St. Louis, MO) unless otherwise stated. MSA-CEP and MSA-CTL2 (50 ug each) were denatured with 6 M guanidine hydrochloride (GuHCl), reduced with 25 mM dithiothreitol (DTT), alkylated with 50 mM iodoacetamide, dialyzed against 50 mM ammonium bicarbonate using 10 kDa MWCO Slide-A-Lyzer cassettes (Thermo Scientific, Rockford, IL). Protein was digested 1 to 50 enzyme to protein with trypsin, chymotrypsin, and trypsin/Glu-C overnight at 37°C; note all enzymes purchased from Roche Diagnostics GMBH, Germany. For HSA-CEP and HSA-CTL1 digestions, 125 μg protein was denatured using ProteaseMAX surfactant (Promega, Madison, WI), reduced with 5 mM DTT, alkylated with 15 mM iodoacetamide and digested with 1 to 50 trypsin to protein overnight at 37°C.

Reverse phase LC-MS/MS analysis. Resulting peptides from MSA-CEP and MSA-CTL2 were analyzed by LC-ESI MS/MS on a Thermo Velos Orbitrap coupled to a Waters nanoACQUITY UPLC (Milford, MA). 70 pmol of digested MSA-CEP and MSA-CTL2 were HPLC separated on column (Waters Acquity HSS T3 1.8 μm beads, 1 × 100 mm at 40°C) at 15 μL/min. The 55 min gradient started 0–3 min, 3% B (B = acetonitrile, 0.1% formic acid), increased to 97% B at 35 min, then 95% B at 37 min, followed by washing and column equilibration. Mass spectrometer parameters included a full scan event using the FTMS analyzer at 100000 resolution from m/z 300–2000 for 10 ms. Collision induced dissociation (CID) MS/MS was conducted on the top seven intense ions (excluding 1+ ions) in the ion trap analyzer, activated at 500 (for first event) and 2000 (for remaining events) signal intensity for 10 ms.

For HSA-CEP and HSA-CTL1 digestion, resulting peptides were analyzed by LC-ESI MS/MS on a Thermo LTQ Orbitrap Discovery coupled to Agilent CapLC (Santa Clara, CA). 10 pmol of digested HSA-CEP and HSA-CTL1 were HPLC separated on column (Waters Acuity BEH C18, 1.7 μm, 1 × 100 mm column at 40°C) at 10 μL/min. The 80 min gradient started 0–1 min, 4%

B, increased to 7% B at 1.1 min, 45% B at 55 min, then 95% B at 63 min, followed by washing and column equilibration. Mass spectrometer parameters included a full scan event using the FTMS analyzer at 30000 resolution from m/z 300–2000 for 30 ms. CID MS/MS was conducted on the top seven intense ions (excluding 1+ ions) in the ion trap analyzer, activated at 500 (for all events) signal intensity for 30 ms.

Data analysis and database searching. All mass spectra were processed in Qual Browser V 2.0.7 (Thermo Scientific). Mascot generic files (mgf) were generated with MS DeconTools (R.D. Smith Lab, PPNL) and searched using Mascot V2.3.01 (Matrix Science Inc., Boston, MA) database search against the SwissProt database, V57, with 513,877 sequences. Search parameters included: enzyme: semitrypsin, chymotrypsin, none, or trypsin/Glu-C, allowed up to two missed cleavages; fixed modification carbamidomethyl on cysteines (when applicable); variable modifications searched: Arg-CEP on arginines, Gln-> pyro-Glu (at N-term glutamine), Lys-CEP on lysines, oxidation on methionine; peptide tolerance: ±25 ppm; MS/MS tolerance: ±0.6 Da. Sequence coverage and CEP modification assessments were evaluated on peptide scores with >95% confidence. High-scoring peptide ions were then selected for manual MS/MS analysis using Qual Browser.

Structural Analyses of CEP Adducts

SDS-PAGE analysis. After boiled for 5 min, 5 μg of total protein mixed with 4X sample buffer (Invitrogen, cat. # NP0007) was loaded on 4–12% NuPAGE Bis-Tris gel (Invitrogen, Cat. # NP0321BOX) with NuPAGE MOPS running buffer (Invitrogen, Cat. # NP0001). SeeBlue Plus2 Protein Ladder (Invitrogen, cat. # LC5925) or BenchMark Protein Ladder (Invitrogen, Cat. # 10747-012) was used to estimate the protein size. Gel was stained with SimpleBlue SafeStain (Invitrogen, Cat. # LC6060) for overnight at 4°C and destained with HPLC water. The gel image was taken by Bio-Rad ChemiDoc XRS+ Imaging System.

Size exclusion chromatography (SEC). Human Serum Albumin (HSA) samples (20 μg) were injected on Shodex KW-803 column with 1 mL/min flow rate, 20 mM Tris, 200 mM NaCl, 0.25 mM TCEP, 3 mM NaN3, pH 7.5 as mobile phase on Agilent 1200 HPLC. UV signal was recorded at 280 nm by Agilent 1260 DAD detector. Mouse Serum Albumin (MSA) samples (50 μg) were injected on a Large S200 Column with GE Superdex 200 10/300GL and at 500 μL/min flow rate, 150 mM NaCl and 0.02% NaN3 in Dulbecco's PBS as mobile phase on Agilent 1260 BioInert HPLC. UV signal was measured at 280 nm by Wyatt TREOS/OptiLab Rex.

Circular dichroism (CD). Protein samples were diluted in 10X diluted PBS (pH 7.4) to achieve similar concentration. Baseline was blanked by 10X diluted PBS (pH 7.4). The CD spectra (average of five scans) of protein samples were collected from 260 nm to 190 nm on a Jasco J-815 CD Spectrometer with 0.02-cm path length quartz cell at 10°C.

In Vitro Assays

Cell-based TLR assays. HEK293 cells expressing TLR2 and NFKB luciferase reporter (gift from Novartis Vaccine, Siena, Italy) were seeded at 30000 per well the night before. HSA-CEP was added and incubated for either 6 hr or 24 hr. Supernatant was collected for IL-8 ELISA (R&D, cat# DY208). NFkB luciferase activity was assayed on remaining cells, using Bright-Glo (Promega, cat# E2610). HEK293 cells in a separate plate with the same treatment were used for cell viability measurement using CellTiter-Glo (Promega, cat# G7570), according to manufacturer's instruction.

Thp1 (ATCC, cat# TIB-202) was primed with 0.5% DMSO for overnight, at 100,000/well, then incubate with HSA-CEP for 24 hr, with TLR ligands (Pam3CSK4, FSL1, R837 and R848 were all from Invivogen; LPS was purchased from Sigma) as controls. Supernatant was collected for IL-8 ELISA (R&D), and the remaining cells were used for cell viability measurement using CellTiter-Glo.

In vitro tube formation assay. The CellPlayer GFP Angiokit-96 by Essen BioScience (Ann Arbor, MI) was used to measure tube formation in vitro. Briefly, GFP-transfected HU-VEC were co-cultured with human fibroblasts in a specially designed medium for 11 days in a 96-well format. Cells were treated on days 1, 2, 5, 7 and 9 with VEGF, TLR agonists, or CEP-adducted or control-treated proteins. Fluorescence measurements (IncuCyte, Essen BioScience) were taken kinetically every 12 hours for the duration of the experiment and average tube length (mm/mm^2) was quantified on the last day of the experiment according to the manufacturer's instructions. Control wells received media alone. Reagents were obtained from the following sources: Human VEGF 165– Peprotech; Pam3CSK4– InvivoGen; LPS -Sigma-Aldrich; Poly (I:C) –InVivoGen.

In Vivo Assays

Animals. All animal experiments were approved by the Animal Care and Use Committee at the Novartis Institutes for Biomedical Research. Upon arrival at the vivarium, mice were acclimated for at least 4 days before any studies were initiated. The animals were fed standard laboratory chow and sterile water ad libitum. Genotyping was performed on genomic DNA obtained from tail snips by standard procedures. All mouse strains were genotyped for the *Crb1* gene, to determine if they carried the *rd8* mutation.

C57BL/6N mice were obtained from Taconic; the *rd8* mutation was present in these mice. C57BL/6J mice were obtained from Jackson; the *rd8* mutation was absent in these mice. *Myd88*-deficient mice lacking exons 2–5 were generated at Novartis Institutes for Biomedical Research. Mice were back-crossed to C57BL/6J mice for at least 10 generations; the *rd8* mutation was absent in these mice. Heterozygous breeding generated littermate pups of each genotype, identified by PCR genotyping. *Myd88* deficiency was also functionally confirmed by the *in vivo* retinal infiltration assay below: mutant vs. littermate *wt* mice with treated with either TLR2 agonists and with TLR4 agonists and the retinal infiltration was measured as described (not shown).

Laser-induced choroidal neovascularization (CNV). CNV was induced by laser injury in age and sex matched on a) C57BL/6N mice and b) C57BL/6J mice. Two in vivo experiments were performed with each mouse strain. After pupil dilation with 1% cylate and 10% phenylephrine, the mice were anesthetized and the retinas were visualized with a slit lamp microscope and a cover slip. The laser (Iridex Oculight GLx 532 nm green laser) was applied at 3 locations with a successful laser shot inducing a vaporization bubble. Laser pulses are applied to both eye yielding 6 CNV area data points per mouse and with 10 mice per group yielding 60 CNV area data points per test condition. Immediately after laser 2.0 µl of test article was injected into the subretinal space of both eyes. A sclerotomy was first made with a 30 gauge needle, and then the test article was injected through the same incision with a 33 gauge blunt tipped needle and a 10 µl Hamilton syringe. Injections were visualized under a surgical microscope with direct observation of a small retinal detachment. 7 days post laser, mice were injected i.v. with a vascular label and then euthanized. Mouse eyes were fixed in 4%

paraformaldehyde; RPE-choroid-scleral complexes were isolated and mounted on microscope slides. Fluorescent images of each laser-induced CNV were captured using a Axiocam MR3 camera on a Axio.Image M1 microscope (Zeiss). The CNV lesion sizes were quantified with Axiovision software (Version 4.5 Zeiss). Inter-group differences were analyzed with an ANOVA with a Dunnett's multiple comparison test on GraphPad Prism 6 for Windows software. Data was masked during image acquisition and data analysis.

Recombinant human VEGF165 (Peprotech), IgG2A (R&D, MAB006) and a proprietary anti-mouse VEGF antibody (4G3) were reconstituted in sterile saline (Hospira) to a concentration of 0.05, 0.5, 2.5 or 3.3 mg/ml respectively. 1.4 or 1.9 mg/ml of CEP-MSA and MSA-CTL2 (control 2, mouse serum albumin treated but not adducted) or the other reagents were injected in to the subretinal space on day 0 immediately after a laser as described. After the application of laser burns and subretinal injections of test reagents, antibiotic ointment (Tobramycin or Neomycin ophthalmic ointment depending on availability) was applied to both eyes. The anti-VEGF antibody, 4G3, is a mouse anti-VEGF IgG1 antibody. It binds to mouse VEGF with an EC50 of 0.047 nM in a sandwich ELISA and neutralizes mouse VEGF binding to human VEGFR-2 with an EC50 of 0.15 nM in a binding assay (ELISA MSD).

Corneal neovascularization (CoNV). Acute CoNV was induced in 7- to 9- week old anesthetized mice by complete removal of the corneal epithelium with mechanical abrasion, as detailed [14]. At the end of the studies, mice were humanely euthanized and the area of CoNV was quantitated as described [14]. Animals were randomized prior to treatment and analysis was performed in a masked fashion. N = 5–10 mice/group.

In studies using *Myd88−/−* mice (**Figure 6**), male knockout (KO) and male wild-type littermate controls (WT) were abraded on day 0 and euthanized at the end of the study on day 21 for analysis. *Myd88*-deficient and littermate wild-type mice are on the C57BL6/J background and are described above.

For other CoNV studies (**Figure S3**), C57BL6/N mice were used. The *Crb1* gene product is expressed in the retina, but not in the cornea [22]. In the study presented in **Figure S3C**, animals (N = 10–12 mice/group) were injected i.p. with PBS (200 µl), control antibody (IgG1, 0.5 mg/kg) or anti-VEGF antibody (4G3, 0.5 mg/kg) on days 0, 3 and 5 post-abrasion and eyes were collected on day 6 for analysis.

In vivo TLR2-mediated retinal leukocyte infiltration. TLR2 ligand, Pam3CSK4, was purchased from Invivogen. Female C57BL/6N mice (7 weeks old, Taconic) were treated with either dipeptide-CEP (400 µg per animal, in PBS) or Pam3CSK4 (25 µg per animal, in PBS) via intraperitoneal injection. Control animals received an intraperitoneal injection of sterile PBS. Eight hours after injection, mice were euthanized. Eyes were enucleated and were fixed in 4% paraformaldehyde. For immunostaining, retinas were dissected out. Macrophages was stained using the F4/80-Alexa 488 conjugated antibody (AbD serotec, Oxford, UK). Neutrophils were stained using a biotiny-lated-Gr-1 antibody (San Diego, CA) and an Alexa Fluor 594 conjugated streptavidin secondary antibody (Molecular Probes, Eugene, OR). After retinas were flat mounted onto glass slides, fluorescent images were taken. And F4/80 and Gr-1 positive cells on the retina were counted using Zeiss AxioVision program.

Supporting Information

Figure S1 Confirmation of CEP Adduction By ^{1}H-NMR and LC-MS/MS. A) *Structure for Dipeptide-CEP*. B) *^{1}H-NMR* *of Dipeptide-CEP*. The signature peaks for CEP, lysine, and glycine are indicated. The CEP peaks were not detected in the unadducted dipeptide (not shown). C) *LC-MS/MS of completely hydrolyzed MSA-CEP*. MSA-CEP was enzymatically hydrolyzed and processed for LC-MS/MS analysis. Only MSA-CEP showed a peak corresponding to lysine-CEP; untreated MSA-CTL1 (not shown) and treated but unadducted MSA-CTL2 (lower panel) did not have the CEP peak. D) *1H-NMR of completely hydrolyzed HSA-CEP*. The signature peaks for CEP, Tyr, and Phe are indicated. The resonances corresponding to CEP were absent in HSA-CTL1 and HSA-CTL2 (not shown).

Figure S2 Peptide Mapping of CEP Adduction by LC-MS/MS. A) *LC-MS/MS Analysis of Trypsinized HSA-CEP*. LC-MS/MS of trypsin digested HSA-CEP showed sequence coverage of 65% where bold residues represent observed peptides. In HSA-CEP 14 sites of CEP adduction were identified by this analysis (shown as underlined amino acids). B) *LC-MS/MS Analysis of Trypsinized MSA-CEP*. MSA-CEP was digested with trypsin, chymotrypsin, and trypsin-gluC yielding a sequence coverage of 92% with bold residues representing observed peptides. In MSA-CEP 40 sites of CEP adduction were identified by this analysis (shown as underlined amino acids). The initial signal and propeptides are not observed in the mature, processed protein sequence for HSA and MSA and are shown as italicized residues.

Figure S3 CoNV Model is VEGF-Driven. A) *Progression of Neovascularization*. Adult C57BL/6N mice (N = 5 animals/group) were subjected to corneal abrasion on Day 0 and dissected corneas were analyzed for neovascularization area at different timepoints after abrasion. Neovascularization area progressively increased and plateaued around 2 weeks after abrasion. Statistical analysis was performed using one-way ANOVA with Dunnet's post-test, comparing each time point to Naïve. Only the statistically significant differences between groups are indicated. B) *Upregulation of VEGFA Transcript*. Total RNA was prepared from dissected corneas from naïve mice and or cornea-abraded mice that were euthanized on Day 1 and Day 6 post-abrasion as indicated (N = 5 to 6 animals/group). First-strand cDNA was generated using the High Capacity RNA-to-cDNA Master Mix (Applied Biosystems). Pre-amplification products were generated using the Taqman PreAmp Master Mix Kit (Applied Biosystems)

and a pool of FAM-labelled Taqman assays on demand (Applied Biosystems). qPCR was performed on diluted pre-amplification products using the same Taqman assays on demand in qPCR singleplex reactions. Relative quantification (RQ) performed using $\Delta\Delta$Ct method and data presented as RQ median with error bars as RQ min and RQ max. *VEGFA*, *PECAM-1*(expressed by vascular endothelial cells), and *β-actin* mRNA expression was normalized by expression of *β-actin* gene and expressed relative to naïve animals. Statistical analysis was performed using one-way ANOVA with Dunnett's post-test. C) *VEGF Ab inhibits CoNV*. Adult C57BL/6N mice (N = 10–12 animals/group) were subjected to corneal abrasion on Day 0 and injected intraperitoneally with the reagents as indicated on Days 0, 3, and 5 post-abrasion. Reagents included PBS (vehicle), control IgG1 Ab, and anti-VEGF antibody (4G3). The antibodies were dosed at 0.5 mg/kg. On Day 6 the animals were euthanized and CoNV area was measured by fluorescence microscopy as described in Materials and Methods. Statistical analysis was performed using one-way ANOVA with Dunnett's post-test.

Table S1 Synthetic CEP Adducts Generated.

File S1 Supplemental Data and Methodology of the Mouse Laser-Induced CNV Assay.

Acknowledgments

Myd88−/− mice were generated by Mueller M, Wirsching J, Lemaistre M, Doll T, Isken A, Kinzel B (Developmental & Molecular Pathways, NIBR, Basel, Switzerland). Littermate *Myd88−/−* and *Myd88+/+* mice were bred and genotyped by Vanessa Davis and John Halupowski (Transgenic Services, NIBR, Cambridge, USA). Shawn Hanks (Ophthalmology, NIBR, Cambridge, USA) helped with some of the statistical analyses.

Author Contributions

Conceived and designed the experiments: JG EM QH KA S-ML SP BEG MM BJ SA. Performed the experiments: JG JSC JZ FC XY HY EM CR YX ADE MC NB YQ EF SS AW AJ RC SG. Analyzed the data: JG JSC JZ EM QH KA S-ML SP BEG MM BJ SA. Contributed reagents/materials/analysis tools: JG EM QH KA S-ML SP BEG SA. Wrote the paper: JG EM QH KA S-ML SP BEG MM BJ SA.

References

1. Miller JW (2013) Age-related macular degeneration revisited–piecing the puzzle: the LXIX Edward Jackson memorial lecture. Am J Ophthalmol 155: 1–35 e13.
2. Kuno N, Fujii S (2011) Dry age-related macular degeneration: recent progress of therapeutic approaches. Curr Mol Pharmacol 4: 196–232.
3. Nguyen DH, Luo J, Zhang K, Zhang M (2013) Current therapeutic approaches in neovascular age-related macular degeneration. Discov Med 15: 343–348.
4. Salomon RG, Hong L, Hollyfield JG (2011) Discovery of carboxyethylpyrroles (CEPs): critical insights into AMD, autism, cancer, and wound healing from basic research on the chemistry of oxidized phospholipids. Chem Res Toxicol 24: 1803–1816.
5. Crabb JW, Miyagi M, Gu X, Shadrach K, West KA, et al. (2002) Drusen proteome analysis: an approach to the etiology of age-related macular degeneration. Proc Natl Acad Sci U S A 99: 14682–14687.
6. Gu X, Meer SG, Miyagi M, Rayborn ME, Hollyfield JG, et al. (2003) Carboxyethylpyrrole protein adducts and autoantibodies, biomarkers for age-related macular degeneration. J Biol Chem 278: 42027–42035.
7. Gu J, Pauer GJ, Yue X, Narendra U, Sturgill GM, et al. (2009) Assessing susceptibility to age-related macular degeneration with proteomic and genomic biomarkers. Mol Cell Proteomics 8: 1338–1349.
8. Ebrahem Q, Renganathan K, Sears J, Vasanji A, Gu X, et al. (2006) Carboxyethylpyrrole oxidative protein modifications stimulate neovascularization: Implications for age-related macular degeneration. Proc Natl Acad Sci U S A 103: 13480–13484.

9. West XZ, Malinin NL, Merkulova AA, Tischenko M, Kerr BA, et al. (2010) Oxidative stress induces angiogenesis by activating TLR2 with novel endogenous ligands. Nature 467: 972–976.
10. Lu L, Gu X, Hong L, Laird J, Jaffe K, et al. (2009) Synthesis and structural characterization of carboxyethylpyrrole-modified proteins: mediators of age-related macular degeneration. Bioorg Med Chem 17: 7548–7561.
11. Gu J (2009) Biomarkers for Age-Related Macular Degeneration [Electronic Thesis or Dissertation.]. Cleveland, Ohio: Case Western Reserve University. 223 p.
12. Mattapallil MJ, Wawrousek EF, Chan CC, Zhao H, Roychoudhury J, et al. (2012) The Rd8 mutation of the Crb1 gene is present in vendor lines of C57BL/6N mice and embryonic stem cells, and confounds ocular induced mutant phenotypes. Invest Ophthalmol Vis Sci 53: 2921–2927.
13. Hollyfield JG, Bonilha VL, Rayborn ME, Yang X, Shadrach KG, et al. (2008) Oxidative damage-induced inflammation initiates age-related macular degeneration. Nat Med 14: 194–198.
14. Sivak JM, Ostriker AC, Woolfenden A, Demirs J, Cepeda R, et al. (2011) Pharmacologic uncoupling of angiogenesis and inflammation during initiation of pathological corneal neovascularization. J Biol Chem 286: 44965–44975.
15. Renganathan K, Gu J, Rayborn ME, Crabb JS, Salomon RG, et al. (2013) CEP Biomarkers as Potential Tools for Monitoring Therapeutics. PLoS One 8: e76325.
16. Cruz-Guilloty F, Saeed AM, Duffort S, Cano M, Ebrahimi KB, et al. (2014) T cells and macrophages responding to oxidative damage cooperate in

pathogenesis of a mouse model of age-related macular degeneration. PLoS One 9: e88201.

17. Organisciak DT, Darrow RM, Rapp CM, Smuts JP, Armstrong DW, et al. (2013) Prevention of retinal light damage by zinc oxide combined with rosemary extract. Mol Vis 19: 1433–1445.

18. Yuan X, Gu X, Crabb JS, Yue X, Shadrach K, et al. (2010) Quantitative proteomics: comparison of the macular Bruch membrane/choroid complex from age-related macular degeneration and normal eyes. Mol Cell Proteomics 9: 1031–1046.

19. Jang G-F, Zhang L, Hong L, Wang H, Salomon RG, et al. (2012) Quantification Of CEP By LC MS/MS. Investigative Ophthalmology & Visual Science 53: 6478.

20. Saeed AM, Duffort S, Ivanov D, Wang H, Laird JM, et al. (2014) The Oxidative Stress Product Carboxyethylpyrrole Potentiates TLR2/TLR1 Inflammatory Signaling in Macrophages. PLoS One 9: e106421.

21. Ahmed N, Argirov OK, Minhas HS, Cordeiro CA, Thornalley PJ (2002) Assay of advanced glycation endproducts (AGEs): surveying AGEs by chromatographic assay with derivatization by 6-aminoquinolyl-N-hydroxysuccinimidyl-carbamate and application to Nepsilon-carboxymethyl-lysine- and Nepsilon-(1-carboxyethyl)lysine-modified albumin. Biochem J 364: 1–14.

22. Alves CH, Pellissier LP, Wijnholds J (2014) The CRB1 and adherens junction complex proteins in retinal development and maintenance. Prog Retin Eye Res 40: 35–52.

17

Global Intracoronary Infusion of Allogeneic Cardiosphere-Derived Cells Improves Ventricular Function and Stimulates Endogenous Myocyte Regeneration throughout the Heart in Swine with Hibernating Myocardium

Gen Suzuki[2,5], **Brian R. Weil**[2,5], **Merced M. Leiker**[2,5], **Amanda E. Ribbeck**[2,5], **Rebeccah F. Young**[2,5], **Thomas R. Cimato**[2,5], **John M. Canty Jr.**[1,2,3,4,5]*

1 Division of Cardiovascular Medicine, Veterans Affairs Western New York Health Care System, Buffalo, New York, United States of America, 2 Department of Medicine, Division of Cardiovascular Medicine, University at Buffalo, Buffalo, New York, United States of America, 3 Department of Physiology & Biophysics, University at Buffalo, Buffalo, New York, United States of America, 4 Department of Biomedical Engineering, University at Buffalo, Buffalo, New York, United States of America, 5 The Clinical and Translational Research Center, University at Buffalo, Buffalo, New York, United States of America

Abstract

Background: Cardiosphere-derived cells (CDCs) improve ventricular function and reduce fibrotic volume when administered via an infarct-related artery using the "stop-flow" technique. Unfortunately, myocyte loss and dysfunction occur globally in many patients with ischemic and non-ischemic cardiomyopathy, necessitating an approach to distribute CDCs throughout the entire heart. We therefore determined whether global intracoronary infusion of CDCs under continuous flow improves contractile function and stimulates new myocyte formation.

Methods and Results: Swine with hibernating myocardium from a chronic LAD occlusion were studied 3-months after instrumentation (n = 25). CDCs isolated from myocardial biopsies were infused into each major coronary artery ($\sim$33$\times$10^6 icCDCs). Global icCDC infusion was safe and while $\sim$3% of injected CDCs were retained, they did not affect ventricular function or myocyte proliferation in normal animals. In contrast, four-weeks after icCDCs were administered to animals with hibernating myocardium, %LADWT increased from 23$\pm$6 to 51$\pm$5% (p<0.01). In diseased hearts, myocyte proliferation (phospho-histone-H3) increased in hibernating and remote regions with a concomitant increase in myocyte nuclear density. These effects were accompanied by reductions in myocyte diameter consistent with new myocyte formation. Only rare myocytes arose from sex-mismatched donor CDCs.

Conclusions: Global icCDC infusion under continuous flow is feasible and improves contractile function, regresses myocyte cellular hypertrophy and increases myocyte proliferation in diseased but not normal hearts. New myocytes arising via differentiation of injected cells are rare, implicating stimulation of endogenous myocyte regeneration as the primary mechanism of repair.

Editor: Yaoliang Tang, Georgia Regents University, United States of America

Funding: This study was supported by NIH HL55324, HL61610, F32 HL114335, American Heart Association SDG3990004, New York State Department of Health NYSTEM CO24351, and the Albert and Elizabeth Rekate Fund in Cardiovascular Medicine. The funders had no role in the study design, data collection and analysis, decision to publish, or preparation of the manuscript.

Competing Interests: The authors have declared that no competing interests exist.

* Email: canty@buffalo.edu

Introduction

A large number of preclinical studies have demonstrated the ability of diverse adult stem cell formulations to prevent post-infarction remodeling [1]. Stimulating resident progenitor cells or administering cell preparations that include exogenous cardiac stem cells may be particularly effective approaches to elicit cardiac repair. Although several stem and progenitor cell populations have been identified, there is compelling evidence that the reparative potential of these cells is maximized when delivered as a heterogeneous mixture of heart-derived subpopulations[2–4]. This notion has been supported by the encouraging results of studies utilizing cardiosphere-derived cells (CDCs), a population of cardiac stromal cells derived from myocardial biopsies that fulfill the criteria for cardiac progenitor cells and have recently been shown to reduce scar mass and increase viable myocardium in patients with left ventricular dysfunction after myocardial infarction [5].

At this stage of therapeutic development, most basic and clinical studies have focused on infusing stem cells down the infarct-related artery with the "stop-flow" technique or injecting them into the peri-infarct tissue in an attempt to replace scar with newly regenerated myocardial tissue. While these approaches may effect repair in the infarcted region, viable dysfunctional myocardium without fibrosis may also be an important target for cardiac repair. Viable dysfunctional regions develop as a consequence of repetitive ischemia as in hibernating myocardium where there is regional myocyte loss and compensatory myocyte hypertrophy [6]. In addition, viable dysfunctional myocardium can also develop in regions that are normally perfused and remote from a large dysfunctional region (viable or infarcted) due to apoptosis-induced myocyte loss and left ventricular remodeling [7,8]. Importantly, despite revascularization procedures, these viable dysfunctional regions comprise a much larger percent of the heart in patients with advanced ischemic cardiomyopathy as infarct volume only averages ~20% of the left ventricle (LV) [9]. Thus, regenerating myocytes globally, including areas of dysfunctional myocardium without scar, may be required to reverse deleterious LV remodeling and optimally improve LV function.

To determine the extent that the reparative actions of CDCs may be independent of scar replacement, we administered cells via intracoronary infusion to the entire heart under continuous flow in a swine model of hibernating myocardium. Myocyte proliferation was assessed in hibernating as well as normally-perfused remote regions and sex-mismatched allogeneic CDCs were used to quantify myocytes arising directly from the injected cells. Our results demonstrate that global intracoronary infusion of CDCs without using the "stop-flow" technique improves myocardial function. These effects are primarily related to actions on endogenous myocytes in hibernating regions as well as normally-perfused remote myocardium with only rare myocytes arising from donor cells.

Methods

Ethics Statement

All procedures and protocols conformed to institutional guidelines for the care and use of animals in research and were approved by the University at Buffalo Institutional Animal Care and Use Committee (Protocol #MED02011Y). All studies were performed under either isoflurane or propofol anesthesia as described below, and all efforts were made to minimize suffering. **Figure 1** outlines the experimental groups and steps in the isolation of CDCs described below.

Cultivation, Expansion and In Vitro Characterization of CDCs

CDCs were cultivated using the techniques described by Smith et al [2] (**Figure 1A**). LV tissue specimens were obtained by needle biopsies (2–5 biopsies from the LV basal free wall, 20–50 mg total) and cut into 1–2 mm pieces [2,10]. After gross connective tissue was removed from fragments, they were washed and partially enzymatically digested in a solution of type IV collagenase for 60 minutes at 37 degrees. Tissue fragments were cultured as "explants" on dishes coated with fibronectin. After ~8 days, a layer of stromal-like cells arose from and surrounded the explants. Over this layer a population of small, round, phase-bright cells migrated. Once confluent, the cells surrounding the explants were harvested by gentle enzymatic digestion. These cardiosphere-forming cells were seeded at 2 to 3×10^4 cells/mL on

poly-D-lysine-coated dishes in cardiosphere medium (20% heat-inactivated fetal calf serum, gentamicin 50 µg/ml, 2 mmol/L L-glutamine, and 0.1 mmol/L 2-mercaptoethanol in Iscove's modified Dulbecco medium). Cardiospheres formed after 4–10 days in culture, detached from the tissue culture surface, and began to slowly grow in suspension. When sufficient in size and number, free-floating cardiospheres were harvested by aspirating them along with media. Cells that remained adherent to the poly-D-lysine-coated dishes were discarded. Detached cardiospheres were then plated on fibronectin-coated flasks where they attached to the culture surface and formed monolayers of "Cardiosphere-Derived Cells" (CDCs) [2]. CDCs were subsequently passaged by trypsinization and splitting at a 1:2 ratio. Up to 100 million CDCs developed within 4–6 weeks of the time that the original cardiac biopsies were obtained. Cells were characterized by flow cytometry and immunohistochemistry with hematopoietic (CD45, cKit, CD133), mesenchymal (CD90, CD105) and cardiac (GATA4, Nkx2.5, cTnT, and cTnI) markers.

Prior to administration, cell suspensions were filtered through a 30–100 µm pore filter to circumvent administering cell aggregates (MACS pre-separation filters, Miltenyi Biotec) and suspended in heparinized HBSS solution (3000 U heparin in 30 ml in total) for intracoronary infusion. Approximately $10–15 \times 10^6$ cells were infused into each of the three proximal coronary arteries including the stump of the occluded left anterior descending artery (LAD). Thus, cells were administered to the entire heart with the hibernating LAD region receiving cells through the stenotic LAD and/or collateral vessels that typically develop in this model. The cell suspensions were each slowly infused over 10-minutes with no untoward hemodynamic changes or electrocardiographic evidence of myocardial ischemia.

In Vivo Studies of Intracoronary CDC (icCDC) Infusion in Swine

The effects of intracoronary CDC infusion on flow, function and myocyte proliferation were assessed in normal animals as well as a series of animals with viable dysfunctional hibernating myocardium. Experimental groups and samples sizes are summarized in **Figure 1B**. All animals were in good health at the time of study and the specific protocols performed in each group are summarized below.

Intracoronary CDC Infusion in Normal Swine (n = 11). To assess the safety of infusing allogeneic intracoronary CDCs throughout the entire heart, the acute electrocardiographic, hemodynamic and functional effects of global icCDC infusion were assessed in normal animals. In six animals, serial echocardiography and ST-T changes were assessed by Holter monitoring and hearts were excised after 24-hours to evaluate postmortem necrosis by pathology and TTC staining. We measured average CDC size immediately prior to cell infusion with a hemocytometer. To exclude minor injury beyond the detection of TTC and pathology, serum cardiac TnI (pig cTnI ELISA Kit, Life Diagnostics) was assessed at baseline, 2 hours and 24 hours after injection. Less than 0.04 ng/ml of serum cTnI was considered normal. Regional and global function was assessed with echocardiography at baseline and 2-hours after icCDC infusion. In an additional 5 pigs, follow-up was carried out for 2-weeks to assess late retention and the effects of icCDC infusion on myocyte nuclear density, morphometry and immunohistochemistry in the completely normal heart. TTC staining was also performed to exclude necrosis.

Intracoronary CDC Infusion in Swine with Hibernating Myocardium (n = 25). In order to study the effects of icCDCs on cardiac myocyte function in a fashion that was independent of

Global Intracoronary Infusion of Allogeneic Cardiosphere-Derived Cells Improves Ventricular Function...

185

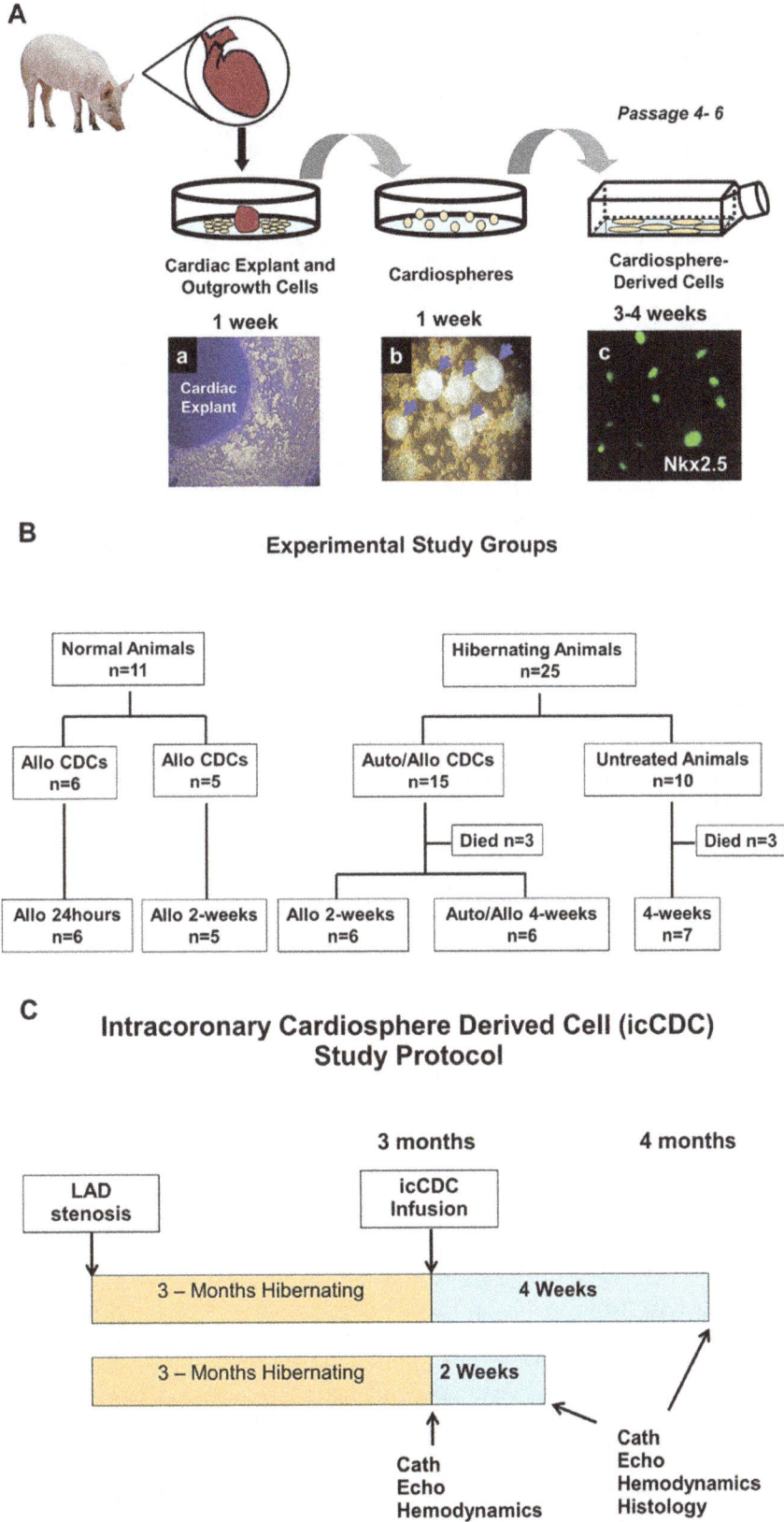

Figure 1. CDC Isolation, Summary of Experimental Study Groups, and Study Protocol Timeline. A.) Myocardial needle biopsies were obtained from the LV free wall at the time of initial instrumentation in juvenile pigs with hibernating myocardium. The tissue fragments were cultured as "explants" on dishes coated with fibronectin. After one week, a layer of stromal-like cells started surrounding the explants (a). A population of small, round and phase-bright cells developed over this layer that subsequently formed cardiospheres (b). These cells were harvested

and began forming cardiosphere-derived cells which are stained in green with the cardiac transcription factor Nkx2.5 (c). Expansion of CDCs to obtain a total injected volume >30 million CDCs typically required six passages in culture. **B.)** Experimental data was acquired from 36 animals. Eleven pigs were normal and used to exclude myocardial necrosis after infusing icCDCs as well as assess CDC retention and the effects of icCDCs on myocyte proliferation in the normal heart. Twenty-five animals had viable dysfunctional myocardium (hibernating animals) and were studied 3-months after the development of a chronic LAD occlusion and the development of chronic hibernating myocardium. Of the animals with hibernating myocardium, 15 were randomized to receive intracoronary CDC infusion and 10 served as untreated controls. Follow-up in animals with icCDCs was at either 24 weeks after icCDC infusion. Sudden cardiac death from lethal ventricular arrhythmias [22] occurred before the final study was completed in six animals with hibernating myocardium (3 in untreated and 3 in icCDC treated). These rates are similar to those previously published. Auto – autologous; Allo – allogeneic. **C.)** Juvenile swine were instrumented with a Delrin LAD stenosis to produce hibernating myocardium with serial physiological studies beginning at 3-months. At that time, a baseline closed-chest study was conducted to assess myocardial function with echocardiography (echo), myocardial perfusion with microspheres at rest and vasodilation, hemodynamics and coronary angiography (Cath). Subsequently, intracoronary CDCs were infused throughout the heart in 15 animals and 10 served as untreated controls. One group of animals underwent repeat study and tissue harvesting after 2-weeks of treatment and the second group was evaluated 4-weeks after icCDC treatment.

myocardial scar a porcine model of viable dysfunctional hibernating myocardium was produced as previously described [11]. Briefly, juvenile pigs were sedated (Telazol 100 mg/ml/xylazine 100 mg/ml, 0.022 mg/kg i.m.), intubated and ventilated with a 0.5–2% isoflurane-oxygen mixture. Through a limited pericardiotomy, the proximal LAD was instrumented with a Delrin occluder (1.5 mm). Antibiotics (cefazolin, 25 mg/kg and gentamicin, 3 mg/kg i.m.) were given 1-hour before surgery and repeated after closing the chest. Analgesia included an intercostal nerve block (0.5% Marcaine) and intramuscular doses of butorphanol (2.2 mg/kg q6h) and flunixin (1–2 mg/kg q.d.). Pigs with hibernating myocardium underwent initial studies 3-months after instrumentation. We have previously demonstrated that reductions in resting flow, function and flow reserve develop in this model after 3-months and remain unchanged between 3- and 5-months after instrumentation producing a stable model of ischemic LV dysfunction [11,12]. In contrast to models of total coronary occlusion, histological and TTC evidence of infarction is absent in this model.

Serial Physiological Studies [12]. The timeline of physiological studies is summarized in **Figure 1C**. Animals with chronic hibernating myocardium underwent an initial baseline physiological study 3-months after initial instrumentation to establish the presence of contractile dysfunction at rest. Sedation was initiated with a Telazol (100 mg/ml)/xylazine (100 mg/ml) mixture (0.037 ml/kg i.m.) and maintained with propofol (5–10 mg/kg/hr i.v.). Under sterile conditions, a 6-Fr introducer was inserted into the left brachial artery. A 5F Millar micromanometer was placed into the LV apex with the lumen used for microsphere injection. The introducer side port was used to monitor aortic pressure and provide a reference blood withdrawal for microspheres. Animals were heparinized (100 U/kg), and hemodynamics allowed to equilibrate for at least 30-minutes. Regional wall-thickening was assessed using off-axis M-Mode echocardiography (GE Vivid 7) employing a right parasternal approach. All hibernating pigs had contractile dysfunction in the distribution of the anterior wall supplied by the LAD. To quantify this, systolic wall-thickening ($\Delta WT = ESWT-EDWT$; $\%WT = \Delta WT/EDWTx100$) was measured in the dysfunctional LAD region as well as in normally-perfused remote regions (posterior wall) of the same heart. All measurements were calculated from echocardiographic dimensions using ASE criteria. After baseline measurements, myocardial perfusion was assessed with microspheres at rest and following pharmacological vasodilation using adenosine (0.9 mg/kg/min iv) while phenylephrine was infused and titrated to maintain mean blood pressure at ~100 mmHg. Subsequently, animals received icCDCs (n = 15) or were untreated (n = 10). Since initial results (n = 6) demonstrated comparable effects of autologous and allogeneic icCDCs, allogeneic icCDCs with cyclosporine A immunosuppression (5 mg/kg/day p.o., Watson Pharma) were

used for the majority of experiments (20/23 icCDC-treated animals). At the end of each study, the catheters were removed and pigs were allowed to recover and returned to the animal housing facility. Two- or four-weeks later, the pigs were brought back to the laboratory for a second physiological study which was performed in a fashion similar to the initial baseline protocol. Once measurements were completed, animals were euthanized under general anesthesia. The LV was rapidly excised, weighed and sectioned into 1-cm rings parallel to the AV groove from apex to base. Thin rings above each major ring were incubated in TTC to assess infarction. Additional tissue samples were taken for quantifying microsphere flow and histology.

Microsphere Flow Measurements [13]. Regional perfusion was assessed using 15 μm microspheres labeled with fluorescing dyes [14,15]. Approximately 3×10^6 microspheres were injected into the LV while a reference sample was withdrawn at 6 ml/min for 90-seconds. At the end of the study, samples were taken from a mid-ventricular ring and divided into twelve circumferential wedges, each of which was cut into 3 transmural layers. Fluorescent dyes were extracted using standard techniques and fluorescence quantified at selected excitation wavelengths. In each animal the circumferential flow distribution during adenosine was analyzed to identify the hibernating risk region as compared to normal regions where flow increased 4–6 fold. From these data, we determined the weighted average flow from samples in the central hypoperfused region (hibernating LAD) or normally-perfused remote region. Samples with intermediate vasodilated flows were considered border regions and excluded from analysis. Using these data, we also assessed relative and absolute coronary flow reserve in animals receiving icCDCs vs. untreated swine with hibernating myocardium using. Relative flow reserve was determined by dividing the flow in LAD regions by the corresponding average full-thickness value from normal myocardium. Absolute coronary flow reserve was assessed by comparing flow in each region during vasodilation to the corresponding values at rest.

Myocardial Histopathology and Flow Cytometry of Circulating Progenitor Cells

Quantitative immunohistochemistry and morphometry were evaluated in icCDC-treated and untreated animals with hibernating myocardium as previously described in detail and summarized below [12,16].

Myocyte Nuclear Density and Morphometry. Samples approximately midway between the base and apex that were immediately adjacent to the LAD (hibernating) and posterior descending arteries (normal) were fixed (10% formalin) and paraffin-embedded. Point-counting of trichrome-stained sections was used to quantify connective tissue [6]. PAS stained sections were used to quantify myocyte diameter. Myocyte diameter was assessed by counting at least 100 cells from the inner and outer half of the LAD and remote regions. Myocytes were included regardless of size as long as myofilaments could be identified surrounding the nucleus. We assessed regional myocyte nuclear density in PAS stained sections as previously described [12]. To evaluate the possibility that changes in nuclear density reflected increases in the number of nuclei per myocyte, longitudinal myocytes (n = 20 in each histological sample) that had boundaries that were clearly visible (intercalated discs and the lateral borders of the myocyte) were used to quantify the number of nuclei per cell in each heart using previously published methodology [17].

Immunohistochemical Assessment of Myocyte Proliferation and Angiogenesis. All of the antibodies we employed have been successfully used in the pig in previous studies by our group [12]. Paraffin-fixed tissue sections with 4 μm thickness were incubated with anti-phospho-histone-H3 rabbit polyclonal antibody (Upstate Biotech, 1:1000) to detect proliferating cells and anti-cTnI (rabbit polyclonal antibody, Santa Cruz, 1:200) to detect myofilaments. For capillary density evaluation, paraffin-fixed tissue sections were incubated with Von Willebrand factor (vWF, Biocare Medical, 1:200). Myocardial levels of CD45 negative resident and bone marrow derived progenitor cells (CD45 antibody, 1:200, AbD serotec) were quantified in frozen tissue sections using the cell surface marker cKit (AbD serotec, 1:200) and CD133 (Miltenyi biotec, 1:200) [12,18]. To optimize identification of cKit and CD133 antigens, we conducted the quantitative analysis using frozen tissue sections. Samples were post-treated with Alexa Fluor 488 conjugated anti-mouse and Alexa Fluor 555 conjugated anti-rabbit antibody (Invitrogen). Nuclei were stained with TO-PRO-3 (Molecular Probes) or DAPI (Vectashield). Image acquisition was performed with a confocal microscope (Zeiss 510 Meta) and AxioImager equipped with ApoTome (Zeiss). Phospho-histone-H3 positive myocytes were counted and evaluated as positive nuclei per million myocyte nuclei as previously described [12]. The number of cKit+ and CD133+ cells in myocardium was also expressed in relation to myocyte nuclear density or cells per million myocytes. Data represent the averages from 462±46 fields examined per slide (area of 72±7 mm^2 per section).

Fluorescence in Situ Hybridization of Sex-mismatched CDC Donor-Recipient Pairs. To track the fate of infused CDCs in the myocardium, male CDCs were injected into female recipients and Fluorescence in Situ Hybridization used to quantify the frequency of Y-chromosomes from donor CDCs (Y-FISH). We hybridized tissue samples with FITC-conjugated porcine Y-chromosome probe (IDLabs, Ontario, Canada) according to the manufacturer's instructions. The nuclear diameter is greater than the section thickness and thus, the Y chromosome is not present in each nucleus sectioned. We therefore also evaluated the frequency of Y-FISH staining in male control cardiac tissue to determine the efficiency of Y chromosome identification. Tissues were incubated with cTnI, vWF and alpha-SMA to detect co-localization of donor derived Y chromosome cells in myocytes, endothelial cells and vascular smooth muscle. Nuclei were stained with DAPI (Vectashield) or TOPRO3 (Molecular Probes).

Flow Cytometry of Circulating Progenitor Cells [12]. We determined whether icCDCs mobilized hematopoietic progenitor cells (cKit+ or CD133+) in hibernating (n = 3) and normal swine (n = 3) as we have previously described. Mononuclear cells were isolated from peripheral blood (30 ml) using the Becton Dickinson CPT cell separation system before, 3 days and 2 weeks after icCDC treatment. Leukocyte counts were performed using an automated hemocytometer while monocyte counts were done using a manual hemocytometer. Approximately 10–20×10^6 mononuclear cells were analyzed by FACS after staining for cKit

(CD117, AbD Serotec), CD133 (PE conjugated, Miltenyi Biotech) and CD 45 (PE-Cy5 conjugated, BD Pharmingen). Isotype controls were used as negative controls. Single stains were also performed to determine quality control and for multi-channel compensation. Data were expressed as progenitor cells (cKit+ and CD45−, CD133+ and CD45−) per million mononuclear cells. All cell counts were corrected for the absolute mononuclear cell count. Immunohistochemical analysis and morphometric analysis of excised tissue is summarized below.

Statistical Analysis

Data are expressed as the mean ± standard error. Differences in physiological parameters at baseline and after treatment with icCDCs as well as comparisons between the hibernating and normally-perfused remote regions of the same heart were assessed using paired t-tests. Differences among icCDC treated animals and age matched untreated animals were assessed using a two-way ANOVA and the post-hoc Holm-Sidak test (Sigma Stat 3.0). For all comparisons, p<0.05 was considered significant.

Results

In Vitro Characterization of Porcine CDCs

Porcine cardiospheres were readily isolated from ~10 mg ventricular needle biopsies. After plating on fibronectin, they expanded to >30 million CDCs after 4–6 passages and 3 to 4 weeks in culture. **Figure 2** summarizes the immunohistochemical and flow cytometry markers of the CDCs at the time of injection. At the time they were infused into swine (passage 6), most CDCs expressed mesenchymal markers (CD90+ and CD105+, **Figure 2A**) but were CD45 negative. The frequency of cKit+ cells declined after plating with 4.7±0.9% of the CDCs remaining cKit+ at passage 6. At this time, virtually all of the CDCs expressed the cardiac transcription factors Nkx2.5 and GATA4 (**Figure 2B**) but they remained cardiac troponin negative. Thus, at the time of intracoronary infusion, our CDCs were primarily a population of early, cardiac-committed progenitors with a low frequency of cKit positivity.

Safety of Global icCDC Infusion and Quantitative Retention of icCDCs in Normal Swine

We initially evaluated the safety of infusing divided doses of allogeneic icCDCs (~38.5±0.9×10^6 cells, 18±0.4 μm diameter, n = 6) into each of the three major coronary arteries (nominal rate 1.3×10^6 icCDCs/minute). Continuous electrocardiographic monitoring demonstrated no ST elevation or arrhythmias during infusion. Echocardiographic evaluation and serum TnI are summarized in **Figure 3**. There was no change in LAD function (LAD % WT 61±7% at baseline to 66±5% at 2-hours, p = ns) and no elevation of cTnI after icCDC infusion (0 to 0.05±0.03 ng/ml at 2-hours, p = ns). When reassessed at 24-hours, function remained unchanged (%LAD wall thickening (%LADWT) 65±2%, p = ns vs. baseline) with a small, insignificant increase in cTnI (0.41±0.24 ng/ml, p = 0.15). There was no TTC evidence of infarction and no histologic increase in connective tissue or light microscopic evidence of necrosis. Thus, global icCDC infusion without employing transient coronary occlusion did not affect myocardial function nor did it produce significant changes in troponin I or pathological evidence of microinfarction.

We quantified retention of CDCs in the heart, by counting myocardial Y-FISH positive cells in female animals 24-hours after male icCDCs (n = 6). There were 5±2 Y-FISH positive cells per million myocytes (1.4±0.5 per cm^2 section) after icCDCs. Since the efficiency of Y-FISH positivity in male control hearts was

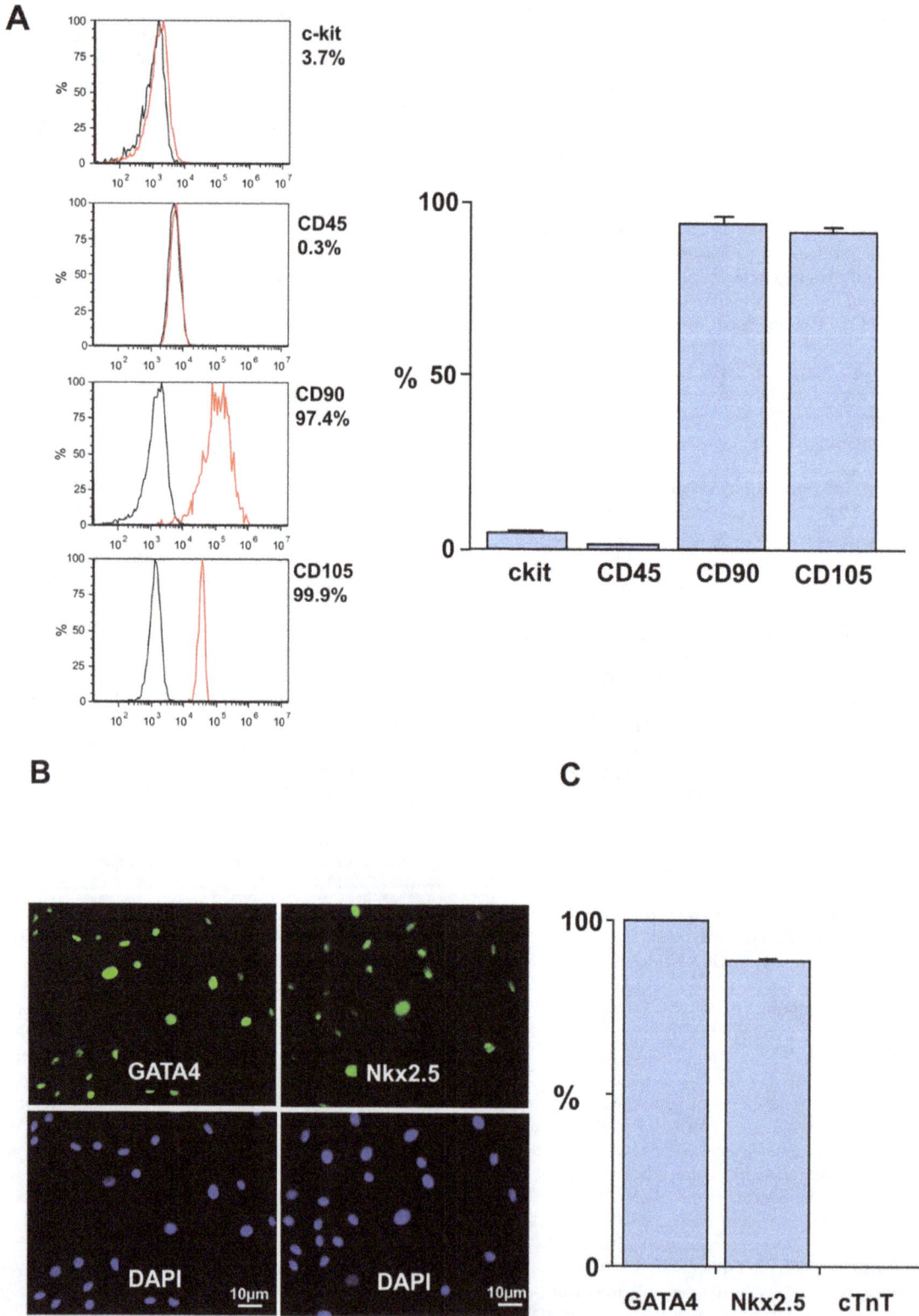

Figure 2. CDC Characterization by Flow Cytometry and Immunohistochemistry. A.) Summary of CDC characterization by flow cytometry at passage 6 (n=6). Most CDCs expressed mesenchymal markers (CD90+ and CD105+) but were CD45 negative. While cKit+ cells were identified in cardiospheres, they declined after plating with 4.7±0.9% of the CDCs remaining cKit positive at the time of icCDC infusion. **B. and C.)** Summary of immunohistochemical characterization of CDCs for markers of commitment to a cardiac lineage (n=6). At passage 6, most CDCs expressed the cardiac transcription factors GATA4 and Nkx2.5 (green stain). In contrast, CDCs did not express cardiac troponin T, suggesting that these cells are at an early stage of cardiac development.

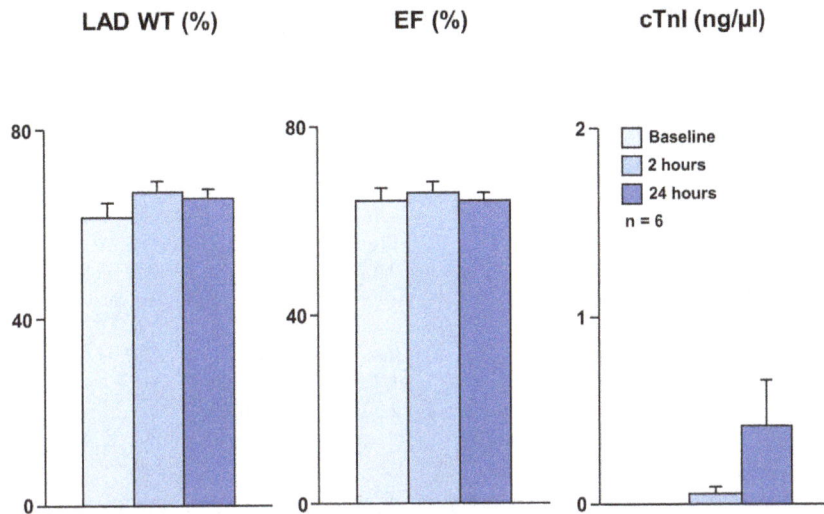

Figure 3. Safety of Global icCDC Infusion under Continuous Flow in Normal Swine. Allogeneic icCDCs (~38.5±0.9×10^6 cells, n = 6) were infused into each of the three major coronary arteries and hearts were harvested 24-hours later for pathological analyses. This figure summarizes echocardiographic function and serum TnI at baseline, 2-hours and 24-hours after icCDCs in normal animals. Two hours after icCDCs, there was no change in regional wall thickening (LAD%WT) or global function (EF). Twenty-four hours after icCDCs, function remained unchanged. There was a small statistically insignificant increase in cTnI. Tetrazolium staining and histology showed no evidence of micro-necrosis. Similar results were seen in normal hearts harvested at 2-weeks (data not shown). These results, along with the absence of hemodynamic and electrocardiographic changes during infusion, indicate that global icCDC infusion is safe, has no effect on myocardial function and produces no pathological evidence of infarction.

40±4%, the frequency of Y-positive nuclei underestimated actual cell number by a factor of 2.5. Correcting for this factor and extrapolating data to the entire heart indicated that approximately 1.1±0.5 million CDCs or CDC derived cells were present in the heart at 24 hours. This represented 3±1% of the original icCDC dose. This low value is similar to short-term cell retention after intracoronary infusion reported by others using the stop-flow technique as well as after intramyocardial injection[19–21]. The number of Y-FISH positive cells (original icCDCs as well as myocardial cells derived from CDCs) remained approximately similar 2-weeks after sex-mismatched icCDC infusion (3.4±0.7% of injected dose, p-ns vs 24-hours). Thus, while viable CDC derived cells remained in the myocardium, there was no significant increase in the number of donor derived cells beyond those present at 24 hours.

Effects of icCDCs on Myocardial Flow and Function in Swine with Hibernating Myocardium

Baseline physiological studies at 3-months confirmed viable dysfunctional (hibernating) myocardium in all animals (n = 25). At this time there was a severe proximal LAD stenosis (99±1%) with total occlusion and collateral-dependent myocardium in most swine (20 of 25 animals). Regional LAD wall thickening (LAD % WT) was reduced in comparison to normal remote regions (29±3% in LAD vs. 76±7% in remote, p<0.05) with reductions in resting perfusion (LAD 0.83±0.08 vs. 1.32±0.11 ml/min/g in remote, p<0.05). Flow during adenosine vasodilation was severely attenuated (LAD 0.92±0.19 vs. 4.50±0.49 ml/min/g in remote, p<0.01). Although VT/VF develops in the absence of infarction in this animal model of hibernating myocardium [22], survival to the final study was similar in animals after receiving icCDCs (12/15; 80% survival) vs. untreated controls (7/10; 70% survival).

Figure 4 and **Table 1** summarize the serial functional effects of icCDCs. There was no effect of icCDCs on heart rate or blood pressure. In untreated animals, LAD % WT was depressed at rest and did not change when evaluated 4-weeks later (LAD % WT

29.2±2.7% vs. 28.1±1.9%, p = ns) as we have previously demonstrated [12]. In contrast, function after icCDCs significantly increased at 2-weeks (LAD % WT 33.9±3.1% to 49.1±4.0%, p< 0.01) and 4-weeks (LAD %WT 22.8±5.6% to 51.0±5.1%, p< 0.01). Interestingly, while there was no effect of icCDCs on %WT in normally-perfused remote myocardium after 2-weeks (Remote %WT 85±7% to 80±9%, p = ns), it significantly increased after 4-weeks (Remote %WT 68±10% to 107±11% at 4-weeks, p< 0.05). Global function was mildly reduced at rest before icCDCs and returned to values similar to normal animals after icCDC infusion (LVEF 56±1% to 64±2% at 2-weeks and 54±2% to 71±4% 4-weeks after icCDCs, p<0.05).

Despite the improvement in function in hibernating myocardium, there was no significant effect of icCDCs on paired serial measurements of coronary flow at rest or during adenosine vasodilation (**Figure 5**). As a result, while LAD flow reserve (adenosine/rest) was critically impaired in hibernating myocardium, it did not increase significantly after icCDCs (LAD adenosine/rest 1.88±0.20 initial vs. 2.03±0.31 final, p = ns). Coronary flow reserve also remained unchanged in normally-perfused remote myocardium (4.9±0.5 initial vs. 5.6±0.6 final, p = ns). Likewise, serial measurements of relative flow during adenosine vasodilation (LAD/remote) remained unchanged after icCDCs (0.18±0.04 initial vs. 0.19±0.08 final, p = ns). While there was no measurable increase in coronary collateral flow, icCDCs stimulated capillary angiogenesis and increased myocardial capillary density (1013±31/mm^2 in untreated animals to 1604±27/mm^2 at 4-weeks after icCDCs, p<0.05 vs. untreated, **Figure 6**). Thus, improvements in regional and global function after icCDCs were not secondary to stimulation of increased coronary collateral flow or arteriogenesis.

Effects of icCDCs on Myocyte Proliferation

Intracoronary CDC infusion increased myocytes in the mitotic phase of the cell cycle throughout the entire hibernating heart. **Figure 7** shows that phospho-histone-H3 (pHH3) positive myo-

Table 1. Effects of icCDCs on Echocardiographic Measurements of Cardiac Function and Hemodynamic Variables in Swine with Hibernating Myocardium.

	n	LAD Δ WT (mm)	Remote Δ WT (mm)	FS (%)	EF (%)	LV$_{Sys}$ (mm Hg)	HR (bpm)	LVdP/dt$_{Max}$ (mmHg/sec)
Untreated Hibernating	7							
Initial		2.8±0.2	6.4±0.7	24.4±1.6	53±2	128±3.0	112±7.5	2171±64
Final		3.2±0.3	7.7±0.7	24.9±1.6	53±3	124±4.0	101±7.6	2035±111
CDCs Hibernating 2-weeks	6							
Initial		2.7±0.2	5.8±0.3	29.2±0.9	56±1	138±6.7	110±9.0	2375±89
Final		4.8±0.3*†	6.9±0.5*	34.9±1.7*	64±2*†	139±4.1	107±6.5	2528±74*†
CDCs Hibernating 4-weeks	6							
Initial		2.0±0.4	5.2±0.9	26.4±2.2	54±2	144±3.4	86±2.3	2370±97
Final		5.0±0.4*†	8.1±0.8*	35.4±2.9*†	71±4*†	150±3.1	83±5.2	2633±148*†

Values are mean ± SEM; *p<0.05 vs. Initial; †vs. Untreated; LAD – Left Anterior Descending Artery; LV – Left Ventricular; WT – Wall Thickening; FS – Fractional Shortening; EF – Ejection Fraction; Δ WT = End-Systolic Wall Thickness – End-diastolic Wall Thickness.

cytes increased in both LAD and remote regions of hibernating hearts and remained significantly elevated at 4-weeks. In contrast, despite a significant retention of icCDCs in normal control hearts, they did not stimulate myocyte mitosis and pHH3 positive myocytes remained low and similar to untreated normal animals. These data indicate that icCDCs stimulated myocyte proliferation in ischemic as well as remote myocardium of the diseased heart with no effect on myocyte proliferation in the normal heart. **Figure 8** summarizes the effects of icCDCs on resident myocardial cKit$^+$ cells. Rare cKit+ cells were identified in untreated animals with hibernating myocardium as well as normals. Total cKit$^+$ cells (as well as cKit$^+$/CD45$^-$ cells) increased from 0.8±0.2/mm^2 (cKit$^+$/CD45$^-$: 0.2±0.04/mm^2) in untreated animals to 4.8±0.3/mm^2 (cKit$^+$/CD45$^-$: 2.5±0.1/mm^2) at 2 weeks after icCDCs (both p<0.05 vs. untreated). The number remained elevated at 2.7±0.3/mm^2 (cKit$^+$/CD45−: 0.7±0.1/mm^2, p<0.05 vs. untreated) at 4 weeks after icCDCs. In contrast, icCDCs did not increase total cKit$^+$ (1.3±0.2/mm^2, p-ns vs. untreated) or cKit$^+$/CD45$^-$ cells (0.8±0.1/mm^2, p-ns vs. untreated) in normal animals.

Since cell cycle markers such as pHH3 measured at a single time point cannot quantify the cumulative number of new myocytes produced, we quantified myocyte nuclear density, nuclear number per myocyte and myocyte diameter from histological sections in animals with and without treatment with icCDCs [12]. Concordant with the global increase in the mitotic marker pHH3, we found that icCDCs increased myocyte nuclear density in LAD as well as remote regions of the hibernating heart. At the same time, cardiomyocyte size was substantially smaller after icCDC infusion (**Figure 9**). Myocyte diameter decreased from 15.7±0.4 to 8.9±0.3 μm in the hibernating LAD region (p<0.05) and from 15.0±0.4 to 9.5±0.4 μm in remote myocardium (p<0.05). The myocyte diamters after icCDCs were also notably smaller than myocyte size in our previous publications of untreated animals with or without hibernating myocardium [12,18,23]. At the same time, LAD myocyte nuclear density which was reduced in the hibernating region in untreated animals (749±56 myocyte nuclei/mm^2) increased to 1472±74 after icCDCs (p<0.05). Likewise, it rose from 981±22 to 1574±145 myocyte nuclei/mm^2 in the remote region (p<0.05). Despite increases in the number and reductions in the size of myocytes after icCDCs, gross pathological analysis confirmed that icCDCs did not increase global LV mass (icCDC treated 2.0±0.2 vs. 2.4±0.4 g/Kg by postmortem analysis, p-ns). Consistent with the lack of effect of icCDCs on mitotic markers in normal hearts, there was no effect of icCDC infusion on myocyte nuclear density or myocyte diameter in normal controls. Interestingly, icCDCs also attenuated myocardial interstitial fibrosis. Interstitial connective tissue in untreated hibernating LAD regions was 6.6±1.4% vs. 4.1±0.8% in remote regions. After icCDCs, LAD connective tissue decreased similar to remote regions (3.4±0.3% vs. 3.7±0.1%, p-ns vs. untreated hibernating myocardium).

To exclude the possibility that icCDCs simply stimulated myocyte nuclear division (with increased nuclear number per cell) without cell division, we assessed average myocyte nuclear number per cell in longitudinally-oriented myocytes in each histological sample (**Figure 9B**). Porcine myocytes in normal hearts were multi-nucleated with 4.8±0.1 nuclei/myocyte, similar to values obtained in isolated porcine myocytes [24]. There was a small reduction in nuclei/myocyte from animals with untreated hibernating myocardium (LAD region 4.1±0.1 nuclei/cell, p<0.05 vs. normal animals) while remote regions remained unchanged (4.5±0.2 nuclei/myocyte vs. 4.6±0.1 nuclei/myocyte in normal animals, p-ns). Four-weeks after global icCDC infusion,

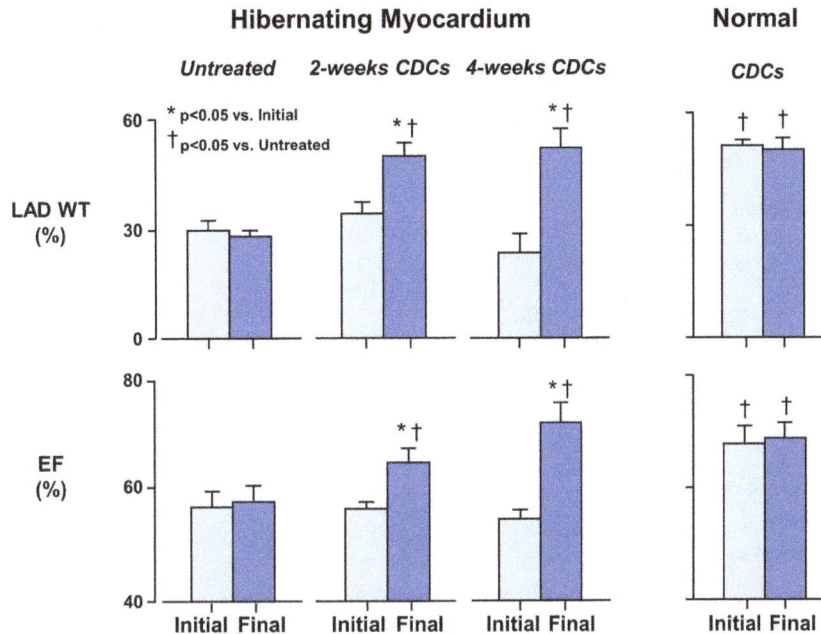

Figure 4. Effects of icCDCs on Regional and Global Function in Hibernating and Normal Hearts. Most animals with hibernating myocardium developed a chronic LAD occlusion in the absence of infarction. Regional LAD wall thickening (% LAD WT) was chronically depressed in hibernating animals as compared to normal remote regions (29±3% vs. 76±7%, p<0.05) with a mild reduction in ejection fraction (EF) vs. normals (57±3% vs. 68±3%, p<0.05). In untreated animals with hibernating myocardium, serial physiological parameters were stable with no change in LAD wall thickening or ejection fraction between studies performed up to 4-weeks later. In contrast, in animals treated with icCDCs, regional LAD wall thickening increased along with a return of ejection fraction to normal. There was no effect of icCDCs on regional or global function in normal animals. Light Blue-Initial; Dark Blue-Final.

the number of nuclei/myocyte in hibernating myocardium increased modestly to values similar to those found in normal hearts (5.0±0.2 nuclei/myocyte in LAD and 4.8±0.2 nuclei/myocyte in remote after icCDCs; p-ns vs. normal hearts). These data indicate that some of the increase in myocyte nuclear density after icCDCs reflected normalization of the reduced nuclear number/myocyte that was present in hibernating myocardium. Nevertheless, the relative increase in nuclei per cell were smaller than the measured increase in myocyte nuclei per mm^2 after icCDCs.

Estimate of Myocytes Derived Directly From icCDCs vs. Endogenous Myocyte Regeneration

To determine the number of new myocytes that were directly derived from icCDCs, we assessed Y-FISH positive cells in the hearts of female animals treated with male donor icCDCS 2-weeks after treatment (n = 5). We found 7±1 Y-FISH positive cells per million myocyte nuclei after icCDCs (1±0.1 per cm^2 tissue section). Importantly, this was two orders of magnitude lower than the frequency of proliferating myocytes (~600 pHH3 positive myocytes per million myocytes or ~98 per cm^2, **Figure 7**). When these measurements were extrapolated to the entire heart, only ~3.4±0.7% of the original icCDC dose (or a total of~1 million CDCs or CDC-derived cells) remained in the heart. Most donor cells were non-myocytes but there were rare examples of Y-FISH positive cardiomyocytes derived from donor CDCs (**Figure 10**). Specifically, 1.0±0.7 Y-FISH positive cells per million myocyte nuclei were also positive for cardiac troponin I, indicating that ~15% of the retained CDCs differentiated into cardiomyocytes. Even if we assumed that each Y-FISH positive cell will differentiate into a cardiomyocyte, CDC-derived myocytes would represent no more than 1 in every 5000 myocytes (or 0.05%).

Thus, the paucity of Y-FISH staining in the face of significant increases in myocyte nuclear density indicates that most of the new myocytes arose from the recipient and not via differentiation of donor CDCs.

Effects of icCDCs on Circulating cKit$^+$ and CD133$^+$ Bone Marrow Progenitor Cells (n = 6)

Since CDCs expressed mesenchymal markers and we previously demonstrated that intracoronary mesenchymal stem cell infusion mobilizes bone marrow progenitor cells which contribute to cardiac repair [12], we evaluated circulating progenitor cells after icCDCs. FACS analysis demonstrated that icCDCs did not increase circulating cKit$^+$ or CD133$^+$ cells when assessed either 3-days or 2-weeks after icCDC infusion. Since the results were similar in hibernating and sham animals, the final results were pooled. Bone marrow mononuclear cells that were cKit+ and CD45- averaged 6.9±1.7% at baseline, 3.2±0.2% 3-days after icCDCs and 4.2±1.5% 2-weeks after icCDCs (all p-ns vs. initial baseline values). The percentage of bone marrow mononuclear cells that were CD133+ but negative for hematopoietic lineage markers (CD45−) averaged 0.02±0.01%, 0.02±0.01% at 3 days and 0.06±0.02% after 2-weeks (all p-ns). Circulating mononuclear cells were not affected by icCDCs with cKit+/CD45− cells averaging 1.2±0.3% at baseline, 0.3±0.04% after 3-days and 1.2±0.4% after 2-weeks (all p-ns). Likewise, circulating CD133+/ CD45- mononuclear cells averaged 0.05±0.02% at baseline, 0.02±0.01% after 3-days and 0.06±0.03% after 2-weeks (all p-ns), and CD133+ cells were rarely observed in myocardial tissue after icCDCs (data not shown). Thus, unlike the intracoronary infusion of autologous mesenchymal stem cells [12], icCDCs did not mobilize bone marrow progenitor cells.

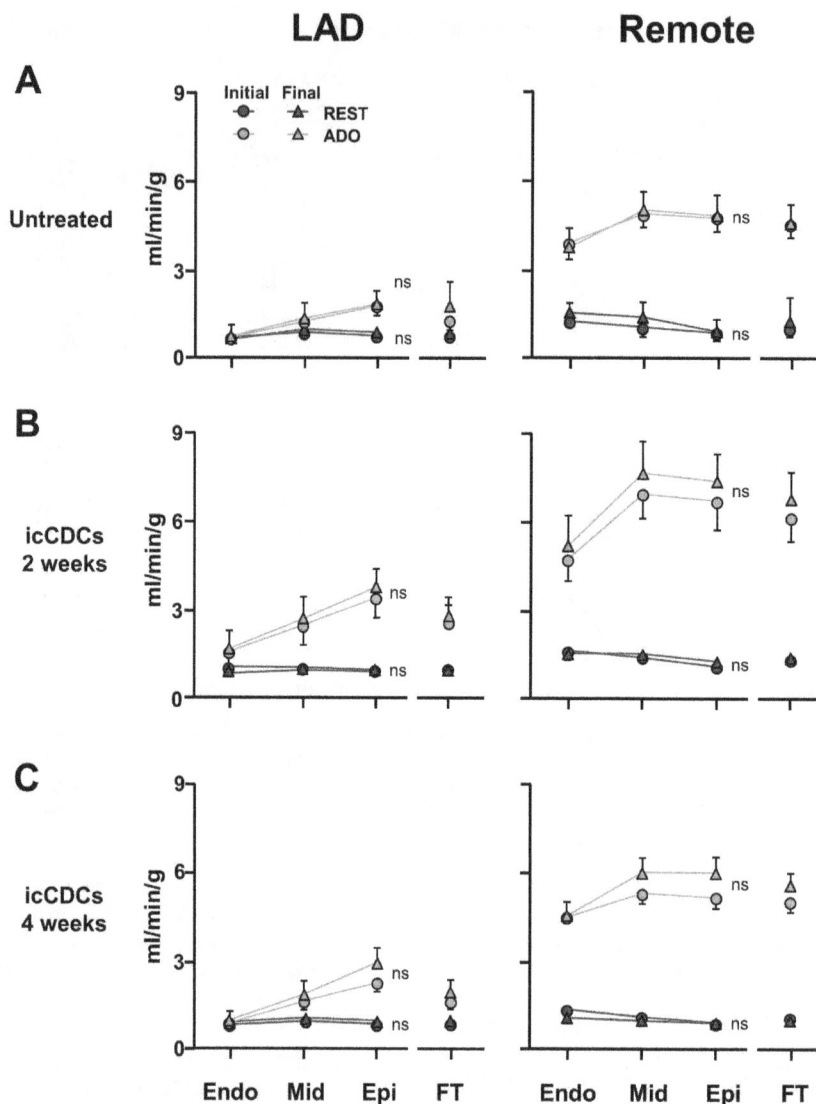

Figure 5. Administration of icCDCs Does Not Alter Myocardial Perfusion. Serial microsphere measurements of transmural myocardial perfusion at rest (solid symbols) and pharmacological vasodilation with adenosine (open symbols) in swine with hibernating myocardium. At the initial baseline study, coronary flow reserve was critically impaired in the LAD region but increased over 4-fold after adenosine in normally perfused remote regions. In untreated animals (**A**), paired analysis of initial and final measurements demonstrated stable myocardial perfusion indicating no spontaneous improvement in coronary collateral flow in this model over time. Paired analysis of initial and final measurements of coronary flow at rest and vasodilation also did not change either 2-weeks (B) or 4-weeks (C) after icCDC infusion. These results indicate that the functional improvement seen after icCDC infusion is not related to an increase in coronary collateral perfusion in swine with hibernating myocardium.

Discussion

Our results demonstrate that global infusion of icCDCs without using the "stop-flow" technique is a feasible, safe and efficacious approach to administer CDCs to the entire heart. Allogeneic icCDCs given with cyclosporine immunosuppression improved global and regional contractile function in swine with hibernating myocardium with no effect when administered to normal controls. This improvement was accompanied by increased myocyte nuclear density and a reduction in myocyte size in the dysfunctional hibernating region as well as in normally-perfused remote regions of the hibernating heart. New myocytes were primarily derived from endogenous sources since Y-FISH staining demonstrated only rare Y-FISH positive myocytes. Collectively, our results in an animal model with contractile dysfunction that is

devoid of infarction support the feasibility of employing global intracoronary infusion of allogeneic CDCs to treat viable dysfunctional myocardium in patients with chronic ischemic heart disease.

Relation to Previous Studies

Autologous icCDCs have previously been infused into the infarct-related artery using the "stop-flow" technique. Using this approach, studies in swine [10] and the Phase I/II CADUCEUS trial [5] confirmed that CDCs regenerate muscle mass and decrease scar volume yet global systolic function assessed from the ejection fraction did not increase. Our findings raise the possibility that this may relate to restricting CDC delivery to the infarct zone. As a result, deleterious myocyte loss and dysfunction in the large amount of remodeled myocardial tissue remote from the infarct is

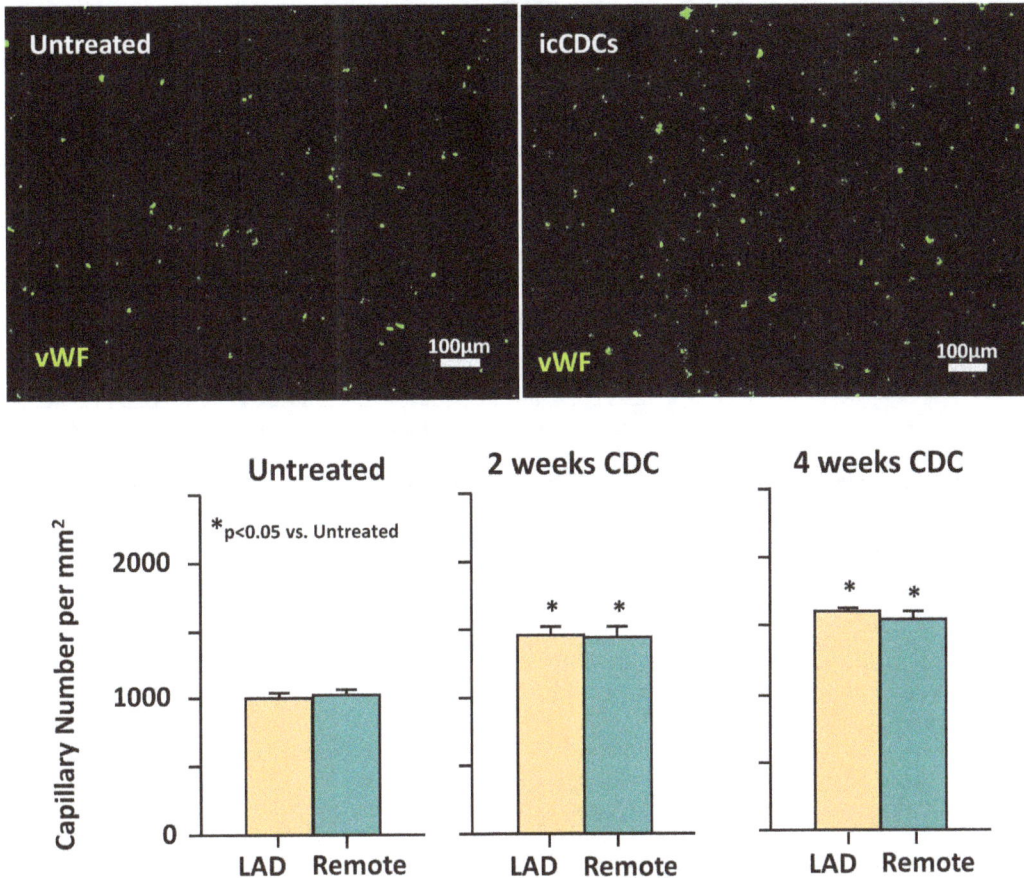

Figure 6. icCDCs Increased Capillary Density in Hibernating Myocardium. Capillary density was quantified using von Willebrand Factor (vWF). Upper images show vWF staining (green) from hibernating myocardium. Animals receiving icCDCs exhibited increased capillary density at 2-weeks and 4-weeks vs. untreated animals with hibernating myocardium (both p<0.05 vs. untreated, lower graph). While icCDCs stimulated capillary angiogenesis, their contribution to coronary vascular resistance was small since there was no functional improvement in coronary collateral perfusion at rest or after pharmacological vasodilation with adenosine. This is consistent with the negligible contribution of the low resistance capillary bed to total coronary vascular resistance.

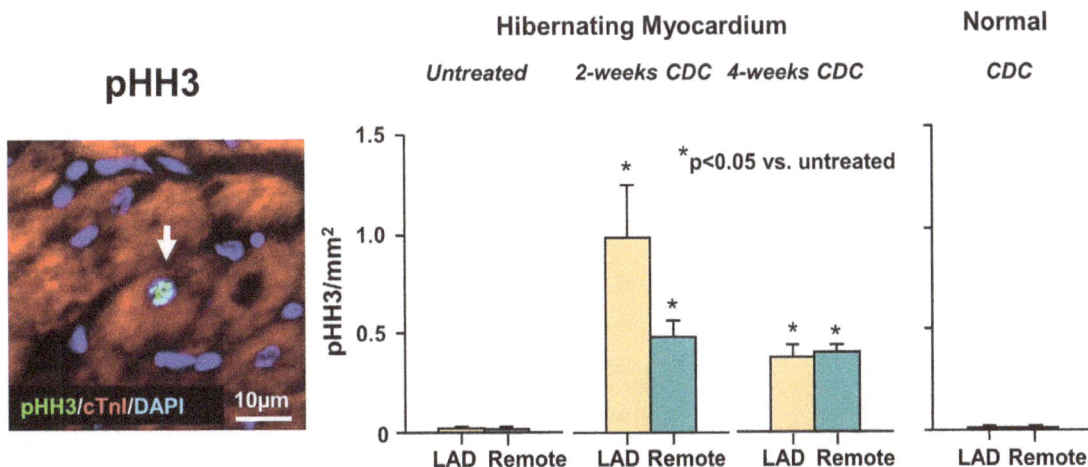

Figure 7. Effects of icCDCs on Cardiomyocyte Markers of Mitosis. Mitotic myocytes were quantified with phospho-histone-H3 (arrow, pHH3 positive myocyte nucleus localized with co-staining for DAPI and TnI). We found that icCDCs increased myocytes in the mitotic phase of the cell cycle in hibernating LAD and remote regions 2-weeks after injection. These declined but remained significantly elevated as compared to untreated animals and normal controls 4-weeks after icCDCs. There was no effect of icCDCs on the number of pHH3 positive myocytes in normal animals.

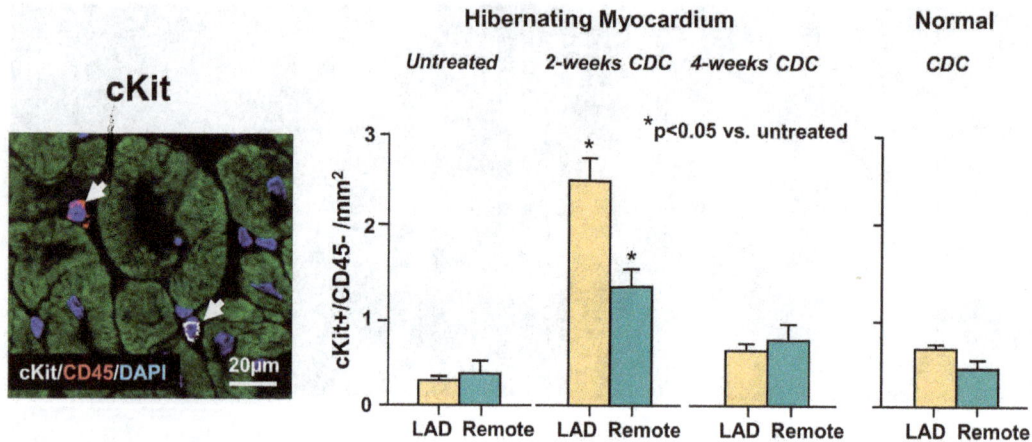

Figure 8. Effects of icCDCs on Myocardial cKit Cells. cKit positive cardiac progenitor cells (white) that did not express CD45(red) were counted and quantified in each heart. Intracoronary CDCs produced a transient increase in cKit+/CD45- cells at 2-weeks which returned to baseline after 4-weeks. Increases in cKit+ cells were global and similar in dysfunctional LAD and remote regions of hibernating hearts. In contrast, there was no effect of icCDCs on cKit+ cells in normal hearts. LAD-yellow bars; Remote-green bars.

not effectively treated [7,8]. With intracoronary infusion under continuous flow, the percentage of injected CDCs residing in the heart at 2-weeks was similar to intracoronary MSCs [12,25] and similar to the CDC retention reported by others using "stop-flow" or direct intramyocardial injection[21,26–28]. At the same time, the troponin I levels we found after global intracoronary infusion were lower than many studies using the "stop-flow" technique to administer cell-therapies into the infarct-related artery [5,10,29,30]. This could be due to the use of a Millipore filter, a lower regional cell dose or preventing in vivo aggregation of cells by infusing them under flow. Based on these findings, global CDC infusion appears to be a safe and clinically relevant alternative approach for intracoronary stem cell delivery. Further clinical studies will be required to establish feasibility in patients.

Source of New Myocytes After icCDCs

We found that icCDCs administered to the entire heart stimulated myocyte proliferation leading to an increase in myocyte nuclear density in a fairly short period of time. At the same time, there was a marked reduction in myocyte size throughout the entire heart. These changes were associated with an elevation in pHH3-positive myocytes indicative of myocyte cell division. Our results indicate that allogeneic icCDCs primarily facilitate endogenous myocyte proliferation of host recipient cells since the frequency of Y-FISH donor cells (including myocytes and non-myocytes) in tissue harvested at postmortem was extremely low. In addition, Y-FISH positive cells were also much lower than myocardial cKit+ cells suggesting that most tissue cKit+ cells likely arose from the host rather than donor CDC population. Although we [12] and others [31] have used cardiac localization of cKit+ cells as a marker of endogenous cardiac stem cell mobilization, it is important to note that the role of endogenous cKit+ cells in forming cardiomyocytes has recently been questioned [32]. Based upon our findings, the majority of newly formed myocytes probably arose from endogenous myocyte cell division but we cannot exclude a contribution arising from the differentiation of endogenous cardiac stem cells since we are unable to discriminate between these two possibilities in the large animal model. Support for both mechanisms is provided by a recent study documenting an approximately equal contribution of cardiomyocyte proliferation and endogenous stem cell recruitment

to new myocyte formation following CDC treatment in mice with myocardial infarction [33].

Regardless of the cellular source of new myocytes, the specific mechanism(s) by which CDCs stimulate endogenous myocyte regeneration remain elusive. Though not tested in the present study, one possibility is that icCDCs synthesize and release growth factors that stimulate endogenous myocyte proliferation. A similar mechanism has been proposed for MSCs [12]. Alternatively, it is plausible that icCDCs interact with resident CSCs or myocytes and effect cell-cell transfer of microRNAs or exosomes that allow myocytes or CSCs to re-enter the proliferative phase of the cell cycle [34]. Recent work in small animal models of myocardial infarction supports this notion. For example, Xie et al. demonstrated that β1 integrin signaling mediates cell-to-cell contact-dependent CDC stimulation of myocyte proliferation [35], while Ibrahim et al. identified secretion of microRNA-rich exosomes as a key mediator of CDC-mediated cardiac regeneration [36]. This exciting progress may facilitate the future development of new strategies to promote myocardial repair without the use of exogenous cells.

Autologous vs. Allogeneic icCDCs

While most previous studies in swine have used autologous CDC preparations [10,37], we found that allogeneic icCDCs were effective at stimulating myocyte proliferation and improving function when given with cyclosporine immunosuppression. Others have recently demonstrated the efficacy of allogeneic CDCs in rodents and more recently confirmed this in swine without using cyclosporine immunosuppression [21,38]. This likely reflects the fact that CDCs are immunoprivileged and do not express class II MHC molecules. It is also compatible with the notion that, like intracoronary mesenchymal stem cells (icMSCs) [12], CDCs facilitate endogenous cardiac regeneration and are not the direct precursor of most of the myocytes formed. Recently, Malliaras and colleagues demonstrated that myocyte regeneration after CDC treatment arises primarily from the host tissue using genetic fate mapping in mice [33] and histological evaluation in a swine model of myocardial infarction [21]. Our results provide additional support for this mechanism in a large animal model of ischemic LV dysfunction without infarction and confirm the feasibility of using allogeneic CDCs with cyclosporine to promote

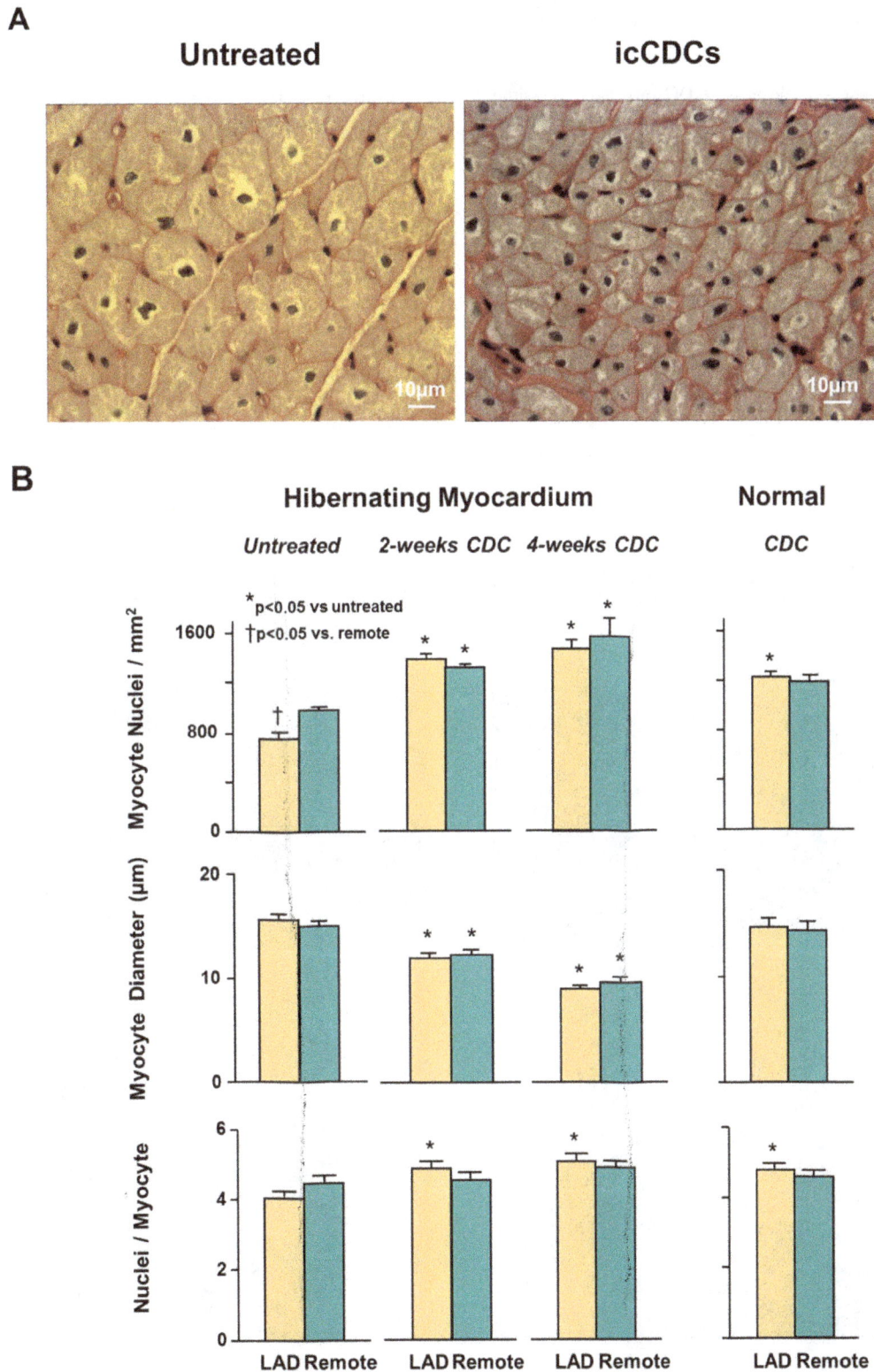

Figure 9. Quantitative Effects of icCDCs on Myocyte Nuclear Density and Myocyte Diameter in Dysfunctional LAD Regions and Normally-Perfused Remote Myocardium. A.) PAS staining demonstrates the typical increase in myocyte nuclear density and reduction of myocyte size found in a dysfunctional LAD region after icCDC treatment. Myocyte diameter in untreated hibernating myocardium was increased due to the loss of myocytes arising from apoptosis during the development of hibernating myocardium as we have previously described [6]. **B).** After icCDC infusion, there was a progressive increase in LAD myocyte nuclear density and a reduction in myocyte diameter that was significant as early as 2-weeks after treatment. Directionally similar changes were observed in remote regions from hibernating hearts. There was no effect of icCDCs on myocyte nuclear density or myocyte diameter in normal animals. Evaluation of longitudinally oriented myocytes showed only small differences in the

number of nuclei per myocyte. While values of nuclei per myocyte were lower in untreated hibernating myocardium, they rose to become no different than normal hearts after icCDCs. LAD-yellow bars; Remote-green bars.

cardiac repair. Although allogeneic CDCs are effective without immunosuppression [21,38], cyclosporine may have enhanced their effects by delaying an immune response to other mismatched donor antigens. Nevertheless, based upon the fact that most of the myocyte regeneration appears to be derived via recipient myocyte proliferation rather than derived from the donor CDCs in our sex mismatched recipients, it seems unlikely that chronic cyclosporine would be necessary once endogenous repair has been stimulated. While further studies could determine whether cyclosporine affords any advantage over using allogeneic CDCs alone, allogeneic icCDC therapy using the stop-flow approach and regional infusion to the infarct related artery without immunosuppression has recently been advanced to a Phase I/II clinical safety trial (NCT01458405, ClinicalTrials.gov).

Comparison of the Effects of icCDCs and icMSCs on Myocyte Regeneration

The effects of allogeneic icCDCs on myocyte regeneration are qualitatively similar to changes we reported after autologous icMSCs in swine with hibernating myocardium [12]. The two cell types are somewhat similar in that they each express mesenchymal surface antigens such as CD90 and CD105 and stimulate myocyte proliferation throughout hearts with viable dysfunctional myocardium. Nevertheless, they differ in other respects and the cellular mechanisms responsible for cardiac repair may differ. For example, although both icCDCs and icMSCs increased myocardial cKit+ cells and increased myocytes in the mitotic phase of the cell cycle, only icMSC treatment resulted in increased myocardial CD133+ bone marrow progenitor cells. Another difference is that porcine icMSCs did not express cKit, although it has recently been argued that the cKit fraction does not contribute to the regenerative efficacy of CDCs [39]. Finally, icMSCs rarely express

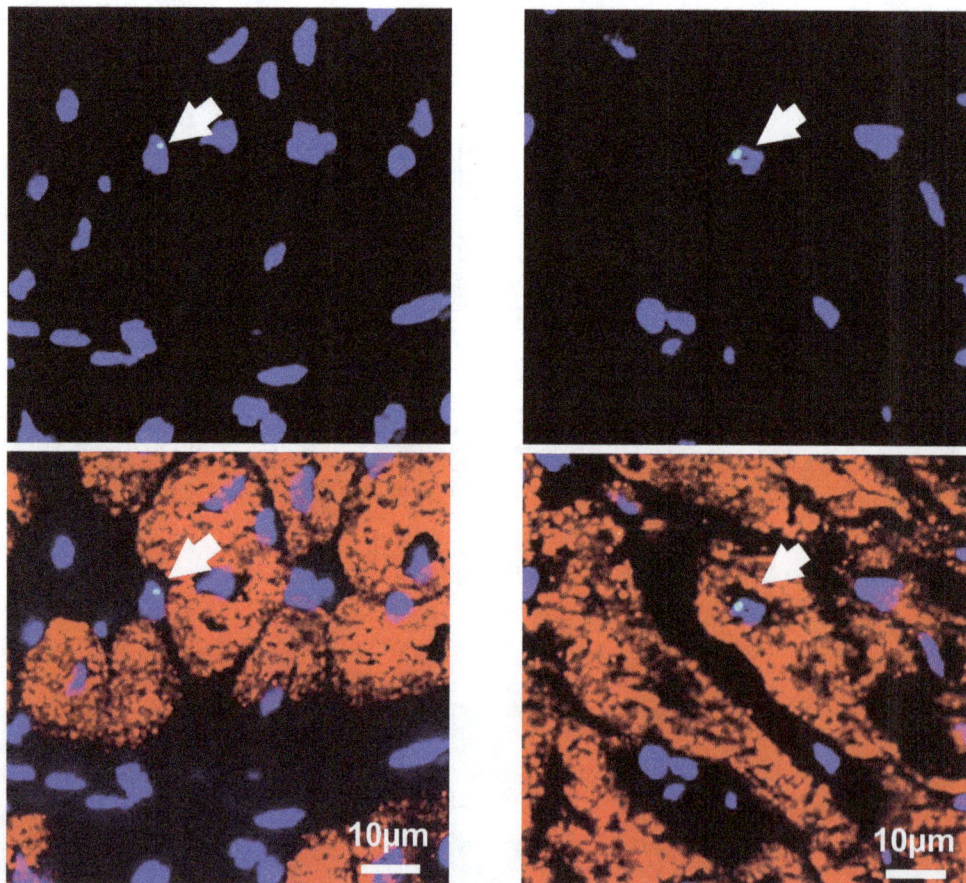

Figure 10. Examples of Sex-Mismatched CDC-Derived Cells in Hibernating Myocardium. To determine the fate of injected CDCs, male donor CDCs were injected into female recipients and Y-chromosomes were detected by fluorescence in situ hybridization (Y-FISH). An average of 3.4% of the total injected CDC dose wad retained in the left ventricle at 2-weeks. Y-chromosome positive cells were primarily found in the interstitial space. The arrows in the pair of photomicrographs in the left upper and lower panels demonstrate a Y-chromosome nucleus (green) in an interstitial cell with the nucleus stained blue (DAPI). This is distinct from the cardiac myocytes which are stained with troponin I in red. While infrequent, we identified rare Y-positive cardiac myocytes as depicted by the arrow on the upper and lower right panel. This indicates the potential for CDCs to differentiate into myocytes. Nevertheless, these constituted less than 1 in 5,000 of the new myocytes formed. None of the Y-FISH positive cells co-localized with cKit.

cardiac transcription factors (GATA4 and Nkx2.5) that were present in nearly all of the icCDCs we injected in the present study. Thus, while both intracoronary stem cell formulations stimulated myocyte regeneration, the precise molecular mechanisms along with the source of new myocytes could differ as could their quantitative impact on myocardial function and the number of new myocytes regenerated. Both autologous MSCs and CDCs have completed early phase I/II clinical trials using approaches that restrict administration to the infarct related artery or employ intramyocardial injection [40]. Since it will be challenging to compare different stem cell therapies in clinical trials, blinded head-to-head comparisons of stem cell formulations in preclinical studies like ours may provide insight into whether one approach is superior to the other to better inform selection of the best clinical platform for cardiac repair.

Experimental Considerations

In contrast to previous studies examining human CDCs [2], [41] the porcine CDCs used in the present study exhibited very high expression of CD90. This is a consistent finding in our laboratory, as we reliably observe ~95–99% CD90$^+$ expression in porcine CDCs between passage 4 and passage 6. It is unclear if this reflects species-related differences in the antigenic profile of CDCs, but this notion is supported by the fact that our cultivation protocol is identical to that utilized in previously published studies [2,5]. This may contribute to the paucity of myocytes derived directly from injected CDCs in light of recent *in vitro* data indicating that the CD90− fraction of human CDCs show superior cardiogenic differentiation potential [42]. Nevertheless, our results demonstrate that icCDC-mediated functional cardiac repair is not dependent on the presence of a significant population of CD90− cells. The use of thin (4 μm) tissue sections for immunohistochemical analyses precludes z-stack confocal imaging to confirm the localization of pHH3 positive and Y-FISH positive nuclei but also reduces the likelihood that a non-myocyte nucleus could be superimposed on top of a myocyte. Thus, we believe that the relative changes in pHH3 positive and Y-FISH positive myocytes between untreated and icCDC-treated animals are indicative of increased myocyte proliferation and CDC engraft-

ment, respectively. Finally, it is important to acknowledge the limitations of two-dimensional trans-thoracic echocardiography in providing precise quantification of LV ejection fraction in swine. Future studies should include advanced, three-dimensional imaging approaches (i.e., multi-detector computed tomography or cardiac magnetic resonance imaging), particularly when indices of global LV function are a primary endpoint.

Translational Implications

The results of the present study suggest that global intracoronary infusion of allogeneic CDCs to the entire heart is safe and efficacious without employing the "stop-flow" technique. Like allogeneic MSCs, allogeneic CDCs would provide an "off-the-shelf" formulation to treat patients with LV dysfunction and myocardial infarction circumventing the need for autologous stem cell harvesting and expansion. This could provide an approach that is easily implemented in any standard catheterization laboratory and broadly available since it would not require advanced instrumentation to inject cells into the myocardium, nor a preexisting coronary stent which is required for the "stop-flow" approach. While further studies are required to determine the most efficacious cell type, the global intracoronary infusion of allogeneic CDCs affords promise as a treatment for regional and global LV dysfunction due to ischemic and non-ischemic LV dysfunction.

Acknowledgments

These studies could not have been completed without the technical assistance of Deana Gretka, Elaine Granica, and Beth Palka. Anne Coe and Marsha Barber assisted with preparation of the manuscript and figures.

Author Contributions

Conceived and designed the experiments: GS JMC. Performed the experiments: GS MML AER RFY TRC BRW. Analyzed the data: GS BRW JMC TRC. Contributed to the writing of the manuscript: GS BRW JMC TRC MML AER RFY.

References

1. Jeevanantham V, Butler M, Saad A, Abdel-Latif A, Zuba-Surma EK, et al. (2012) Adult bone marrow cell therapy improves survival and induces long-term improvement in cardiac parameters: a systematic review and meta-analysis. Circulation 126: 551–568.
2. Smith RR, Barile L, Cho HC, Leppo MK, Hare JM, et al. (2007) Regenerative potential of cardiosphere-derived cells expanded from percutaneous endomyocardial biopsy specimens. Circulation 115: 896–908.
3. Marban E, Cheng K (2010) Heart to heart: The elusive mechanism of cell therapy. Circulation 121: 1981–1984.
4. Simpson DL, Mishra R, Sharma S, Goh SK, Deshmukh S, et al. (2012) A strong regenerative ability of cardiac stem cells derived from neonatal hearts. Circulation 126: S46–53.
5. Makkar RR, Smith RR, Cheng K, Malliaras K, Thomson LE, et al. (2012) Intracoronary cardiosphere-derived cells for heart regeneration after myocardial infarction (CADUCEUS): a prospective, randomised phase 1 trial. Lancet 379: 895–904.
6. Lim H, Fallavollita JA, Hard R, Kerr CW, Canty JM Jr (1999) Profound apoptosis-mediated regional myocyte loss and compensatory hypertrophy in pigs with hibernating myocardium. Circulation 100: 2380–2386.
7. Beltrami CA, Finato N, Rocco M, Feruglio GA, Puricelli C, et al. (1994) Structural basis of end-stage failure in ischemic cardiomyopathy in humans. Circulation 89: 151–163.
8. Abbate A, Narula J (2012) Role of apoptosis in adverse ventricular remodeling. Heart Fail Clin 8: 79–86.
9. Fallavollita JA, Heavey BM, Luisi AJ Jr, Michalek SM, Baldwa S, et al. (2014) Regional myocardial sympathetic denervation predicts the risk of sudden cardiac arrest in ischemic cardiomyopathy. J Am Coll Cardiol 63: 141–149.
10. Johnston PV, Sasano T, Mills K, Evers R, Lee ST, et al. (2009) Engraftment, differentiation, and functional benefits of autologous cardiosphere-derived cells in porcine ischemic cardiomyopathy. Circulation 120: 1075–1083.
11. Fallavollita JA, Logue M, Canty JM Jr (2001) Stability of hibernating myocardium in pigs with a chronic left anterior descending coronary artery stenosis: Absence of progressive fibrosis in the setting of stable reductions in flow, function and coronary flow reserve. J Am Coll Cardiol 37: 1989–1995.
12. Suzuki G, Iyer V, Lee TC, Canty JM Jr (2011) Autologous mesenchymal stem cells mobilize cKit+ and CD133+ bone marrow progenitor cells and improve regional function in hibernating myocardium. Circ Res 109: 1044–1054.
13. Fallavollita JA, Perry BJ, Canty JM Jr (1997) ^{18}F-2-deoxyglucose deposition and regional flow in pigs with chronically dysfunctional myocardium: Evidence for transmural variations in chronic hibernating myocardium. Circulation 95: 1900–1909.
14. Malm BJ, Suzuki G, Canty JM Jr, Fallavollita JA (2002) Variability of contractile reserve in hibernating myocardium: Dependence on the method of stimulation. Cardiovasc Res 56: 422–433.
15. Glenny RW, Bernard S, Brinkley M (1993) Validation of fluorescent-labeled microspheres for measurement of regional organ perfusion. J Appl Physiol 74: 2585–2597.
16. Kajstura J, Zhang X, Reiss K, Szoke E, Li P, et al. (1994) Myocyte cellular hyperplasia and myocyte cellular hypertrophy contribute to chronic ventricular remodeling in coronary artery narrowing-induced cardiomyopathy in rats. Circ Res 74: 383–400.
17. Bruel A, Nyengaard JR (2005) Design-based stereological estimation of the total number of cardiac myocytes in histological sections. Basic Res Cardiol 100: 311–319.
18. Suzuki G, Iyer V, Cimato T, Canty JM Jr (2009) Pravastatin improves function in hibernating myocardium by mobilizing CD133+ and cKit+ hematopoietic

progenitor cells and promoting myocytes to reenter the growth phase of the cardiac cell cycle. Circ Res 104: 255–264.

19. Perin EC, Silva GV, Assad JA, Vela D, Buja LM, et al. (2008) Comparison of intracoronary and transendocardial delivery of allogeneic mesenchymal cells in a canine model of acute myocardial infarction. J Mol Cell Cardiol 44: 486–495.

20. Hou D, Youssef EA, Brinton TJ, Zhang P, Rogers P, et al. (2005) Radiolabeled cell distribution after intramyocardial, intracoronary, and interstitial retrograde coronary venous delivery: implications for current clinical trials. Circulation 112: I150–156.

21. Malliaras K, Smith RR, Kanazawa H, Yee K, Seinfeld J, et al. (2013) Validation of contrast-enhanced magnetic resonance imaging to monitor regenerative efficacy after cell therapy in a porcine model of convalescent myocardial infarction. Circulation 128: 2764–2775.

22. Canty JM Jr, Suzuki G, Banas MD, Verheyen F, Borgers M, et al. (2004) Hibernating myocardium: Chronically adapted to ischemia but vulnerable to sudden death. Circ Res 94: 1142–1149.

23. Suzuki G, Lee TC, Fallavollita JA, Canty JM Jr (2005) Adenoviral gene transfer of FGF-5 to hibernating myocardium improves function and stimulates myocytes to hypertrophy and reenter the cell cycle. Circ Res 96: 767–775.

24. Spinale FG, Zellner JL, Tomita M, Crawford FA, Zile MR (1991) Relation between ventricular and myocyte remodeling with the development and regression of supraventricular tachycardia-induced cardiomyopathy. Circ Res 69: 1058–1067.

25. Leiker M, Suzuki G, Iyer VS, Canty JM Jr, Lee TC (2008) Assessment of a nuclear affinity labeling method for tracking implanted mesenchymal stem cells Cell Transpl 17: 911–922.

26. Li Z, Lee A, Huang M, Chun H, Chung J, et al. (2009) Imaging survival and function of transplanted cardiac resident stem cells. J Am Coll Cardiol 53: 1229–1240.

27. van der Bogt KE, Sheikh AY, Schrepfer S, Hoyt G, Cao F, et al. (2008) Comparison of different adult stem cell types for treatment of myocardial ischemia. Circulation 118: S121–129.

28. Terrovitis J, Kwok KF, Lautamaki R, Engles JM, Barth AS, et al. (2008) Ectopic expression of the sodium-iodide symporter enables imaging of transplanted cardiac stem cells in vivo by single-photon emission computed tomography or positron emission tomography. J Am Coll Cardiol 52: 1652–1660.

29. Chugh AR, Beache GM, Loughran JH, Mewton N, Elmore JB, et al. (2012) Administration of cardiac stem cells in patients with ischemic cardiomyopathy: the SCIPIO trial: surgical aspects and interim analysis of myocardial function and viability by magnetic resonance. Circulation 126: S54–64.

30. Bolli R, Tang XL, Sanganalmath SK, Rimoldi O, Mosna F, et al. (2013) Intracoronary deliver of autologous cardiac stem cells improves cardiac function

in a porcine model of chronic ischemic cardiomyopathy. Circulation 128: 122–131.

31. Xiong Q, Ye L, Zhang P, Lepley M, Swingen C, et al. (2012) Bioenergetic and functional consequences of cellular therapy: activation of endogenous cardio-vascular progenitor cells. Circ Res 111: 455–468.

32. van Berlo JH, Kanisicak O, Maillet M, Vagnozzi RJ, Karch J, et al. (2014) c-kit+ cells minimally contribute cardiomyocytes to the heart. Nature 509: 337–341.

33. Malliaras K, Zhang Y, Seinfeld J, Galang G, Tseliou E, et al. (2013) Cardiomyocyte proliferation and progenitor cell recruitment underlie therapeu-tic regeneration after myocardial infarction in the adult mouse heart. EMBO Mol Med 5: 191–209.

34. Hosoda T, Zheng H, Cabral-da-Silva M, Sanada F, Ide-Iwata N, et al. (2011) Human cardiac stem cell differentiation is regulated by a mircrine mechanism. Circulation 123: 1287–1296.

35. Xie Y, Ibrahim A, Cheng K, Wu Z, Liang W, et al. (2014) Importance of cell-cell contact in the therapeutic benefits of cardiosphere-derived cells. Stem Cells 32: 2397–2406.

36. Ibrahim AG, Cheng K, Marban E (2014) Exosomes as critical agents of cardiac regeneration triggered by cell therapy. Stem Cell Reports 2: 606–619.

37. Lee ST, White AJ, Matsushita S, Malliaras K, Steenbergen C, et al. (2011) Intramyocardial injection of autologous cardiospheres or cardiosphere-derived cells preserves function and minimizes adverse ventricular remodeling in pigs with heart failure post-myocardial infarction. J Am Coll Cardiol 57: 455–465.

38. Malliaras K, Li TS, Luthringer D, Terrovitis J, Cheng K, et al. (2012) Safety and efficacy of allogeneic cell therapy in infarcted rats transplanted with mismatched cardiosphere-derived cells. Circulation 125: 100–112.

39. Cheng K, Ibrahim A, Hensley MT, Shen D, Sun B, et al. (2014) Relative Roles of CD90 and c-Kit to the Regenerative Efficacy of Cardiosphere-Derived Cells in Humans and in a Mouse Model of Myocardial Infarction. J Am Heart Assoc 3: e001260.

40. Williams AR, Trachtenberg B, Velazquez DL, McNiece I, Altman P, et al. (2011) Intramyocardial stem cell injection in patients with ischemic cardiomy-opathy: functional recovery and reverse remodeling. Circ Res 108: 792–796.

41. Li TS, Cheng K, Malliaras K, Smith RR, Zhang Y, et al. (2012) Direct comparison of different stem cell types and subpopulations reveals superior paracrine potency and myocardial repair efficacy with cardiosphere-derived cells. J Am Coll Cardiol 59: 942–953.

42. Gago-Lopez N, Awaji O, Zhang Y, Ko C, Nsair A, et al. (2014) THY-1 Receptor Expression Differentiates Cardiosphere-Derived Cells with Divergent Cardiogenic Differentiation Potential. Stem Cell Reports 2: 576–591.

18

More than Just Two Sexes: The Neural Correlates of Voice Gender Perception in Gender Dysphoria

Jessica Junger[1,2]*, Ute Habel[1,2], Sabine Bröhr[1], Josef Neulen[3], Christiane Neuschaefer-Rube[4], Peter Birkholz[4], Christian Kohler[5], Frank Schneider[1,2], Birgit Derntl[1,2], Katharina Pauly[1,2]

1 Department of Psychiatry, Psychotherapy and Psychosomatics, Medical School, RWTH Aachen University, Aachen, Germany, 2 Jülich Aachen Research Alliance-Translational Brain Medicine, Jülich, Germany, 3 Department of Gynaecological Endocrinology and Reproductive Medicine, Medical School, RWTH Aachen University, Aachen, Germany, 4 Department of Phoniatrics, Pedaudiology and Communication Disorders, Medical School, RWTH Aachen University, Aachen, Germany, 5 Department of Psychiatry, Neuropsychiatry Division, University of Pennsylvania School of Medicine, Philadelphia, Pennsylvania, United States of America

Abstract

Gender dysphoria (also known as "transsexualism") is characterized as a discrepancy between anatomical sex and gender identity. Research points towards neurobiological influences. Due to the sexually dimorphic characteristics of the human voice, voice gender perception provides a biologically relevant function, e.g. in the context of mating selection. There is evidence for a better recognition of voices of the opposite sex and a differentiation of the sexes in its underlying functional cerebral correlates, namely the prefrontal and middle temporal areas. This fMRI study investigated the neural correlates of voice gender perception in 32 male-to-female gender dysphoric individuals (MtFs) compared to 20 non-gender dysphoric men and 19 non-gender dysphoric women. Participants indicated the sex of 240 voice stimuli modified in semitone steps in the direction to the other gender. Compared to men and women, MtFs showed differences in a neural network including the medial prefrontal gyrus, the insula, and the precuneus when responding to male vs. female voices. With increased voice morphing men recruited more prefrontal areas compared to women and MtFs, while MtFs revealed a pattern more similar to women. On a behavioral and neuronal level, our results support the feeling of MtFs reporting they cannot identify with their assigned sex.

Editor: Bernhard Fink, University of Goettingen, Germany

Funding: This study was supported by the German Research Foundation (DFG: HA 3202/7-1 and IRTG 1328) and the Brain Imaging Facility of the Interdisciplinary Centre for Clinical Research of the Faculty of Medicine at RWTH Aachen University, Germany. The funders had no role in study design, data collection and analysis, decision to publish, or preparation of the manuscript.

Competing Interests: The authors have declared that no competing interests exist.

* Email: jjunger@ukaachen.de

Introduction

The dichotomous classification of gender is based on the fact that the two sexes are essential to the survival of our [1] and other [2] species. A combination of chromosome complement and genital and gonadal phenotypes also contributes to the conventional dichotomy [3]. However, non-binary gender diversity abounds in both humans and animals [4]. Primary sexual characteristics such as testes and ovaries develop in early pregnancy prior to the sexual differentiation of the brain. The latter is associated with sexually distinctive behavior and appears to be continuously influenced by sex hormones. The two processes are independent of each other and can take different routes [5]. Therefore, increasing evidence is linking their disparate differentiation to gender dysphoria [6] also known as transsexualism [7,8]. According to DSM-5, gender dysphoria describes the experience of a marked incongruence between one's assigned and one's expressed gender [9]. Gender denotes one's public role as male or female with biological and psychological factors interacting with gender development. Expressed gender includes several alternative identities beyond binary stereotypes [10]. This discrepancy, accompanied by distress or impairment in social and occupational

functioning, can result in mental problems calling for clinical interventions [11,12,13]. Individuals who suffer from gender dysphoria often aspire to adapt their assigned to their expressed gender with hormonal treatment or sex-reassignment surgery [14]. Those individuals are often referred to with assigned-to-target sex terms such as male-to-female (MtFs) or female-to-male (FtMs).

Although the etiopathogenesis of gender dysphoria is yet to be adequately defined, recent findings point towards neurobiological mechanisms involved [15]. No detectable influences of hormonal status [16] on gender dysphoria have been found so far. Concerning genetic factors the evidence is inconsistent: Some studies lacked to find genetic influences [17,18,19], whereas Hare and colleagues [20] identified an association between gender dysphoria and the androgen receptor allele. This is consistent with studies suggesting a much higher prevalence of gender dysphoria in monozygotic than in dizygotic twins [21] as well as in non-twin siblings than in the general population [22]. The experience of gender identity is linked to the sexual differentiation of the brain and may therefore be detectable in brain structure and function [8,23]. Post-mortem anatomical studies have consistently revealed a similarity in the bed nucleus of stria terminalis (BSTc), one of the regions with marked gender differences in volume, between MtFs and women [24,25] as well as between one FtM and men [26]. However, generalization from these findings is limited by the constraints of small sample sizes in post-mortem studies.

In contrast, magnetic resonance imaging (MRI) is a sensitive tool to investigate the cerebral basis of sexual differentiation in vivo. There are only a few structural studies examining gender dysphoria: Simon and colleagues [27] reported grey matter volume differences between gender dysphoric individuals and a mixed gender control group, Luders and colleagues [28] as well as Savic and Arver [29] provided evidence for differences in grey matter volumes between individuals with gender dysphoria and both men and women. Rametti and colleagues [30] reported an intermediate position for hormonally untreated MtFs between male and female brains by measure of white matter microstructure. Further, studies found female-like white matter structures in FtMs [31] as well as changes in white matter structure in FtMs [32,33] and cortical thickness in FtMs and MtFs [33] through hormonal treatment. Thus, structural evidence points towards differences between gender dysphoric individuals and their biological and partly also the opposite sex as well as changes through the intake of hormones.

Functional imaging provides a more directed pattern: Greater similarity was found between an untreated FtM and females compared to males in functional connectivity maps involving the lingual gyrus and the precuneus [34]. However, in the light of a single case study, results should be taken with caution. In contrast, to the best of our knowledge other functional studies, so far, rather found male-like brain activation in FtMs [35] and female-like cerebral activity in MtFs in language and mental rotation tasks [36,37,38] as well as during the presentation of erotic stimuli [39] and, therefore, in tasks especially sensitive to neural sex differences in neuronal activity [40,41,42]. Thus in sum, findings from functional studies point toward increased similarities between gender dysphoric individuals and their aspired gender. However, the amount of research that has been done in this area is rather small, and differences in sample size and methodology have led to an ambiguous integration of gender dysphoria into neurobiological research. Further studies with larger sample sizes are needed to explore additional biological and social factors influencing gender-specific experience and behavior [43,44].

The human voice is an important secondary sexual characteristic directly linked to gender identification. It plays a key role in social interaction [45,46] and mate selection [47,48,49]. The extra-linguistic parameters, fundamental frequency (F0; i.e. vocal fold vibration perceived as pitch) and formant frequencies (i.e. vocal tract resonances) are of major relevance to sex identification [50,51,52,53]. While the average voice pitch varies about one octave (i.e. 12 semitones) between men and women, there is an overlapping, "gender-ambiguous" range in which the decision for a male or female voice depends especially on formants and other contextual parameters such as visual information or prosodic characteristics [51,54,55].

Voices are first processed in the superior temporal gyri, the anterior superior temporal sulci and the middle temporal gyri [46,56,57,58,59,60]. Further there is evidence for an involvement of the inferior frontal gyrus and the cerebellum in voice gender perception [61]. Thereby right inferior frontal gyrus seems to reflect the processing of F0 modulation, which is the main acoustic correlate of prosody [62]. In a male sample, the differentiation of male and female voices resulted in greater activation in the right anterior superior temporal gyrus while listening to female voices, and in the right precuneus while perceiving male voices [52]. In a gender-mixed sample, stronger neural activity was observed in

response to female as compared to male voices in the right supratemporal plane, the right posterior superior temporal gyrus, left postcentral gyrus, the bilateral inferior parietal lobes and the insula [63]. In a previous study [64], we found the first evidence of an interaction between voice gender and the listener's sex; namely, stronger activation in men compared to women in the left middle temporal, left orbital and right medial prefrontal cortices for the processing of female compared to male voices. Further, increased voice gender morphing in the direction of the opposite sex resulted in stronger activation in the superior and middle frontal gyri in men compared to women.

A possible explanation for this may come from animal research, which indicated a neural basis for sex differences in the auditory perception of mating calls [65,66]. Detection of gender distinguishing signals in general is important for the identification of adequate mates and results in a preference as well as stronger emotional and attentional reactions to signals of the opposite sex [67,68,69]. Thus, sexual dimorphism of vocalization and its perception in both animals and humans are evolutionary relevant for reproductive needs. Therefore, we expected that a voice gender perception task should be highly sensitive to sex differences in neuronal activity in areas involved in auditory, affective, attentional and evaluative processes.

In the present study, we aimed at investigating the neural correlates of voice gender processing in MtFs in comparison to men and women. Based on the few structural and functional findings, so far, we speculate to find at least distinct neural activation in MtFs compared to men and women if not a tendency to a more female-like activation pattern in voice-selective brain regions, such as the superior and middle temporal gyri and prefrontal areas. Further, we hypothesized that hormonally treated (contrasted to untreated) MtFs are more equal to the aspired sex. Results are discussed with respect to an integration of gender-variant identities into our understanding of the sexes.

Materials and Methods

Participants

17 hormonally untreated MtFs, 16 hormonally treated MtFs, 21 non-gender dysphoric men and 20 non-gender dysphoric women participated in this study. MtFs were recruited in self-help groups at the Department of Phoniatrics, Pedaudiology and Communication Disorders of the RWTH Aachen University Hospital and by word-of-mouth recommendation. MtFs identified themselves as gender dysphoric, expressed a strong sense of belonging to the opposite sex, and lived the desired role in everyday life. MtFs taking hormones were treated following the German transsexual law [70] for at least 3 months and therefore had overcome the first phase of endocrinological adjustment. Untreated MtFs declared their intention of undergoing cross-sex hormone therapy in the future. The German version of the Structured Clinical Interview of the fourth edition of the Diagnostic and Statistical Manual of Mental Disorders (DSM-IV) [71] was used to ensure the exclusion of participants with mental disorders of axis I unrelated to gender dysphoria. Further exclusion criteria were neurological disorders and other medical conditions affecting the cerebral metabolism as well as first degree relatives with a history of mental diseases. All participants were native speakers of German and right-handed aside from one left-handed participant in each group. Handedness was assessed by means of the Edinburgh Handedness Inventory [72].

The hormonal status was obtained on the day of testing from all participants, except 3 from whom no blood samples could be taken and 5 from whom some blood parameters were not available due to technical problems. The number of hetero- and homosexual participants was equal in both MtF samples. (Sexual orientation in MtFs was defined according to their anatomical sex, i.e. homosexual MtFs prefer male partners).

Four participants were excluded due to excessive movement in the scanner. Hence, data from 71 participants (16 untreated MtFs, 16 treated MtFs, 19 women, 20 men) were included in the final analyses (Table 1). Data pertaining to the controls have already been reported [64]. Kolmogorov-Smirnov tests showed that data were normally distributed regarding age, years of education or crystallized verbal intelligence [73] estimation. Using analyses of variance (ANOVA) group comparisons showed no significant differences in all three measures. Since hormonal data did not fulfill the assumptions of normal distribution, nonparametric Kruskal-Wallis tests were used for analyzing the latter (Table 1).

The local Institutional Ethics Committee of the Medical Faculty of RWTH Aachen University approved the study (reference: EK 088/09). All participants were financially reimbursed and gave their written informed consent.

Stimuli and procedure

A more detailed description of the stimulus preparation and presentation appears in Junger et al. [64]. In brief, the voices of 10 men and 10 women were recorded while reading 3 out of 6 emotionally neutral 3-syllable nouns. To guarantee a natural and consistent prosody and pronunciation, target words were spoken in the context of the carrier sentence "I said ..." and then cut out subsequently. The resulting 30 original male and 30 original female voice stimuli were further modified (morphed) in 2, 4 and 6 semitone steps (st) in the direction of the other gender. The software Praat [74] was used to shift pitch contour and formant structure accordingly. Correspondingly, the final task consisted of

8 experimental conditions in a 2×4 event-related design with the factors voice gender and morphing level: 1.) original/0st male voice, 2.) male voice morphed by 2st, 3.) male voice morphed by 4st, 4.) male voice morphed by 6st, 5.) original/0st female voice, 6.) female voice morphed by 2st, 7.) female voice morphed by 4st, and finally 8.) female voice morphed by 6st. Every condition was presented ten times and comprised 3 different nouns spoken by 3 different voices of the same sex. This resulted in a total of 240 voice stimuli presented in a pseudo-randomized order.

Stimulus presentation was done via electrostatic headphones with Presentation 14.2 software (http://www.neurobs.com) individually adapted for loudness. Participants were asked to indicate the gender of each speaker by button press as fast as possible.

Analysis of the behavioral data

Behavioral data were analyzed using SPSS 18.0.0 (SPSS Inc., Chicago, IL). Since data were not normally distributed, nonparametric Friedman rank tests [75] were performed to detect mean rank differences between conditions regarding the amount of hits and reaction times (RT). Due to the two-alternative discrimination task, inclusion of the error rates only revealed a pattern opposite to the hit rates and thus provided no additional information.

To decompose significant effects regarding voice gender (male/female) and stimulus morphing (0st/2st/4st/6st) for hits and RT, Wilcoxon tests were calculated for within-group comparisons in the whole group and each group separately and Mann-Whitney-U tests for the between-groups comparisons.

For each morphing level, discrimination sensitivity (d-prime - d') and potential response biases (log ß) [76] were evaluated for two-alternative discrimination tasks [77]. Because of their normal distribution, one-sample t-tests were calculated for each group separately in order to analyze if d-prime was above chance level

Table 1. Characteristics of the sample (mean and standard deviations for age, education, IQ, hormonal level and sexual orientation) and group comparisons.

	Men	Women	MtF untreated	MtF treated	P (ANOVA)
Age	32.35 (10.27)	33.16 (12.34)	36.38 (14.02)	30.19 (10.95)	0.528
Education	15.00 (2.92)	14.95 (3.21)	14.50 (3.06)	13.81 (3.04)	0.646
IQ	112.45 (14.72)	112.21 (16.14)	113.13 (13.57)	104.25 (6.81)	0.204
Sexual orientation[D]					P (Chi square)
Heterosexual	18	18	7	5	
Homosexual	1	0	7	9	
Bisexual	0	0	2	2	
					<0.001*
Hormonal level					P (Kruskal-Wallis)
17-ß-Estradiol (pmol/l)	88.62 (34.89)	136.04 (134.52)	91.67 (50.92)	1400.10 (3137.44)[A]	<0.001*
FSH (U/l)	5.35 (6.08)	13.4 (24.97)	4.62 (2.77)	5.57 (9,99)	0.055
LH (U/l)	5.76 (1.99)	10.32 (15.26)	5.30 (3.08)	4.14 (9.72)	0.011
Progesterone (nmol/l)	2.32 (1.01)	4.46 (8.69)	2.06 (1.04)	1.7 (0.80)	0.267
Prolactin (mU/l)	166.16 (58.41)	195.42 (57.12)	165.50 (65.25)	571.46 (538.28)[A]	0.001*
Sex steroid binding globulin (SHBG; nmol/l)	27.88 (11.05)[C]	127.83 (65.95)[B]	40.28 (40.01)[C]	91.19 (71.76)	<0.001*
Free testosterone (pmol/l)	40.71 (14.32)[C]	3.74 (2.33)[B]	32.85 (13.64)[C]	4.48 (5.97)	<0.001*

Significant differences are marked in asterisk.
[A]significant difference with respect to all three other groups, $p = 0.008$ Bonferroni corrected.
[B]significant differences with respect to men and MtF untreated, $p = 0.008$ Bonferroni corrected.
[C]significant differences with respect to MtF treated, $p = 0.008$ Bonferroni corrected.
[D]respective data are missing in one man and one woman.

and if potential biases reached significance. To compare groups, ANOVAs were computed and decomposed by post hoc two-sample t-tests where significant.

All post-hoc tests were Bonferroni corrected for multiple comparisons.

All data points used to determine averages and summary statistics in Table 1 and Table 2 as well as in the behavioral results section can be found in Table S4–Table S7 in File S1.

fMRI data acquisition and pre-processing

Functional imaging data were acquired at the Department of Psychiatry, Psychotherapy and Psychosomatics of the RWTH Aachen University Hospital on a 3 T Siemens Trio MR Scanner (Siemens Medical Systems, Erlangen, Germany). Echo-planar imaging (EPI) sensitive to blood-oxygen-level-dependent (BOLD) contrast were used (T2*, voxel size: $3.1 \times 3.1 \times 3.1$ mm^3, distance factor 15%, GAP 0.5 mm, 64×64 matrix, FoV: 200×200 mm^2, TR = 2s, TE = 30 ms, $\alpha = 76°$) with 36 slices covering the entire brain. To avoid magnetic field saturation effects, image acquisition was preceded by 6 dummy scans which were discarded before preprocessing. The resulting 785 volumes per subject were analyzed using SPM8 (http://www.fil.ion.ucl.ac.uk/spm) implemented in MATLAB 2010b (Mathworks, Sherborn, MA). Images were realigned to the first volume. Spatial normalization into MNI space was accomplished by means of the unified segmentation approach [78]; an 8 mm FWHM Gaussian kernel was used for smoothing. A 128 Hz high pass filter removed effects of low frequency noise.

Analysis of the fMRI data

On the first level, regressors were modeled for each of the eight experimental conditions (i.e. the four morphing levels of male and female voices) for each subject and subsequently entered into the second level (data available from the Dryad Digital Repository: http://doi.org/10.5061/dryad.48tj0). Three different flexible factorial designs were calculated for the group analyses applying a general linear mixed-effects model (GLM) approach. Participants were entered as random effects and conditions as fixed effects. Movement parameter regressors and individual mean RTs were included as nuisance covariates. The first model contained 4 groups, namely men, women, and untreated and treated MtFs. However, no significant activation differences were found between the two MtF groups. To further analyze the effect of sexual orientation, we ran a GLM with the 4 groups (men, women, heterosexual MtFs and homosexual MtFs) and found no significant differences between hetero- and homosexual MtF groups in our investigated contrasts. Therefore, we pooled the data of all MtFs resulting in a GLM with 3 groups.

Based on this third model, main effects for original voices were calculated to investigate the neural correlates of voice perception in each group separately.

In order to compare our findings with previous results, contrasts were computed for interactions between the listeners' sex and the original voice gender [64] for the comparison between MtFs and the control groups: 1) [men 0 w > men 0 m] > [MtF 0 w > MtF 0 m], 2) [women 0 w > women 0 m] > [MtF 0 w > MtF 0 m], 3) [men 0 m > men 0 w] > [MtF 0 m > MtF 0 w], and 4) [women 0 m > women 0 w] > [MtF 0 m > MtF 0 w]. (Due to the subtraction method in the construction of contrasts, these interactions are arithmetically equivalent to 1) [MtF 0 m > MtF 0 w] > [men 0 m > men 0 w], 2) [MtF 0 m > MtF 0 w] > [women 0 m > women 0 w], 3) [MtF 0 w > MtF 0 m] > [men 0 w > men 0 m], and 4) [MtF 0 w > MtF 0 m] > [women 0 w > women 0 m]). To decompose the underlying origin of the effects, mean beta values and standard errors of each peak voxel were extracted in each group separately and for both voice conditions.

Table 2. Behavioral outcome measures.

	Men	Women	MtF	Men	Women	MtF
	Mean (SD)	Mean (SD)	Mean (SD)	Mean (SD)	Mean (SD)	Mean (SD)
Hits (%)						
Male voice				*Female voice*		
0st	97.00 (3.73)	99.29 (1.78)	98.02 (2.79)	97.66 (2.88)	94.21 (6.29)	93.44 (7.83)
2st	90.66 (7.48)	96.84 (4.07)	93.33 (7.43)	91.16 (10.83)	80.35 (11.75)	87.19 (9.20)
4st	63.00 (15.02)	83.15 (12.78)	75.94 (12.64)	81.66 (9.64)	65.78 (15.98)	72.19 (11.87)
6st	36.50 (17.18)	62.63 (15.65)	52.19 (17.47)	58.66 (16.27)	36.49 (15.93)	50.94 (15.48)
Reaction time						
Male voice				*Female voice*		
0st	1.22 (0.19)	1.06 (0.15)	1.14 (0.18)	1.25 (0.23)	1.07 (0.16)	1.18 (0.2)
2st	1.32 (0.29)	1.14 (0.14)	1.25 (0.22)	1.32 (0.27)	1.18 (0.24)	1.32 (0.22)
4st	1.51 (0.29)	1.22 (0.13)	1.39 (0.23)	1.34 (0.25)	1.27 (0.21)	1.32 (0.21)
6st	1.65 (0.37)	1.35 (0.16)	1.53 (0.30)	1.47 (0.29)	1.38 (0.27)	1.39 (0.24)
D-prime				*Log β*		
0st	3.87 (0.34)	3.78 (0.41)	3.58 (0.55)	−0.14 (0.74)	0.63 (0.75)	0.49 (0.84)
2st	3.03 (0.48)	2.84 (0.40)	2.88 (0.58)	−0.20 (1.29)	1.22 (1.08)	0.57 (1.07)
4st	1.35 (0.32)	1.53 (0.43)	1.40 (0.39)	−0.36 (0.53)	0.37 (0.49)	0.05 (0.49)
6st	−0.15 (0.45)	−0.03 (0.29)	0.09 (0.42)	−0.07 (0.30)	0.37 (0.49)	0.03 (0.21)

Mean percentage of hits and reaction times (in seconds) for correct responses, discrimination sensitivity (d-prime) and answering bias (log ß) in response to male and female voices of the different morphing steps in semitones (st) for men, women and MtFs.

To determine the stimulus effects of gender morphing, male and female voices were weighted linearly ascending (0st*-3 < 2st*-1 < 4st*1 < 6st*3) and mean centered according to their morphing level. Resulting effects were compared between all sex groups.

For direct comparisons of men and women, see Text S1 in File S1 and Table S2 in File S1 for the interactions between the listeners' sex and the original voice gender, and Text S2 in File S1 and Table S3 in File S1 for the linear increase of voice morphing).

For all fMRI analyses, a Monte Carlo corrected threshold was applied using AlphaSim by Ward (2000) implemented in AFNI 2011 [79]. Assuming an uncorrected per voxel probability threshold of $p = 0.001$ and by entering the measurement parameters (e.g. voxel size, smoothing kernel), after 1.000 simulations a cluster size extent threshold of 20 contiguous resampled voxels was indicated to correct for multiple comparisons at $p < 0.05$. Effects still significant after a more conservative multiple comparison cluster level correction implemented in SPM (single-voxel threshold of $p < 0.001$ and cluster-level threshold of $p < 0.05$, family-wise error (FWE) corrected for multiple comparisons across the whole brain) and after Bonferroni correction for multiple interaction post hoc tests ($p < 0.0125$) are marked by asterisks.

Results

Behavioral results

Since we found no significant behavioral differences between treated and untreated MtFs in percentage of hits (male voices $p = 0.396$, female voices $p = 0.720$), RTs (male voices $p = 0.572$, female voices $p = 0.940$), d-prime (all $p > 0.100$) and log ß (all $p > 0.345$), both groups were pooled for further analyses.

Hit rates. Friedman tests indicated significant differences in the percentage of hits across all groups, voice gender and morphing conditions ($x^2(7) = 370.69$, $p < 0.001$).

Stimulus morphing decreased percentage of hits in the whole group and in all groups separately (all $p < 0.001$) without group differences ($p \leq 0.008$).

An effect of voice gender was found in men ($z = -2.47$, $p = 0.014$) and women ($z = -3.02$, $p = 0.003$) with more correct answers for voices of the opposite sex. In contrast, MtFs revealed no such effect ($z = -1.13$, $p = 0.258$).

Directly comparing men and MtFs as well as men and women, differences were observed with MtFs and women performing better in response to male voices (men vs. MtFs: $z = -2.672$, $p = 0.008$; men vs. women: $z = -3.921$, $p < 0.001$) and men in response to female voices (men vs. MtFs: $z = -2.776$, $p = 0.006$; men vs. women: $z = -3.584$, $p < 0.001$). Comparing women and MtFs, differences regarding male and female voices (both $z > -2.045$, $p < 0.05$) did not survive Bonferroni correction ($p \leq 0.008$).

None of the comparisons of each of the 8 sub-conditions between men and MtFs as well as women and MtFs did survive Bonferroni correction ($p \leq 0.002$) though trends ($p < 0.05$) become obvious. In contrast, comparisons between men and women revealed significant differences (see Table 2, Figure 1).

Discrimination and response bias. One-sample t-tests revealed discrimination ability above chance level for all conditions ($p < 0.001$) except 6st (men: $p = 0.147$, women: $p = 0.608$, MtFs: $p = 0.243$). As expected, discrimination sensitivity declined as the degree of morphing increased (Figure 1). Comparisons of d' for each morphing level revealed no significant group differences (Figure 1, Table 2).

The repeated measures ANOVA for log ß revealed significant effects of morphing ($F = 8.642$, df = 2.1, 138.5; $p < 0.001$) and

group ($F = 10.483$, df = 2, 67; $p < 0.001$) as well as an interaction of morphing and group ($F = 4.932$, df = 4.1, 138.5; $p = 0.001$).

While men revealed no significant position bias (all $p \geq 0.293$) except a trend to indicate female voices for 4st ($p = 0.006$), women were biased towards indicating male voices (all $p \leq 0.004$) except in the condition most morphed into the female direction ($p = 0.493$). Also MtFs revealed a bias towards indicating male voices for 0st ($p = 0.002$) and by trend for 2st ($p = 0.005$). Group comparisons regarding log ß reflected significant differences between men and women and by trend between men and MtFs for all morphing levels except 6st, Women and MtFs only differed by trend in the 4st condition (Figure 1, Table 2).

Reaction times. Friedman tests indicated significant differences in RT across all groups, voice gender and morphing conditions ($x^2(7) = 304.277$, $p < 0.001$). RT increased with morphing level (all $p < 0.001$) with increases in RT for increased morphing. Participants as a whole group reacted faster to female compared to male voices ($z = -2.848$, $p < 0.004$).

Direct group comparisons indicated women reacted faster than men in response to male voices ($z = -2.894$, $p = 0.003$) and to each morphing level (all $p \leq 0.022$). Further, women by trend reacted faster than MtFs in response to male voices ($z = -2.046$, $p = 0.041$). There was no difference between men and MtFs in any of the morphing levels (all $p > 0.207$) or regarding the sex of the voices (both $p > 0. 175$; Table 2).

Due to the reported group differences, we included RT as covariate into the functional imaging analyses.

Functional imaging results

Similar to the behavioral data, we found no significant brain activation differences between treated and untreated MtFs in the processing of male vs. female original voices as well as regarding gender morphing. Therefore both groups were pooled for further analyses.

Effects of original voice perception. One-sample analyses revealed bilateral activation in voice-selective areas such as the bilateral superior temporal gyri and bilateral cingulate cortex (for details see Figure S1 in File S1 and Table S1 in File S1), confirming the validity of the paradigm.

Effects of listener sex and original voice gender. Interaction analyses yielded stronger activation in *men* compared to MtFs for the processing of male vs. female original voices in the right hemispheric area triangularis, insula, and cuneus, the bilateral lingual gyrus extending to the calcarine gyrus and to the parahippocampus on the left side (Table 3, Figure 2).

The comparison between *women* and MtFs revealed enhanced activation for the processing of male vs. female original voices in the bilateral MPFC extending into the rostral anterior cingulate cortex (rACC), the right superior temporal gyrus (STG), precentral gyrus, cuneus, cerebellum, thalamus and the left precuneus extending into the paracentral lobe (Table 3, Figure 2).

There was no significant activation difference in a) MtFs or women compared to men in response to male vs. female voices or b) men or women compared to MtFs in response to female vs. male voices.

For a detailed description of the comparison between men and women see Junger and colleagues [64] and Text S1 in File S1 and Table S2 in File S1.

Effects of voice gender morphing. With respect to the parametric weighting of the linearly increasing morphing degree, stronger activation in the right superior frontal gyrus (SFG, MNI 27 −1 46, k = 21, t = 3.51) was found in men compared to MtFs (Figure 3). Parameter estimates reflected an overall increase in activation with higher degrees of morphing, whereas MtFs did not

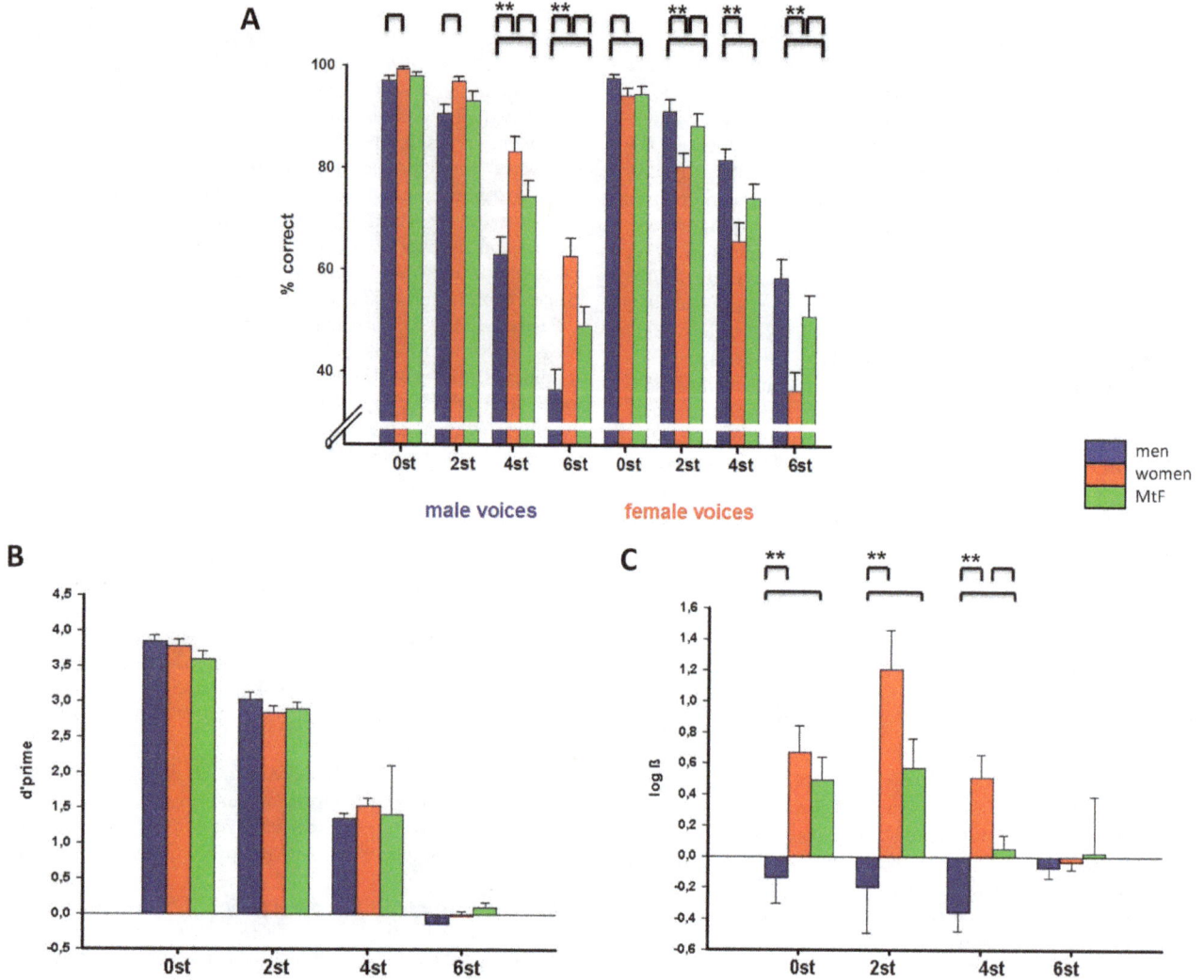

Figure 1. Behavioral performance. A. Performance (% hits with standard error bars) in response to male (left) and female (right) voices of the different morphing steps in semitones (st) in men (blue), women (red) and MtFs (green). Significant differences between groups are marked by bars (p<0.05) or asterisks (p≤0.002 Bonferroni corrected). B. Gender discrimination sensitivity (d-prime with standard error bars). C. Response bias (log ß with standard error bars) for each morphing step in men, women and MtFs with positive values representing bias to choose male voices and negative values representing bias to choose female voices. Significant differences are marked by bars (p<0.05) or asterisks (p≤0.002 Bonferroni corrected).

show such a pattern in this region. No activation difference was found for the reverse contrast (MtFs > men). There was no difference in either contrasts between women and MtFs.

Discussion

Hormonal level and sexual preference in MtFs

In line with the observations of Haraldsen et al. [80] and Wisnewski et al. [81], but contrary to those of Van Goozen et al. [82,83], we found no differences between gender dysphoric individuals untreated or treated with cross-sex hormones. Differences in findings might arise from the usage of different paradigms, such as more complex visuospatial and verbal fluency tasks, known to robustly reflect gender differences on a behavioral level. However, Van Goozen et al. [84] also did not find a effect of treatment when comparing MtFs and FtMs tested prior to and 14 weeks after cross-sex hormonal treatment. The notion of possible task effects is further corroborated by the study of Miles et al. [85], who found hormone effects in a paired association learning task,

but not in verbal memory and other cognitive tasks [86]. Overall, hormonal effects seem to be rather subtle both in behavioral and neurofunctional terms.

Although there is evidence from vaginal responses to visual erotic stimuli indicating differences between homo- and heterosexual MtFs [87,88], we did not find differences in brain activation between these groups in our sample. This is possibly due to the different physiological measure and stimuli used. Further, structural brain-imaging data revealed subtle differences in the bed nucleus of the stria terminalis when comparing heterosexual men with MtFs, but not in the comparison of hetero- and homosexual men [25]. According to that, our resulting pattern measured in a sample with mixed sexual orientation is in line with various functional studies exploring different biologically relevant domains in heterosexual [89], homosexual [27,30,31,37,90] and mixed [91] gender dysphoric samples in comparison to heterosexual men and women as well as in samples with no information on sexual orientation [28,36,39]. This might indicate that differences

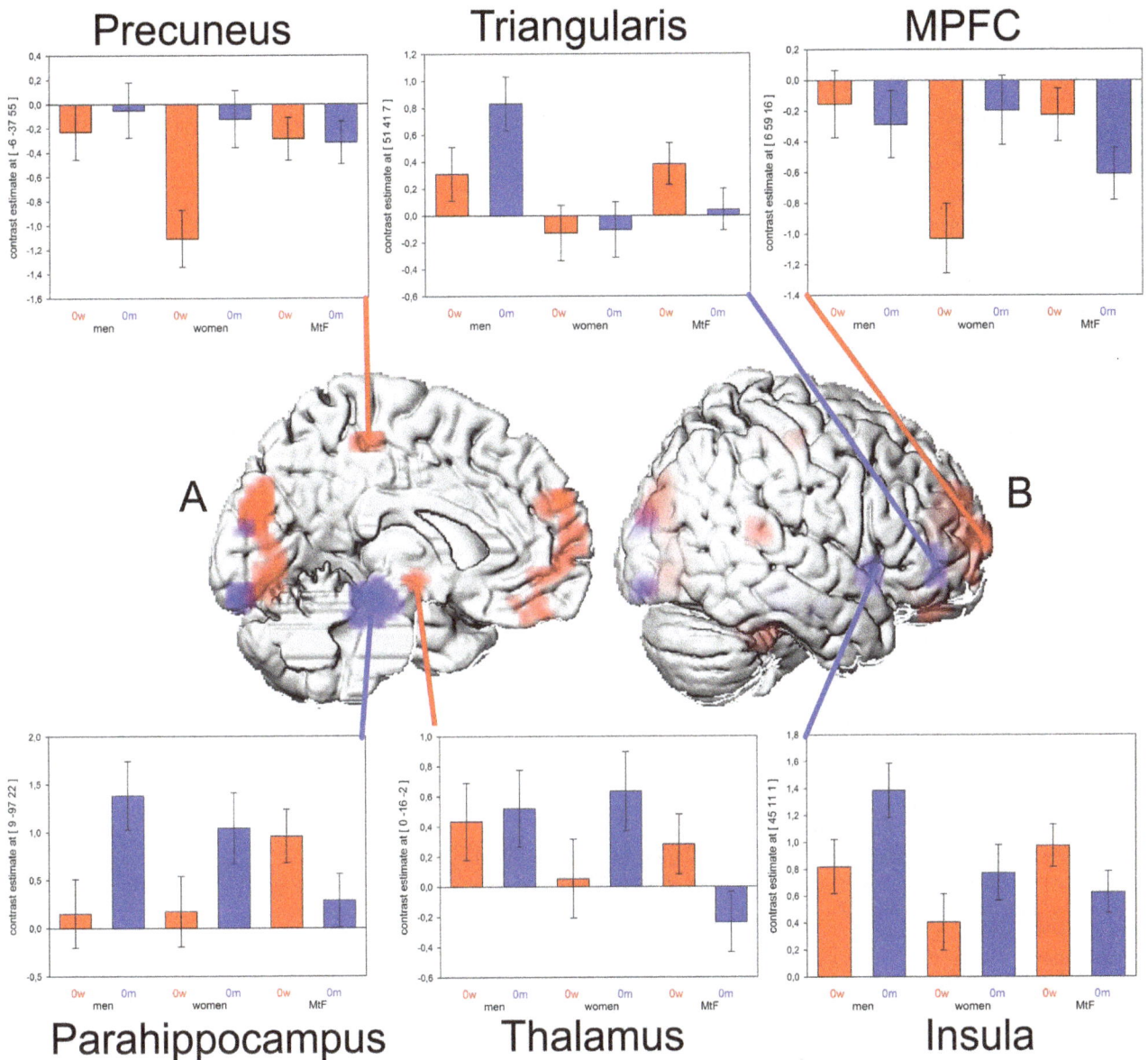

Figure 2. Interaction between original voice sex and group (p<0.05 Monte Carlo corrected, extent threshold =20 voxels). Male vs. female voices in men compared to MtFs (blue) and in women compared to MtFs (red).Parameter estimates are shown separately for male (0 m) and female (0 w) voices for men, women and MtFs A: left hemisphere, B: right hemisphere.

between individuals with gender dysphoria and their biological sex are more pronounced than differences related to their sexual orientation. However, small sub-group sample sizes might have disguised group differences based on sexual orientation within our group of individuals with gender dysphoria.

Differences in voice gender processing

In contrast to men and women, who showed a significantly better performance in response to voices of the opposite sex [64,92,93], MtFs displayed similar performance accuracy for male and female voices. Moreover, while men and women differed regarding RTs, MtFs exhibited an intermediate position with no significant differences compared to men and women. In line with the latter observation, Cohen-Kettenis and colleagues [94] proposed an intermediate position for MtFs and FtMs, placing both groups between men and women, in spatial cognition and

verbal memory tasks; i.e. the respective response patterns of the groups reflected incongruence with their biological sex. Similar to these behavioural results, brain activation in MtFs differed from the other two groups when listening to male (as compared to female) voices, supporting the notion of an intermediate position between men and women.

Activation differences between men and MtFs. MtFs showed less activation in the parahippocampal gyrus, IFG and insula than men when contrasting male and female voices. The parahippocampus is associated with emotional auditory processing [95], related to automatic matching of incoming meaningful sounds to stored representations [96] and was found to be less active in good compared to weak learners in an auditory memory task [97]. The insula is involved in paralinguistic information processing, such as vocal identity [98] and in the creation of an acoustic "mean voice" representation. [99]. Increased IFG

Table 3. Comparisons between groups and voice gender.

Brain region	L/R	x	y	z	k	t
a) men > MtFs						
Lingual gyrus extending to parahippocampal gyrus*	L	−12	−37	−11	311	4.64
Cuneus	R	9	−97	22	29	4.43
Lingual gyrus extending to calcarine gyrus	L	−9	−97	−8	50	4.02
Insula	R	45	11	1	33	3.76
Calcarine gyrus extending to lingual gyrus	R	18	−94	−5	31	3.60
Area tringularis (IFG)	R	51	41	7	26	3.59
b) women > MtFs						
Cuneus*	R	6	−91	28	348	4.78
MPFC extending to rACC*	R	6	59	16	278	4.66
MPFC extending to rACC*	L	−15	53	7	100	4.14
Cerebellum	R	12	−40	−29	49	3.92
STG	R	48	−43	16	31	3.76
Precuneus, paracentral lobe	L	−6	−37	55	35	3.73
Thalamus	-	0	−16	−2	22	3.61
Precentral gyrus	R	21	−25	58	20	3.45

Stronger activation/less deactivation in a) men compared to MtFs and b) women compared to MtFs for the processing of male vs. female original voices with no significant results for the opposite interactions ([MtF 0m > MtF 0w] > [men/women 0m > men/women 0w]; MNI coordinates, p<0.05 Monte Carlo corrected, k = cluster extension).
*significant at SPM cluster level (p<0.0125 Bonferroni corrected).

activation in response to a voice discrimination task has been related to increased processing demand [100,101] and, accordingly, indicates augmented processing demands in the identification of male vs. female voices in men compared to MtFs. In line with this, the smaller activation differences in MtFs in response to male as compared to female voices may reflect a good stored representation of both male and female voices leading to a similar demand when comparing them to learned "mean voice" representations. Due to the fact that MtFs grew up with male voices and trained themselves to sound more feminine, they possess extensive experience and expertise with respect to the voices of both genders.

Activation in the IFG triangularis in particular correlated negatively with implicitly perceived vocal attractiveness [102]. Moreover, MtFs exhibited less activation than men in the lingual and calcarine gyri when processing male as compared to female voices. Given that it was an auditory task, the observed activation differences in the visual system seem surprising. However, there is evidence of activation in these areas during auditory word and pseudoword processing [103] and passive speech listening [104]. This indicates the involvement of other primary sensory cortices in auditory word processing besides the auditory cortex. In line with von Kriegstein and colleagues [59], who reported that activation in the cuneus, the lingual and calcarine gyri results from attention to the verbal content of an auditory presented sentence, we propose that activation in these regions reflects that differences were based on a differential focus of attentional processes on the semantic content rather than voice characteristics. Accordingly, this difference between male and female voice processing seems to be more pronounced in men than in MtFs.

In light of a mating-related opposite-sex effect in voice gender perception in men [64], the stronger activity in all mentioned areas (in men compared to MtFs) may reflect men's relative indifference regarding male voices.

Thus, neuronal differences between men and MtFs may be related to both mating behavior and voice gender expertise, the former resulting in less attention to masculine voices in men and the latter resulting in generally decreased activation in MtFs compared to men and women (see below). In sum, our observation provides evidence for distinct cerebral activation patterns in MtFs different from their biological sex.

Activation differences between women and MtFs. During the evaluation of male as compared to female voices, women revealed stronger activation than MtFs in the right STG, which is linked to affective and identity information processing inherent in vocal stimuli [56]. Moreover, voice recognition accuracy was correlated with activation in the right superior temporal area [105]. Since we found no differences in accuracy between women and MtFs, decreased activation in MtFs might suggest that they need less effort to achieve levels of performance similar to women. This might be due to the fact that MtFs are more attuned to issues related to voice gender perception in everyday life.

Women also revealed less deactivation in the precuneus. Gur and colleagues [106] demonstrated increased deactivation when contrasting target detection with novelty detection, explaining it in terms of greater attentional demands. Thus, this might indicate women pay more attention to male voices than MtFs, while MtFs revealed similar responses to male and female voices. This was also reflected in similar accuracy and RT found in MtFs for both stimulus types. The anterior region of the precuneus is known to be involved in self-centered mental imagery. Interestingly, women reveal stronger connectivity than men between the precuneus and the thalamus [107] as well as between the medial dorsal-anterior precuneus and the ACC during attentional processes [108]. This fits nicely to our observation of increased activation in all mentioned areas connected to the precuneus in women.

The ACC, finally, is linked to reward anticipation [109] and the reflection on subjective preferences [110]. In both men and

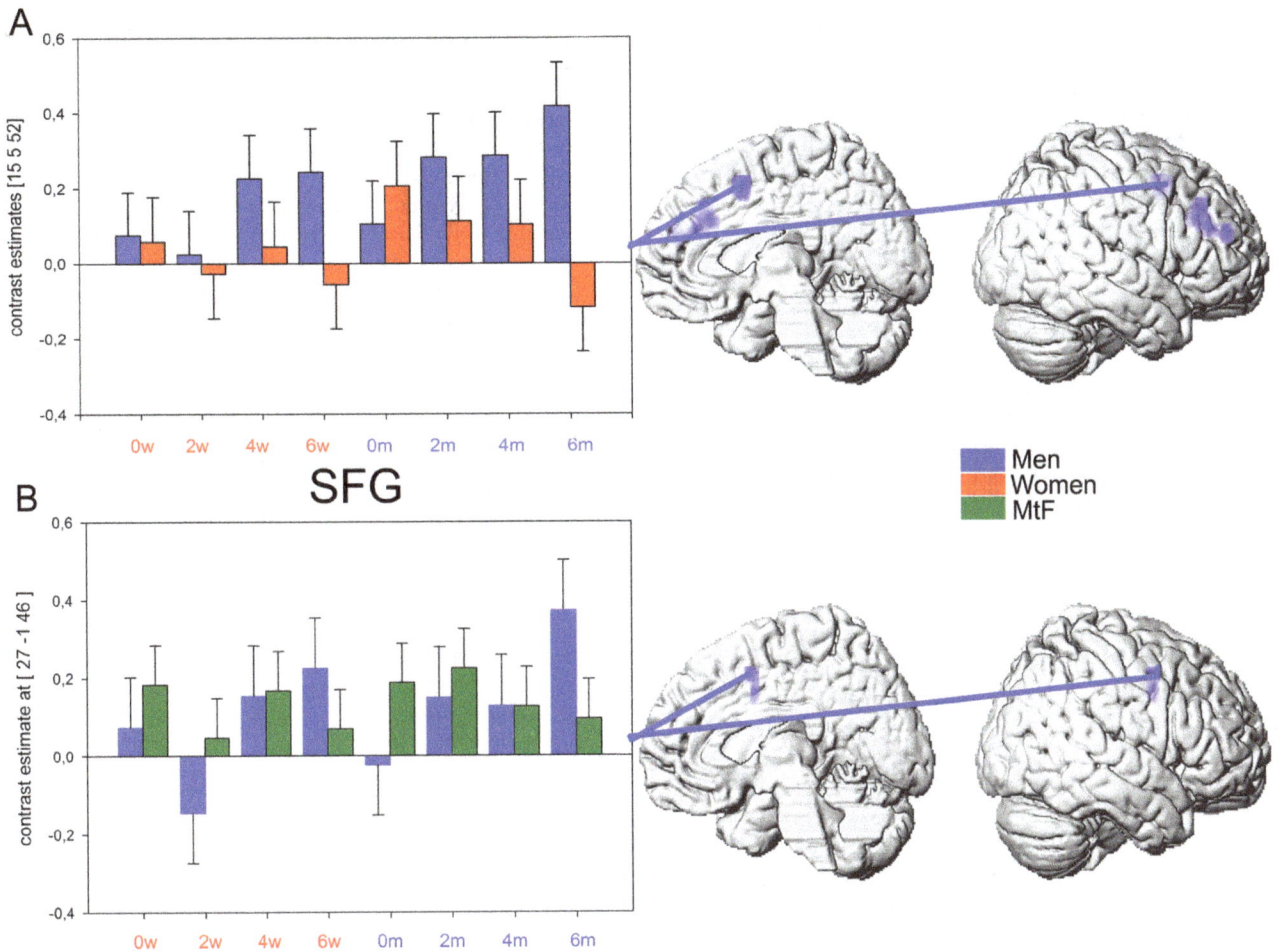

Figure 3. Graduate voice gender morphing. A. Contrast estimates of stronger activation in men as compared to women (blue) in the right SFG (peak voxel 15 5 52) for increasing morphing degree plotted for all 8 conditions. B. Contrast estimates of stronger activation in men as compared to MtFs (blue) in the right SFG (peak voxel 27 −1 46) for increasing morphing degree plotted for all 8 conditions (p<0.05 Monte Carlo corrected, extent threshold = 20 voxels).

women, it revealed enhanced activation in response to opposite-sex stimuli, which were construed as having greater salience [111]. Similarly, we could show its increased down-regulation in response to female voices in women and in response to male voices in men [64] with MtFs revealing activation patterns more similar to the male control sample.

Thus, parts of the reported behavioral and brain activation patterns underline an exceptional position of MtFs with qualitative differences from both men and women.

Differences in gradual voice gender morphing

In line with Cohan and Forget [112], who could show that women and hormonally treated MtFs performed similarly in two auditory tasks, we found a response bias in MtFs more similar to their aspired gender than to their biological sex, i.e. a tendency towards indicating male voices, at least for 0st and 2st voices. Further, in the direct group comparison, MtFs as well as women performed better than men in response to male voices underlining a certain overlap with the aspired gender.

As reported before [64], we found no sex-dependent activation patterns in typical voice-selective auditory areas in response to increased morphing. Instead, we observed increased activation in the SFG in men compared to women and to MtFs. Stronger

activation in the SFG has been associated with greater top-down control or cognitive effort to fulfil tasks in men compared to women [113,114]. Men seem to require greater attentional resources and cognitive control in order to discriminate voices with increasing difficulty in contrast to women and MtFs.

Concerning increased voice gender morphing, MtFs resemble women more closely than men on both behavioral and neuronal levels. This observation is in line with other fMRI studies which found a more female-like activation pattern in MtFs [39] [89].

Conclusions

We found sex-specific brain networks in men, women and MtFs when identifying gender by means of vocal sounds [64]. It seems that sex differences in voice gender perception are reflected in a widespread network involved in auditory, attentional, emotional and retrieval processes.

In contrast to men and women, MtFs showed no opposite-sex performance effect in voice perception. However, in direct comparison MtFs and women performed better in response to male voices and men when evaluating female voices. Further, there was a difference in RT between men and women, but none between MtFs and the two other groups. Both performance measures suggest a distinction between MtFs and men as well as women.

In line with the behavioral results, MtFs showed differences (compared to men and women) in neuronal response patterns with respect to male vs. female voices. Presumably, a different strategy is used in MtFs' voice gender identification due to early processing differences. They also might more intensively examine their own and aspired vocal characteristics during gender alignment, resulting in a certain expertise. In this sense, attentional differences due to automatized processing could lead to less brain activation in MtFs.

By morphing original voices into an ambiguous gender range, we have shown additional sex differences in terms of increased activation in prefrontal areas related to higher cognitive task demands in men compared to MtFs and women, as well as a more female-like response bias in MtFs.

In sum, our data support gender dysphoric individuals' lack of identification with their biological sex. Brain activation patterns of MtFs differed from those of men and, partly, also from those of women. Thus, differences between MtFs, men and women in voice gender processing reflect a qualitative difference in behavioral and neuronal processing that cannot be easily located on a linear continuum between men and women.

Further research is necessary to replicate our results in other tasks and samples of FtMs. Facilitating the removal of the stigma associated with gender dysphoria and providing access to care to individuals experiencing severe distress due to gender nonconformity should be considered the primary goal of research in this area.

Supporting Information

File S1 This file contains supporting information, including Figure S1, Text S1, Text S2, and Table S1–Table S7. Figure S1, Brain activation in men, women and MtFs (from top to bottom) for original voices (p<0.05 Monte Carlo corrected, extent threshold = 20 voxels) revealing activation in typical voice-related areas including the bilateral superior temporal gyri. Table S1, Brain activation in men, women and MtFs for original voices (p<0.05 Monte Carlo corrected, extent threshold = 20 voxels). Text S1, As described in Junger and colleagues (2013), men revealed stronger activation compared to women for the processing of female vs. male original voices mainly

in prefrontal areas but also in the left middle temporal gyrus (MTG). Table S2, Stronger activation/less deactivation in men compared to women for the processing of female vs. male original voices ([men 0 w > men 0 m] > [women 0 w > women 0 m]) with no significant results for the opposite contrast ([women 0 w > women 0 m] > [men 0 w > men 0 m]); (MNI coordinates, p< 0.05 Monte Carlo corrected, k = cluster extension). Text S2, As described in Junger and colleagues (2013) analyzing the parametric weighting of the linearly increasing morphing degree yielded stronger activation in right superior and middle frontal gyri in men compared to women (Table S3) with increased activation with increasing morphing degree only in men. Table S3, Activation peaks (MNI coordinates) and cluster extension (k) for a linear increase of voice morphing regarding gender identity for men contrasted to women; p<.05 Monte Carlo corrected (with no significant results for the opposite contrast). Table S4, Data points used to determine averages and summary statistics in Table 1. Table S5, Data points used to determine averages and summary statistics in Table 2 for correct responses (hits) in response to male and female voices of the different morphing steps (0, 2, 4, 6 semitones (st)). Table S6, Data points used to determine averages and summary statistics in Table 2 for reaction times (RT; in milliseconds) in response to male and female voices of the different morphing steps (0, 2, 4, 6 semitones (st)). Table S7, Data points used to determine averages and summary statistics in Table 2 for discrimination sensitivity (d-prime) and answering bias (log ß) in response to male and female voices of the different morphing steps (0, 2, 4, 6 semitones).

Acknowledgments

The authors thank Anna Pohl, Cordula Kemper, Maria Peters and Thilo Kellermann for their assistance and support. In particular, authors want to thank all self-help groups, web forums and participants, whose great endorsement and perseverance made this study possible.

Author Contributions

Conceived and designed the experiments: JJ UH KP CN PB BD. Performed the experiments: JJ KP SB. Analyzed the data: JJ KP JN. Contributed reagents/materials/analysis tools: PB KP FS. Wrote the paper: JJ. Revised the manuscript: UH JN CN PB CK BD FS KP.

References

1. Ellison PT (2009) On Fertile Ground: A Natural History of Human Reproduction: Harvard University Press.
2. Clutton-Brock T (2007) Sexual selection in males and females. Science 318: 1882–1885.
3. Joel D (2012) Genetic-gonadal-genitals sex (3G-sex) and the misconception of brain and gender, or, why 3G-males and 3G-females have intersex brain and intersex gender. Biol Sex Differ 3: 27.
4. Bullough VLR, Bullough B (1994) Human sexuality: an encyclopedia: Garland Publishing, Incorporated.
5. Garcia-Falgueras A, Swaab DF (2010) Sexual hormones and the brain: an essential alliance for sexual identity and sexual orientation. Endocr Dev 17: 22–35.
6. Cohen-Kettenis PT (2009) The DSM Diagnostic Criteria for Gender Identity Disorder in Adolescents and Adults. Arch Sex Behav.
7. Swaab DF (2007) Sexual differentiation of the brain and behavior. Best Pract Res Clin Endocrinol Metab 21: 431–444.
8. Bao AM, Swaab DF (2011) Sexual differentiation of the human brain: relation to gender identity, sexual orientation and neuropsychiatric disorders. Front Neuroendocrinol 32: 214–226.
9. APA. (2013) Diagnostic and statistical manual of mental disorders: DSM-V (5th ed.). Washington, DC: American Psychiatric Association.
10. APA. (2013) Gender Dysphoria. American Psychiatric Association.
11. Shechner T (2010) Gender identity disorder: a literature review from a developmental perspective. Isr J Psychiatry Relat Sci 47: 132–138.
12. Hoshiai M, Matsumoto Y, Sato T, Ohnishi M, Okabe N, et al. (2010) Psychiatric comorbidity among patients with gender identity disorder. Psychiatry Clin Neurosci 64: 514–519.
13. Hepp U, Kraemer B, Schnyder U, Miller N, Delsignore A (2005) Psychiatric comorbidity in gender identity disorder. J Psychosom Res 58: 259–261.
14. Gires (2008) Gender Variance (dysphoria). Gender Identity Research and Education Society http://www.gires.org.uk/assets/gdev/gender-dysphoria.pdf.
15. Swaab DF, Garcia-Falgueras A (2009) Sexual differentiation of the human brain in relation to gender identity and sexual orientation. Funct Neurol 24: 17–28.
16. Gooren L (2006) The biology of human psychosexual differentiation. Horm Behav 50: 589–601.
17. Bentz EK, Hefler LA, Kaufmann U, Huber JC, Kolbus A, et al. (2008) A polymorphism of the CYP17 gene related to sex steroid metabolism is associated with female-to-male but not male-to-female transsexualism. Fertil Steril 90: 56–59.
18. Lombardo F, Toselli L, Grassetti D, Paoli D, Masciandaro P, et al. (2013) Hormone and Genetic Study in Male to Female Transsexual Patients. J Endocrinol Invest.
19. Ujike H, Otani K, Nakatsuka M, Ishii K, Sasaki A, et al. (2009) Association study of gender identity disorder and sex hormone-related genes. Prog Neuropsychopharmacol Biol Psychiatry 33: 1241–1244.

20. Hare L, Bernard P, Sanchez FJ, Baird PN, Vilain E, et al. (2009) Androgen receptor repeat length polymorphism associated with male-to-female transsexualism. Biol Psychiatry 65: 93–96.

21. Heylens G, De Cuypere G, Zucker KJ, Schelfaut C, Elaut E, et al. (2012) Gender identity disorder in twins: a review of the case report literature. J Sex Med 9: 751–757.

22. Gomez-Gil E, Esteva I, Almaraz MC, Pasaro E, Segovia S, et al. (2010) Familiality of gender identity disorder in non-twin siblings. Arch Sex Behav 39: 546–552.

23. Corsello SM, Di Donna V, Senes P, Luotto V, Ricciato MP, et al. (2011) Biological aspects of gender disorders. Minerva Endocrinol 36: 325–339.

24. Kruijver FP, Zhou JN, Pool CW, Hofman MA, Gooren LJ, et al. (2000) Male-to-female transsexuals have female neuron numbers in a limbic nucleus. J Clin Endocrinol Metab 85: 2034–2041.

25. Zhou JN, Hofman MA, Gooren LJ, Swaab DF (1995) A sex difference in the human brain and its relation to transsexuality. Nature 378: 68–70.

26. Garcia-Falgueras A, Swaab DF (2008) A sex difference in the hypothalamic uncinate nucleus: relationship to gender identity. Brain 131: 3132–3146.

27. Simon L, Kozak LR, Simon V, Czobor P, Unoka Z, et al. (2013) Regional grey matter structure differences between transsexuals and healthy controls-a voxel based morphometry study. PLOS One 8: e83947.

28. Luders E, Sanchez FJ, Gaser C, Toga AW, Narr KL, et al. (2009) Regional gray matter variation in male-to-female transsexualism. Neuroimage 46: 904–907.

29. Savic I, Arver S (2011) Sex dimorphism of the brain in male-to-female transsexuals. Cereb Cortex 21: 2525–2533.

30. Rametti G, Carrillo B, Gomez-Gil E, Junque C, Zubiarre-Elorza L, et al. (2011) The microstructure of white matter in male to female transsexuals before cross-sex hormonal treatment. A DTI study. J Psychiatr Res 45: 949–954.

31. Rametti G, Carrillo B, Gomez-Gil E, Junque C, Segovia S, et al. (2010) White matter microstructure in female to male transsexuals before cross-sex hormonal treatment. A diffusion tensor imaging study. J Psychiatr Res.

32. Rametti G, Carrillo B, Gomez-Gil E, Junque C, Zubiaurre-Elorza L, et al. (2012) Effects of androgenization on the white matter microstructure of female-to-male transsexuals. A diffusion tensor imaging study. Psychoneuroendocrinology 37: 1261–1269.

33. Zubiaurre-Elorza L, Junque C, Gomez-Gil E, Guillamon A (2014) Effects of Cross-Sex Hormone Treatment on Cortical Thickness in Transsexual Individuals. J Sex Med.

34. Santarnecchi E, Vatti G, Dettore D, Rossi A (2012) Intrinsic Cerebral Connectivity Analysis in an Untreated Female-to-Male Transsexual Subject: A First Attempt Using Resting-State fMRI. Neuroendocrinology 96: 188–193.

35. Ye Z, Kopyciok R, Mohammadi B, Kramer UM, Brunnlieb C, et al. (2011) Androgens Modulate Brain Networks of Empathy in Female-to-Male Transsexuals: An fMRI Study. Zeitschrift Fur Neuropsychologie 22: 263–277.

36. Sommer IE, Cohen-Kettenis PT, van Raalten T, Vd Veer AJ, Ramsey LE, et al. (2008) Effects of cross-sex hormones on cerebral activation during language and mental rotation: An fMRI study in transsexuals. Eur Neuropsychopharmacol 18: 215–221.

37. Carrillo B, Gomez-Gil E, Rametti G, Junque C, Gomez A, et al. (2010) Cortical activation during mental rotation in male-to-female and female-to-male transsexuals under hormonal treatment. Psychoneuroendocrinology 35: 1213–1222.

38. Schoning S, Engelien A, Bauer C, Kugel H, Kersting A, et al. (2010) Neuroimaging differences in spatial cognition between men and male-to-female transsexuals before and during hormone therapy. J Sex Med 7: 1858–1867.

39. Gizewski ER, Krause E, Schlamann M, Happich F, Ladd ME, et al. (2009) Specific cerebral activation due to visual erotic stimuli in male-to-female transsexuals compared with male and female controls: an fMRI study. J Sex Med 6: 440–448.

40. Thomsen T, Hugdahl K, Ersland L, Barndon R, Lundervold A, et al. (2000) Functional magnetic resonance imaging (fMRI) study of sex differences in a mental rotation task. Med Sci Monit 6: 1186–1196.

41. Semrud-Clikeman M, Fine JG, Bledsoe J, Zhu DC (2012) Gender differences in brain activation on a mental rotation task. Int J Neurosci 122: 590–597.

42. Lykins AD, Meana M, Strauss GP (2008) Sex differences in visual attention to erotic and non-erotic stimuli. Arch Sex Behav 37: 219–228.

43. Cross SE, Madson L (1997) Models of the self: self-construals and gender. Psychol Bull 122: 5–37.

44. Cahill L (2006) Why sex matters for neuroscience. Nat Rev Neurosci 7: 477–484.

45. Fitch WT (2000) The evolution of speech: a comparative review. Trends Cogn Sci 4: 258–267.

46. Belin P, Fecteau S, Bedard C (2004) Thinking the voice: neural correlates of voice perception. Trends Cogn Sci 8: 129–135.

47. Hodges-Simeon CR, Gaulin SJ, Puts DA (2011) Voice correlates of mating success in men: examining "contests" versus "mate choice" modes of sexual selection. Arch Sex Behav 40: 551–557.

48. Hodges-Simeon CR, Gaulin SJ, Puts DA (2010) Different Vocal Parameters Predict Perceptions of Dominance and Attractiveness. Hum Nat 21: 406–427.

49. Bruckert L, Bestelmeyer P, Latinus M, Rouger J, Charest I, et al. (2010) Vocal attractiveness increases by averaging. Curr Biol 20: 116–120.

50. Poon S, Ng ML (2011) Contributions of voice fundamental frequency and formants to the identification of speakè's gender. Proceedings of the 17th International Congress of Phonetic Sciences.

51. Titze IR (1989) Physiologic and acoustic differences between male and female voices. J Acoust Soc Am 85: 1699–1707.

52. Sokhi DS, Hunter MD, Wilkinson ID, Woodruff PW (2005) Male and female voices activate distinct regions in the male brain. Neuroimage 27: 572–578.

53. van Dommelen WA (1990) Acoustic parameters in human speaker recognition. Lang Speech 33 (Pt 3): 259–272.

54. Oates J, Dacakis G (1997) Voice change in transsexuals. Venereology 10: 178–187.

55. Gelfer MP, Schofield KJ (2000) Comparison of acoustic and perceptual measures of voice in male-to-female transsexuals perceived as female versus those perceived as male. J Voice 14: 22–33.

56. Belin P, Zatorre RJ, Ahad P (2002) Human temporal-lobe response to vocal sounds. Brain Res Cogn Brain Res 13: 17–26.

57. Belin P, Zatorre RJ, Lafaille P, Ahad P, Pike B (2000) Voice-selective areas in human auditory cortex. Nature 403: 309–312.

58. von Kriegstein K, Smith DR, Patterson RD, Kiebel SJ, Griffiths TD (2010) How the human brain recognizes speech in the context of changing speakers. J Neurosci 30: 629–638.

59. von Kriegstein K, Eger E, Kleinschmidt A, Giraud AL (2003) Modulation of neural responses to speech by directing attention to voices or verbal content. Brain Res Cogn Brain Res 17: 48–55.

60. Formisano E, De Martino F, Bonte M, Goebel R (2008) "Who" is saying "what"? Brain-based decoding of human voice and speech. Science 322: 970–973.

61. Joassin F, Maurage P, Campanella S (2011) The neural network sustaining the crossmodal processing of human gender from faces and voices: an fMRI study. Neuroimage 54: 1654–1661.

62. Hesling I, Clement S, Bordessoules M, Allard M (2005) Cerebral mechanisms of prosodic integration: evidence from connected speech. Neuroimage 24: 937–947.

63. Lattner S, Friederici AD (2003) Talker's voice and gender stereotype in human auditory sentence processing-evidence from event-related brain potentials. Neurosci Lett 339: 191–194.

64. Junger J, Pauly K, Brohr S, Birkholz P, Neuschaefer-Rube C, et al. (2013) Sex matters: Neural correlates of voice gender perception. Neuroimage 79: 275–287.

65. Narins PM, Capranica RR (1976) Sexual differences in the auditory system of the tree frog Eleutherodactylus coqui. Science 192: 378–380.

66. Ryan MJ, Perrill SA, Wilczynski W (1992) Auditory Tuning and CAll Frequency Predict Population-Based MAting Preferences in the Cricket Frog, Acris crepitans. American Naturalist 139: 1370–1383.

67. Duncan LA, Park JH, Faulkner J, Schaller M, Neuberg SL, et al. (2007) Adaptive Allocation of Attention: Effects of Sex and Sociosexuality on Visual Attention to Attractive Opposite-Sex Faces. Evol Hum Behav 28: 359–364.

68. Hofmann SG, Suvak M, Litz BT (2006) Sex differences in face recognition and influence of facial affect. Personality and Individual Differences 40: 1683–1690.

69. Proverbio AM, Riva F, Martin E, Zani A (2010) Neural markers of opposite-sex bias in face processing. Front Psychol 1: 169.

70. Schneider F, Frister H, Olzen D (2010) Transsexuellengesetz. Begutachtung psychischer Störungen: Springer Berlin Heidelberg. pp. 247–261.

71. Wittchen HU, Zaudig M, Fydrich T (1997) Strukturiertes Klinisches Interview für DSM-IV. Göttingen: Hogrefe.

72. Oldfield RC (1971) The assessment and analysis of handedness: the Edinburgh inventory. Neuropsychologia 9: 97–113.

73. Lehrl S (1999) Mehrfachwahl-Wortschatz-Intelligenztest: MWT-B: Spitta.

74. Boersma P, Weenink D (2010) Praat: doing phonetics by computer [Computer program]. Version 5313, Available: http://wwwpraatorg/. Accessed 2012 Apr 11.

75. Friedman M (1937) The use of ranks to avoid the assumption of normality implicit in the analysis of variance. Journal of the American Statistical Association 32: 675–701.

76. D'Ausilio A, Bufalari I, Salmas P, Busan P, Fadiga L (2011) Vocal pitch discrimination in the motor system. Brain Lang 118: 9–14.

77. Wickens T (2002) Elementary Signal Detection Theory. Los Angeles: Oxford University Press.

78. Ashburner J, Friston KJ (2005) Unified segmentation. Neuroimage 26: 839–851.

79. Cox RW (2012) AFNI: what a long strange trip it's been. Neuroimage 62: 743–747.

80. Haraldsen IR, Egeland T, Haug E, Finset A, Opjordsmoen S (2005) Cross-sex hormone treatment does not change sex-sensitive cognitive performance in gender identity disorder patients. Psychiatry Res 137: 161–174.

81. Wisniewski AB, Prendeville MT, Dobs AS (2005) Handedness, functional cerebral hemispheric lateralization, and cognition in male-to-female transsexuals receiving cross-sex hormone treatment. Arch Sex Behav 34: 167–172.

82. Van Goozen SH, Cohen-Kettenis PT, Gooren LJ, Frijda NH, Van de Poll NE (1995) Gender differences in behaviour: activating effects of cross-sex hormones. Psychoneuroendocrinology 20: 343–363.

83. Slabbekoorn D, van Goozen SH, Megens J, Gooren LJ, Cohen-Kettenis PT (1999) Activating effects of cross-sex hormones on cognitive functioning: a study

of short-term and long-term hormone effects in transsexuals. Psychoneuroendocrinology 24: 423–447.

84. van Goozen SH, Slabbekoorn D, Gooren LJ, Sanders G, Cohen-Kettenis PT (2002) Organizing and activating effects of sex hormones in homosexual transsexuals. Behav Neurosci 116: 982–988.

85. Miles C, Green R, Sanders G, Hines M (1998) Estrogen and memory in a transsexual population. Horm Behav 34: 199–208.

86. Miles C, Green R, Hines M (2006) Estrogen treatment effects on cognition, memory and mood in male-to-female transsexuals. Horm Behav 50: 708–717.

87. Chivers ML, Rieger G, Latty E, Bailey JM (2004) A sex difference in the specificity of sexual arousal. Psychol Sci 15: 736–744.

88. Lawrence AA, Latty EM, Chivers ML, Bailey JM (2005) Measurement of sexual arousal in postoperative male-to-female transsexuals using vaginal photoplethysmography. Arch Sex Behav 34: 135–145.

89. Berglund H, Lindstrom P, Dhejne-Helmy C, Savic I (2008) Male-to-female transsexuals show sex-atypical hypothalamus activation when smelling odorous steroids. Cereb Cortex 18: 1900–1908.

90. Nawata H, Ogomori K, Tanaka M, Nishimura R, Urashima H, et al. (2010) Regional cerebral blood flow changes in female to male gender identity disorder. Psychiatry Clin Neurosci 64: 157–161.

91. Luders E, Sanchez FJ, Tosun D, Shattuck DW, Gaser C, et al. (2011) Increased Cortical Thickness in Male-to-Female Transsexualism. Journal of Behavioral and Brain Science.

92. Feinberg DR (2008) Are human faces and voices ornaments signaling common underlying cues to mate value? Evolutionary Anthropology 17: 112–118.

93. Jones BC, Feinberg DR, Debruine LM, Little AC, Vukovic J (2008) Integrating cues of social interest and voice pitch in men's preferences for women's voices. Biol Lett 4: 192–194.

94. Cohen-Kettenis PT, van Goozen SH, Doorn CD, Gooren LJ (1998) Cognitive ability and cerebral lateralisation in transsexuals. Psychoneuroendocrinology 23: 631–641.

95. Koelsch S (2005) Investigating emotion with music: neuroscientific approaches. Ann N Y Acad Sci 1060: 412–418.

96. Engelien A, Stern E, Isenberg N, Engelien W, Frith C, et al. (2000) The parahippocampal region and auditory-mnemonic processing. Ann N Y Acad Sci 911: 477–485.

97. Gaab N, Gaser C, Schlaug G (2006) Improvement-related functional plasticity following pitch memory training. Neuroimage 31: 255–263.

98. Remedios R, Logothetis NK, Kayser C (2009) An auditory region in the primate insular cortex responding preferentially to vocal communication sounds. J Neurosci 29: 1034–1045.

99. Andics A, McQueen JM, Petersson KM, Gal V, Rudas G, et al. (2010) Neural mechanisms for voice recognition. Neuroimage 52: 1528–1540.

100. Tesink CM, Petersson KM, van Berkum JJ, van den Brink D, Buitelaar JK, et al. (2009) Unification of speaker and meaning in language comprehension: an FMRI study. J Cogn Neurosci 21: 2085–2099.

101. Hutchison ER, Blumstein SE, Myers EB (2008) An event-related fMRI investigation of voice-onset time discrimination. Neuroimage 40: 342–352.

102. Bestelmeyer PE, Latinus M, Bruckert L, Rouger J, Crabbe F, et al. (2012) Implicitly perceived vocal attractiveness modulates prefrontal cortex activity. Cereb Cortex 22: 1263–1270.

103. Xiao Z, Zhang JX, Wang X, Wu R, Hu X, et al. (2005) Differential activity in left inferior frontal gyrus for pseudowords and real words: an event-related fMRI study on auditory lexical decision. Hum Brain Mapp 25: 212–221.

104. Hwang JH, Wu CW, Chen JH, Liu TC (2006) The effects of masking on the activation of auditory-associated cortex during speech listening in white noise. Acta Otolaryngol 126: 916–920.

105. Phan KL, Wager T, Taylor SF, Liberzon I (2002) Functional neuroanatomy of emotion: a meta-analysis of emotion activation studies in PET and fMRI. Neuroimage 16: 331–348.

106. Gur RC, Turetsky BI, Loughead J, Waxman J, Snyder W, et al. (2007) Hemodynamic responses in neural circuitries for detection of visual target and novelty: An event-related fMRI study. Hum Brain Mapp 28: 263–274.

107. Cavanna AE, Trimble MR (2006) The precuneus: a review of its functional anatomy and behavioural correlates. Brain 129: 564–583.

108. Zhang S, Li CS (2012) Functional connectivity mapping of the human precuneus by resting state fMRI. Neuroimage 59: 3548–3562.

109. Hare TA, Camerer CF, Knoepfle DT, Rangel A (2010) Value computations in ventral medial prefrontal cortex during charitable decision making incorporate input from regions involved in social cognition. J Neurosci 30: 583–590.

110. McClure SM, Li J, Tomlin D, Cypert KS, Montague LM, et al. (2004) Neural correlates of behavioral preference for culturally familiar drinks. Neuron 44: 379–387.

111. Spreckelmeyer KN, Rademacher L, Paulus FM, Grunder G (2012) Neural activation during anticipation of opposite-sex and same-sex faces in heterosexual men and women. Neuroimage 66C: 223–231.

112. Cohen H, Forget H (1995) Auditory cerebral lateralization following cross-gender hormone therapy. Cortex 31: 565–573.

113. Gauthier CT, Duyme M, Zanca M, Capron C (2009) Sex and performance level effects on brain activation during a verbal fluency task: a functional magnetic resonance imaging study. Cortex 45: 164–176.

114. Allendorfer JB, Lindsell CJ, Siegel M, Banks CL, Vannest J, et al. (2012) Females and males are highly similar in language performance and cortical activation patterns during verb generation. Cortex 48: 1218–1233.

Permissions

The contributors of this book come from diverse backgrounds, making this book a truly international effort. This book will bring forth new frontiers with its revolutionizing research information and detailed analysis of the nascent developments around the world.

We would like to thank all the contributing authors for lending their expertise to make the book truly unique. They have played a crucial role in the development of this book. Without their invaluable contributions this book wouldn't have been possible. They have made vital efforts to compile up to date information on the varied aspects of this subject to make this book a valuable addition to the collection of many professionals and students.

This book was conceptualized with the vision of imparting up-to-date information and advanced data in this field. To ensure the same, a matchless editorial board was set up. Every individual on the board went through rigorous rounds of assessment to prove their worth. After which they invested a large part of their time researching and compiling the most relevant data for our readers.

The editorial board has been involved in producing this book since its inception. They have spent rigorous hours researching and exploring the diverse topics which have resulted in the successful publishing of this book. They have passed on their knowledge of decades through this book. To expedite this challenging task, the publisher supported the team at every step. A small team of assistant editors was also appointed to further simplify the editing procedure and attain best results for the readers.

Apart from the editorial board, the designing team has also invested a significant amount of their time in understanding the subject and creating the most relevant covers. They scrutinized every image to scout for the most suitable representation of the subject and create an appropriate cover for the book.

The publishing team has been an ardent support to the editorial, designing and production team. Their endless efforts to recruit the best for this project, has resulted in the accomplishment of this book. They are a veteran in the field of academics and their pool of knowledge is as vast as their experience in printing. Their expertise and guidance has proved useful at every step. Their uncompromising quality standards have made this book an exceptional effort. Their encouragement from time to time has been an inspiration for everyone.

The publisher and the editorial board hope that this book will prove to be a valuable piece of knowledge for researchers, students, practitioners and scholars across the globe.

List of Contributors

Maria E. Teves, Wei Li and Lauren van Reesema
Department of Obstetrics and Gynecology, Virginia Commonwealth University, Richmond, Virginia, United States of America

Patrick R. Sears
Cystic Fibrosis Center, University of North Carolina, Chapel Hill, North Carolina, United States of America

Zhengang Zhang
Department of Obstetrics and Gynecology, Virginia Commonwealth University, Richmond, Virginia, United States of America
Department of Infectious Diseases, Tongji Medical College, Huazhong University of Science and Technology, Wuhan, Hubei, China

Waixing Tang
Department of Otorhinolaryngology, University of Pennsylvania, Philadelphia, Pennsylvania, United States of America

Richard M. Costanzo
Department of Physiology and Biophysics, Virginia Commonwealth University, Richmond, Virginia, United States of America

C. William Davis and Michael R. Knowles
Department of Cell & Molecular Physiology of Medicine, University of North Carolina, Chapel Hill, North Carolina, United States of America

Jerome F. Strauss III and Zhibing Zhang
Department of Obstetrics and Gynecology, Virginia Commonwealth University, Richmond, Virginia, United States of America
Department of Biochemistry and Molecular Biology, Virginia Commonwealth University, Richmond, Virginia, United States of America

Lindsay J. Henderson
Department of Neurobiology, Physiology and Behavior, University of California Davis, Davis, California, United States of America
College of Medical, Veterinary & Life Sciences, The University of Glasgow, Glasgow, United Kingdom

Aileen Adams and Neil P. Evans
College of Medical, Veterinary & Life Sciences, The University of Glasgow, Glasgow, United Kingdom

Britt J. Heidinger
College of Medical, Veterinary & Life Sciences, The University of Glasgow, Glasgow, United Kingdom
Department of Biological Sciences, North Dakota State University, Fargo, North Dakota, United States of America

Kathryn E. Arnold
College of Medical, Veterinary & Life Sciences, The University of Glasgow, Glasgow, United Kingdom
Environment Department, The University of York, York, United Kingdom

Ken Harata, Yasuyuki Kubo
Laboratory of Plant Pathology, Graduate School of Life and Environmental Sciences, Kyoto Prefectural University, Kyoto, Japan

Min Da, Bo Qian, Xuming Mo, Yuanli Hu and Yu Feng0
Department of Cardiothoracic Surgery, The Affiliated Nanjing Children's Hospital of Nanjing Medical University, Nanjing, Jiangsu, P.R. China

Jing Xu
Department of Thoracic and Cardiovascular Surgery, The First Affiliated Hospital of Nanjing Medical University, Nanjing, Jiangsu, P.R. China

Yuan Lin, Bixian Ni and Zhibin Hu
Department of Epidemiology and Biostatistics, State Key Laboratory of Reproductive Medicine, Nanjing Medical University, Nanjing, Jiangsu, P.R. China

Sara L. Montgomery, Daria Vorojeikina and Matthew D. Rand
Department of Environmental Medicine, University of Rochester School of Medicine and Dentistry, Rochester, New York, United States of America

Wen Huang, Trudy F. C. Mackay and Robert R. H. Anholt
Department of Biological Sciences, Genetics Program, and W. M. Keck Center for Behavioral Biology, North Carolina State University, Raleigh, North Carolina, United States of America

Veronica Colangelo, Stéphanie François and Raffaella Meneveri
Department of Health Sciences, University of Milano-Bicocca, Monza, Italy

Giulia Soldà , Francesca Roma and Enrico Ginelli
Department of Medical Biotechnology and Translational Medicine, University of Milan, Milan, Italy

Raffaella Picco
Department of Medical and Biological Sciences, University of Udine, Udine, Italy

Di Yu, Yu Feng, Lei Yang, Min Da, Changfeng Fan, Song Wang and Xuming Mo
Department of Cardiothoracic Surgery, Nanjing Children's Hospital, Nanjing Medical University, Nanjing, China

Thibaut Larcher, Helicia Goubin, Maéva Dutilleul, Lydie Guigand and Yan Cherel
INRA, UMR703 APEX, Oniris, Atlantic Gene Therapies, Université de Nantes, Oniris, École nationale vétérinaire, agro-alimentaire et de l'alimentation, Nantes, France

Thibaut Larcher, Helicia Goubin, Maéva Dutilleul, Lydie Guigand and Yan Cherel
INRA, UMR703 APEX, Oniris, Atlantic Gene Therapies, Université de Nantes, Oniris, École nationale vétérinaire, agro-alimentaire et de l'alimentation, Nantes, France

Aude Lafoux, Corinne Huchet and Gilles Toumaniantz
INSERM, UMR 1087/CNRS 6291 Institut du Thorax, Université de Nantes, Faculté des Sciences et des Techniques, Nantes, France

Laurent Tesson, Séverine Remy, Virginie Thepenier and Ignacio Anegon
INSERM, UMR 1064-Center for Research in Transplantation and Immunology, ITUN, CHU Nantes, Université de Nantes, Faculté de Médecine, Nantes, France

Virginie François
INSERM, UMR 1089, Atlantic Gene Therapies, Thérapie génique pour les maladies de la rétine et les maladies neuromusculaires, Université de Nantes, Faculté de Médecine, Nantes, France

Caroline Le Guiner
INSERM, UMR 1089, Atlantic Gene Therapies, Thérapie génique pour les maladies de la rétine et les maladies neuromusculaires, Université de Nantes, Faculté de Médecine, Nantes, France
Genethon, Evry, France

Anne De Cian, Charlotte Boix, Jean-Baptiste Renaud, Carine Giovannangeli and Jean-Paul Concordet
INSERM, U1154, CNRS, UMR 7196, Muséum National d'Histoire Naturelle, Paris, France

Noelia Díaz, Laia Ribas and Francesc Piferrer
Institut de Ciències del Mar, Consejo Superior de Investigaciones Científicas (CSIC), Barcelona, Spain

Md. Rafiqul Islam, Miyako Takayanagi, Hirofumi Yasumuro, Wataru Inami, Ailidana Kunahong and Roman M. Casco-Robles
Graduate School of Life and Environmental Sciences, University of Tsukuba, Tsukuba, Ibaraki, Japan

Kenta Nakamura and Chikafumi Chiba
Faculty of Life and Environmental Sciences, University of Tsukuba, Tsukuba, Ibaraki, Japan

Fubito Toyama
Graduate School of Engineering, Utsunomiya University, Utsunomiya, Tochigi, Japan

Ponnalagu Murugeswari and Veerappan Muthukkaruppan
Department of Immunology and Cell Biology, Aravind Medical Research Foundation, Dr.G.Venkataswamy Eye Research Institute, Madurai, India

Dhananjay Shukla, Ramasamy Kim and Perumalsamy Namperumalsamy
Vitreous and Retina Service, Aravind Eye Care System, Madurai, India,

Alan W. Stitt
Centre for Experimental Medicine, Queens University, Belfast, United Kingdom

Marlos R. Domingues
Sports Department, Federal University of Pelotas, Pelotas, Rio Grande do Sul, Brazil

Alicia Matijasevich
Postgraduate Programme in Epidemiology, Federal University of Pelotas, Pelotas, Rio Grande do Sul, Brazil

Department of Preventive Medicine, School of Medicine, University of Sao Paulo, Sao Paulo, Brazil

Aluísio J. D. Barros, Iná S. Santos and Bernardo L. Horta
Postgraduate Programme in Epidemiology, Federal University of Pelotas, Pelotas,
Rio Grande do Sul, Brazil

Pedro C. Hallal
Sports Department, Federal University of Pelotas, Pelotas, Rio Grande do Sul, Brazil
Postgraduate Programme in Epidemiology, Federal University of Pelotas, Pelotas,
Rio Grande do Sul, Brazil

Xinyu Tang, Todd G. Nick, Ming Li and Jingyun Li
Biostatistics Program, Department of Pediatrics, College of Medicine, University of Arkansas for Medical Sciences, Little Rock, Arkansas, United States of America

Mario A. Cleves, Stewart L. MacLeod and Charlotte A. Hobbs
Division of Birth Defects Research, Department of Pediatrics, College of Medicine, University of Arkansas for Medical Sciences, Little Rock, Arkansas, United States of America

Stephen W. Erickson
Department of Biostatistics, College of Public Health, University of Arkansas for Medical Sciences, Little Rock, Arkansas, United States of America

Ning Li, Nina A. Lehr and Francesc López-Giráldez
Department of Ecology and Evolutionary Biology, Yale University, New Haven, Connecticut, United States of America

Zheng Wang
Department of Ecology and Evolutionary Biology, Yale University, New Haven, Connecticut, United States of America
Department of Biostatistics, Yale University, New Haven, Connecticut, United States of America

David A. Hewitt
Department of Botany, Academy of Natural Sciences, Philadelphia, Pennsylvania, United States of America

Wagner Free Institute of Science, Philadelphia, Pennsylvania, United States of America

Frances Trail
Department of Plant Biology, Michigan State University, East Lansing, Michigan, United States of America
Department of Plant, Soil and Microbial Sciences, Michigan State University, East Lansing, Michigan, United States of America

Jeffrey P. Townsend
Department of Ecology and Evolutionary Biology, Yale University, New Haven, Connecticut, United States of America
Department of Biostatistics, Yale University, New Haven, Connecticut, United States of America
Program in Computational Biology and Bioinformatics, Yale University, New Haven, Connecticut, United States of America
Program in Microbiology, Yale University, New Haven, Connecticut, United States of America

Yu Feng, Di Yu, Lei Yang, Min Da, Changfeng Fan, Song Wang and Xuming Mo
Department of Cardiothoracic Surgery, The Affiliated Children's Hospital of Nanjing Medical University, Nanjing, Jiangsu, China

Tao Chen and Jin Liu
Department of Epidemiology and Biostatistics, School of Public Health, Nanjing Medical University, Nanjing, Jiangsu, China

Xing Tong
Atherosclerosis Research Center, Key Laboratory of Cardiovascular Disease and
Molecular Intervention, Nanjing Medical University, Nanjing, Jiangsu, China

Shutong Shen
Department of Cardiology, The First Affiliated Hospital of Nanjing Medical University, Nanjing, Jiangsu, China

John Gounarides, Jennifer S. Cobb, Jing Zhou, Frank Cook, Xuemei Yang, Hong Yin and Chang Rao
Analytical Sciences, Novartis Institutes for Biomedical Research, Cambridge, MA, United States of America

Erik Meredith and Muneto Mogi
Global Discovery Chemistry, Novartis Institutes for Biomedical Research, Cambridge, MA, United States of America

Qian Huang and YongYao Xu
Developmental and Metabolic Pathways, Novartis Institutes for Biomedical Research, Cambridge, MA, United States of America

Karen Anderson, Andrea De Erkenez, Sha-Mei Liao, Maura Crowley, Natasha Buchanan, Stephen Poor, Yubin Qiu, Elizabeth Fassbender, Siyuan Shen, Amber Woolfenden, Amy Jensen, Rosemarie Cepeda, Bijan Etemad-Gilbertson, Shelby Giza, Bruce Jaffee, Sassan Azarian
Ophthalmology, Novartis Institutes for Biomedical Research, Cambridge, MA, United States of America

Gen Suzuki, Brian R. Weil, Merced M. Leiker, Amanda E. Ribbeck, Rebeccah F. Young and Thomas R. Cimato
Department of Medicine, Division of Cardiovascular Medicine, University at Buffalo,
Buffalo, New York, United States of America
The Clinical and Translational Research Center, University at Buffalo, Buffalo, New York,
United States of America

John M. Canty Jr.
Division of Cardiovascular Medicine, Veterans Affairs Western New York Health Care System, Buffalo, New York, United States of America
Department of Medicine, Division of Cardiovascular Medicine, University at Buffalo, Buffalo, New York, United States of America

Department of Physiology & Biophysics, University at Buffalo, Buffalo, New York, United States of America
Department of Biomedical Engineering, University at Buffalo, Buffalo, New York, United States of America
The Clinical and Translational Research Center, University at Buffalo, Buffalo, New York, United States of America

Jessica Junger, Ute Habel, Frank Schneider, Birgit Derntl and Katharina Pauly
Department of Psychiatry, Psychotherapy and Psychosomatics, Medical School, RWTH Aachen University, Aachen, Germany
Jülich Aachen Research Alliance-Translational Brain Medicine, Jülich, Germany

Sabine Bröhr
Department of Psychiatry, Psychotherapy and Psychosomatics, Medical School, RWTH Aachen University, Aachen, Germany

Josef Neulen
Department of Gynaecological Endocrinology and Reproductive Medicine, Medical School, RWTH Aachen University, Aachen, Germany

Christiane Neuschaefer-Rube and Peter Birkholz
Department of Phoniatrics, Pedaudiology and Communication Disorders, Medical School, RWTH Aachen University, Aachen, Germany

Christian Kohler
Department of Psychiatry, Neuropsychiatry Division, University of Pennsylvania School of Medicine, Philadelphia, Pennsylvania, United States of America

Index

www.ingramcontent.com/pod-product-compliance
Lightning Source LLC
Chambersburg PA
CBHW082042190326
41458CB00010B/3434